Seismology of Azimuthally Anisotropic Media and Seismic Fracture Characterization

Ilya Tsvankin and Vladimir Grechka

Geophysical References Series No. 17
Ken Larner and Michael Slawinski, volume editors
Sergey Fomel, managing editor

Society of Exploration Geophysicists
The international society of applied geophysics
Tulsa, Oklahoma, U.S.A

ISBN 978-0-931830-47-1 (Series)
ISBN 978-1-56080-228-0 (Volume)

Society of Exploration Geophysicists
P.O. Box 702740
Tulsa, OK 74170-2740

Published 2011
Printed in the United States of America

Library of Congress Cataloging-in-Publication Data
Tsvankin, I. D.
Seismology of azimuthally anisotropic media and seismic fracture characterization / Ilya Tsvankin and Vladimir Grechka.
p. cm. – (Geophysical references series ; no. 17)
Includes bibliographical references and index.
ISBN 978-1-56080-228-0 (volume : alk. paper) – ISBN 978-0-931830-47-1 (series : alk. paper)
1. Seismic waves. 2. Seismology. 3. Anisotropy. 4. Rock deformation–Measurement. I. Grechka, V. IU II. Title.
QE538.5.T884 2011
551.22–dc22
2011002794

Contents

About the authors

Ilya Tsvankin received his M.S. (1978) and Ph.D. (1982) degrees in geophysics from Moscow State University in Russia. From 1978 to 1989 he worked at the Institute of Physics of the Earth in Moscow and was deputy head of the laboratory "Geophysics of Anisotropic Media." After moving to the United States in 1990, Ilya became a consultant to the Amoco Production Research Center in Tulsa. Since 1992, he has been on the faculty of Colorado School of Mines, where currently he is professor of geophysics and co-leader of the Center for Wave Phenomena. Ilya's research has focused on seismic modeling, inversion, and processing for anisotropic media, fracture characterization, nonlinear elasticity, and time-lapse seismic. In 1996, he received the Virgil Kauffman Gold Medal Award from SEG for his pioneering work in seismic anisotropy. Ilya's monograph *Seismic Signatures and Analysis of Reflection Data in Anisotropic Media*, published in 2001 and reprinted in 2005, is regarded as a major text in the field.

Vladimir Grechka received his M.S. (1984) in geophysical exploration from Novosibirsk State University, Russia, and a Ph.D. (1990) in geophysics from the Institute of Geophysics, Novosibirsk, Russia. He worked in the same institute from 1984 to 1994 as a research scientist. He was a graduate student at the University of Texas at Dallas from 1994 to 1995. Then Vladimir joined the Department of Geophysics, Colorado School of Mines, where he was an associate research professor and a co-leader of the Center for Wave Phenomena. Since 2001, he is a senior geophysicist at Shell Exploration & Production Company. Vladimir's research is focused on theory of seismic wave propagation in anisotropic media, velocity analysis, fracture characterization, and microseismic. Vladimir received J. Clarence Karcher Award from SEG (1997) and the East European Award from the European Geophysical Society (1992). Vladimir and Ilya are teaching a two-day course on anisotropy as part of the SEG Continuing Education Program.

Preface

As has been known for some time, velocities of elastic waves propagating in most subsurface formations exhibit directional dependence, or seismic anisotropy. On their way to exploration targets, seismic waves pass through sedimentary strata, whose effective anisotropy is caused by such physical mechanisms as the intrinsic structure of shales and clays, thin layering, natural fracture systems, and nonhydrostatic stress. Given the ubiquity of seismic anisotropy, ignoring it in data processing often leads to many serious distortions: migrated images are blurred, reflectors are misplaced both laterally and in depth, and interpretation of reflection amplitudes is incorrect. Accounting for anisotropy, however, requires complicated data-processing workflows and involves estimation of additional, sometimes poorly constrained medium parameters. Despite the severity of this challenge, the industry has been making steady progress in implementing ever more complex and realistic anisotropic models.

Indeed, this book was written during a period of accelerating transition toward anisotropy-based methods in seismic exploration and reservoir characterization. In particular, implementation of reverse time migration and other powerful algorithms for 3D prestack depth migration has convincingly demonstrated the advantages of anisotropic P-wave imaging. Application of transversely isotropic models with a vertical and tilted symmetry axis (VTI and TTI, respectively) significantly improves well ties, spatial positioning of reflectors for a wide range of dips, focusing of steep reflectors, fault resolution, etc. Despite the higher computational cost of anisotropic migration, it is already used in the majority of 3D imaging projects in the Gulf of Mexico and elsewhere. Still, the nonuniqueness of parameter estimation might seriously hamper velocity model-building, even if the symmetry-axis orientation is assumed to be known (e.g., orthogonal to reflectors). In the book we present velocity-analysis algorithms for layered TI media and study the influence of the tilt of the symmetry axis on the inversion results.

A development with far-reaching implications for seismic processing and inversion is routine acquisition of wide-azimuth marine data with the main goal of better illuminating subsalt targets. Azimuthal velocity variation (i.e., azimuthal anisotropy) observed in those surveys can provide valuable constraints for anisotropic parameter estimation. Wide-azimuth acquisition is likely to expose the limitations of TTI models, which often represent an oversimplified symmetry for dipping beds. We expect that in many cases processing of wide-azimuth data will require employing more complicated, but more realistic, orthorhombic velocity models. Orthorhombic symmetry has not yet become part of industry practice, although it is typical for naturally fractured formations and rocks under nonhydrostatic stress. One of the main goals of the book is to give an in-depth description of parameter-estimation methods for wide-azimuth P-wave and multicomponent data from layered orthorhombic media.

Shear waves and mode conversions (PS- and SP-waves) are particularly sensitive to anisotropy, in part because they split into the fast and slow modes with orthogonal polarizations. The S-wave splitting coefficient, usually estimated by applying Alford (or Alford-style) rotation to the horizontal displacement components, can serve as an indicator of near-vertical sets of aligned fractures. Also, the parameters responsible for shear-wave velocity anisotropy typically are larger than their P-wave counterparts. Depth misties between PP and PS sections observed in ocean-bottom surveys provide unambiguous evidence of the influence of anisotropy on reflection traveltimes and indicate the high potential of combining P- and S-waves in anisotropic velocity analysis. Because P-to-S mode conversions often represent the only available source of shear-wave information, the book includes a detailed discussion of joint processing and inversion of P- and PS-waves. Our results demonstrate that wide-azimuth, multicomponent data should play a major role in robust parameter estimation for azimuthally anisotropic models, especially when the symmetry is lower than TI.

One of the most important applications of anisotropic processing is in characterization of naturally fractured reservoirs because vertical or dipping fracture sets create azimuthal anisotropy on the scale of seismic wavelength. Although the velocities and amplitudes of vertically traveling shear modes give a direct estimate of the S-wave splitting coefficient, it has become common to infer fracture parameters from the azimuthal variation of traveltimes and prestack amplitudes of higher-quality P-waves (sometimes combined with PS-waves). Extraction of fracture-related attributes from reflection and vertical seismic profiling data represents an example of increased emphasis on "looking at the anisotropy" (as opposed to "looking past anisotropy" for imaging purposes) in seismic methods. We introduce effective medium theories to obtain the parameters of fractured formations on a seismic scale and describe seismic fracture-characterization methods for orthorhombic and HTI (TI with a horizontal symmetry axis) symmetry operating with different sets of input data.

While the number of anisotropy-related journal articles is almost staggering, there is only a handful of books (Helbig, 1994; MacBeth, 2002; Rüger, 2002; Tsvankin, 2005) and expanded course notes (Thomsen, 2002; Grechka, 2009) devoted to applications of anisotropic models in seismic processing and reservoir characterization. In addition, a broad but brief overview of the state of the art in applied seismic anisotropy is given in the paper by Tsvankin et al. (2010). For an historical analysis and major milestones of anisotropy research, we refer the reader to Helbig and Thomsen (2005).

This book can serve as a sequel to the monograph of Tsvankin (2005; the first edition was published in 2001), which reviews the basics of anisotropic wave propagation and discusses inversion and processing of reflection data from anisotropic (mostly transversely isotropic) media. Here, we summarize research developments over the past decade in employing compressional and multicomponent seismic data from azimuthally anisotropic media for purposes of velocity analysis and fracture characterization. It should be mentioned that Chapters 7 and 9 represent revised and expanded versions of the corresponding parts of the book by Grechka (2009).

Chapter 1 is devoted to analytic treatment of the most stable, hyperbolic portion of azimuthally varying reflection moveout using the general concept of the NMO ellipse. After giving an explicit representation of the NMO ellipse for a homogeneous medium, we develop the generalized Dix equation for a stack of horizontal, arbitrarily anisotropic layers above a dipping reflector. When the overburden is laterally heterogeneous, the effective NMO ellipse can be built by an efficient Dix-type averaging procedure that involves *NMO-velocity surfaces* obtained by plotting $V_{\rm nmo}$ as the radius-vector in 3D space. An alternative way of computing NMO ellipses in heterogeneous media at a comparable computational cost is based on dynamic ray tracing of the zero-offset ray. These modeling methods are orders of magnitude faster than multioffset, multiazimuth kinematic ray tracing. Whereas most of the chapter is devoted to pure (PP or SS) reflections, we also analyze normal moveout of converted waves for multiple horizontal layers with a horizontal symmetry plane. PS-wave traveltime in such models is an even function of offset, and the NMO ellipses of pure and converted waves are related by a simple Dix-type equation.

The results of Chapter 1 provide the foundation for processing and inversion of wide-azimuth, conventional-spread P-wave data discussed in **Chapter 2**. We present a complete processing flow for NMO ellipses in horizontally layered media that includes a correction for lateral velocity variation. This methodology is applied to P-wave data acquired over a fractured reservoir in the Powder River Basin, Wyoming, to compute anisotropy-induced interval NMO ellipses and estimate depth-varying fracture orientation. Then we extend moveout inversion to models composed of homogeneous transversely isotropic layers separated by plane dipping or curved interfaces. The presence of dips in the overburden can mitigate trade-offs between the TI parameters and help constrain reflector depth, as illustrated by synthetic and physical-modeling examples. To perform interval parameter estimation for multilayered VTI media, we develop an inversion technique, *stacking-velocity tomography*, that operates with P-wave NMO ellipses, zero-offset traveltimes, and reflection time slopes. When the symmetry axis is tilted but is known to be orthogonal to a dipping reflector, the NMO ellipse can be inverted for the parameters V_{P0} and δ, while ϵ remains unconstrained by conventional-spread moveout. For layered orthorhombic models, we use joint inversion of the NMO ellipses of horizontal and dipping events to estimate the parameter set responsible for time-domain processing of P-wave data.

In **Chapter 3** we extend moveout analysis to long-spread P-wave reflection traveltimes from azimuthally anisotropic media. Deviations from hyperbolic moveout are described by an exact expression for the quartic moveout coefficient A_4 valid for arbitrary anisotropy and heterogeneity. To gain insight into the behavior of nonhyperbolic moveout, we simplify this equation for horizontal and dipping layers with TI and orthorhombic symmetry. The azimuthal variation of A_4 for an orthorhombic layer can have multiple lobes and is controlled by the anellipticity parameters $\eta^{(1)}$, $\eta^{(2)}$, and $\eta^{(3)}$. Then we demonstrate that a highly accurate traveltime approximation for models composed of horizontal orthorhombic, HTI, and VTI layers is provided by the generalized version of the Alkhalifah-Tsvankin nonhyperbolic moveout equation.

The effective moveout parameters are estimated using a global 3D algorithm that maximizes semblance for the full range of offsets and azimuths. To perform interval parameter estimation using long-spread data, we employ the velocity-independent layer-stripping (VILS) method, which is much more stable than Dix-type techniques. Applications to wide-azimuth P-wave data from Weyburn field in Canada and Rulison field in Colorado, USA, illustrate the high potential of nonhyperbolic moveout inversion in fracture characterization and time-domain anisotropic velocity analysis.

Starting with **Chapter 4**, we turn our attention to multicomponent seismic and joint processing/inversion of P- and S-wave reflections. The most common source of information about shear-wave velocities is mode-converted PS-waves, whose properties are analyzed in the beginning of the chapter. To avoid problems caused by the moveout asymmetry, conversion-point dispersal, and polarity reversals of mode conversions, we introduce the so-called "PP+ PS = SS" method designed to construct data (ΨS-waves) kinematically equivalent to pure SS primary reflections. The generated ΨS events can be processed by any velocity-analysis technique developed for pure modes. The method needs no information about the velocity field and anisotropy parameters (except for that used in PP-PS event correlation) and can be applied to wide-azimuth data from arbitrarily anisotropic, heterogeneous media. In addition to the kinematic PP+ PS = SS technique that operates with traveltime picks, we describe the full-waveform version of the method based on convolution of PP and PS traces. The accuracy and requirements of the algorithm are illustrated on synthetic and field-data examples.

Chapter 5 explores the possibilities of combining PP-wave reflection moveout with traveltimes of mode-converted (PS) and shear (SS) waves in parameter estimation for TI media. Despite the information provided by mode conversions, PP and PS traveltimes from horizontally layered VTI media do not constrain the Thomsen parameters and reflector depth, even for uncommonly large spreadlength-to-depth ratios. We show that multicomponent moveout inversion generally becomes better posed in the presence of reflector dip, and extend stacking-velocity inversion (tomography) described in Chapter 2 to the combination of PP and ΨS (computed using the PP+ PS = SS method) data from layered TTI media with dipping or curved interfaces. The kinematic PP+ PS = SS method and multicomponent stacking-velocity inversion are applied to 2D PP and PS data from the North Sea to build an anisotropic velocity model and use it for converted-wave processing. In particular, taking anisotropy into account greatly improves the image of the reservoir and overburden faults on PS time sections. Another case study, based on the full-waveform version of the PP+ PS = SS method, reveals substantial anisotropy in sedimentary layers in the Gulf of Mexico.

Multicomponent velocity analysis is even more essential for lower-symmetry orthorhombic and monoclinic media, as discussed in **Chapter 6**. PP and PS (fast PS_1 and slow PS_2) reflection traveltimes for horizontally layered orthorhombic models can be used to compute the NMO ellipses of the split shear waves S_1S_1 and S_2S_2. Then we invert the interval NMO ellipses of PP and SS reflections, supplemented by known reflector depths, for the interval anisotropy parameters. This approach is validated by a physical-modeling experiment, which simulates a wide-azimuth reflection survey over an orthorhombic layer. To invert wide-azimuth PP and SS (i.e., ΨS_1 and ΨS_2) traveltimes from multiple orthorhombic layers separated by dipping or curved interfaces, we supplement stacking-velocity tomography with depth constraints. Moveout analysis for the more complex monoclinic symmetry is facilitated by introducing a Thomsen-style notation that captures the combinations of the stiffness elements responsible for the kinematics of reflected waves. We demonstrate that the interval parameters of horizontally layered monoclinic media can be obtained from wide-azimuth PP, PS_1, and PS_2 reflection traveltimes and depth information using the same algorithm as that for orthorhombic media.

Although the book is largely focused on reflection seismology, the topic of **Chapter 7** is anisotropic inversion of data acquired in vertical seismic profiling (VSP) surveys. The unique feature of VSP geometry is the potential to obtain a complete (triclinic) local stiffness tensor at the receiver location without making a priori assumptions about the symmetry (provided the model is structurally simple). We present synthetic and field examples of successful estimation of the triclinic stiffness tensor from multicomponent, wide-azimuth, walkaway VSP data. In the case study from Vacuum field (New Mexico, USA), the inverted stiffnesses are well-described by an orthorhombic model with a subhorizontal symmetry plane. This methodology, however, is not applicable when the overburden is laterally heterogeneous, and the input data are limited to polarization vectors and vertical slownesses. We express the P-wave slowness-of-polarization function for VTI and orthorhombic models through appropriately modified anisotropy parameters. Applications to VSP surveys from a subsalt play in the deepwater Gulf of Mexico and from a tight-gas land reservoir confirm that these parameters are well resolved and contain useful information about lithology and fracturing.

The first seven chapters deal primarily with arrival times, which generally represent the best-resolved component of seismic wavefields. Reflection amplitudes, analyzed in **Chapter 8**, tend to be more noisy but can provide superior vertical resolution and sensitive physical attributes. Using weak-contrast, weak-anisotropy approximations for plane-wave reflection coefficients, we review the azimuthal amplitude-variation-with-offset (AVO) response for orthorhombic and HTI media (for more details, see Rüger, 2002). Robust reconstruction of reflection coefficients requires correction for geometrical spreading in AVO processing, especially when the overburden is anisotropic. We describe the moveout-based anisotropic spreading correction (MASC), which accurately removes the geometrical-spreading factor from wide-azimuth P-wave reflections in layered orthorhombic media. The performance of P-wave azimuthal AVO and moveout analyses is illustrated on a case study for a fractured gas-sand reservoir at Rulison field. The advanced anisotropic processing sequence helps delineate pronounced azimuthal anomalies of the AVO gradient, which provide insight into the orientation and spatial distribution of natural fractures. Long-offset AVO response, seldom included in conventional processing, can be used to identify the critical reflections from high-velocity layers (this method is called *seismic critical-angle reflectometry*). We give concise expressions for critical angles in VTI and orthorhombic media and discuss the potential of critical-angle reflectometry in anisotropic parameter estimation. As established by experimental studies, attenuation anisotropy often is much stronger than velocity anisotropy, and attenuation coefficients might serve as fracture-detection attributes. In the last section of the chapter we develop an analytic framework for describing and inverting the attenuation coefficients of P- and S-waves in anelastic TI and orthorhombic media. Treatment of attenuation coefficients is greatly simplified by employing dimensionless attenuation-anisotropy parameters introduced by generalizing Thomsen notation.

The book concludes with **Chapter 9**, in which we use effective media theories to examine anisotropy caused by small-scale aligned fractures. Comparison of two popular theories proposed by J. Hudson and M. Schoenberg shows that, even on a qualitative level, Schoenberg's formalism is expected to be more accurate. This conclusion is confirmed by modeling the effective elasticity of so-called digital (i.e., numerically simulated) rocks. The synthetic study also demonstrates that multiple sets of realistic irregularly shaped, intersecting fractures produce the same effective stiffnesses as three orthogonal sets of isolated, penny-shaped cracks. Hence, the symmetry of the effective medium is nearly orthorhombic (for an isotropic background matrix), with fewer independent parameters than for general orthorhombic models. Whereas this discovery shows that microstructural information carried by long seismic waves is inherently limited, it underscores the importance of orthorhombic symmetry and paves the way for devising practical fracture-characterization techniques. We discuss application of one such technique to wide-azimuth, multicomponent reflection data from the fractured reservoir at Rulison field. Joint inversion of the NMO ellipses of pure-mode P, S_1, and S_2 reflections yields an effective orthorhombic reservoir model without relying on borehole information and provides evidence for multiple fracture sets in part of the study area.

The subject of azimuthal anisotropy and fracture characterization is so broad that some topics inevitably had to be left out. Much of the book is devoted to parameter estimation, which represents the main challenge in anisotropic processing and imaging. We focus primarily on inversion of prestack data, but do not address technical problems related to migration velocity analysis and migration algorithms for azimuthally anisotropic (in practice, still largely limited to TTI) models. Also, the book does not discuss rotation techniques and polarization layer stripping needed to separate split PS- or SS-waves used in multicomponent inversion methods. Several key references on shear-wave polarization analysis can be found in Tsvankin et al. (2010). An interesting application of anisotropy-related attributes, not covered here, is in time-lapse seismic monitoring, especially for compacting reservoirs. Another promising methodology of reservoir characterization, which is outside the scope of the book, is microseismic. Passive seismic measurements can be used to evaluate the permeability of fracture networks and calibrate fracture-characterization methods operating with VSP and surface reflection data.

In organizing the book, we kept the complexity of the subject in mind and generally adopted the recipe of Tsvankin's (2005) monograph. The main text is designed primarily to provide physical insights into anisotropic signatures and processing methods, with essential mathematical derivations given in the appendices. For a number of more complex mathematical developments, we refer the reader to the original papers. The purpose of the summary sections at the end of each chapter is not just to list the main conclusions but, in some cases, also to describe relevant results that could not be included in the book.

While we strived to be comprehensive in analyzing each topic, most of the material is based on our publications. We are deeply indebted to the following colleagues, including our former and current students, who collaborated with one or both of us on the papers used in the book: Abdulfattah Al-Dajani, Tariq Alkhalifah, Andrey Bakulin, Jyoti Behura, Jack Cohen, Pedro Contreras, Thomas Davis, Pawan Dewangan, Gerardo Franco, Carol Gentry, Baoniu Han, Jan Ove Hansen, Patsy Jorgensen, Mark Kachanov, Martin Landrø, Yves Le Stunff, Jorge Lopez, Albena Mateeva, Andres Pech, Andreas Rüger, Claude Signer, Stephen Theophanis, Leon Thomsen, Ivan Vasconcelos, Xiaoxiang Wang, Xiaoxia Xu, and Yaping Zhu. Fruitful discussions with other colleagues, too numerous to be mentioned here by name, have been indispensable in improving our understanding of the subject. We are particularly grateful to the A(nisotropy)-Team of the Center for Wave Phenomena (CWP) at the Colorado School of Mines (CSM), which has pioneered many groundbreaking developments in anisotropic processing and inversion methods. Special thanks to Ken Larner and Michael Slawinski, who made an extraordinary time commitment to provide comprehensive reviews of the book and many insightful suggestions.

The LaTeX style files for the book were set up by John Stockwell of CWP. We also appreciate the assistance of Barbara McLenon (CWP) and the SEG staff in preparing the manuscript for publication. The research program led by I. Tsvankin at CSM has been supported by the Consortium Project on Seismic Inverse Methods for Complex Structures at CWP, Office of Basic Energy Sciences of the U.S. Department of Energy, Petroleum Research Fund of the American Chemical Society, and Research Partnership to Secure Energy for America (RPSEA).

Chapter 1

Normal-moveout (NMO) ellipse and generalized Dix equation

Traveltimes of reflected waves (reflection moveout) in heterogeneous anisotropic media are usually modeled by multioffset and multiazimuth ray tracing (e.g., Gajewski and Pšenčík, 1987). Whereas anisotropic ray-tracing codes are sufficiently fast for forward modeling, their application in moveout inversion requires repeated generation of azimuthally-dependent traveltimes around many common-midpoint (CMP) locations, which makes the inversion procedure extremely time-consuming. Also, purely numerical solutions do not give insight into the influence of anisotropy on reflection traveltimes.

This chapter is devoted to analytic treatment of conventional-spread reflection moveout in anisotropic media. For models with moderate structural complexity and spreadlength-to-depth ratios close to unity, traveltimes in CMP geometry are well-described by normal-moveout (NMO) velocity defined in the zero-spread limit (Tsvankin and Thomsen, 1994; Tsvankin, 2005). Even in the presence of nonhyperbolic moveout, NMO velocity ($V_{\rm nmo}$) is still responsible for the most stable, conventional-offset portion of the moveout curve. The description of $V_{\rm nmo}$ given here provides an analytic basis for moveout inversion, helps evaluate the contribution of the anisotropy parameters to reflection traveltimes, and leads to a significant increase in the efficiency of traveltime modeling/inversion methods.

A theoretical framework for 3D anisotropic moveout analysis was proposed by Grechka and Tsvankin (1998a), who showed that NMO velocity of pure (i.e., non-converted) modes varies with azimuth as an *ellipse*, even if the medium is arbitrarily anisotropic and heterogeneous. This conclusion breaks down only for subsurface models in which CMP traveltime does not increase with offset (i.e., the case of reverse moveout) or cannot be described by a series expansion in offset. The orientation of the NMO ellipse and the values of its semiaxes can be expressed through the spatial derivatives of the slowness vector, which are determined by both the direction of the reflector normal and the medium properties above the reflector. The main results of Grechka and Tsvankin (1998a), including explicit expressions for the NMO ellipse in homogeneous transversely isotropic (TI) and orthorhombic layers, are discussed in Tsvankin's (2005) monograph.

The elliptical azimuthal dependence of the NMO-velocity function has been used to develop algorithms for stacking-velocity analysis and moveout correction of 3D wide-azimuth data. Even more important, the equation of the NMO ellipse provides a foundation for moveout-velocity inversion in anisotropic media described in Chapters 2, 5, and 6.

We briefly review the concept of the NMO ellipse at the beginning of this chapter. For a homogeneous anisotropic layer of arbitrary symmetry, the NMO ellipse is expressed as an explicit function of the components of the slowness vector. Then we present the generalized Dix equation for the effective NMO velocity in models that include a stack of horizontal, homogeneous, arbitrarily anisotropic layers above a horizontal or dipping reflector. While this equation is structured similarly to the conventional Dix formula, it is based on averaging of the 2×2 matrices that define the interval NMO ellipses.

To extend this approach to laterally heterogeneous media, we introduce the NMO-velocity *surface* obtained by plotting NMO velocity as the radius-vector along all possible directions in 3D space. The NMO ellipse at the surface and conventional-spread moveout as a whole can be modeled by Dix-type averaging of specifically oriented cross sections of NMO-velocity surfaces along the zero-offset reflection raypath. This procedure is particularly simple to implement for a stack of homogeneous anisotropic layers separated by plane dipping boundaries.

For general heterogeneous media, we present an efficient methodology to compute the NMO ellipse using dynamic ray tracing of a single (zero-offset) ray. We show that the quantities needed to obtain the geometrical spreading (e.g., Červený, Molotkov and Pšenčík, 1977; Kendall and Thomson, 1989; Červený, 2001) provide sufficient information to build the NMO ellipse and, therefore, model reflection moveout without tracing a large family of rays.

In addition to pure (PP or SS) modes, we analyze conventional-spread moveout of mode-converted waves in layer-cake media with a horizontal symmetry plane. Because of the mirror symmetry with respect to the interfaces, the traveltime series for PS-waves contains only even terms in offset and, for conventional spreadlengths, reduces to a hyperbolic equation parameterized by NMO velocity. We show that the azimuthally dependent V_{nmo} for converted waves has the same elliptical form as that for pure modes and can be found by a Dix-type summation of the matrices responsible for the NMO ellipses of PP- and SS-waves.

1.1 Equation of the NMO ellipse

Suppose that the traveltimes (*reflection moveout*) of a pure-mode (PP or SS) reflected wave have been recorded in a number of CMP gathers with different azimuthal orientation but the same midpoint location (Figure 1.1). If the medium is anisotropic and heterogeneous, the dependence of long-offset reflection traveltimes on the azimuth α of the CMP line can become rather complicated (see Chapter 3). For spreadlengths close to the distance between the common midpoint and the reflector, however, move-

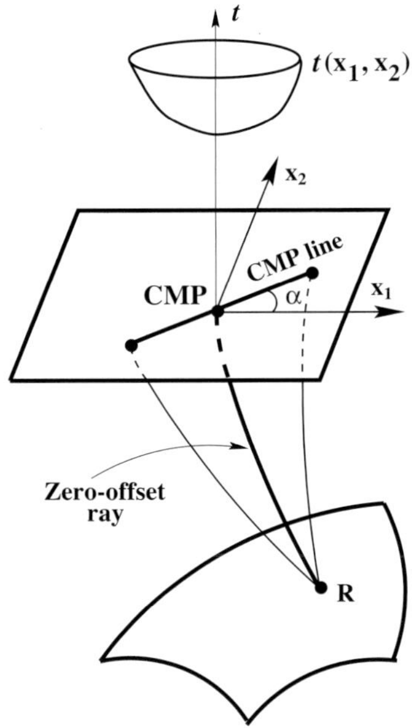

Figure 1.1: Normal-moveout velocity is calculated on CMP lines with different azimuths and a fixed midpoint location (Grechka et al., 1999b).

out in CMP geometry can be approximated by a hyperbolic equation,

$$t^2(\alpha) = t_0^2 + \frac{x^2}{V_{\rm nmo}^2(\alpha)}, \tag{1.1}$$

where t_0 is the zero-offset reflection time, x is the source-receiver offset, and $V_{\rm nmo}(\alpha)$ is the normal-moveout velocity defined as

$$V_{\rm nmo}^2(\alpha) = \lim_{x \to 0} \frac{d[x^2(\alpha)]}{d[t^2(\alpha)]}. \tag{1.2}$$

According to equation 1.2, NMO velocity determines the initial slope of the $t^2(x^2)$ curve at zero offset. For any realistic subsurface model, the function $t^2(x^2)$ deviates from a straight line because of the influence of heterogeneity and/or anisotropy, but equation 1.1 usually remains sufficiently accurate for conventional moderate offsets limited by the distance between the CMP and reflector (Tsvankin and Thomsen, 1994; Tsvankin, 2005).

The analysis below is based on the general result of Grechka and Tsvankin (1998a), who showed that NMO velocity is described by the following simple quadratic form:

$$V_{\rm nmo}^{-2}(\alpha) = W_{11}\cos^2\alpha + 2W_{12}\sin\alpha\cos\alpha + W_{22}\sin^2\alpha, \tag{1.3}$$

where **W** is a symmetric matrix,

$$W_{ij} = \tau_0 \left. \frac{\partial^2 \tau}{\partial x_i \, \partial x_j} \right|_{\mathbf{x}_{\rm CMP}} = \tau_0 \left. \frac{\partial p_i}{\partial x_j} \right|_{\mathbf{x}_{\rm CMP}} , \qquad (i, j = 1, 2) . \tag{1.4}$$

Here, $\tau(x_1, x_2)$ is the *one-way* traveltime from the zero-offset reflection point to the location $\mathbf{x} = \{x_1, x_2\}$ at the surface, τ_0 is the one-way zero-offset traveltime, p_i are the components of the slowness vector corresponding to the ray emerging at point $\mathbf{x}$, and $\mathbf{x}_{\rm CMP}$ is the CMP location. In Figure 1.1, the common midpoint coincides with the origin of the coordinate system; however, shifting the origin to any other location does not change the derivatives in equation 1.4. The one-way traveltimes appear in equation 1.4 because reflection-point dispersal has no influence on NMO velocity of pure modes, and rays recorded at relatively small offsets can be assumed to propagate through the reflection point of the zero-offset ray (Hubral and Krey, 1980; Tsvankin, 1995a).

It is convenient to use the eigenvectors of the matrix **W** as the auxiliary horizontal axes rotated by the angle β with respect to the original coordinate frame:

$$\beta = \tan^{-1} \left[\frac{W_{22} - W_{11} + \sqrt{(W_{22} - W_{11})^2 + 4W_{12}^2}}{2W_{12}} \right] . \tag{1.5}$$

Equation 1.3 then takes the form:

$$V_{\rm nmo}^{-2}(\alpha) = \lambda_1 \cos^2(\alpha - \beta) + \lambda_2 \sin^2(\alpha - \beta) , \tag{1.6}$$

where $\lambda_{1,2}$ are the eigenvalues of the matrix **W**. Grechka and Tsvankin (1998a) conclude that for positive λ_1 and λ_2 the NMO velocity from equation 1.3, plotted as the radius-vector in each azimuthal direction, defines a centered *ellipse*. A negative eigenvalue implies that $V_{\rm nmo}^2 < 0$ in certain azimuthal directions where the CMP traveltime decreases with offset. Although such reverse moveout can exist in some cases (e.g., for turning waves, as described by Hale et al., 1992), typically both λ_1 and λ_2 are positive, and the azimuthal dependence of NMO velocity is indeed elliptical. This conclusion is valid for arbitrarily anisotropic, heterogeneous media provided the traveltime field is sufficiently smooth to be adequately approximated by a Taylor series expansion.

1.2 NMO ellipse for a homogeneous layer

1.2.1 Arbitrary anisotropic symmetry

To obtain NMO velocity for any given model from equations 1.3 and 1.4, we need to evaluate the spatial derivatives of the slowness vector at the CMP location. As demonstrated by Grechka et al. (1999b; section 1.2 follows their results), for the model of a single homogeneous layer this can be done by using the relationship between the group-velocity and slowness vectors (see Appendix 1A):

$$\mathbf{W} = \frac{p_1 q_{,1} + p_2 q_{,2} - q}{q_{,11} q_{,22} - q_{,12}^2} \begin{pmatrix} q_{,22} & -q_{,12} \\ -q_{,12} & q_{,11} \end{pmatrix}, \tag{1.7}$$

$$\begin{aligned} V_{\mathrm{nmo}}^{-2}(\alpha) &\equiv V_{\mathrm{nmo}}^{-2}(\alpha, p_1, p_2) \\ &= \frac{p_1 q_{,1} + p_2 q_{,2} - q}{q_{,11} q_{,22} - q_{,12}^2} \left[q_{,22} \cos^2\alpha - 2q_{,12} \sin\alpha \cos\alpha + q_{,11} \sin^2\alpha \right]; \end{aligned} \tag{1.8}$$

$q \equiv q(p_1, p_2) \equiv p_3$ is the vertical slowness as a function of the horizontal slownesses p_1 and p_2, $q_{,i} \equiv \partial q / \partial p_i$, and $q_{,ij} \equiv \partial^2 q / (\partial p_i \partial p_j)$. The derivatives in equation 1.8 are evaluated for the zero-offset ray.

Equation 1.8 is valid for pure modes reflected from a horizontal or curved interface beneath an arbitrarily anisotropic, homogeneous layer with any strength of anisotropy (i.e., for any magnitude of the anisotropy parameters). The NMO velocity is fully determined by the azimuth α of the CMP line and by the slowness surface near the slowness direction corresponding to the zero-offset ray. According to Snell's law, the slowness vector of the zero-offset ray is normal to the reflecting interface at the reflection point. Therefore, the slowness components p_1, p_2 and q can be found by solving the Christoffel equation for the slowness (phase) direction normal to the reflector. Because this equation is cubic with respect to the squared phase velocity, it yields explicit expressions for the phase and slowness vectors.

The derivatives of the vertical slowness q can be found directly from the Christoffel equation as well. The slowness components satisfy the equation $\Gamma(q, p_1, p_2) = 0$, where Γ is (in general) a sixth-order polynomial with respect to q. For anisotropic models with a horizontal symmetry plane, Γ becomes a *cubic* polynomial with respect to q^2. In either case, the derivatives $q_{,i}$ and $q_{,ij}$ can be obtained as

$$q_{,i} = -\frac{\Gamma_{p_i}}{\Gamma_q} \tag{1.9}$$

and

$$q_{,ij} = -\frac{\Gamma_{p_i p_j} + \Gamma_{p_i q}\, q_{,j} + \Gamma_{p_j q}\, q_{,i} + \Gamma_{qq}\, q_{,i}\, q_{,j}}{\Gamma_q}, \tag{1.10}$$

where $\Gamma_{p_i} \equiv \partial\Gamma/\partial p_i$, $\Gamma_q \equiv \partial\Gamma/\partial q$, $\Gamma_{p_i p_j} \equiv \partial^2\Gamma/(\partial p_i \partial p_j)$, $\Gamma_{p_i q} \equiv \partial^2\Gamma/(\partial p_i \partial q)$, and $\Gamma_{qq} \equiv \partial^2\Gamma/\partial q^2$. Therefore, all terms in equation 1.8 can be obtained *explicitly* from the Christoffel equation.

Equation 1.8 can also be used to develop weak-anisotropy approximations for NMO velocity by linearizing q and its derivatives in the dimensionless anisotropy parameters or in perturbations of the stiffness coefficients. These analytic approximations provide valuable insight into the influence of the anisotropy parameters on normal moveout (e.g., Tsvankin, 1995a, 2005; Cohen, 1998). There is no need,

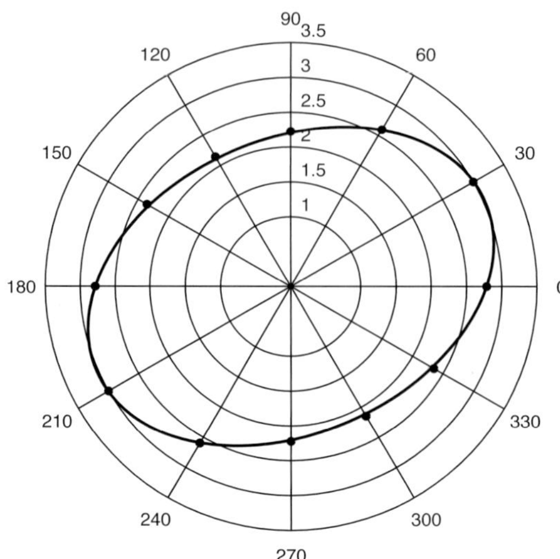

Figure 1.2: Comparison of the P-wave NMO velocity from equation 1.8 (solid line) with the moveout (stacking) velocity (dots) obtained by least-squares fitting of a hyperbola to the exact traveltimes computed for spreadlength equal to the distance between the CMP and the reflector (Grechka et al., 1999b). The model contains a homogeneous orthorhombic layer (with the vertical symmetry planes at azimuths 0° and 90°) above a plane dipping reflector; the dip and azimuth of the reflector are equal to 30° (azimuth is shown on the perimeter). The relevant medium parameters [in Tsvankin's (1997a; 2005) notation; see Appendix 1B] are V_{P0}=2.0 km/s, $\epsilon^{(1)}$=0.110, $\delta^{(1)}$=−0.035, $\epsilon^{(2)}$=0.225, $\delta^{(2)}$=0.100, and $\delta^{(3)}$=0. The vertical symmetry plane at zero azimuth has the properties of the VTI model of Dog Creek shale, while the orthogonal vertical symmetry plane is equivalent to Taylor sandstone; both models are described in Thomsen (1986).

however, to substitute weak-anisotropy approximations for the exact equations in numerical modeling and inversion.

Therefore, equation 1.8 gives a simple and numerically efficient recipe to model azimuthally-dependent reflection moveout in a homogeneous layer. The example in Figure 1.2, generated for an orthorhombic layer[1] above a dipping reflector, illustrates the high accuracy of the hyperbolic moveout equation parameterized by the analytic NMO velocity (equation 1.8) in describing conventional-spread reflection traveltimes. Despite the presence of anisotropy-induced nonhyperbolic moveout, the P-wave NMO velocity is close to the moveout (stacking) velocity estimated from the exact traveltimes on six CMP lines with different orientation. The maximum difference between V_{nmo} (solid line) and the finite-spread moveout velocity (dots) is just 1.4%, which is even less than the corresponding value (2.7%) for the same model, but with a

[1]See Helbig (1994) or Tsvankin (2005) for a detailed description of symmetry types. The main properties of orthorhombic models are also described in section 2.5 and Appendix 1B.

horizontal reflector (see Grechka and Tsvankin, 1998a). Therefore, the magnitude of nonhyperbolic moveout for this model decreases with reflector dip (see Chapter 3); the same observation was made by Tsvankin (1995a) for vertical transverse isotropy. Note that although the azimuth of the dip plane of the reflector is equal to 30°, the major axis of the $V_{\rm nmo}(\alpha)$ ellipse has the azimuth close to 24° because of the influence of the azimuthal anisotropy above the reflector.

1.2.2 Special cases

Model with a vertical symmetry plane

Next, consider a model in which the dip plane of the reflector coincides with a vertical symmetry plane of the layer. The medium can be, for instance, transversely isotropic, orthorhombic, or monoclinic. The mirror symmetry of the model as a whole with respect to the dip plane implies that one of the axes of the NMO ellipse points in the dip direction, formally shown below.

It is convenient to align the x_1-axis with the azimuth of the dip plane, while the x_2-axis will point in the strike direction. Evidently, the zero-offset ray should lie in the vertical symmetry plane $x_2=0$, and its slowness component p_2 goes to zero. As another consequence of the mirror symmetry with respect to the dip plane, $\partial p_2/\partial x_1=0$ (i.e., rays recorded in the symmetry plane $x_2=0$ cannot have a nonzero p_2-component), so the cross-terms W_{12} (equation 1.3) and $q_{,12}$ (equation 1.8) vanish, and the NMO velocity in equation 1.8 simplifies to

$$V_{\rm nmo}^{-2}(\alpha, p_1) = \frac{p_1 q_{,1} - q}{q_{,11} q_{,22}} \left[q_{,22} \cos^2\alpha + q_{,11} \sin^2\alpha\right]. \tag{1.11}$$

Equation 1.11 describes an ellipse with the semiaxes in the dip ($\alpha=0$) and strike ($\alpha=\pi/2$) directions:

$$V_{\rm nmo}^2(\alpha=0, p_1) = \frac{q_{,11}}{p_1 q_{,1} - q}, \tag{1.12}$$

$$V_{\rm nmo}^2(\alpha=\frac{\pi}{2}, p_1) = \frac{q_{,22}}{p_1 q_{,1} - q}. \tag{1.13}$$

The dip-line NMO velocity 1.12 was originally obtained by Tsvankin (1995a) as a function of the in-plane phase velocity V expressed through the phase angle θ with the vertical:

$$V_{\rm nmo}(0,\phi) = \frac{V(\phi)}{\cos\phi} \frac{\sqrt{1+\frac{1}{V(\phi)}\frac{d^2V}{d\theta^2}\Big|_{\theta=\phi}}}{1-\frac{\tan\phi}{V(\phi)}\frac{dV}{d\theta}\Big|_{\theta=\phi}}; \tag{1.14}$$

ϕ is the reflector dip. $V_{\rm nmo}^2(0,p_1)$ in the form 1.12 was first given by Cohen (1998). Equation 1.13 provides a similar representation for the NMO velocity in the strike direction.

Equations 1.12 and 1.13 are always valid for vertical transverse isotropy because of the mirror symmetry with respect to *any* vertical plane in that model. The vertical slowness for VTI media can be represented as $q(p_1, p_2) \equiv q\left(\sqrt{p_1^2 + p_2^2}\right)$, so for $p_2 = 0$, $q_{,22} = q_{,1}/p_1$. Then equation 1.13 for the strike-line NMO velocity reduces to the expression obtained by Grechka and Tsvankin (1998a),

$$[V_{\rm nmo}^{\rm VTI}]^2\,(\alpha = \frac{\pi}{2}, p_1) = \frac{q_{,1}}{p_1\,(p_1 q_{,1} - q)}\,. \tag{1.15}$$

Grechka and Tsvankin (1998a) also gave an equivalent form of equation 1.15 in terms of the phase-velocity function and linearized it in the anisotropy parameters. Due to the axial symmetry of the VTI model, both the dip-line (equation 1.12) and strike-line (equation 1.15) NMO velocities depend on the derivatives of q with respect to the single horizontal (in-plane) slowness component (p_1). The equation for $q^2(p_1^2)$ in VTI media is particularly easy to solve because it splits into a quadratic equation for P- and SV-waves and a linear equation for SH-waves.

Finally, in isotropic media the vertical slowness can be directly expressed through the reflector dip ϕ:

$$q^{\rm ISO} = \sqrt{V^{-2} - p_1^2} = \frac{\cos\phi}{V}\,,$$

and equations 1.12 (or 1.14) and 1.15 yield the well-known relationships published by Levin (1971):

$$V_{\rm nmo}^{\rm ISO}\,(\alpha = 0) = \frac{V}{\cos\phi}\,, \tag{1.16}$$

$$V_{\rm nmo}^{\rm ISO}\,(\alpha = \frac{\pi}{2}) = V\,. \tag{1.17}$$

Horizontal reflector

For a horizontal reflector ($p_1 = p_2 = 0$), equation 1.8 reduces to

$$V_{\rm nmo}^{-2}(\alpha, 0, 0) = -\,\frac{q}{q_{,11}\,q_{,22} - q_{,12}^2}\left[q_{,22}\cos^2\alpha - 2q_{,12}\sin\alpha\cos\alpha + q_{,11}\sin^2\alpha\right], \tag{1.18}$$

where q and $q_{,ij}$ should be evaluated for the vertical slowness vector.

Further simplification can be achieved for a medium with a vertical symmetry plane. Aligning one of the horizontal coordinate axes with the symmetry-plane direction and substituting $q_{,12} = 0$ into equation 1.18 (or $p_1 = 0$ into equation 1.11), we find

$$V_{\rm nmo}^{-2}(\alpha) = -\,\frac{q}{q_{,11}\,q_{,22}}\left[q_{,22}\cos^2\alpha + q_{,11}\sin^2\alpha\right] \tag{1.19}$$

$$= \frac{\sin^2\alpha}{\left[V_{\rm nmo}^{(1)}\right]^2} + \frac{\cos^2\alpha}{\left[V_{\rm nmo}^{(2)}\right]^2}\,, \tag{1.20}$$

where $V_{\rm nmo}^{(1)}$ and $V_{\rm nmo}^{(2)}$ are the semiaxes of the NMO ellipse:

$$V_{\rm nmo}^{(1)} = \sqrt{-\frac{q_{,22}}{q}}\,, \tag{1.21}$$

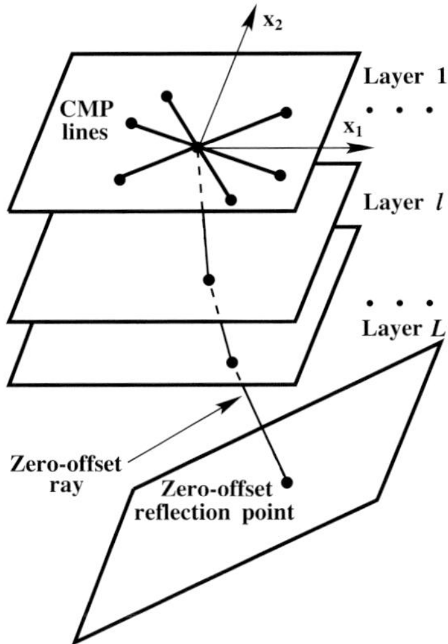

Figure 1.3: Dipping reflector beneath a horizontally layered overburden. The NMO velocity in this model can be obtained from the generalized Dix equation (Grechka et al., 1999b).

$$V_{\rm nmo}^{(2)} = \sqrt{-\frac{q_{,11}}{q}} \,. \tag{1.22}$$

For example, equations 1.19 – 1.22 are valid for an orthorhombic layer with a horizontal symmetry plane because the other two symmetry planes are vertical (the symmetry planes of orthorhombic media are mutually orthogonal). Then, as discussed in sections 2.5 and 6.1, $V_{\rm nmo}^{(1)}$ and $V_{\rm nmo}^{(2)}$ represent the symmetry-plane NMO velocities, which can be found for each mode (P, S_1, or S_2) by analogy with VTI media (Tsvankin, 1997a; Grechka et al., 1999b).

1.3 NMO ellipse for a layer-cake medium above a dipping reflector

1.3.1 Generalized Dix equation

Here, we show that the NMO ellipse for vertically heterogeneous arbitrarily anisotropic media above a dipping reflector (Figure 1.3) can be obtained by Dix-type averaging of the matrices **W** responsible for the interval NMO ellipses. Our derivation, based on the work by Grechka et al. (1999b), generalizes the approach employed by Alkhalifah and Tsvankin (1995) to obtain a 2D Dix-type NMO equation for vertical symmetry planes. We make no assumptions about the mutual orientation of the CMP line and reflector strike, and take full account of out-of-plane phenomena associated with both model geometry and depth-varying anisotropy.

The effective NMO ellipse recorded at the surface can be obtained from the matrix **W** defined in equation 1.4:

$$W_{ij}(L) = \tau(L) \frac{\partial p_i}{\partial x_j(L)}, \qquad (i, j = 1, 2), \tag{1.23}$$

where $\tau(L)$ is the total one-way zero-offset traveltime and $x_i(L)$ is the horizontal ray displacement between the zero-offset reflection point located at the Lth (generally dipping) interface and the surface (Figure 1.3). Due to the continuity of the ray, both $\tau(L)$ and $x_i(L)$ are equal to the sum of the respective interval values:

$$\tau(L) = \sum_{\ell=1}^{L} \tau_\ell, \tag{1.24}$$

$$x_i(L) = \sum_{\ell=1}^{L} x_{i,\ell}, \qquad (i = 1, 2). \tag{1.25}$$

(Note that here and below in the section on layered media, the comma in the subscripts separates the layer index and *does not* denote differentiation.)

It is convenient to introduce an auxiliary matrix,

$$Y_{ij}(L) \equiv \frac{\partial x_i(L)}{\partial p_j}, \qquad (i, j = 1, 2), \tag{1.26}$$

with the derivatives evaluated for the ray parameters[2] p_1 and p_2 of the zero-offset ray. Then,

$$\mathbf{W} \equiv \mathbf{W}(L) = \tau(L)\, \mathbf{Y}^{-1}(L). \tag{1.27}$$

Because the medium above the reflector is laterally homogeneous, the horizontal components p_1 and p_2 of the slowness vector remain constant along any given ray between the reflection point and the surface. Therefore, substituting equation 1.25 into equation 1.26, we find

$$Y_{ij}(L) \equiv \frac{\partial x_i(L)}{\partial p_j} = \sum_{\ell=1}^{L} \frac{\partial x_{i,\ell}}{\partial p_j} \equiv \sum_{\ell=1}^{L} Y_{ij,\ell}. \tag{1.28}$$

Equation 1.28 explains the reason for introducing the effective matrix $\mathbf{Y}(L)$: unlike the matrix **W**, it can be decomposed into the sum of the matrices $\mathbf{Y}_\ell$ for the individual layers. Since all intermediate boundaries are horizontal, the ray displacements $x_{i,\ell}$ in any layer coincide with the values that should be used in computing the matrix

[2]Throughout the book, "ray parameter" is used as a synonym for "horizontal slowness," whether or not the horizontal slowness is preserved along the ray. In an alternative formulation, ray parameter is defined as a quantity that remains constant along the ray, which helps obtain analytic expressions for traveltime, ray trajectory, and polarization in anisotropic heterogeneous media (Slawinski, 2010, chapter 14; Epstein and Slawinski, 2009; Slawinski and Webster, 1999).

W and the interval NMO velocity for this particular layer. Hence, we can apply equation 1.27 to layer ℓ:

$$\mathbf{W}_\ell = \tau_\ell \, \mathbf{Y}_\ell^{-1} \tag{1.29}$$

and

$$\mathbf{Y}_\ell = \tau_\ell \, \mathbf{W}_\ell^{-1} \,. \tag{1.30}$$

Substituting equations 1.30 and 1.28 into equation 1.27 leads to the final result:

$$\mathbf{W}^{-1}(L) = \frac{1}{\tau(L)} \sum_{\ell=1}^{L} \tau_\ell \, \mathbf{W}_\ell^{-1} \,. \tag{1.31}$$

The interval matrices $\mathbf{W}_\ell$ in terms of the components of the slowness vector are given by equation 1.7, whereas the traveltimes τ_ℓ should be obtained from kinematic ray tracing (i.e., by computing group velocity) of the zero-offset ray. Because the eigenvalues of the matrices $\mathbf{W}_\ell$ and $\mathbf{W}(L)$ usually are positive (under the assumptions discussed above), these matrices are nonsingular and can be inverted.

Equation 1.31 performs Dix-type averaging of the interval matrices $\mathbf{W}_\ell$ to obtain the effective matrix $\mathbf{W}(L)$ and the effective NMO ellipse $V_{\text{nmo}}(\alpha, L)$ (equation 1.3). It should be emphasized that the interval NMO velocities $V_{\text{nmo},\ell}(\alpha)$ (or the interval matrices $\mathbf{W}_\ell$) in equation 1.31 are computed for the horizontal components of the slowness vector of the zero-offset ray. As mentioned above, this vector is perpendicular to the reflector at the reflection point. This means that the interval matrices $\mathbf{W}_\ell$ correspond to the generally *nonexistent* reflectors that are orthogonal to the slowness vector of the zero-offset ray in each layer.

Rewriting equation 1.31 in the "Dix differentiation" form gives

$$\mathbf{W}_\ell^{-1} = \frac{\tau(\ell) \, \mathbf{W}^{-1}(\ell) - \tau(\ell - 1) \, \mathbf{W}^{-1}(\ell - 1)}{\tau(\ell) - \tau(\ell - 1)} \,. \tag{1.32}$$

Equations 1.31 and 1.32 generalize the Dix (1955) formula for horizontally layered, arbitrarily anisotropic media above a dipping reflector. Formally, this extension looks relatively straightforward: the squared NMO velocities in the Dix formula are simply replaced by the inverse matrices $\mathbf{W}^{-1}$. Also, the robustness of the generalized Dix differentiation is subject to the same limitation as that of its conventional counterpart: the thickness of the layer of interest should not be too small relative to the layer's depth.

In contrast to the conventional Dix equation, however, the matrix $\mathbf{W}^{-1}(\ell - 1)$ in equation 1.32 *cannot* be obtained from seismic data directly because the corresponding reflector usually does not exist in the subsurface. Therefore, layer stripping by means of equation 1.32 involves recalculating each interval matrix $\mathbf{W}_\ell$ from one value of the slowness vector (corresponding to a certain real reflector in a given layer) to another – that of the zero-offset ray. If the medium is anisotropic, this procedure requires knowledge of the interval anisotropy parameters, as discussed for the 2D case by Alkhalifah and Tsvankin (1995), Alkhalifah (1997), and Tsvankin (2005).

Only in the simplest special case of a *horizontal* reflector, does the slowness vector of the zero-offset ray remain straight (vertical) all the way to the surface, and the interval matrices $\mathbf{W}_\ell$ correspond to the NMO velocities from horizontal interfaces, which can be measured from reflection data. Although such a model is laterally homogeneous, the zero-offset *ray* is not necessarily vertical if the medium does not have a horizontal symmetry plane.

1.3.2 Model with a vertical symmetry plane

Here, we consider the same special case as for the single-layer model – a medium in which all layers have a common vertical symmetry plane that coincides with the dip plane of the reflector (e.g., each layer may be VTI). If the x_1-axis lies in the symmetry plane, the interval matrices $\mathbf{W}_\ell$ are diagonal (see above), and

$$W_{12,\ell} = 0\,. \tag{1.33}$$

Consequently, the off-diagonal elements of the effective matrix $\mathbf{W}(L)$ (equation 1.31) vanish as well:

$$W_{12}(L) = 0\,. \tag{1.34}$$

When the matrix $\mathbf{W}$ is diagonal, its two components directly determine the semiaxes of the NMO ellipse (see equation 1.3):

$$W_{kk,\ell} = [V^{(k)}_{\mathrm{nmo},\ell}]^{-2} \tag{1.35}$$

and

$$W_{kk}(L) = [V^{(k)}_{\mathrm{nmo}}(L)]^{-2}\,, \qquad (k = 1, 2)\,, \tag{1.36}$$

where $V^{(1)}_{\mathrm{nmo}} \equiv V_{\mathrm{nmo}}(\alpha = 0)$ and $V^{(2)}_{\mathrm{nmo}} \equiv V_{\mathrm{nmo}}(\alpha = \pi/2)$ are the NMO velocities measured in the dip and strike directions, respectively.

Substitution of equations 1.33 – 1.36 into equations 1.31 and 1.32 yields more conventional Dix-type averaging and differentiation formulas for the dip and strike components of the NMO velocity:

$$[V^{(k)}_{\mathrm{nmo}}(L)]^2 = \frac{1}{\tau(L)} \sum_{\ell=1}^{L} \tau_\ell\, [V^{(k)}_{\mathrm{nmo},\ell}]^2 \tag{1.37}$$

and

$$[V^{(k)}_{\mathrm{nmo},\ell}]^2 = \frac{\tau(\ell)\,[V^{(k)}_{\mathrm{nmo}}(\ell)]^2 - \tau(\ell-1)\,[V^{(k)}_{\mathrm{nmo}}(\ell-1)]^2}{\tau(\ell) - \tau(\ell-1)}\,, \qquad (k = 1, 2)\,. \tag{1.38}$$

Equations 1.37 and 1.38 for the dip-line ($k = 1$) NMO velocity coincide with the 2D Dix-type equations (Alkhalifah and Tsvankin, 1995; Tsvankin, 2005). Our derivation shows that the same Dix-type formalism can be applied to the *strike* component ($k = 2$) of the NMO velocity, which determines the second semiaxis of the NMO ellipse. Despite the close resemblance of expressions 1.37 and 1.38 to the conventional Dix equation, one should keep in mind that the interval NMO velocities in equations 1.37 and 1.38 correspond to the *nonexistent* reflectors perpendicular to the slowness vector of the zero-offset ray in each layer.

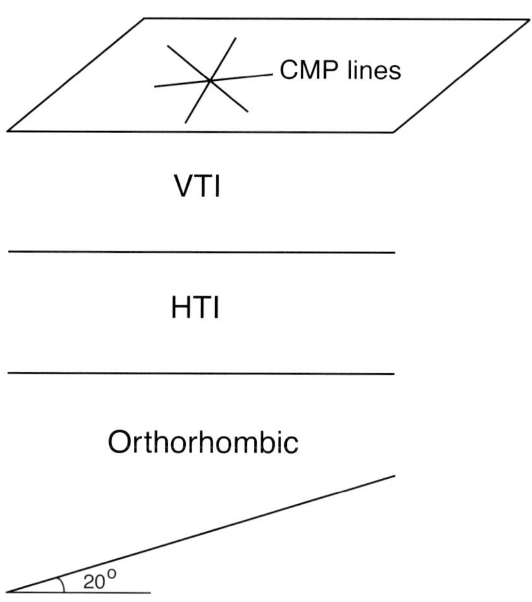

Figure 1.4: Model used in Figure 1.5 to check the accuracy of the generalized Dix equation (Grechka et al., 1999b). Layer 1 is VTI with $V_{P0,1}$=2.5 km/s, ϵ_1=0.2, and δ_1=0.1. Layer 2 is HTI (TI with a horizontal symmetry axis) with the symmetry-axis azimuth β_2=30°, $V_{P0,2}$=3.0 km/s, $\epsilon_2^{(2)}=\epsilon_2^{(V)}=-0.2$, and $\delta_2^{(2)}=\delta_2^{(V)}=-0.15$ (notation for HTI and orthorhombic media is described in Appendix 1B). Layer 3 is orthorhombic with $V_{P0,3}$=3.5 km/s, $\epsilon_3^{(1)}$=0.2, $\delta_3^{(1)}$=0.15, $\epsilon_3^{(2)}=-0.3$, $\delta_3^{(2)}=-0.2$, and $\delta_3^{(3)}$=0.05; the azimuth of the $[x_1, x_3]$ symmetry plane is β_3=60°. The interface depths are z_1=1.0 km, z_2=2.0 km, and z_3=3.0 km. The reflector dip is 20° and the dip-plane azimuth is 0°.

1.3.3 Synthetic example

The accuracy of the single-layer NMO equation was discussed above (see Figure 1.2). Figure 1.5 illustrates the performance of the generalized Dix equation 1.31 for a model that includes three anisotropic layers with different symmetry above a dipping reflector (Figure 1.4). P-wave reflection traveltimes were computed by ray tracing in six azimuthal directions with an increment of 30°. The moveout velocities (dots in Figure 1.5a) were estimated by fitting a hyperbola to the exact traveltimes. Despite the complexity of the model, the best-fit ellipse found from the finite-spread moveout velocities (dashed line) is sufficiently close to the theoretical NMO ellipse (solid) computed from equations 1.31 and 1.3. The small difference between the ellipses is caused by the influence of nonhyperbolic moveout associated with both anisotropy and vertical heterogeneity. From Figure 1.5b, the contribution of nonhyperbolic moveout typically becomes substantial only at source-receiver offsets that exceed the distance between the CMP and the reflector.

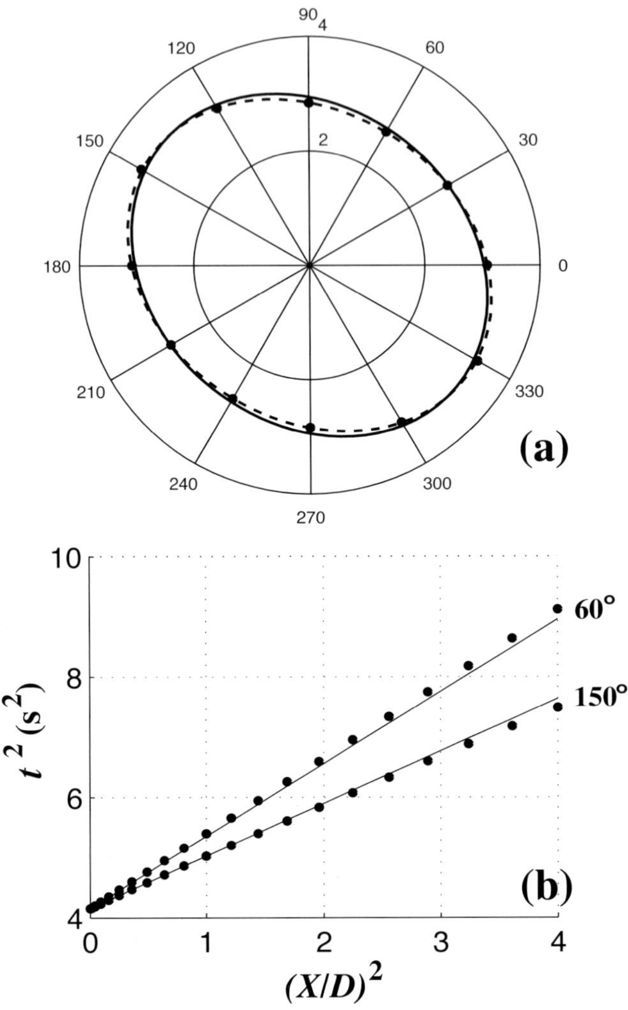

Figure 1.5: (a) Comparison between the P-wave NMO ellipse calculated from the generalized Dix equation 1.31 (solid curve) and moveout velocities obtained from ray-traced traveltimes (dots) for spreadlength equal to the distance between the CMP and the reflector (Grechka et al., 1999b). The model is shown in Figure 1.4; the dashed line marks the best-fit ellipse found from the finite-spread moveout velocities. (b) The hyperbolic moveout curve parameterized by the exact NMO velocity (solid lines) and ray-traced traveltimes (dots) in two azimuthal directions (60° and 150°); D is the CMP-to-reflector distance.

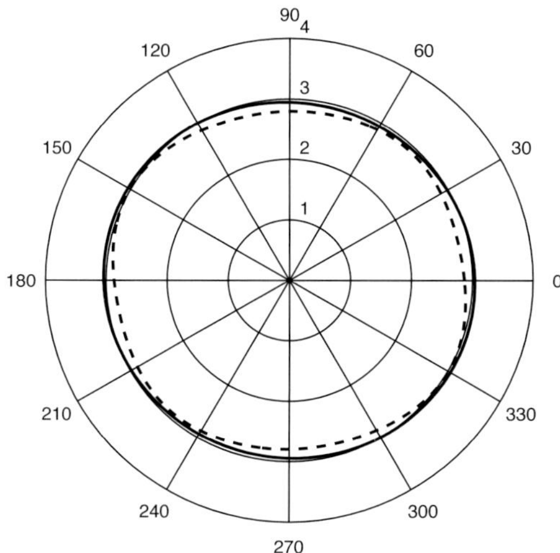

Figure 1.6: Comparison of the exact P-wave NMO ellipse (solid line) and an approximate NMO velocity obtained by the Dix-type averaging (e.g., equation 1.37) of the interval NMO velocities (dashed line) for each azimuth (Grechka et al., 1999b). The model contains three horizontal orthorhombic layers with a horizontal ($[x_1, x_2]$) symmetry plane. The azimuth of the $[x_1, x_3]$ symmetry plane (also, the direction of one of the axes of the interval NMO ellipse) in the first (top) layer is β_1=0°, in the second layer β_2=45°, and in the third layer β_3=60°. The vertical P-wave velocities are $V_{P0,1}$=2.0 km/s, $V_{P0,2}$=3.0 km/s, and $V_{P0,3}$=3.5 km/s; the interval zero-offset traveltimes are equal to one another ($\tau_1=\tau_2=\tau_3$=1.0 s). The relevant anisotropy parameters are (subscripts denote the layer number): $\delta_1^{(1)}$=0.25 and $\delta_1^{(2)}=-0.15$ (layer 1); $\delta_2^{(1)}=-0.20$ and $\delta_2^{(2)}$=0.20 (layer 2); and $\delta_3^{(1)}$=0.25 and $\delta_3^{(2)}=-0.15$ (layer 3).

1.3.4 Accuracy of rms averaging of NMO velocities

Although the generalized Dix equation 1.31 operates with the matrices $\mathbf{W}_\ell^{-1}$, we demonstrated that Dix-type averaging can be applied to the dip- and strike-components of the NMO velocity in a model that has a common (that is, throughgoing) vertical symmetry plane aligned with the dip plane of the reflector. Also, from the results of the previous section rms averaging of the interval NMO velocities is valid in any azimuthal direction, if all interval NMO ellipses degenerate into circles. Hence, the error of this more conventional averaging procedure depends on the elongation of the interval ellipses, a quantity controlled by both azimuthal anisotropy and reflector dip. Grechka et al. (1999b) show that this error increases rather slowly as the interval ellipses deviate from a circle because the rms averaging of the interval velocities computed for any given azimuth provides a *linear* approximation (in terms of the "eccentricity" coefficient) to the exact NMO velocity for the same azimuth.

To quantify this conclusion, we consider two numerical examples. Figure 1.6 shows the azimuthally dependent P-wave NMO velocity in an orthorhombic medium

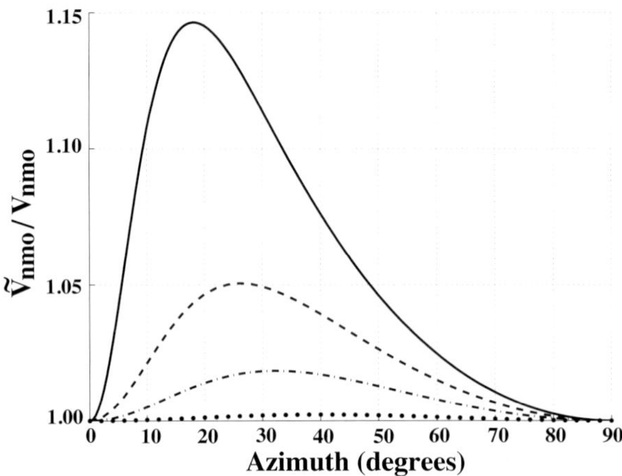

Figure 1.7: Rms-averaged NMO velocity $\tilde{V}_{\text{nmo}}$ for a given azimuth normalized by the exact V_{nmo} computed from the generalized Dix equation 1.31 (Grechka et al., 1999b). The model contains three isotropic layers above a dipping reflector; the interval NMO velocity can be computed from equations 1.16 and 1.17 (Levin, 1971). The interval velocities are V_1=2.0 km/s, V_2=3.0 km/s, and V_2=3.5 km/s; the interval zero-offset traveltimes are τ_1=τ_2=τ_3=1.0 s. The reflector dip is ϕ=40° (dotted line), ϕ=60° (dash-dotted), ϕ=70° (dashed), and ϕ=80° (solid); the azimuth is measured with respect to the dip plane of the reflector.

consisting of three horizontal layers with strong azimuthal anisotropy (i.e., with a large difference between the parameters $\delta^{(1)}$ and $\delta^{(2)}$). While the exact NMO ellipse (solid line) happens to be close to a circle, the approximate, rms-averaged normal-moveout velocity (dashed line) has a nonelliptical shape because the interval NMO ellipses deviate significantly from circles. The maximum error of the rms averaging is about 6.3%, which will lead to much higher errors in the interval velocities after application of the Dix differentiation (equation 1.38). Therefore, for this model it is necessary to use the exact NMO equation, which properly accounts for the influence of azimuthal anisotropy on normal moveout.

For a horizontal reflector and moderate azimuthal anisotropy above it (i.e., with interval NMO-velocity variation limited by 10% – 20%), the accuracy of rms averaging of NMO velocities is much higher. This implies that for such media it is possible to obtain a close approximation for the interval NMO velocity by conventional Dix differentiation in any given azimuthal direction. In the special case of horizontally layered HTI media, this conclusion was made by Al-Dajani and Tsvankin (1998).

Another example, in which the interval NMO ellipses differ from circles due to the influence of reflector dip in a purely isotropic layered model, is shown in Figure 1.7. For isotropic media the dip plane of the reflector always represents a symmetry plane, and one axis of all interval NMO ellipses is parallel to the dip direction. Then rms averaging of the interval NMO velocities (equation 1.37) becomes exact for the dip (azimuth $\alpha = 0°$) and strike ($\alpha = 90°$) lines (Figure 1.7). Although for all other azimuthal directions the NMO velocity has to be found from the generalized Dix equation, Figure 1.7 indicates that rms averaging is quite accurate for a wide range of dips, with the error reaching only 1.85% for the dip $\phi = 60°$. The error increases with dip as the interval NMO ellipses become more elongated.

For dipping reflectors, however, the Dix differentiation cannot be applied in the standard fashion, even if the dip is mild and the rms-averaging equation provides sufficient accuracy. Indeed, the interval NMO velocities of dipping events are calculated for nonexistent reflectors and cannot be found directly from the data. For isotropic media it is straightforward to obtain these interval velocities from V_{nmo} for the corresponding horizontal interfaces. In the presence of anisotropy, interval parameter estimation using dipping events involves a layer-stripping procedure that requires reconstruction of the NMO ellipses in the overburden (see Chapter 2).

On the whole, we recommend to use the generalized Dix equation for any azimuthally anisotropic model, provided the azimuthal data coverage is sufficient to reconstruct the function $V_{\text{nmo}}(\alpha)$. Operating with the NMO ellipses rather than individual azimuthal moveout measurements gives the additional advantage of smoothing the azimuthal variation of NMO velocity, which helps eliminate "outliers" and stabilize interval parameter estimation. A field-data application of the generalized Dix equation is discussed in Chapter 2.

1.4 NMO-velocity surface

The generalized Dix equation discussed above is restricted to laterally homogeneous models above the reflector. Next, using the results of Grechka and Tsvankin (2002a), we introduce the concept of *NMO-velocity surfaces* and use it to develop a Dix-type averaging scheme for more complicated, laterally heterogeneous media.

To define the NMO-velocity surface, let us examine the normal-moveout velocity measured in common-midpoint geometry along an arbitrary direction $\boldsymbol{\mathcal{L}}$ in 3D space. (One can imagine, for instance, recording reflection arrivals along an oblique or vertical borehole.) If the vectors $V_{\text{nmo}}(\boldsymbol{\mathcal{L}})$ are plotted from the common-midpoint location, their ends form the NMO-velocity *surface*, while the NMO ellipse is the intersection of this surface with the horizontal plane. Even though under normal circumstances we cannot count on measuring V_{nmo} along many different directions in space, this formalism naturally leads to an efficient Dix-type procedure for computing NMO ellipses in laterally heterogeneous, anisotropic media.

1.4.1 General formulation

Suppose a pure-mode reflected wave is recorded along an arbitrarily oriented CMP line $\boldsymbol{\mathcal{L}}$. We assume that reflection traveltime is uniquely defined for each (moderate compared to the reflector depth) source-receiver offset. If, instead, the traveltime function becomes multivalued, as in the vicinity of shear-wave cusps, a more elaborate approximation than the hyperbolic moveout equation is required.

The exact function $V_{\rm nmo}(\boldsymbol{\mathcal{L}})$ is derived in Appendix 1C (equation 1.109) as

$$V_{\rm nmo}^{-2}(\boldsymbol{\mathcal{L}}) = \boldsymbol{\mathcal{L}}\, \mathbf{U}\, \boldsymbol{\mathcal{L}}^{\rm T}, \tag{1.39}$$

where $\boldsymbol{\mathcal{L}} = [\mathcal{L}_1, \mathcal{L}_2, \mathcal{L}_3]$ is a unit row vector, $\boldsymbol{\mathcal{L}}^{\rm T}$ is a unit column vector, and $\mathbf{U}$ is a 3×3 symmetric matrix with elements

$$U_{km} = \tau_0 \frac{\partial^2 \tau(\mathbf{x})}{\partial x_k \partial x_m} \equiv \tau_0 \frac{\partial p_k(\mathbf{x})}{\partial x_m}, \qquad (k, m = 1, 2, 3). \tag{1.40}$$

Here, $\tau(\mathbf{x})$ is the one-way traveltime from the zero-offset reflection point to location $\mathbf{x}$, $\tau_0 = \tau(\mathbf{x}_{\rm CMP})$ corresponds to the common midpoint, and $\mathbf{p}(\mathbf{x}) = [p_1(\mathbf{x}), p_2(\mathbf{x}), p_3(\mathbf{x})]$ is the slowness vector for the ray excited at the zero-offset reflection point and recorded at location $\mathbf{x}$. The derivatives in equation 1.40 are evaluated at the common midpoint.

Azimuthally-dependent NMO velocity in the horizontal plane (usually an ellipse, as discussed above) can be viewed as the intersection of the NMO-velocity surface with the horizontal plane. For a horizontal unit vector $\boldsymbol{\mathcal{L}}^{\rm hor} = [\cos\alpha, \sin\alpha, 0]$, general expression 1.39 yields the NMO ellipse (equation 1.3):

$$V_{\rm nmo}^{-2}(\alpha) = U_{11} \cos^2\alpha + 2\, U_{12} \sin\alpha \cos\alpha + U_{22} \sin^2\alpha\,. \tag{1.41}$$

Therefore, the 2×2 matrix $\mathbf{W}$, which defines the NMO ellipse in equation (1.3), coincides with the upper left submatrix of $\mathbf{U}$:

$$W_{ij} = \tau_0 \frac{\partial p_i(\mathbf{x})}{\partial x_j} = U_{ij}, \qquad (i, j = 1, 2). \tag{1.42}$$

1.4.2 Possible shapes of NMO-velocity surfaces

Equation 1.39 indicates that the function $V_{\rm nmo}(\boldsymbol{\mathcal{L}})$ defines a *centered quadratic* surface in 3D space. The shape of this surface is determined by the eigenvalues of the matrix $\mathbf{U}$, which have to be real because $\mathbf{U}$ is real and symmetric. Using equation 1.42, $\mathbf{U}$ can be written in the form

$$\mathbf{U} = \begin{pmatrix} W_{11} & W_{12} & U_{13} \\ W_{12} & W_{22} & U_{23} \\ U_{13} & U_{23} & U_{33} \end{pmatrix}. \tag{1.43}$$

Therefore, if the NMO ellipse $\mathbf{W}$ has been estimated from moveout data, one needs to determine only three additional quantities U_{k3} to reconstruct the whole NMO-velocity surface $\mathbf{U}$. The elements U_{k3} can be found by differentiating the Christoffel equation at the CMP location (see below). For homogeneous media it is possible to obtain explicit expressions for the elements U_{k3} using the approach applied to the matrix $\mathbf{W}$ (equation 1.7).

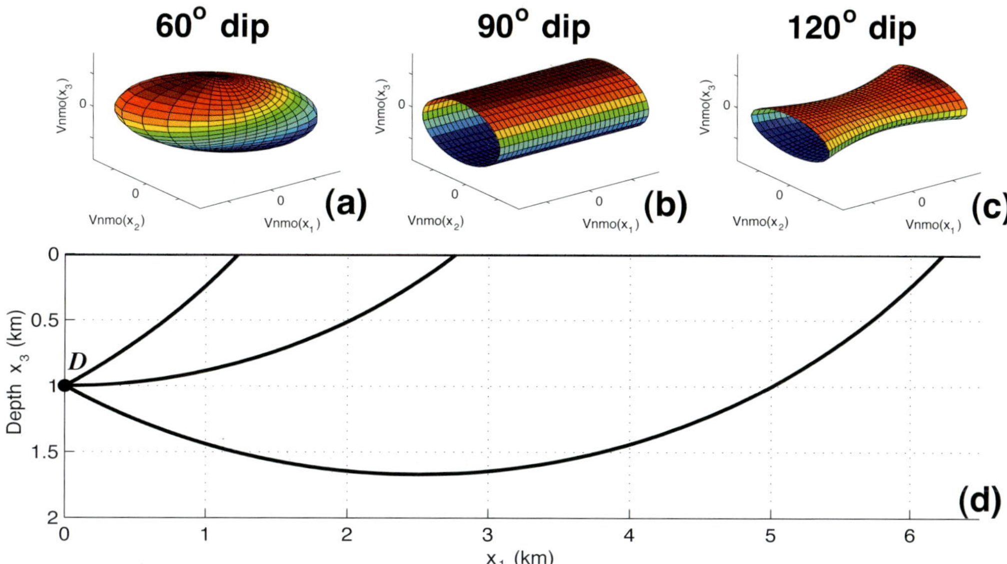

Figure 1.8: NMO-velocity surfaces (a,b,c) and trajectories of the zero-offset rays (d) in an isotropic medium with a constant vertical-velocity gradient (Grechka and Tsvankin, 2002a). Reflector dip is (a) 60°; (b) 90°; and (c) 120°. The zero-offset reflection point D is located at a depth of 1 km; the parameters of the velocity function $V(x_3)=V_0+\zeta x_3$ are $V_0=2.0$ km/s and $\zeta=0.6$ s^{-1}.

As mentioned above, the intersection of the NMO-velocity surface $\mathbf{U}$ with the horizontal plane is elliptical because the matrix $\mathbf{W}$ typically defines an ellipse in the horizontal plane. There are only three distinct types of quadratic surfaces that have elliptical cross sections symmetric with respect to the common midpoint: an *ellipsoid*, an *elliptical cylinder*, and a *one-sheeted hyperboloid* (Figure 1.8). (A hyperboloid and a cylinder may also have a nonelliptical intersection with the horizontal plane.) Note that if the NMO-velocity surface has the form of a cylinder, V_{nmo} along the axis of the cylinder is infinite, which implies that the CMP traveltime in this direction does not change with offset. A numerical example below shows that both elliptical cylinders and one-sheeted hyperboloids can be encountered in realistic subsurface models.

In principle, a quadratic surface might have other shapes including a two-sheeted hyperboloid, an imaginary elliptical cylinder, and a hyperbolic cylinder. These shapes correspond to a matrix $\mathbf{U}$ with at least two nonpositive eigenvalues, which means that reflection traveltime does not increase with offset in two or three mutually orthogonal directions in space. Although such cases are not prohibited by the theory, their occurrence is expected to be rare.

To illustrate the shape of NMO-velocity surfaces for typical seismological models, we computed the matrix $\mathbf{U}$ for a vertically heterogeneous isotropic medium with a constant vertical-velocity gradient (i.e., the velocity function is defined as $V(x_3) = V_0 + \zeta x_3$). The one-way traveltime $\tau(\mathbf{x})$ from the origin of the coordinate system to point $\mathbf{x} = [x_1, x_2, x_3]$ can be found analytically (Slotnick, 1959):

$$\tau(\mathbf{x}) = \frac{1}{\zeta} \cosh^{-1} \left[1 + \frac{\zeta^2 \,(x_1^2 + x_2^2 + x_3^2)}{2\, V_0 \,(V_0 + \zeta x_3)} \right]. \tag{1.44}$$

Substitution of equation 1.44 into equation 1.40 yields explicit expressions for the matrix $\mathbf{U}$ in terms of V_0, ζ, and the depth and dip of the reflector.

Figure 1.8 displays the NMO-velocity surfaces, along with computed ray trajectories (circular arcs), for reflectors beneath this constant-gradient isotropic medium. Depending on reflector dip, the surface can take any of the three shapes (an ellipsoid, a cylinder, and a hyperboloid) discussed above. For isotropic models in which velocity monotonically increases with depth, NMO-velocity ellipsoids correspond to reflector dips below 90° (Figure 1.8a), cylinders to vertical reflectors (Figure 1.8b), and hyperboloids to dips exceeding 90° (overhangs; Figure 1.8c).

1.4.3 NMO-velocity surface recorded in homogeneous media

In addition to the submatrix W_{ij} responsible for the NMO ellipse, the matrix $\mathbf{U}$ (equation 1.43) includes the elements U_{3k}, which depend on the spatial derivatives of the vertical slowness p_3. The components of the slowness vector $\mathbf{p}$ near spatial location $\mathbf{x}$ are related to each other by the Christoffel equation,

$$\Gamma(\mathbf{p}, \mathbf{x}) = 0\,. \tag{1.45}$$

In general, equation 1.45 contains a separate contribution of the coordinates $\mathbf{x}$ because of the spatial variation of the stiffness coefficients c_{ij} in heterogeneous media. Grechka and Tsvankin (2002a) show that equation 1.45 can be used to reconstruct the NMO-velocity surface from the NMO ellipse $\mathbf{W}(\mathbf{x})$ by specifying the slowness vector of the zero-offset ray and the stiffnesses $c_{ij}(\mathbf{x})$ (expressed according to the Voigt recipe) in the vicinity of the common midpoint.

Hereafter, we assume that the medium near the CMP can be treated as locally homogeneous, which significantly simplifies the representation of the NMO-velocity surface (Appendix B in Grechka and Tsvankin, 2002a):

$$\mathbf{U}^{\text{hom}} = \begin{pmatrix} W_{11} & W_{12} & q_{,1} W_{11} + q_{,2} W_{12} \\ \bullet & W_{22} & q_{,1} W_{12} + q_{,2} W_{22} \\ \bullet & \bullet & q_{,1}^2 W_{11} + 2 q_{,1} q_{,2} W_{12} + q_{,2}^2 W_{22} \end{pmatrix}, \tag{1.46}$$

where, as before, $q \equiv q(p_1, p_2) \equiv p_3$ is the vertical slowness, and $q_{,i} \equiv \partial q / \partial p_i$; the derivatives are evaluated for the slowness vector of the zero-offset ray. The bullets denote the elements $U_{21} = U_{12}$, $U_{31} = U_{13}$, and $U_{32} = U_{23}$. The NMO-velocity surface in equation 1.46 can be obtained from the NMO ellipse (i.e., from the matrix $\mathbf{W}$) and the Christoffel equation expressed in terms of the slownesses p_1, p_2, and q (see above).

To find the shape of the NMO-velocity surface for homogeneous media, note that the third column of the matrix $\mathbf{U}^{\rm hom}$ is a linear combination of the first two columns:

$$q_{,1}\, U_{k1}^{\rm hom} + q_{,2}\, U_{k2}^{\rm hom} = U_{k3}^{\rm hom}\,, \qquad (k = 1, 2, 3)\,. \tag{1.47}$$

As follows from equation 1.47,

$$\det \mathbf{U}^{\rm hom} = 0\,. \tag{1.48}$$

Because the first and the second columns are generally independent, the matrix $\mathbf{U}^{\rm hom}$ has one zero eigenvalue, so the surface defined by $\mathbf{U}^{\rm hom}$ has to be a *cylinder.* For models in which the matrix $\mathbf{W}$ describes an ellipse in the horizontal plane, the NMO-velocity surface is an *elliptical cylinder.*

The axis of the NMO-velocity cylinder is parallel to the eigenvector $\mathbf{e} = [e_1, e_2, e_3]$ corresponding to the zero eigenvalue of the matrix $\mathbf{U}^{\rm hom}$. Substituting the eigenvector $\mathbf{e}$ into the first two rows of $\mathbf{U}^{\rm hom}$ (equation 1.46) yields

$$\begin{cases} W_{11}\,\dfrac{e_1}{e_3} + W_{12}\,\dfrac{e_2}{e_3} = -q_{,1} W_{11} - q_{,2} W_{12}\,, \\[2ex] W_{12}\,\dfrac{e_1}{e_3} + W_{22}\,\dfrac{e_2}{e_3} = -q_{,1} W_{12} - q_{,2} W_{22}\,. \end{cases} \tag{1.49}$$

Therefore,

$$\frac{e_i}{e_3} = -q_{,i}\,, \qquad (i = 1, 2)\,. \tag{1.50}$$

The orientation of the vector $\mathbf{e}$ can be found from equation 1.71 for the components of the group-velocity vector $\boldsymbol{\mathcal{G}}$:

$$\frac{\mathcal{G}_i}{\mathcal{G}_3} = -q_{,i}\,, \qquad (i = 1, 2)\,. \tag{1.51}$$

Comparison of equations 1.50 and 1.51 shows that $\mathbf{e}$ is parallel to the vector $\boldsymbol{\mathcal{G}}$ at the CMP location. In other words, the axis of the NMO-velocity cylinder in homogeneous media of any symmetry points in the direction of the zero-offset ray. According to the geometrical meaning of the NMO-velocity surface, this result implies that the NMO velocity on a CMP line parallel to the zero-offset ray is infinite. Indeed, if sources and receivers are placed on the (straight) zero-offset ray, the reflected rays travel along the acquisition line; consequently, the two-way reflection traveltime in CMP geometry is independent of offset (i.e., $V_{\rm nmo} = \infty$).

These conclusions are valid for any pure-mode reflection recorded at a common midpoint located in a homogeneous, arbitrarily anisotropic medium. Whereas the stiffness coefficients are assumed to be constant near the CMP, they can vary elsewhere along the raypath. For example, the model can be composed of homogeneous layers separated by plane or curved interfaces. In the special case of a *single* homogeneous layer, the matrix $\mathbf{W}$ was found as an explicit function of the slowness components and the derivatives $q_{,i}$ and $q_{,ij}$ (equation 1.7). Then the whole NMO-velocity surface can be obtained by solving the Christoffel equation for the slowness vector of the zero-offset ray and substituting the results into equation 1.46.

1.4.4 P-wave NMO-velocity cylinder in a VTI layer

According to equation 1.46, if the NMO-velocity cylinder $\mathbf{U}^{\rm hom}$ has been reconstructed from seismic data, it should be possible to find the derivatives $q_{,i}$ in addition to the NMO ellipse. Because for some models $q_{,i}$ might depend on medium parameters not constrained by the NMO ellipse, the NMO-velocity surface can provide valuable information for anisotropic inversion. This point is illustrated here for the P-wave NMO-velocity cylinder $\mathbf{U}^{\rm VTI}$ from a plane dipping reflector beneath a homogeneous VTI layer.

To simplify the derivation of the matrix $\mathbf{U}^{\rm VTI}$, we linearize it in the parameters ϵ and δ under the assumption of weak anisotropy. The P-wave NMO ellipse (i.e., the elements W_{ij} in equation 1.46 expressed through the horizontal slowness components p_1 and p_2) in VTI media is fully controlled by the NMO velocity from a horizontal reflector,

$$V_{{\rm nmo},P} = V_{P0}\,\sqrt{1+2\,\delta}\,, \tag{1.52}$$

and the anellipticity parameter η defined as

$$\eta \equiv \frac{\epsilon-\delta}{1+2\,\delta}\,, \tag{1.53}$$

where V_{P0} is the P-wave vertical velocity (Alkhalifah and Tsvankin, 1995; Tsvankin, 2005). Hence, it is instructive to express our results in terms of $V_{{\rm nmo},P}$, η, and one of the Thomsen parameters (e.g., δ) instead of the more conventional parameter set $[V_{P0},\,\epsilon,\,\delta]$.

Selecting the coordinate frame in which the reflector normal lies in the vertical plane $[x_1,\,x_3]$, so that the zero-offset slowness component $p_2=0$ (Figure 1.9a), equations 1.46 and 1.7 can be used to obtain

$$U_{11}^{\rm VTI} \equiv W_{11}^{\rm VTI} = \frac{1}{V_{{\rm nmo},P}^2} - p_1^2\left[1+2\eta\,(4y^2-9y+6)\right], \tag{1.54}$$

$$U_{12}^{\rm VTI} \equiv W_{12}^{\rm VTI} = 0\,, \tag{1.55}$$

$$U_{22}^{\rm VTI} \equiv W_{22}^{\rm VTI} = \frac{1}{V_{{\rm nmo},P}^2} - 2\eta\,p_1^2\,(2-y)\,, \tag{1.56}$$

$$U_{13}^{\rm VTI} = -\frac{p_1\sqrt{1-y}}{V_{{\rm nmo},P}}\left[1+\delta-\eta\,y\,\frac{8y^2-15y+8}{1-y}\right], \tag{1.57}$$

$$U_{23}^{\rm VTI} = 0\,, \tag{1.58}$$

$$U_{33}^{\rm VTI} = p_1^2\left[1+2\delta-4\eta\,y\,(1-2y)\right], \tag{1.59}$$

where $y \equiv p_1^2 V_{{\rm nmo},P}^2$.

Equations 1.54 – 1.56 describe the NMO ellipse ($\mathbf{W}^{\rm VTI}$) with axes pointing in the dip and strike directions of the reflector. The dip component of the P-wave NMO velocity is defined by $W_{11}^{\rm VTI}$ in equation 1.54, while equation 1.56 for $W_{22}^{\rm VTI}$ yields

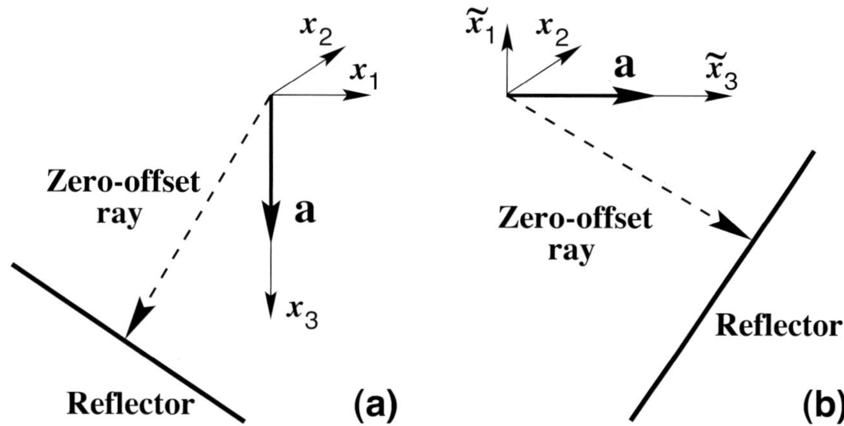

Figure 1.9: Dipping reflectors beneath (a) VTI and (b) HTI media (Grechka and Tsvankin, 2002a). The HTI model is obtained by rotating the symmetry axis **a** of the VTI model by 90° around the strike direction x_2.

the strike component (the exact NMO ellipse is defined by equations 1.12 and 1.15). Clearly, the NMO ellipse $\mathbf{W}^{\rm VTI}$ as a whole is governed by $V_{{\rm nmo},P}$ and η, with no dependence on δ; this conclusion holds for strong anisotropy as well (Grechka and Tsvankin, 1998a; Tsvankin, 2005).

Equations 1.57 – 1.59, which specify the additional components of $\mathbf{U}^{\rm VTI}$ needed to build the NMO-velocity cylinder, indicate that $V_{\rm nmo}$ in nonhorizontal directions depends on δ in addition to $V_{{\rm nmo},P}$ and η. This result also follows from the equation of the NMO ellipse in TI media with a *horizontal* symmetry axis (HTI) given by Contreras et al. (1999). Note that the vertical symmetry axis **a** in Figure 1.9a becomes horizontal after rotating the whole plot by 90° around the coordinate axis x_2. Hence, this rotation transforms the VTI model in Figure 1.9a into the HTI model in Figure 1.9b. The quantities $U_{22}^{\rm VTI}$ and $U_{33}^{\rm VTI}$ determine the NMO ellipses in both the vertical $[x_2, x_3]$-plane of the original VTI model and the horizontal $[x_2, \tilde{x}_3]$-plane of the new HTI model. The results of Contreras et al. (1999) show that the NMO ellipse $\mathbf{W}^{\rm HTI}$ for a dipping HTI layer is a function of δ as well as of $V_{{\rm nmo},P}$ and η.

Although the discussion of NMO velocities measured outside the horizontal plane may seem purely academic (unless vertical or oblique boreholes are available), intersections of the NMO-velocity surface with nonhorizontal planes play an important role in Dix-type averaging of NMO velocities in anisotropic media with dipping interfaces in the overburden (discussed next). As we demonstrate in Chapter 2, the information contained in the matrix elements U_{k3} can be extracted from $V_{\rm nmo}$ at the surface in the presence of lateral heterogeneity (interface dips) above the reflector. The only exception is layered elliptically anisotropic models, for which normal moveout is independent of the interval parameter $\epsilon = \delta$ (Dellinger and Muir, 1988).

1.4.5 NMO ellipse for models with intermediate dipping interfaces

The NMO-velocity surface incorporates the influence of the medium properties along the whole raypath between the zero-offset reflection point and the CMP location. Typically, we are not interested in building the whole effective matrix $\mathbf{U}$ at the earth's surface. However, as shown by Grechka and Tsvankin (2002a), NMO-velocity surfaces provide a convenient tool to obtain the effective NMO ellipse in the presence of lateral heterogeneity. The discussion below is limited to the practically important special case of homogeneous, anisotropic layers or blocks separated by plane dipping interfaces.

Dix-type averaging procedure

As discussed above, NMO-velocity surfaces in piecewise homogeneous media always have a cylindrical shape (equation 1.46). Figure 1.10 schematically illustrates the 3D procedure developed by Grechka and Tsvankin (2002a) to construct NMO-velocity cylinders and the effective NMO ellipse in layered media with plane dipping interfaces. Assuming that the interval slowness vectors $\mathbf{p}_\ell$ and traveltimes $\tau_{0,\ell}$ for the zero-offset ray have been obtained from ray tracing, we build the effective NMO ellipse as follows:

Step 1. Using equations 1.46 and 1.7, compute the NMO-velocity cylinder $\mathbf{U}_1$ in the layer $\ell = 1$ immediately above the reflector; the slowness vector $\mathbf{p}_1$ of the zero-offset ray is parallel to the reflector normal. In the first layer, the interval cylinder $\mathbf{U}_1$ coincides with the effective cylinder $\mathbf{U}(1)$. If the layer number $\ell > 1$, the cylinder $\mathbf{U}(\ell)$ (dashed lines in Figure 1.10a) is obtained from the continuation procedure described below.

Step 2. Find the intersection $\mathbf{W}(\ell)$ (magenta line in Figure 1.10a) of the cylinder $\mathbf{U}(\ell)$ with the ℓth interface using equation 1.46 (Grechka and Tsvankin, 2002a).

Step 3. Compute the interval cylinder $\mathbf{U}_{\ell+1}$ (dashed lines in Figure 1.10b) for the slowness vector $\mathbf{p}_{\ell+1}$ from equations 1.46 and 1.7. Find the intersection $\mathbf{W}_{\ell+1}$ (magenta line in Figure 1.10b) of $\mathbf{U}_{\ell+1}$ with the ℓth interface.

Step 4. Apply the generalized Dix equation 1.31 to obtain the intersection $\mathbf{W}(\ell+1)$ (magenta line in Figure 1.10c) of the effective cylinder $\mathbf{U}(\ell+1)$ with a plane parallel to the ℓth interface:

$$[\mathbf{W}(\ell+1)]^{-1} = \frac{\tau_0(\ell)\,[\mathbf{W}(\ell)]^{-1} + \tau_{0,\ell+1}\,[\mathbf{W}_{\ell+1}]^{-1}}{\tau_0(\ell+1)}, \tag{1.60}$$

where $\tau_0(\ell) = \sum_{j=1}^{\ell} \tau_{0,j}$. $\mathbf{W}(\ell+1)$ is computed at the top of the $\ell+1$th layer, with the ℓth interface playing the role of the horizontal interfaces used in the derivation of equation 1.31.

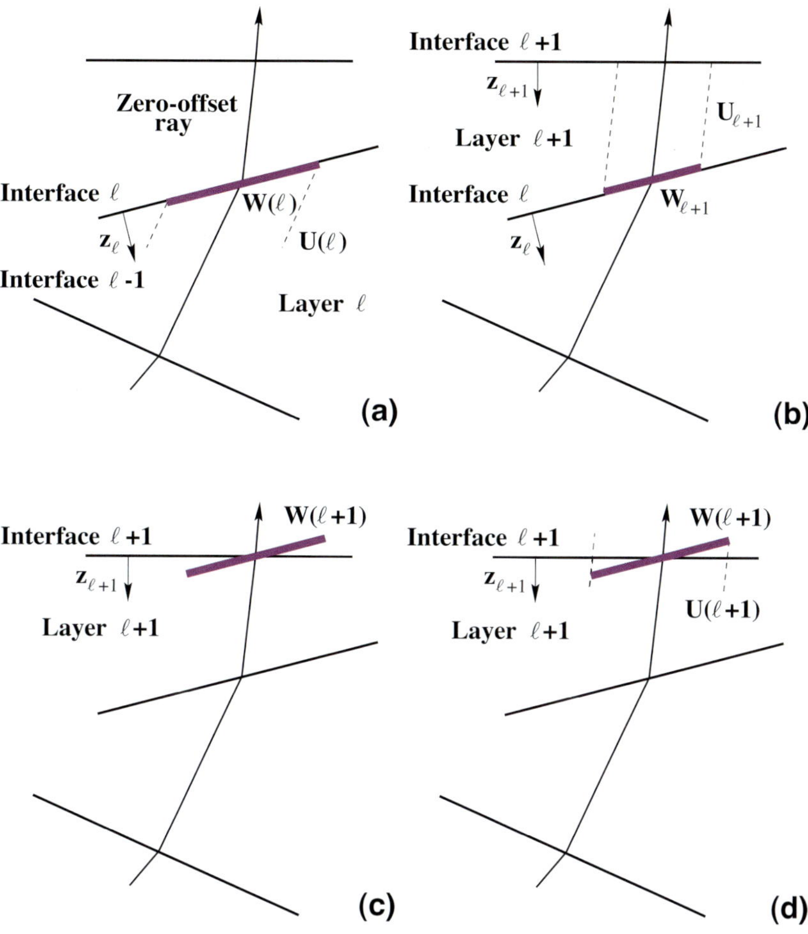

Figure 1.10: Dix-type procedure to obtain the effective NMO ellipse for a stack of homogeneous, anisotropic layers separated by plane dipping interfaces (Grechka and Tsvankin, 2002a). The intersection $\mathbf{W}(\ell)$ of the effective NMO-velocity cylinder $\mathbf{U}(\ell)$ with the top of the ℓth layer (a) and the interval NMO ellipse $\mathbf{W}_{\ell+1}$ (b) in the $(\ell+1)$th layer are substituted into equation 1.60 to produce the cross section $\mathbf{W}(\ell+1)$ (c) of the effective cylinder $\mathbf{U}(\ell+1)$ (d). Note that $\mathbf{W}(\ell+1)$ is the intersection of $\mathbf{U}(\ell+1)$ with a plane parallel to the ℓth interface. According to Snell's law, the projection of the zero-offset slowness vector $\mathbf{p}$ onto each interface is preserved (i.e., $\mathbf{p}_\ell \times \mathbf{z}_\ell = \mathbf{p}_{\ell+1} \times \mathbf{z}_\ell$ at the ℓth interface with unit normal $\mathbf{z}_\ell$).

Step 5. Reconstruct the cylinder $\mathbf{U}(\ell + 1)$ (dashed lines in Figure 1.10d) from its cross section $\mathbf{W}(\ell + 1)$, as described in Appendix D of Grechka and Tsvankin (2002a).

Step 6. Repeat **Step 2** for the next $[(\ell + 1)\text{th}]$ layer.

Hence, the effective NMO ellipse and conventional-spread reflection traveltimes for layered media with plane dipping interfaces can be computed without time-consuming multioffset, multiazimuth ray tracing. This Dix-type averaging procedure is employed in Chapters 2, 5, and 6 to devise efficient algorithms for stacking-velocity inversion in piecewise-homogeneous TI and orthorhombic models.

Numerical example

To verify the accuracy of Dix-type averaging of NMO-velocity surfaces, we computed the effective NMO ellipse at the top of a model composed of three dipping TI layers with a tilted symmetry axis (Table 1.1). Figure 1.11 displays the NMO ellipses for the reflection from the bottom of the model determined from the Dix-type averaging procedure (solid curve) and 3D anisotropic ray tracing (dotted). The ellipses almost coincide, confirming that the Dix-type equations give an adequate description of reflection moveout on conventional-length spreads. The small difference of up to 1.6% between the theoretical and ray-traced ellipses in Figure 1.11 can be attributed to the influence of nonhyperbolic moveout. As mentioned above, typically the deviation from hyperbolic moveout is small if the offset does not exceed the distance between the CMP and the reflector. This conclusion holds for P-wave data in a wide variety of anisotropic models of moderate structural complexity (Tsvankin, 2005).

1.5 Computation of the NMO ellipse for heterogeneous media

The concept of the NMO ellipse helps model reflection traveltimes on conventional CMP spreads without performing ray tracing for each source-receiver pair. According to equation 1.3, the NMO ellipse and conventional-spread moveout as a whole are fully defined by only three quantities – the matrix elements W_{11}, W_{12}, and W_{22}. For longer offsets exceeding the distance between the CMP and the reflector it is necessary to take nonhyperbolic moveout into account.

In principle, estimating $V_{\text{nmo}}(\alpha)$ from traveltimes computed in three distinct azimuthal directions is sufficient to reconstruct the NMO ellipse and find the NMO velocity for all other azimuths. Although this modeling procedure is much more efficient than multiazimuth ray tracing, it still time-consuming and does not take advantage of the explicit expressions for the parameters of the NMO ellipse. It is much more attractive to build the NMO ellipse directly from the spatial derivatives of the ray parameter $\partial p_i/\partial x_j$ at the CMP location (equations 1.3 and 1.4). Below, we follow Grechka et

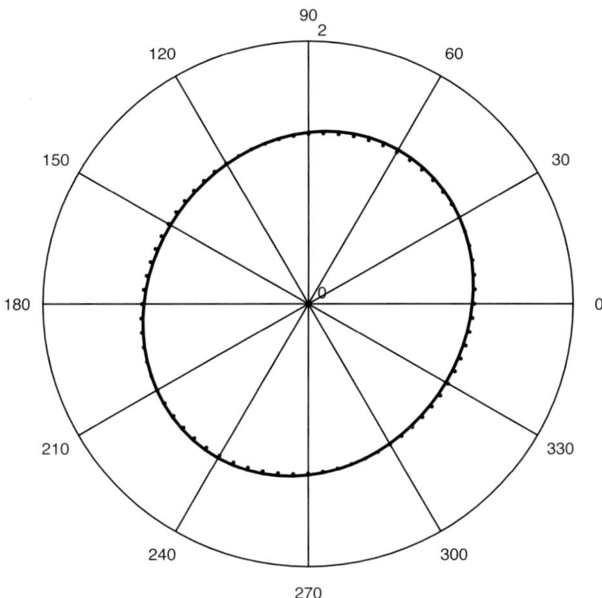

Figure 1.11: NMO ellipses computed for the P-wave reflection from the bottom of the layered TI model from Table 1.1 (Grechka and Tsvankin, 2002a). The solid line is the ellipse calculated using the Dix-type averaging procedure discussed above. The dotted ellipse is reconstructed from the best-fit moveout velocities obtained from ray-traced traveltimes in six azimuths separated by 30°; the maximum offset (3 km) is equal to the distance between the CMP and the reflector.

Reflector depth (km)	V_{P0} (km/s)	ϵ	δ	ν	β	ϕ	ψ
1.0	0.5	0.20	0.10	10.0	60.0	20.0	20.0
2.0	1.0	0.10	0.07	20.0	50.0	40.0	60.0
3.0	2.0	0.15	0.10	30.0	40.0	30.0	0.0

Table 1.1: Relevant parameters of the three-layer tilted TI model used in Figure 1.11. The layers are separated by plane dipping interfaces; ϕ and ψ are the dip and azimuth of the bottom of a layer (in degrees). The orientation of the symmetry axis in each layer is described by the angles ν (tilt from the vertical) and β (azimuth). The parameters ϵ and δ are defined with respect to the symmetry axis.

al. (1999b), who suggested an efficient method of computing these derivatives using dynamic ray-tracing equations for the zero-offset ray.

To describe rays propagating from the zero-offset reflection point to the surface, it is convenient to use the ray coordinates $(\gamma_1, \gamma_2, \tau)$. The parameter τ is defined as the traveltime along the ray, whereas γ_1 and γ_2 uniquely determine the raypath and can be chosen, for instance, as the horizontal components of the slowness vector (p_1 and p_2). Here, we use another option suggested by Kashtan (1982) and Kendall and Thomson (1989) and define γ_1 and γ_2 as the polar and azimuthal angles of the slowness vector.

The spatial derivatives of the vector $\mathbf{p}$, needed to calculate $V_{\text{nmo}}(\alpha)$, can be formally written as

$$\frac{\partial p_i}{\partial x_j} = \frac{\partial p_i}{\partial \gamma_1}\frac{\partial \gamma_1}{\partial x_j} + \frac{\partial p_i}{\partial \gamma_2}\frac{\partial \gamma_2}{\partial x_j} + \frac{\partial p_i}{\partial \tau}\frac{\partial \tau}{\partial x_j}. \tag{1.61}$$

Using the matrix notation

$$\mathbf{P} = \left[\frac{\partial \mathbf{p}}{\partial \gamma_1}, \frac{\partial \mathbf{p}}{\partial \gamma_2}, \frac{\partial \mathbf{p}}{\partial \tau}\right], \tag{1.62}$$

$$\mathbf{X} = \left[\frac{\partial \mathbf{x}}{\partial \gamma_1}, \frac{\partial \mathbf{x}}{\partial \gamma_2}, \frac{\partial \mathbf{x}}{\partial \tau}\right], \tag{1.63}$$

and the fact that the inverse matrix $\mathbf{X}^{-1}$ contains the rows

$$\mathbf{X}^{-1} = \left[\begin{array}{c} \partial \gamma_1/\partial \mathbf{x} \\ \partial \gamma_2/\partial \mathbf{x} \\ \partial \tau/\partial \mathbf{x} \end{array}\right],$$

equation 1.61 can be represented in the form

$$\frac{\partial p_i}{\partial x_j} = \left[\mathbf{P}\,\mathbf{X}^{-1}\right]_{ij}. \tag{1.64}$$

Hence, if the matrices in equations 1.62 and 1.63 have been calculated for the zero-offset ray at the CMP (surface) location, the derivatives $\partial p_i/\partial x_j$, $(i, j = 1, 2)$ can be determined as the upper-left 2×2 submatrix of the 3×3 matrix in equation 1.64. Both $\mathbf{p}$ and $\mathbf{x}$ should be computed for rays emanating from an imaginary source located at the reflection point of the zero-offset ray.

The third columns of the matrices $\mathbf{P}$ and $\mathbf{X}$ (i.e., the derivatives $\partial \mathbf{p}/\partial \tau$ and $\partial \mathbf{x}/\partial \tau$) can be obtained directly from kinematic ray-tracing equations. To find the first and second columns [i.e., the derivatives $\partial \mathbf{p}/\partial \gamma_n$ and $\partial \mathbf{x}/\partial \gamma_n$, $(n = 1, 2)$], one needs to integrate dynamic ray-tracing equations (e.g., Červený, Molotkov and Pšenčík, 1977; Kendall and Thomson, 1989). In fact, the derivatives that control NMO velocity are exactly the same as those required to compute the geometrical spreading along the zero-offset ray. This result is not surprising because NMO velocity is related to the wavefront curvature at the CMP location (Shah, 1973), which, in turn, determines geometrical spreading.

Thus, the NMO ellipse in heterogeneous, arbitrarily anisotropic media can be computed by employing kinematic and dynamic ray-tracing equations for the one-way zero-offset ray and substituting the results into equations 1.64, 1.4, and 1.3. Because this approach involves a single (zero-offset) ray, it is orders of magnitude less time consuming than is the tracing of hundreds or thousands of reflected rays for different azimuths and source-receiver offsets.

In the special case of a medium composed of homogeneous layers (or blocks) separated by smooth interfaces, the ray trajectory is piecewise linear, and integration of kinematic ray-tracing equations reduces to summation along straight ray segments. Integration of dynamic ray-tracing equations becomes relatively straightforward as well, because it requires only the continuation of the derivatives $\partial \mathbf{x}/\partial \gamma_n$ and $\partial \mathbf{p}/\partial \gamma_n$ across homogeneous layers and smooth interfaces (Kashtan, 1982). For plane interfaces, $\partial \mathbf{x}/\partial \gamma_n$ and $\partial \mathbf{p}/\partial \gamma_n$ can be expressed in terms of the slowness, polarization, and group-velocity vectors used or the traveltime calculation, making this algorithm especially simple.

The methodology based on ray tracing was tested on a model that includes VTI, HTI, and TTI layers separated by dipping interfaces (Figure 1.12). Since the interfaces are plane, the effective NMO ellipse could be found from the Dix-type averaging procedure discussed above, but this time we employed the ray-tracing equations for the zero-offset ray. The modeled NMO ellipse provides an excellent approximation to the finite-spread moveout velocity for the full range of azimuths, with a maximum error of less than 1.5%. In addition to confirming the high accuracy of the algorithm, this example demonstrates again that the analytic NMO velocity adequately describes P-wave reflection traveltimes on conventional-length spreads.

1.6 NMO ellipse of mode-converted waves

In general, reflection traveltime of mode-converted (PS or SP) waves does not remain the same when the source and receiver are interchanged. This moveout asymmetry, discussed in more detail in Chapter 4, adds odd (linear, cubic, etc.) terms in offset to the traveltime series $t(x)$. Therefore, conventional hyperbolic and higher-order moveout equations developed for pure-mode reflections cannot be applied to converted waves.

In this section, however, the discussion is restricted to horizontally layered media with a horizontal symmetry plane (Appendix 1D). The layers, for instance, can be orthorhombic, monoclinic (in both models one of the symmetry planes should be horizontal) or transversely isotropic with a vertical (VTI) or horizontal (HTI) symmetry axis. As shown in Appendix 1D, since the slowness and group-velocity surfaces in such models are symmetric with respect to the horizontal plane, traveltime of mode-converted waves is an even function of offset. Hence, converted-wave moveout on conventional-length spreads is governed primarily by NMO velocity obtained as a

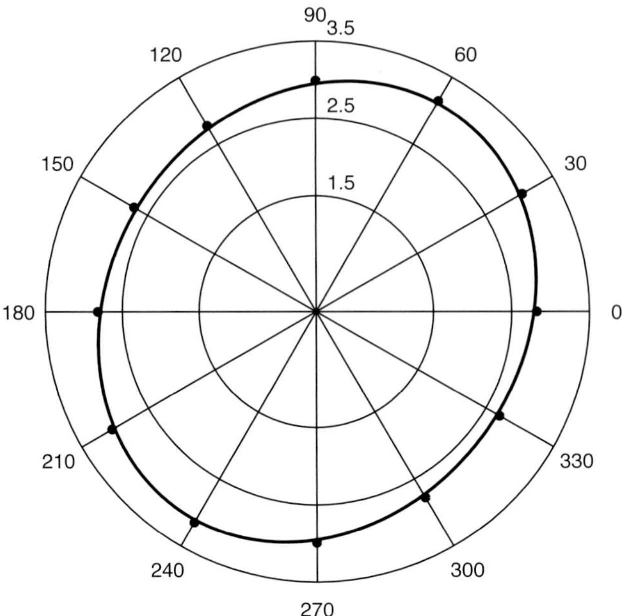

Figure 1.12: Comparison of the P-wave NMO ellipse computed from equations 1.3, 1.4, and 1.64 (solid curve) with the finite-spread moveout velocity (dots) estimated for the spreadlength equal to the CMP-reflector distance (Grechka et al., 1999b). The model includes three TI layers above a dipping reflector. The first (shallowest) layer is VTI with $V_{P0,1}$=2.0 km/s, ϵ_1=0.2, and δ_1=0.1. The second layer is HTI (the azimuth of the symmetry axis is 30°) with $V_{P0,2}$=2.4 km/s, ϵ_2=0.15, and δ_2=0. The third layer is TI with a tilted symmetry axis (the azimuth is 60°, the tilt is 30°) and $V_{P0,3}$=3 km/s, ϵ_3=0.25, and δ_3=0.08. The dip ϕ and azimuth ψ of the bottom of the first layer are ϕ_1=10° and ψ_1=70°; for the second layer, ϕ_2=15° and ψ_2=20°; and for the third layer (reflector), ϕ_3=35° and ψ_3=50°. The normal distances between the CMP and the interfaces are 1, 2, and 3 km.

function of azimuth from the pure-mode equation 1.3:

$$V_{\rm nmo}^{-2}(\alpha) = W_{11}\cos^2\alpha + 2W_{12}\sin\alpha\cos\alpha + W_{22}\sin^2\alpha\,. \tag{1.65}$$

NMO equation 1.65 can be expressed through the eigenvalues λ_1 and λ_2 of the matrix $\mathbf{W}$ (equation 1.6). Typically, λ_1 and λ_2 are positive (otherwise, traveltime decreases with offset in some azimuthal directions), and $V_{\rm nmo}$ traces out an ellipse in the horizontal plane. As with pure modes, representation of conventional-spread moveout in terms of NMO velocity breaks down when offset-dependent traveltime cannot be approximated by a Taylor series expansion (e.g., near shear-wave cusps) and for models with anomalously strong nonhyperbolic moveout. It should be mentioned that the magnitude of nonhyperbolic moveout for fixed spreadlength is generally larger for converted waves than for pure-mode reflections (e.g., Tsvankin, 2005).

While equation 1.65 of the NMO ellipse is valid for both pure and converted reflection events, the matrix $\mathbf{W}$ for pure modes depends on the spatial derivatives of the *one-way* traveltime from the zero-offset reflection point to the surface. For a homogeneous medium, these derivatives can be expressed through the slowness components of the zero-offset ray, which leads to a concise representation of the NMO velocity for PP- and SS-waves (equation 1.7).

In contrast, the matrix $\mathbf{W}$ for converted waves is more difficult to relate to the model parameters because it depends on both P- and S-legs of the PS-ray. A convenient way to include shear information in moveout inversion is to obtain the NMO velocity of SS-waves (in anisotropic media there are two split shear waves, S_1 and S_2) by combining PP and PS reflections. When the medium is laterally heterogeneous, computation of pure-mode SS-wave traveltimes from PP and PS data can be accomplished using the PP+ PS = SS method introduced in Chapter 4.

For layer-cake models with a horizontal symmetry plane, however, the NMO ellipses of pure and converted waves are related by a simple expression based on the generalized Dix equation 1.31 (Appendix 1D). The matrix $\mathbf{W}^{(PS)}$ of the converted PS-wave can be found from the corresponding matrices of the pure PP-waves ($\mathbf{W}^{(P)}$) and SS-waves ($\mathbf{W}^{(S)}$) as follows (equation 1.114):

$$2t_{PS0}\left[\mathbf{W}^{(PS)}\right]^{-1} = t_{P0}\left[\mathbf{W}^{(P)}\right]^{-1} + t_{S0}\left[\mathbf{W}^{(S)}\right]^{-1}, \tag{1.66}$$

where t_{P0} and t_{S0} are the two-way zero-offset PP and SS traveltimes, and t_{PS0} is the zero-offset traveltime of the PS-wave $[t_{PS0} = (t_{P0} + t_{S0})/2]$.

For layered media, $\mathbf{W}^{(P)}$ and $\mathbf{W}^{(S)}$ represent effective quantities expressed through the interval matrices using the generalized Dix equation 1.31:

$$\left[\mathbf{W}^{(P)}\right]^{-1} = \frac{1}{t_{P0}} \sum_{\ell=1}^{N} t_{P0,\ell}\left[\mathbf{W}_\ell^{(P)}\right]^{-1}, \tag{1.67}$$

$$\left[\mathbf{W}^{(S)}\right]^{-1} = \frac{1}{t_{S0}} \sum_{\ell=1}^{N} t_{S0,\ell}\left[\mathbf{W}_\ell^{(S)}\right]^{-1}, \tag{1.68}$$

where $t_{P0,\ell}$ and $t_{S0,\ell}$ are the two-way interval zero-offset traveltimes.

This formalism makes it possible to compute interval SS-wave NMO ellipses in layered anisotropic media from PP and PS data. The effective matrices $\mathbf{W}^{(S)}$ of SS-waves for available reflectors can be found from the corresponding NMO ellipses of PP- and PS-waves using equation 1.66. Then the interval shear-wave matrices $\mathbf{W}_\ell^{(S)}$ can be calculated from equation 1.68. Note that for layer-cake media Dix-type differentiation of pure-mode NMO ellipses is a straightforward operation that involves only recorded (horizontal) reflection events. Application of equation 1.66 to data of multicomponent physical modeling for a homogeneous orthorhombic block is discussed in Chapter 6.

1.7 Summary

Azimuthally-dependent NMO velocity is described by a simple quadratic form and usually defines an *elliptical* curve in the horizontal plane. The orientation and semi-axes of the NMO ellipse are determined by the medium properties and the direction of the reflector normal at the zero-offset reflection point.

For a homogeneous, arbitrarily anisotropic medium above a horizontal or dipping reflector, the NMO ellipse was found explicitly as a function of the slowness vector corresponding to the zero-offset ray. The vertical slowness component and its derivatives with respect to the horizontal slownesses, needed to compute the NMO ellipse, can be obtained from the Christoffel equation. This result can be effectively used in moveout inversion, as well as in developing weak-anisotropy approximations for various anisotropic symmetries.

When the medium above the reflector is horizontally layered, the effective NMO ellipse can be obtained by Dix-type averaging of the matrices that describe the interval NMO ellipses. In Chapter 2, we present a field-data application of this generalized Dix equation to fracture detection in azimuthally anisotropic media. It should be emphasized that if the reflector is dipping, the interval NMO ellipses have to be computed for nonexistent reflectors orthogonal to the slowness vector of the zero-offset ray in each layer. Therefore, the generalized Dix differentiation of dipping events entails full-scale layer stripping and cannot be performed using just the reflections from the top and bottom of the target layer.

If the dip plane of the reflector coincides with a plane of symmetry in all layers (e.g., the medium above the reflector is VTI), the semiaxes of the effective NMO ellipse are obtained by rms averaging of the dip-line and strike-line interval NMO velocities. In general, however, the exact V_{nmo} differs from the rms average of the interval NMO velocities computed for the azimuth of the CMP line. Although this deviation becomes significant only in the presence of pronounced azimuthal anisotropy and/or large reflector dip, it is preferable to apply the generalized Dix equation to any wide-azimuth data set. In addition to providing higher accuracy, the averaging of NMO ellipses has the important advantage of smoothing estimated moveout velocities using their correct (elliptical) dependence on azimuth, which reduces instability in the Dix differentiation.

To describe reflection moveout for laterally heterogeneous media above the reflector, we introduced the notion of the NMO-velocity surface obtained by plotting V_{nmo} as the radius-vector from the CMP location. The effective NMO ellipse of a pure (PP or SS) mode can be obtained by Dix-type averaging of specifically oriented cross sections of NMO-velocity surfaces along the zero-offset ray. This modeling procedure is implemented in Chapters 2, 5, and 6 to develop stacking-velocity tomography for piecewise-homogeneous TI and orthorhombic media. The NMO-velocity surface in each anisotropic layer or block encountered by the ray often depends on more parameters than does the intersection of this surface with a horizontal plane (i.e., the NMO ellipse). In the presence of dipping interfaces or other types of lateral heterogeneity, these additional parameters contribute to the effective NMO ellipse and, potentially, can be estimated from reflection data (see Chapter 2). For example, the influence of intermediate dipping interfaces makes P-wave NMO velocity in VTI media dependent on the interval vertical velocity V_{P0} and anisotropy parameters ϵ and δ, not just on V_{nmo} and η.

For arbitrarily heterogeneous media, the NMO ellipse and conventional-spread reflection moveout can be obtained from the results of kinematic and dynamic ray tracing for a single (zero-offset) ray. Although modeling of geometrical spreading requires solving a system of differential equations, this algorithm is orders of magnitude faster than multioffset, multiazimuth kinematic ray tracing. Furthermore, if the medium consists of homogeneous layers or blocks separated by plane dipping interfaces, the NMO ellipse can be expressed through the slowness, polarization, and group-velocity vectors calculated for the zero-offset ray. This method for modeling NMO ellipses is comparable in terms of computational efficiency to the Dix-type averaging procedure based on NMO-velocity surfaces.

Whereas most of the chapter is devoted to pure (PP or SS) reflections, we also analyzed normal moveout of converted (PS or SP) waves for models composed of horizontal anisotropic layers with a horizontal symmetry plane. Converted-wave traveltime in such models is *reciprocal* with respect to source and receiver positions (i.e., it remains the same if the source and receiver are interchanged) and represents an even function of offset. Furthermore, azimuthally-varying moveout of mode conversions on conventional-length spreads is described by the NMO ellipse. Application of the generalized Dix equation yields a simple relationship between the NMO ellipses of pure and converted waves that provides a basis for obtaining shear-wave information from PP and PS data.

Although NMO velocity is defined in the zero-spread limit, our numerical examples for a range of anisotropic models demonstrate that the hyperbolic moveout equation parameterized by the NMO ellipse accurately describes reflection traveltimes for offset-to-depth ratios not exceeding unity. For larger offsets, the influence of anisotropy and/or heterogeneity makes reflection moveout nonhyperbolic, and its azimuthal variation becomes more complicated. An analytic description of P-wave nonhyperbolic moveout in layered azimuthally anisotropic media is given in Chapter 3.

Appendices for Chapter 1

1A NMO ellipse in a homogeneous layer

Here, we obtain an exact expression for the NMO velocity from a dipping reflector beneath a homogeneous, arbitrarily anisotropic layer. The derivation is based on the general equations 1.3 and 1.4 describing the NMO ellipse and follows the work by Grechka et al. (1999b).

To evaluate the derivatives $\partial x_i/\partial p_j$, we need to relate the horizontal ray displacements (x_1, x_2) between the zero-offset reflection point and the surface to the horizontal components of the slowness vector (p_1, p_2). We start by introducing the group-velocity vector $\boldsymbol{\mathcal{G}}$,

$$x_i = \mathcal{G}_i \, \tau \,, \qquad (i = 1, 2, 3) \,, \tag{1.69}$$

where τ is the one-way traveltime. Using the fact that the projection of the group-velocity vector onto the slowness direction is equal to phase velocity, we can write

$$\mathbf{p} \cdot \boldsymbol{\mathcal{G}} = p_1 \, \mathcal{G}_1 + p_2 \, \mathcal{G}_2 + p_3 \, \mathcal{G}_3 = 1 \,. \tag{1.70}$$

Differentiating equation 1.70 with respect to p_1 and p_2 and taking into account that the vertical slowness p_3 can be considered as a function of the horizontal slownesses yields

$$\mathcal{G}_i = -\frac{\partial p_3}{\partial p_i} \, \mathcal{G}_3 - \mathbf{p} \cdot \frac{\partial \boldsymbol{\mathcal{G}}}{\partial p_i} \,, \qquad (i = 1, 2) \,.$$

Because the slowness vector $\mathbf{p}$ is normal to the group-velocity surface (wavefront) $\boldsymbol{\mathcal{G}}(p_1, p_2)$, while the vectors $\partial \boldsymbol{\mathcal{G}}/\partial p_i$ are tangent to this surface, $\mathbf{p} \cdot (\partial \boldsymbol{\mathcal{G}}/\partial p_i) = 0$. Hence,

$$\mathcal{G}_i = -q_{,i} \, \mathcal{G}_3 \,, \qquad (i = 1, 2) \,, \tag{1.71}$$

where $q \equiv q(p_1, p_2) \equiv p_3$ denotes the vertical slowness, and $q_{,i} \equiv \partial q/\partial p_i$. Substitution of equation 1.71 into equation 1.70 gives a representation of $\mathcal{G}_3$ that will be needed later in the derivation:

$$\mathcal{G}_3 = \frac{1}{q - p_1 q_{,1} - p_2 q_{,2}} \,. \tag{1.72}$$

Using equation 1.71, the horizontal ray displacements x_i ($i = 1, 2$) from equations 1.69 can be rewritten as

$$x_i = -q_{,i} \, \mathcal{G}_3 \, \tau \,, \qquad (i = 1, 2) \,. \tag{1.73}$$

Note that the product $\mathcal{G}_3 \tau$ is the depth of the zero-offset reflection point, which is independent of the slowness components p_1 and p_2. Therefore, differentiating equation 1.73, we obtain

$$Y_{ij} \equiv \frac{\partial x_i}{\partial p_j} = -q_{,ij} \, \mathcal{G}_3 \, \tau \,, \tag{1.74}$$

where $q_{,ij} \equiv \partial^2 q/(\partial p_i \partial p_j)$ is a symmetric matrix of the second derivatives of the vertical slowness.

The NMO ellipse is determined by the matrix $\mathbf{W}$ (equation 1.4),

$$\mathbf{W} = \tau_0 \, \mathbf{Y}^{-1}, \tag{1.75}$$

where the inverse matrix $\mathbf{Y}^{-1}$ should be evaluated for the horizontal slowness components of the zero-offset ray.

Substituting Y_{ij} from equation 1.74 into equation 1.75 and using expression 1.72 for $\mathcal{G}_3$, we find

$$\mathbf{W} = \tau_0 \, \mathbf{Y}^{-1} = \frac{p_1 q_{,1} + p_2 q_{,2} - q}{q_{,11} q_{,22} - q_{,12}^2} \begin{pmatrix} q_{,22} & -q_{,12} \\ -q_{,12} & q_{,11} \end{pmatrix}. \tag{1.76}$$

With the matrix $\mathbf{W}$ from equation 1.76, the NMO ellipse in a homogeneous layer (equation 1.3) takes the following form:

$$\begin{aligned} V_{\text{nmo}}^{-2}(\alpha) &\equiv V_{\text{nmo}}^{-2}(\alpha, p_1, p_2) \\ &= \frac{p_1 q_{,1} + p_2 q_{,2} - q}{q_{,11} q_{,22} - q_{,12}^2} \left[q_{,22} \cos^2\alpha - 2 q_{,12} \sin\alpha \cos\alpha + q_{,11} \sin^2\alpha \right]. \end{aligned} \tag{1.77}$$

1B Notation for orthorhombic and HTI media

Orthorhombic (or orthotropic) media have three mutually orthogonal planes of mirror symmetry in which the Christoffel equation has the same form as for transverse isotropy. This equivalence was used by Tsvankin (1997a, 2005) to extend the principle of Thomsen's (1986) VTI notation to orthorhombic symmetry. Tsvankin's notation includes two "isotropic" reference velocities and seven dimensionless anisotropy parameters similar to ϵ, δ, and γ. Below we give the definitions of these parameters in terms of the stiffnesses c_{ij} and density ρ; the coordinate planes are aligned with the symmetry planes of the medium (Figure 2.36).

- V_{P0} is the P-wave vertical velocity:

$$V_{P0} \equiv \sqrt{\frac{c_{33}}{\rho}}. \tag{1.78}$$

- V_{S0} is the vertical velocity of the S-wave polarized in the x_1-direction:

$$V_{S0} \equiv \sqrt{\frac{c_{55}}{\rho}}. \tag{1.79}$$

- $\epsilon^{(1)}$ is the VTI parameter ϵ in the symmetry plane $[x_2, x_3]$ normal to x_1-axis (this explains the superscript "1"):

$$\epsilon^{(1)} \equiv \frac{c_{22} - c_{33}}{2 c_{33}}. \tag{1.80}$$

- $\delta^{(1)}$ is the VTI parameter δ in the $[x_2, x_3]$-plane:

$$\delta^{(1)} \equiv \frac{(c_{23}+c_{44})^2 - (c_{33}-c_{44})^2}{2c_{33}\,(c_{33}-c_{44})}\,. \tag{1.81}$$

- $\gamma^{(1)}$ is the VTI parameter γ in the $[x_2, x_3]$-plane:

$$\gamma^{(1)} \equiv \frac{c_{66}-c_{55}}{2c_{55}}\,. \tag{1.82}$$

- $\epsilon^{(2)}$ is the VTI parameter ϵ in the $[x_1, x_3]$-plane:

$$\epsilon^{(2)} \equiv \frac{c_{11}-c_{33}}{2c_{33}}\,. \tag{1.83}$$

- $\delta^{(2)}$ is the VTI parameter δ in the $[x_1, x_3]$-plane:

$$\delta^{(2)} \equiv \frac{(c_{13}+c_{55})^2 - (c_{33}-c_{55})^2}{2c_{33}\,(c_{33}-c_{55})}\,. \tag{1.84}$$

- $\gamma^{(2)}$ is the VTI parameter γ in the $[x_1, x_3]$-plane:

$$\gamma^{(2)} \equiv \frac{c_{66}-c_{44}}{2c_{44}}\,. \tag{1.85}$$

- $\delta^{(3)}$ is the VTI parameter δ in the $[x_1, x_2]$-plane (x_1 plays the role of the symmetry axis):

$$\delta^{(3)} \equiv \frac{(c_{12}+c_{66})^2 - (c_{11}-c_{66})^2}{2c_{11}\,(c_{11}-c_{66})}\,. \tag{1.86}$$

Vertical and horizontal transverse isotropy can be considered as special cases of an orthorhombic medium with a horizontal symmetry plane. Orthorhombic symmetry reduces to VTI if all vertical planes have identical properties, and the velocity of each mode in the $[x_1, x_2]$-plane (the isotropy plane) is independent of angle. Therefore, the VTI constraints are $\epsilon^{(1)} = \epsilon^{(2)} = \epsilon$, $\delta^{(1)} = \delta^{(2)} = \delta$, $\gamma^{(1)} = \gamma^{(2)} = \gamma$, and $\delta^{(3)} = 0$; ϵ, δ, and γ are Thomsen's (1986) VTI parameters.

If the medium is HTI with the symmetry axis parallel to the x_1-direction, the axes x_2 and x_3 form the isotropy plane, and $\epsilon^{(1)} = \delta^{(1)} = \gamma^{(1)} = 0$. The anisotropy parameter $\delta^{(3)}$ in this model is not independent (Tsvankin, 1997a, 2005). Therefore, wave propagation in HTI media is fully described by the vertical velocities V_{P0} and V_{S0} and the anisotropy parameters $\epsilon^{(2)}$, $\delta^{(2)}$, and $\gamma^{(2)}$ defined in the $[x_1, x_3]$-plane. Because the parameters $\epsilon^{(2)}$, $\delta^{(2)}$ and $\gamma^{(2)}$ correspond to the VTI model equivalent to the $[x_1, x_3]$-plane of HTI media, they are often denoted by $\epsilon^{(V)}$, $\delta^{(V)}$, and $\gamma^{(V)}$, respectively.

The shear-wave splitting parameter at vertical incidence for both orthorhombic and HTI media is described by the fractional difference between c_{44} and c_{55} (assuming $c_{44} > c_{55}$):

$$\gamma^{(S)} \equiv \frac{c_{44} - c_{55}}{2c_{55}} = \frac{\gamma^{(1)} - \gamma^{(2)}}{1 + 2\gamma^{(2)}} \approx \frac{V_{S1} - V_{S0}}{V_{S0}}, \tag{1.87}$$

where $V_{S1} \equiv \sqrt{c_{44}/\rho}$ is the vertical velocity of the fast shear wave. The parameter $\gamma^{(S)}$, which represents a direct measure of the time delay between two split shear waves at vertical incidence, coincides with the generic Thomsen's parameter γ for the HTI medium with the symmetry axis in the x_1-direction.

As discussed in the main text, P-wave time-domain signatures in orthorhombic media are governed by symmetry-plane NMO velocities of horizontal events and the anellipticity parameters $\eta^{(1,2,3)}$ defined below by analogy with the Alkhalifah-Tsvankin VTI parameter η.

- $\eta^{(1)}$ – the parameter η in the $[x_2, x_3]$-plane:

$$\eta^{(1)} \equiv \frac{\epsilon^{(1)} - \delta^{(1)}}{1 + 2\delta^{(1)}} \tag{1.88}$$

- $\eta^{(2)}$ – the parameter η in the $[x_1, x_3]$-plane:

$$\eta^{(2)} \equiv \frac{\epsilon^{(2)} - \delta^{(2)}}{1 + 2\delta^{(2)}} \tag{1.89}$$

- $\eta^{(3)}$ – the parameter η in the $[x_1, x_2]$-plane:

$$\eta^{(3)} \equiv \frac{\epsilon^{(1)} - \epsilon^{(2)} - \delta^{(3)}\,(1 + 2\epsilon^{(2)})}{(1 + 2\epsilon^{(2)})\,(1 + 2\delta^{(3)})} \tag{1.90}$$

For weak anisotropy, the exact definitions of the anellipticity parameters can be linearized in $\epsilon^{(i)}$ and $\delta^{(i)}$:

$$\eta^{(1)} \approx \epsilon^{(1)} - \delta^{(1)}, \tag{1.91}$$

$$\eta^{(2)} \approx \epsilon^{(2)} - \delta^{(2)}, \tag{1.92}$$

$$\eta^{(3)} \approx \epsilon^{(1)} - \epsilon^{(2)} - \delta^{(3)}. \tag{1.93}$$

1C Derivation of the NMO-velocity surface

Here, we develop a small-offset approximation for the squared reflection traveltime t^2 recorded on a straight common-midpoint (CMP) line parallel to an arbitrary unit vector $\boldsymbol{\mathcal{L}}$. The derivation is based on expanding the traveltime in a Taylor series in half-offset h near the CMP location ($h = 0$). The traveltime field is assumed to be smooth enough for all needed derivatives to exist at zero offset.

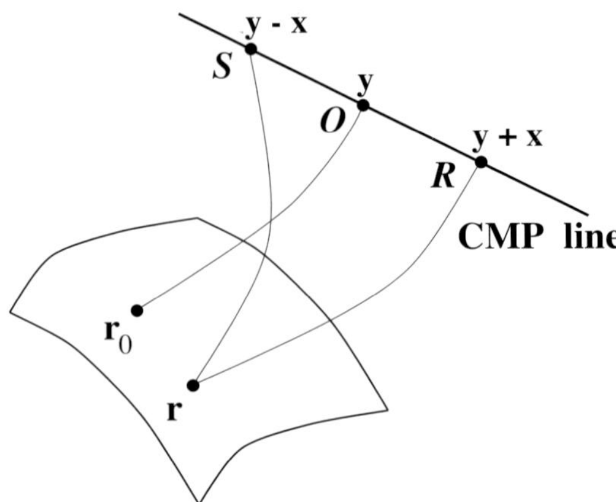

Figure 1.13: Reflected rays recorded on a common-midpoint line in 3D space; $\mathbf{r}_0$ is the reflection point of the zero-offset ray originated at the CMP location O (Grechka and Tsvankin, 2002a). The ray excited at S and emerging at R is reflected at a different point $\mathbf{r}$, but the reflection-point dispersal has no influence on NMO velocity.

If the coordinate of the common midpoint O is denoted by $\mathbf{y}$ (Figure 1.13), the coordinates of the source S and receiver R are $\mathbf{y} - \mathbf{x}$ and $\mathbf{y} + \mathbf{x}$, where

$$\mathbf{x} \equiv [x_1, \, x_2, \, x_3] = h \, \boldsymbol{\mathcal{L}} \equiv h \, [\mathcal{L}_1, \, \mathcal{L}_2, \, \mathcal{L}_3] \, . \tag{1.94}$$

The pure-mode two-way reflection traveltime t depends on the positions of the source and receiver and on the coordinate $\mathbf{r} = \mathbf{r}(\mathbf{y}, \mathbf{x})$ of the reflection point. Summing up the one-way traveltimes τ corresponding to the downgoing and upgoing rays yields

$$t(\mathbf{y}, \mathbf{x}, \mathbf{r}) = \tau(\mathbf{y} - \mathbf{x}, \mathbf{r}) + \tau(\mathbf{y} + \mathbf{x}, \mathbf{r}) \, . \tag{1.95}$$

Since reflection-point dispersal (i.e., the deviation of $\mathbf{r}$ from $\mathbf{r}_0$ in Figure 1.13 for $|\mathbf{x}| \neq 0$) does not change NMO velocity (e.g., Hubral and Krey, 1980; Goldin, 1986), we assume that nonzero-offset rays are reflected at point $\mathbf{r}_0$:

$$t(\mathbf{y}, \mathbf{x}, \mathbf{r}) = t(\mathbf{y}, \mathbf{x}, \mathbf{r}_0) = \tau(\mathbf{y} - \mathbf{x}, \mathbf{r}_0) + \tau(\mathbf{y} + \mathbf{x}, \mathbf{r}_0) \, .$$

Hence, for a given reflection event and a fixed CMP location (constant $\mathbf{y}$ and $\mathbf{r}_0$), t is a function of the one-way traveltime from the zero-offset reflection point:

$$t(\mathbf{x}) = \tau(-\mathbf{x}) + \tau(\mathbf{x}) \, . \tag{1.96}$$

Applying the chain rule to equation 1.96 and taking into account equation 1.94, the first derivative of the traveltime with respect to the half-offset h is obtained as

$$\frac{dt}{dh} = \sum_{k=1}^{3} \frac{\partial t}{\partial x_k} \, \mathcal{L}_k \, . \tag{1.97}$$

Using equation 1.96, we find the derivative 1.97 at zero offset:

$$\left.\frac{dt}{dh}\right|_{h=0} = \sum_{k=1}^{3}\left[-\frac{\partial \tau}{\partial x_k} + \frac{\partial \tau}{\partial x_k}\right]\mathcal{L}_k = 0\,. \tag{1.98}$$

Differentiating equation 1.97 again yields

$$\frac{d^2 t}{dh^2} = \sum_{k,m=1}^{3} \frac{\partial^2 t}{\partial x_k \partial x_m}\,\mathcal{L}_k\,\mathcal{L}_m\,. \tag{1.99}$$

Therefore, at zero offset

$$\left.\frac{d^2 t}{dh^2}\right|_{|\mathbf{x}|=h=0} = \sum_{k,m=1}^{3} \left.\frac{\partial^2 t}{\partial x_k \partial x_m}\right|_{h=0} \mathcal{L}_k\,\mathcal{L}_m = 2\sum_{k,m=1}^{3} \left.\frac{\partial^2 \tau}{\partial x_k \partial x_m}\right|_{h=0} \mathcal{L}_k\,\mathcal{L}_m\,, \tag{1.100}$$

To obtain the equation for NMO velocity along the CMP line $\boldsymbol{\mathcal{L}}$, we expand the traveltime $t(h)$ in a Taylor series truncated after the quadratic term:

$$t(h,\boldsymbol{\mathcal{L}}) = t_0 + \left.\frac{dt}{dh}\right|_{h=0} h + \left.\frac{d^2 t}{dh^2}\right|_{h=0} \frac{h^2}{2}\,, \tag{1.101}$$

where $t_0 = 2\tau_0$ is the two-way zero-offset traveltime. Substituting equations 1.98 and 1.100 into equation 1.101 leads to

$$t(h,\boldsymbol{\mathcal{L}}) = t_0 + h^2 \sum_{k,m=1}^{3} \left.\frac{\partial^2 \tau}{\partial x_k \partial x_m}\right|_{h=0} \mathcal{L}_k\,\mathcal{L}_m\,. \tag{1.102}$$

Squaring equation 1.102 and keeping quadratic and lower-order terms with respect to h yields

$$t^2(h,\boldsymbol{\mathcal{L}}) = t_0^2 + 2\,t_0\,h^2 \sum_{k,m=1}^{3} \left.\frac{\partial^2 \tau}{\partial x_k \partial x_m}\right|_{h=0} \mathcal{L}_k\,\mathcal{L}_m\,. \tag{1.103}$$

Introducing the source-receiver offset

$$X = 2h\,, \tag{1.104}$$

we rewrite equation 1.103 in its final form

$$t^2(X,\boldsymbol{\mathcal{L}}) = t_0^2 + (\boldsymbol{\mathcal{L}}\,\mathbf{U}\,\boldsymbol{\mathcal{L}}^{\mathrm{T}})\,X^2\,. \tag{1.105}$$

Here the superscript "T" denotes transposition, the 3×3 symmetric matrix $\mathbf{U}$ is defined as

$$U_{km} \equiv \tau_0 \left.\frac{\partial^2 \tau}{\partial x_k \partial x_m}\right|_{h=0} = \tau_0 \left.\frac{\partial p_k}{\partial x_m}\right|_{h=0} \qquad (k,m = 1,2,3)\,, \tag{1.106}$$

and

$$p_k = \frac{\partial \tau}{\partial x_k} \qquad (k = 1, 2, 3) \tag{1.107}$$

are the components of the slowness vector $\mathbf{p} = [p_1, p_2, p_3]$.

Comparing equation 1.105 with the conventional definition of the normal-moveout velocity $V_{\rm nmo}(\boldsymbol{\mathcal{L}})$ on the CMP line $\boldsymbol{\mathcal{L}}$,

$$t^2(X, \boldsymbol{\mathcal{L}}) = t_0^2 + \frac{X^2}{V_{\rm nmo}^2(\boldsymbol{\mathcal{L}})}, \tag{1.108}$$

we conclude that

$$\frac{1}{V_{\rm nmo}^2(\boldsymbol{\mathcal{L}})} = \boldsymbol{\mathcal{L}}\, \mathbf{U}\, \boldsymbol{\mathcal{L}}^{\rm T}. \tag{1.109}$$

1D Reciprocal properties and the NMO ellipse of converted waves

First, we show that the reflection traveltime of a converted wave in a horizontal, homogeneous layer with a horizontal symmetry plane is *reciprocal* with respect to the source and receiver positions (i.e., it remains the same if the source and receiver are interchanged). Such reciprocity implies the absence of odd terms in the traveltime series and helps extend the concept of the NMO ellipse to mode conversions. Then we employ the generalized Dix equation 1.31 to find a simple relationship between the NMO ellipses of pure and converted waves.

1D.1 Reciprocity of converted-wave traveltime

Consider a reflected PS-wave traveling along the ray SRG and build a "reciprocal" PS ray G_1R_1G, with G being in the middle of SG_1 (Figure 1.14). Our goal is to prove that the traveltimes along the rays SRG and G_1R_1G are equal to each other.

Because the horizontal plane is a plane of symmetry, the rays of upgoing and downgoing waves with the same horizontal slowness components p_1 and p_2 will be symmetric with respect to it; also, the group velocities along these rays are equal to each other. The same symmetry holds for two downgoing rays with horizontal slownesses of the same magnitude but opposite signs. Hence, to find the reflected PS-wave excited at G_1 and recorded at G, we generate a downgoing P-ray G_1R_1 (Figure 1.14) with the horizontal slownesses $(-p_1)$ and $(-p_2)$, where p_1 and p_2 correspond to ray SR. Then G_1R_1 will represent a mirror image of SR with respect to the horizontal plane, and the traveltimes along these two rays will be identical. Also, as illustrated by the plan view in Figure 1.14, the triangles SRG and G_1R_1G are equal because SR and G_1R_1 are parallel to each other and have the same length. Therefore, the shear-wave rays R_1G and RG are symmetric with respect to the horizontal plane as well. This implies that R_1G corresponds to the ray parameters $(-p_1)$ and $(-p_2)$ and, hence, represents the S-wave reflected (converted) at R_1 (the horizontal slowness should be

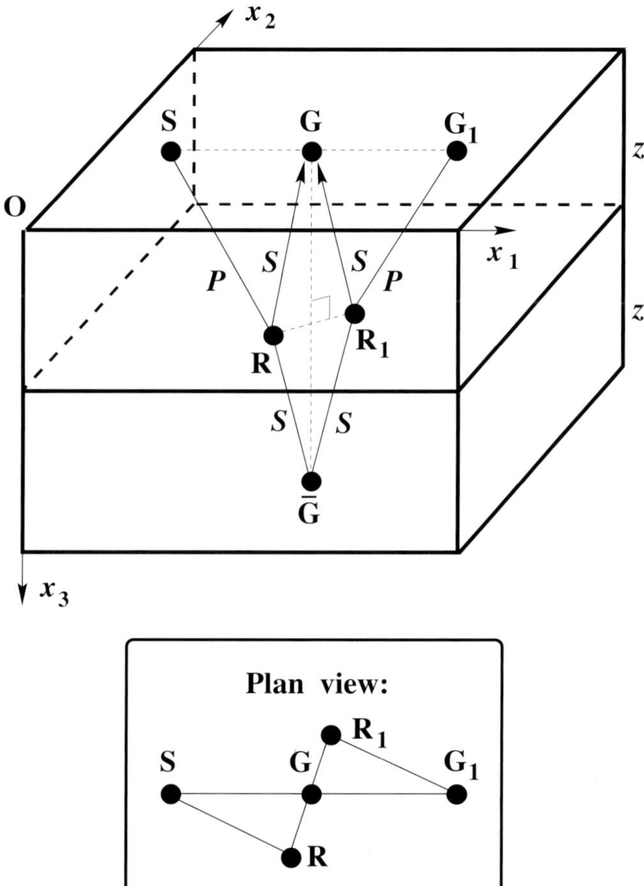

Figure 1.14: Ray geometry for a mode-converted PS-wave in a horizontal layer with a horizontal symmetry plane (Grechka et al., 1999a). The raypaths SRG and G_1R_1G ($SG=GG_1$) represent two "reciprocal" PS rays reflected from the bottom of the layer ($x_3=z$). The layer between $x_3=z$ and $x_3=2z$ (identical to the first layer) is added to obtain the NMO ellipse of the PS-wave from the generalized Dix equation. The inset shows the projections of points S, G, G_1, R, and R_1 onto the horizontal plane.

preserved during reflection/transmission). Thus, both the P and S segments of the PS reflection G_1R_1G represent mirror images with respect to the horizontal plane of the corresponding segments of the ray SRG. As a result, the traveltimes along SRG and G_1R_1G coincide with each other, and the PS-wave moveout is reciprocal with respect to the source and receiver positions.

1D.2 Relationship between the NMO velocities of pure and converted waves

Next, we use the reciprocal properties of the PS-wave moveout to extend the concept of the NMO ellipse to mode conversions. The absence of moveout asymmetry implies that the PS-wave traveltime series contains only even terms in offset, and conventional-spread moveout of converted waves is described by NMO velocity.

To express PS-wave NMO velocity through NMO velocities of the corresponding pure modes, we add a second identical layer to the model in Figure 1.14 and construct a PSSP reflected wave from the bottom of this layer. The intermediate interface represents a symmetry plane, so the transmitted shear-wave ray $R\bar{G}$ is a mirror image of the ray RG (the P-to-S conversion may be caused by a density contrast). The shear-wave reflection $\bar{G}R_1$ should coincide with RG and, therefore, hit point R_1. Finally, R_1G_1 represents the transmitted P-wave in the first layer excited by the S-wave $\bar{G}R_1$ at R_1 because it is a P-ray with the horizontal slowness components p_1 and p_2, which should be preserved along the whole raypath of the PSSP-wave.

Hence, the PS-wave traveltime along the ray SRG can be replaced with the traveltime along $SR\bar{G}$, which is equal to one-half of the reflection traveltime along $SR\bar{G}R_1G_1$. Because the offset $SG = SG_1/2$, the NMO velocities of the PS-wave SRG and the reflected PSSP-wave $SR\bar{G}R_1G_1$ are equal to each other. Kinematically, the PSSP-wave is equivalent to a pure-mode reflection in the two-layer model, so its azimuthally varying NMO velocity is given by the quadratic form 1.3:

$$[V_{\rm nmo}^{(PSSP)}(\alpha)]^{-2} = W_{11}^{(PSSP)} \cos^2\alpha + 2W_{12}^{(PSSP)} \sin\alpha\, \cos\alpha + W_{22}^{(PSSP)} \sin^2\alpha\,. \quad (1.110)$$

The matrix $\mathbf{W}^{(PSSP)}$, which typically describes an ellipse, can be determined from the generalized Dix equation 1.31:

$$t_0^{(PSSP)} \left[\mathbf{W}^{(PSSP)}\right]^{-1} = 2\tau_0^{(P)} \left[\mathbf{W}^{(P)}\right]^{-1} + 2\tau_0^{(S)} \left[\mathbf{W}^{(S)}\right]^{-1}\,, \quad (1.111)$$

where $\tau_0^{(P)}$ and $\tau_0^{(S)}$ are the one-way zero-offset traveltimes of P- and S-waves in the layer, and the matrices $\mathbf{W}^{(P)}$ and $\mathbf{W}^{(S)}$ correspond to the pure PP- and SS-wave reflections. According to equation 1.4,

$$W_{ij}^{(P)} = \tau_0^{(P)} \left.\frac{\partial^2 \tau^{(P)}}{\partial x_i \partial x_j}\right|_{\substack{x_1=0\\x_2=0}}\,. \quad (1.112)$$

$$W_{ij}^{(S)} = \tau_0^{(S)} \left.\frac{\partial^2 \tau^{(S)}}{\partial x_i \partial x_j}\right|_{\substack{x_1=0\\x_2=0}}\,. \quad (1.113)$$

Because the NMO velocities of the PS- and PSSP-waves are equal to each other, the matrix $W_{ij}^{(PSSP)}$ describes the PS-wave NMO ellipse and can be denoted by $W_{ij}^{(PS)}$. Taking into account that $t_0^{(PSSP)} = 2t_0^{(PS)}$ ($t_0^{(PS)}$ is the PS-wave zero-offset traveltime in the layer), we find the matrix $\mathbf{W}^{(PS)}$ as

$$t_0^{(PS)} \left[\mathbf{W}^{(PS)}\right]^{-1} = \tau_0^{(P)} \left[\mathbf{W}^{(P)}\right]^{-1} + \tau_0^{(S)} \left[\mathbf{W}^{(S)}\right]^{-1} . \tag{1.114}$$

Although the derivations in this appendix were carried out for a single layer, they remain entirely valid for a stack of horizontal, homogeneous layers with a horizontal symmetry plane. Therefore, equation 1.114 can be used to obtain the NMO ellipse of mode-converted waves in layered anisotropic media.

Chapter 2

Inversion of P-wave NMO ellipses

In this chapter, the formalism introduced in Chapter 1 is used to develop inversion and processing techniques for conventional-spread P-wave data from azimuthally anisotropic media. Stacking-velocity analysis for wide-azimuth 3D surveys often ignores the azimuthal dependence of normal moveout from horizontal reflectors, which may lead to distortions in the processing results (Lynn et al., 1996). A single value of stacking (NMO) velocity applied to the whole CMP gather causes underestimation of V_{nmo} for source-receiver azimuths near the "fast" direction of the NMO ellipse and overestimation near the "slow" direction. Hence, mixing of different azimuths impairs the performance of the moveout correction and, therefore, the quality of the stacked section. These distortions can be avoided by reconstructing the best-fit NMO ellipse and applying the correct stacking velocity for all azimuthal directions.

The inversion of the NMO ellipse is a much more complicated issue because the ellipticity can be caused not just by azimuthal anisotropy, but also by reflector dip and lateral heterogeneity. If the medium above the reflector is laterally homogeneous, reflector dip manifests itself through the reflection slope on the zero-offset section and can be accounted for in the parameter-estimation procedure. Correcting for the influence of lateral heterogeneity in the overburden is much less straightforward. Because waves recorded in different azimuthal directions propagate through different volumes of rock, lateral velocity variation can make NMO velocity azimuthally dependent even for a horizontal reflector beneath isotropic or VTI media. Anisotropy-induced NMO ellipticity is seldom larger than 10%, so it may well be comparable to distortions caused by mild lateral velocity variation.

The first section in this chapter is devoted to NMO-velocity inversion for horizontally layered, azimuthally anisotropic media. Following the results of Grechka (1998) and Grechka and Tsvankin (1999a), we present a data-driven correction of NMO ellipses for mild lateral velocity variation. We also describe a complete processing flow for moveout inversion of wide-azimuth reflection data that includes the generalized Dix equation introduced in Chapter 1. This methodology is applied to P-wave data acquired over a fractured reservoir in the Powder River Basin, Wyoming, to reconstruct interval NMO ellipses and evaluate depth-varying fracture orientation over the survey area.

Then we analyze the influence of dipping intermediate interfaces on the inversion of conventional-spread P-wave moveout. For VTI media, the presence of dip above the reflector makes P-wave traveltimes dependent on the interval values of the parameters ϵ and δ. Using a simple 2D synthetic model with a curved interface separating VTI and isotropic layers, we show that in special cases it may be possible to invert surface P-wave data for all three relevant VTI parameters (V_{P0}, ϵ, and δ) and, therefore, estimate the depth scale of the model. The dip of an intermediate interface can also mitigate some of the trade-offs between the anisotropy parameters of TI media with a tilted symmetry axis (TTI). We illustrate this point by processing a physical-modeling data set designed to simulate the reflection signature of overthrust areas in the Canadian Foothills.

For multilayered VTI media with dipping or curved interfaces, we introduce a 3D technique for depth-domain inversion of P-wave NMO ellipses and zero-offset traveltimes. This so-called *stacking-velocity tomography* is implemented using an efficient forward-modeling algorithm introduced in Chapter 1 that allows us to compute the hyperbolic portion of reflection moveout by tracing just one (zero-offset) ray. Singular value decomposition and inversion of noise-contaminated data help evaluate the feasibility of estimating the interval parameters V_{P0}, ϵ, and δ and reconstructing the medium interfaces. In particular, for some VTI models with smooth curved interfaces velocity analysis in depth can be performed without any a priori information.

Moveout inversion for tilted TI media generally requires combining the NMO ellipses for two reflectors with different orientations. We show that if one reflector is horizontal, the inversion procedure for a homogeneous TTI layer becomes stable when the dip of the second reflector and the tilt of the symmetry axis exceed 30°. For the important special case of the symmetry axis orthogonal to a dipping reflector, the NMO ellipse can be used to estimate the parameters V_{P0} and δ, but ϵ is not constrained by conventional-spread P-wave moveout. Interval parameter estimation for stratified TTI media is based on the generalized Dix equation and requires the presence of a dipping interface (e.g., a fault plane) in each layer.

For models with orthorhombic (orthotropic) symmetry, P-wave kinematic signatures are controlled by the vertical velocity V_{P0} and five anisotropy parameters. While P-wave reflection traveltimes do not provide enough information to reconstruct the velocity field in depth, we employ joint inversion of the NMO ellipses of horizontal and dipping events to estimate the parameter set responsible for time-domain processing. The feasibility of this inversion procedure depends on the maximum available dip and the relative orientation of the vertical symmetry planes and the dip plane of the structure.

2.1 Interval moveout inversion for horizontally layered models

Many fractured reservoirs, in particular those in the continental United States, are located in relatively simple structural settings without substantial dips or strong lateral velocity gradients. Here, using the results of Grechka and Tsvankin (1999a), we discuss NMO-velocity inversion of wide-azimuth P-wave data acquired over horizontally layered anisotropic media with mild lateral velocity variation in each layer. Imposing these restrictions helps avoid complications in the generalized Dix differentiation of NMO ellipses. Indeed, in the absence of reflector dip estimation of the interval NMO ellipse from equation 1.32 involves only measured NMO velocities of horizontal events and the corresponding zero-offset traveltimes.

2.1.1 Influence of weak lateral heterogeneity on NMO ellipses

Single layer with lateral velocity variation

Let us consider the simplest model of a single anisotropic layer above a horizontal reflector. The horizontal plane is assumed to be a plane of symmetry, which implies that the medium may be monoclinic, orthorhombic, or transversely isotropic with a vertical (VTI) or horizontal (HTI) symmetry axis. In general, the type of symmetry can change laterally provided the slowness surface remains symmetric with respect to the horizontal plane.

Suppose the NMO ellipse (equation 1.3) for the reflection from the bottom of the reference homogeneous layer (with the same parameters as those at the CMP location) is described by the matrix $\mathbf{W}^{\text{hom}}$ (equation 1.4). Whereas the azimuthal variation of NMO velocity in the presence of lateral velocity variation remains elliptical, the corresponding matrix $\mathbf{W}^{\text{het}}$ is different from $\mathbf{W}^{\text{hom}}$. If lateral velocity variation (or lateral heterogeneity, LH) is weak on the scale of the CMP gather (i.e., over the distance equal to the maximum source-receiver offset), NMO velocity can be linearized in the spatial derivatives of the velocity function. Then, as proved in Appendix 2A, the two matrices are related by (equation 2.87)

$$W_{ij}^{\text{het}} = W_{ij}^{\text{hom}} - \frac{\tau_0^2}{3V_0} \frac{\partial^2 V_0}{\partial y_i \partial y_j}\bigg|_{\mathbf{y}=\mathbf{y}_{\text{CMP}}}, \qquad (i, j = 1, 2), \tag{2.1}$$

where $\tau_0 = \tau_0(\mathbf{y})$ is the one-way zero-offset (vertical) traveltime and $V_0 = V_0(\mathbf{y})$ is the vertical velocity at the CMP location $\mathbf{y} = (y_1, y_2)$. Equation 2.1 indicates that in the linear approximation the influence of lateral velocity variation on the NMO ellipse is proportional to the curvature of the vertical-velocity surface $V_0(\mathbf{y})$ at the common midpoint.

If the reflector has dip (ignored here), equation 2.1 will also include the first spatial derivatives of the velocity $V_0(\mathbf{y})$. These derivatives, however, make a noticeable contribution only for substantial dip ϕ because they are multiplied with the cubic and higher powers of $\sin\phi$ (Grechka, 1998). In the case study from the Powder River Basin discussed below, the dips are on the order of a few degrees.

The high accuracy of equation 2.1 is confirmed by the test in Figure 2.1 for an HTI layer with moderate lateral vertical-velocity variation. Ray-traced P-wave reflection traveltimes were generated for the laterally heterogeneous model, as well as

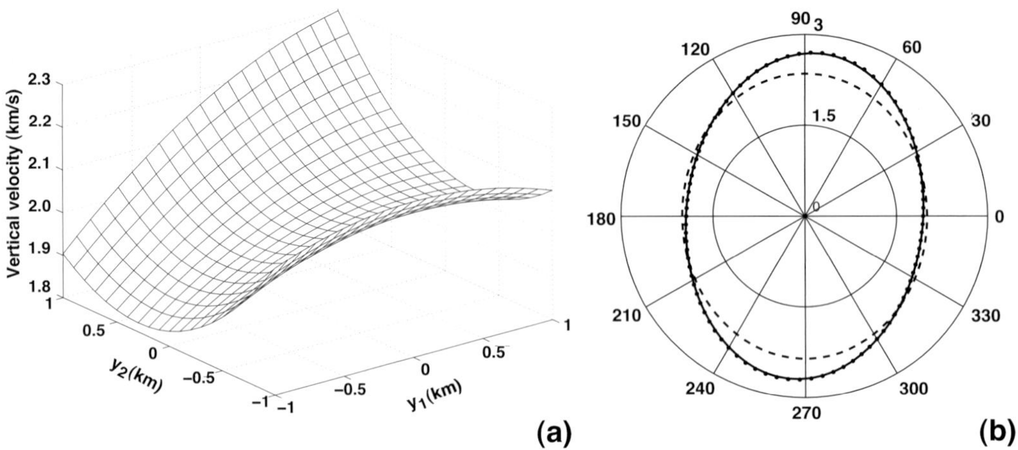

Figure 2.1: (a) Lateral variation of the P-wave vertical velocity in an HTI layer with the symmetry axis parallel to the y_1-direction. (b) The P-wave NMO ellipses for the reference homogeneous layer (dashed) and for the heterogeneous layer (dotted and solid) (Grechka and Tsvankin, 1999a). The dotted NMO ellipse is reconstructed from ray-traced traveltimes on spreadlength equal to the thickness of the layer (1.75 km); the solid ellipse is computed from equation 2.1. The anisotropy parameters (defined with respect to the symmetry axis) are $\epsilon=0$ and $\delta=0.2$. The azimuth 0° on plot (b) corresponds to the y_1-axis on plot (a).

for a reference homogeneous layer with the parameters corresponding to those at the common midpoint located at $y_1 = y_2 = 0$ (Figure 2.1a). After calculating traveltimes along four differently oriented CMP lines, we estimated the best-fit moveout velocities on spreadlength equal to the layer's thickness and reconstructed the NMO ellipses (Figure 2.1b). The difference between the ellipses for the homogeneous (dashed line) and LH (dotted) models can be understood by analyzing the saddle-shaped vertical-velocity surface $V_0(\mathbf{y})$ in Figure 2.1a. The positive curvature of $V_0(\mathbf{y})$ near the y_2-direction increases the velocity for rays and leads to a higher NMO velocity for azimuths close to 90° (Figure 2.1b). Likewise, the small decrease in the NMO velocity at azimuths near 0° results from the negative curvature of $V_0(\mathbf{y})$ in the y_1-direction.

To test the analytic correction for lateral velocity variation, we calculated the derivatives $\partial^2 V_0/(\partial y_i \partial y_j)$ at the CMP location and used equation 2.1 with the exact W_{ij}^{hom} to compute the NMO velocity in the LH layer (solid line in Figure 2.1b). This theoretical ellipse is virtually identical to the one reconstructed numerically (dotted), with the maximum difference less than 0.5%. Note that the linear component of the increase in V_0 in the y_1-direction (Figure 2.1a) has almost no influence on NMO velocity because V_{nmo} is controlled by the curvature of $V_0(\mathbf{y})$ at the common midpoint. This observation agrees with the result of Grechka (1998) who showed that, in the linear approximation, NMO velocity for VTI media is independent of a constant lateral velocity gradient.

Even though equation 2.1 provides an adequate approximation for the NMO ellipse in an LH layer, it is inconvenient to use in data processing because the vertical velocity V_0 in anisotropic media is poorly constrained by P-wave reflection traveltimes (e.g., Tsvankin, 2005). (One exception is the HTI model in which one of the semiaxes of the NMO ellipse is equal to the true vertical velocity.) Because, however, the layer is horizontal and $V_0\,\tau_0 = \text{const}$, the spatial derivatives of the vertical velocity V_0 can be replaced with those of the vertical traveltime τ_0. Differentiating the product $(V_0\,\tau_0)$ twice with respect to y_i and y_j yields

$$\tau_0\,\frac{\partial^2 V_0}{\partial y_i \partial y_j} + V_0\,\frac{\partial^2 \tau_0}{\partial y_i \partial y_j} = 0\,. \tag{2.2}$$

The terms $[(\partial V_0/\partial y_i)\,(\partial \tau_0/\partial y_j)]$ were dropped from equation 2.2 because they are quadratic in the small quantities related to lateral heterogeneity. Substituting equation 2.2 into equation 2.1 gives the following representation of $\mathbf{W}^{\text{hom}}$:

$$W_{ij}^{\text{hom}} = W_{ij}^{\text{het}} - \frac{\tau_0}{3}\,\frac{\partial^2 \tau_0}{\partial y_i \partial y_j}\Bigg|_{\mathbf{y}=\mathbf{y}_{\text{CMP}}}\,, \qquad (i,j=1,2)\,. \tag{2.3}$$

Evaluating the correction term in equation 2.3 usually requires smoothing the time surface $\tau_0(y_1, y_2)$, as illustrated by the case study below.

Stack of horizontal layers

This methodology can be extended to the reflector beneath any number of horizontal anisotropic layers with a horizontal symmetry plane. For multilayered media, however, not only lateral velocity variation, but also azimuthal anisotropy is assumed to be weak; otherwise, the correction term involves the interval vertical velocities.

If the model consists of two horizontal layers, the NMO ellipse for the reflection from the bottom of the second layer can be represented as (Grechka and Tsvankin, 1999a)

$$W_{ij}^{\text{hom}} = W_{ij}^{\text{het}} - \frac{\tau_0}{3}\left[k^2\,\frac{\partial^2 \tau_0}{\partial y_i \partial y_j} + (1+k)\,\frac{\partial^2 \tau_{01}}{\partial y_i \partial y_j}\right], \qquad (i,j=1,2)\,, \tag{2.4}$$

with

$$k = 1 - \frac{\tau_{01} V_{\text{cir1}}^2}{\tau_0 V_{\text{cir}}^2}\,, \tag{2.5}$$

where τ_{01} and τ_0 are the one-way zero-offset traveltimes for the reflections from the bottom of the first and the second layer, respectively; all derivatives are evaluated at the CMP location $\mathbf{y} = \mathbf{y}_{\text{CMP}}$. V_{cir1} and V_{cir} are the "circular" approximations of the NMO ellipses for the reflections from the bottom of the first and second layer. They are obtained by averaging the azimuthally dependent NMO velocity (equation 1.3) in the following way:

$$V_{\text{cir}}^{-2} = \frac{1}{2\pi}\int_0^{2\pi} V_{\text{nmo}}^{-2}(\alpha)\,d\alpha = \frac{W_{11}^{\text{het}} + W_{22}^{\text{het}}}{2}\,. \tag{2.6}$$

To examine how lateral velocity variation in a certain interval contributes to the correction term in equation 2.4, consider a model with a laterally homogeneous top layer:

$$\frac{\partial^2 \tau_{01}}{\partial y_i \partial y_j} = 0\,. \tag{2.7}$$

Then, equation 2.4 simplifies to

$$W_{ij}^{\text{hom}} = W_{ij}^{\text{het}} - \frac{k^2 \tau_0}{3} \frac{\partial^2 \tau_0}{\partial y_i \partial y_j}\,, \qquad (i,j=1,2)\,. \tag{2.8}$$

As can be expected, the correction resulting from lateral velocity variation becomes smaller if a laterally homogeneous layer is added on top of an LH layer. Indeed, the difference between the correction terms in equation 2.8 and in the single-layer equation 2.3 is given by the factor k^2, which is always smaller than unity. This fact can be explained by considering the lateral position of reflected rays in a CMP gather: as the LH layer moves deeper, the rays crossing it sample a smaller vicinity of the common midpoint and, therefore, are less sensitive to the lateral velocity variation.

Furthermore, equation 2.8 shows that if $V_{\text{cir1}}^2 = V_{\text{cir}}^2$ so $k^2 = (1 - \tau_{01}/\tau_0)^2$, the correction for LH is proportional to the *squared* relative thickness of the LH layer. For example, when the relative thickness of the LH layer $[(\tau_0 - \tau_{01})/\tau_0]$ is equal to 0.5, the coefficient k^2 reduces to 0.25 compared to unity for a single layer. Such a nonlinear dependence means that the influence on the effective NMO ellipse of a laterally heterogeneous layer with a fixed thickness rapidly decreases with the layer's depth.

2.1.2 Processing flow for wide-azimuth P-wave data

The generalized Dix equation combined with the correction for lateral velocity variation provides an analytic basis for interval moveout inversion of wide-azimuth data in horizontally layered anisotropic media. A robust processing flow for conventional-spread P-wave reflection events includes the following main steps:

1. 3D semblance analysis to obtain the effective NMO ellipses for each CMP gather or superbin[1].
2. Spatial smoothing of the effective NMO ellipses.
3. Correction for lateral velocity variation.
4. Generalized Dix differentiation to obtain the interval NMO ellipses.

Below, we use a case study for a fractured reservoir to describe each processing step in more detail. Whereas semblance analysis represents the most common tool for estimating reflection moveout from seismic data, Jenner (2001) and Jenner et al. (2001) developed an alternative, windowed crosscorrelation technique for reconstructing effective NMO ellipses. Also, a variety of smoothing algorithms could be applied to NMO ellipses prior to their subsequent processing.

[1] A composite CMP gather that includes all source-receiver pairs with close midpoints.

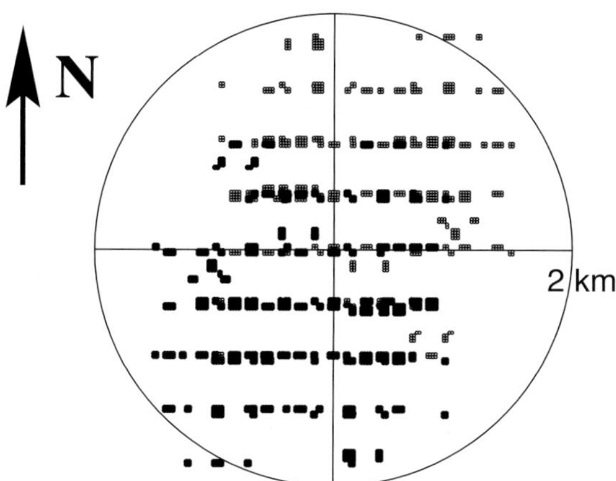

Figure 2.2: Plan view of the source and receiver positions (squares) for a typical superbin from the Powder River Basin data set (Grechka and Tsvankin, 1999a). The superbin contains approximately 400 source-receiver pairs with a common-midpoint scatter of up to about 80 m (2% of the maximum offset). The spreadlength-to-depth ratio for the deepest reflector is close to unity.

2.1.3 Field-data example

Here, we discuss the main results of Grechka and Tsvankin (1999a) who carried out interval moveout inversion of a 3D data set acquired for purposes of fracture detection in the Powder River Basin, Wyoming. The button-patch geometry of the survey was designed to provide full azimuthal coverage in the conventional range of source-receiver offsets. To enhance the signal-to-noise ratio, the data were collected into 169 superbins, each with a somewhat random distribution of azimuths and offsets (Figure 2.2). A description of the geology of the area and data acquisition can be found in Corrigan et al. (1996) and Withers and Corrigan (1997). The processing sequence was designed to recover the interval P-wave NMO ellipses associated with azimuthal anisotropy and use them to characterize natural fracture systems.

3D semblance analysis

The effective NMO ellipses were estimated for several prominent reflectors over the survey area. In principle, one can divide the data for a given superbin into several azimuthal sectors (e.g., Grimm et al., 1999) and perform conventional 2D hyperbolic moveout analysis for source-receiver pairs within each sector. Then, the best-fit moveout (stacking) velocities $V_{\text{nmo}}(\alpha)$ for a given reflection event are approximated with an ellipse using equation 1.3. While azimuthal sectoring makes it possible to use conventional software for semblance analysis, the number and size of the sectors may seriously distort the velocity-estimation results because the distribution of offsets and azimuths is not completely random (e.g., Lynn, 2007).

It is preferable to treat all azimuths simultaneously and perform semblance analysis for the whole superbin at each zero-offset time τ_0. This "global" semblance analysis is based on the hyperbolic moveout equation 1.1 and requires scanning over the three components of the matrix $\mathbf{W}$ responsible for the NMO ellipse (equation 1.3). Because of the limited range of offsets the moveout for the events of interest was close to hyperbolic. Estimation of the NMO ellipse on spreadlengths exceeding reflector depth requires either muting out long offsets or applying nonhyperbolic moveout equations (see Chapter 3).

To make this search more efficient and avoid a full-scale 3D semblance scan, it is convenient to use an equivalent representation of the NMO ellipse in terms of the average velocity $V_{\rm cir}$ (equation 2.6) and two dimensionless quantities, E_1 and E_2:

$$V_{\rm nmo}^{-2}(\alpha) = V_{\rm cir}^{-2}\,(1 + E_1 \cos 2\alpha + E_2 \sin 2\alpha)\,; \tag{2.9}$$

the parameters E_1 and E_2 control the deviation of the NMO ellipse from an average "NMO circle." Combining equations 1.3, 2.6, and 2.9 yields

$$E_1 = \frac{W_{11}^{\rm het} - W_{22}^{\rm het}}{W_{11}^{\rm het} + W_{22}^{\rm het}} \quad \text{and} \quad E_2 = \frac{2\,W_{12}^{\rm het}}{W_{11}^{\rm het} + W_{22}^{\rm het}}\,. \tag{2.10}$$

The introduction of E_1 and E_2 allows us to speed up the semblance analysis by dividing it into two stages. Because typically E_1 and E_2 are small compared to unity, first we assume that $E_1 = E_2 = 0$ and carry out a conventional semblance scan over $V_{\rm cir}$. This procedure, routinely applied in 3D processing, ignores the azimuthal dependence of NMO velocity and yields an average NMO circle that can be treated as an initial guess for the NMO ellipse. Then the coefficients E_1 and E_2 are obtained by deforming the circle into the best-fit NMO ellipse 2.9 that provides the highest semblance value. This search is performed by an efficient minimization technique (Powell's method, see Press et al., 1987) that usually converges in $5-10$ iterations, making an extensive 3D semblance scan unnecessary.

Figure 2.3 shows typical semblance curves obtained from conventional and azimuthal moveout analysis for one of the superbins. By conventional analysis we mean the first stage of our semblance search that provides the best-fit velocity $V_{\rm cir}$. While the two curves are close to each other over most of the time interval (i.e., the NMO ellipses are close to circles), for the reflections at the vertical times 1.54 s, 1.84 s, and, especially, 2.57 s, the azimuthal velocity analysis improves the fit to reflection moveout and provides higher semblance values. For these reflection events, the conventional algorithm smears the azimuthal velocity variation and produces a distorted value of the moveout velocity for any given azimuth, thus leading to a lower quality of stack. Moreover, the velocity $V_{\rm cir}$ cannot be used to obtain information about azimuthal anisotropy.

According to Withers and Corrigan (1997), the reflection at a two-way vertical time of 2.14 s corresponds to the bottom of the Frontier/Niobrara formations, and

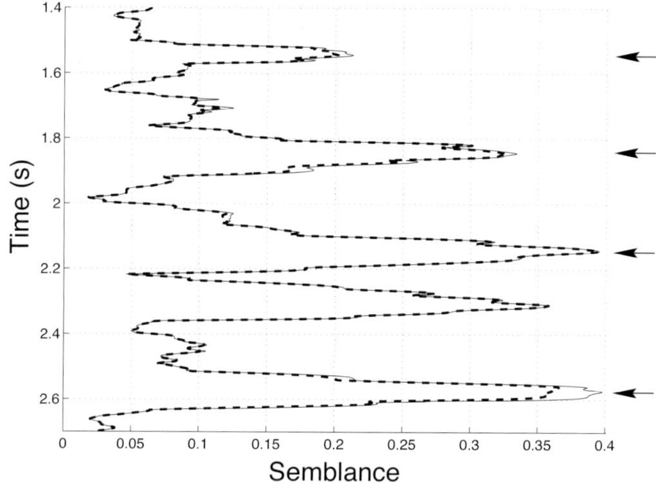

Figure 2.3: Semblance curves computed by conventional moveout analysis (dashed), which ignores the azimuthal dependence of moveout velocity, and by the azimuthal moveout analysis (solid) (Grechka and Tsvankin, 1999a). The arrows mark the reflections used in the generalized Dix differentiation of NMO ellipses.

the event at 2.57 s is the basement reflection. The producing fractured reservoir (the main target of the survey) approximately covers the time interval between 1.84 s and 2.14 s.

The results of the azimuthal moveout analysis for the reflections marked in Figure 2.3 are shown in Figure 2.4 over the entire survey area. Each tick corresponds to the effective NMO ellipse at a certain superbin location. The direction of a tick indicates the azimuth of the major axis of the NMO ellipse, while the tick's length is proportional to the ellipticity E (i.e., the eccentricity of the NMO ellipse). We define E as the fractional difference between the semiaxes $v_{e\ell 2}$ and $v_{e\ell 1}$,

$$E = 2\,\frac{v_{e\ell 2} - v_{e\ell 1}}{v_{e\ell 2} + v_{e\ell 1}}\,. \tag{2.11}$$

The maximum effective ellipticity for the whole survey area does not exceed 0.05. We will see, however, that the interval NMO ellipses in certain layers have much higher values of E. The patterns in Figure 2.4 are sufficiently close to the results of Withers and Corrigan (1997), who used a somewhat different (interactive) algorithm for azimuthal velocity analysis of this data set.

Smoothing of the effective NMO ellipses

An important issue is to what extent one can trust the rapid spatial variations of the parameters of the NMO ellipses in Figure 2.4. The size of the whole survey area does not differ much from the maximum offset for a single superbin (compare

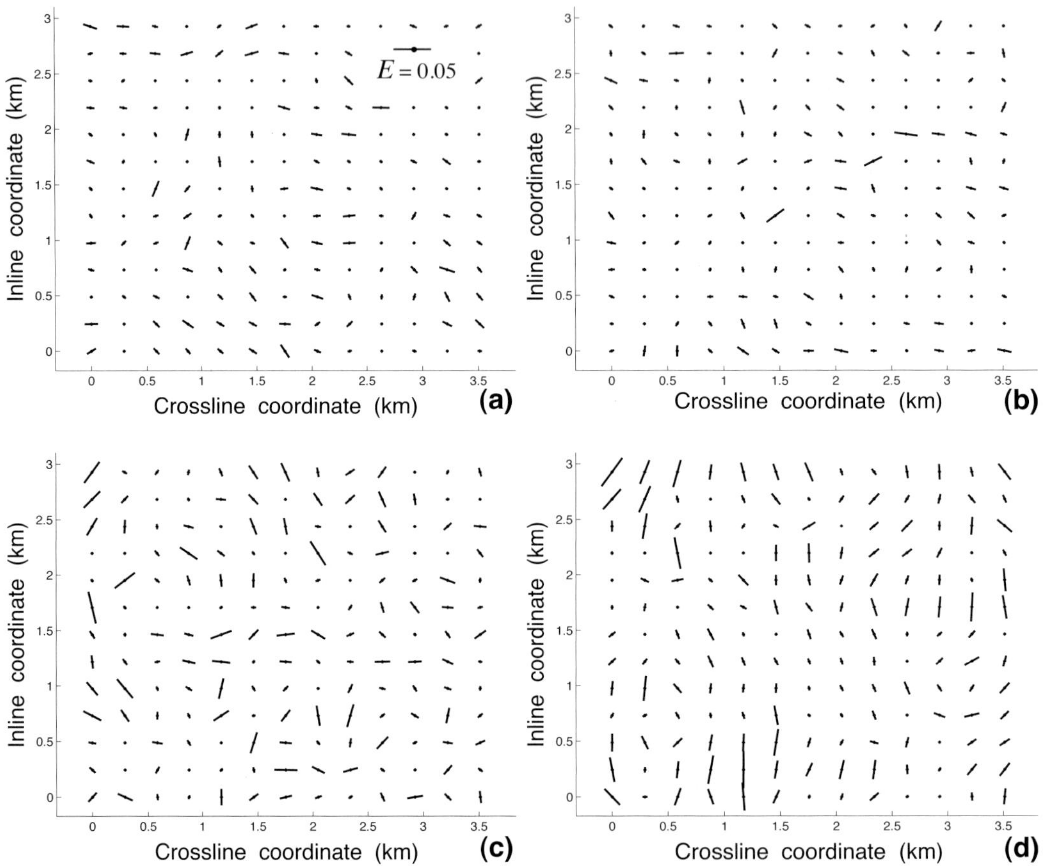

Figure 2.4: Raw effective NMO ellipses for the reflections at the following zero-offset times: (a) 1.54 s, (b) 1.84 s, (c) 2.14 s, and (d) 2.57 s (Grechka and Tsvankin, 1999a).

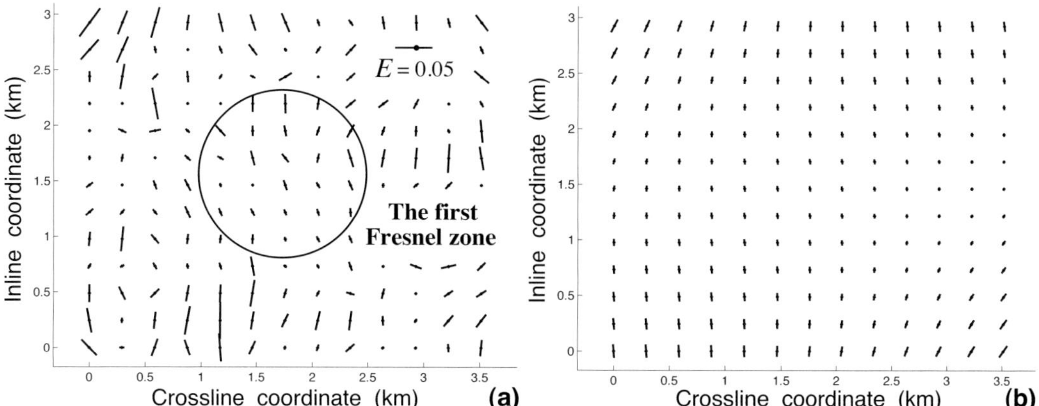

Figure 2.5: (a) Raw effective NMO ellipses for the reflection at 2.57 s with the circle corresponding to the first Fresnel zone at the reflector. (b) The ellipses from plot (a) after spatial smoothing (Grechka and Tsvankin, 1999a).

the scales in Figures 2.2 and 2.4). This means that reflected rays corresponding to adjacent superbins propagate through almost the same subsurface volume. Hence, a 90° rotation of the major axis of the NMO ellipse over a small distance (e.g., between the point with crossline coordinate $y_1 = 0.6$ km and inline coordinate $y_2 = 2.0$ km and an adjacent point with $y_1 = 0.6$ km and $y_2 = 2.2$ km in Figure 2.4d) can most likely be attributed to noise in the input data that leads to errors in azimuthal velocity analysis. Clearly, reliable estimation of the interval moveout is impossible without spatial smoothing of the effective NMO ellipses.

Our design of the smoothing procedure is based on the size of the first Fresnel zone at the reflector, estimated for the central frequency in the data. Figure 2.5 shows that the whole survey area is only about six times larger than the Fresnel zone for the deepest reflector. Because it is reasonable to assume that each Fresnel zone can yield a single NMO ellipse, the spatial variation in the smoothed matrix $\mathbf{W}^{\text{het}}(\mathbf{y})$ should be represented by a function with six independent parameters. In essence, by applying spatial smoothing we suppress short-wavelength spatial variations in NMO velocity that cannot be resolved from the data.

Thus, we seek $\mathbf{W}^{\text{het}}(\mathbf{y})$ as a quadratic polynomial

$$W_{ij}^{\text{het}}(y_1, y_2) = \sum_{k,l=0}^{k+l \leq 2} \mathcal{W}_{ij}^{(kl)} \, y_1^k \, y_2^l \,, \qquad (i, j = 1, 2) \,, \tag{2.12}$$

where the six coefficients $\mathcal{W}_{ij}^{(kl)}$ (for each element W_{ij} of the matrix $\mathbf{W}^{\text{het}}$) are found by least-squares fitting of equation (2.12) to the raw NMO ellipses from Figure 2.4. The choice of the quadratic polynomial means that the NMO ellipses are approximated by surfaces $\mathbf{W}^{\text{het}}(\mathbf{y})$ of constant curvature over the whole survey area.

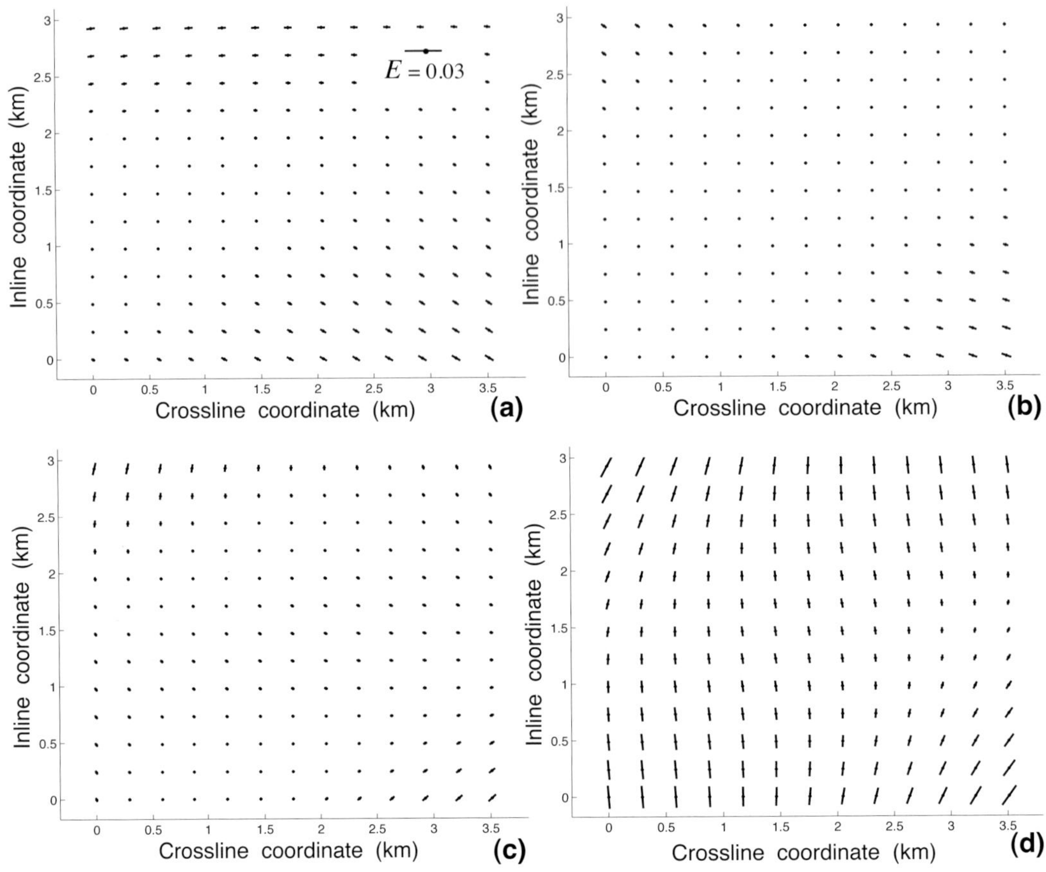

Figure 2.6: Smoothed effective NMO ellipses for the reflections at (a) 1.54 s, (b) 1.84 s, (c) 2.14 s, and (d) 2.57 s (Grechka and Tsvankin, 1999a).

The effective NMO ellipses after the spatial smoothing are shown in Figures 2.5b and 2.6. The smoothing led to an overall decrease of the ellipticity for the events at 1.54, 1.84, and 2.14 s. For the reflection at 2.57 s, the smoothed ellipticity is larger and has a predominant north-south orientation. The results of smoothing indicate that the level of errors in picking azimuthally-dependent NMO velocities is about 1% – 2%, and small effective ellipticities on the order of 2%, quite common for all four reflection events, might not carry useful information about the fracture direction.

Correction for lateral velocity variation

Reflector dips in the survey area are small, and the subsurface model can be adequately represented by a stack of horizontal layers. Therefore, equations 2.3 and 2.4 can be used to correct the effective NMO ellipses for lateral velocity variation. Ap-

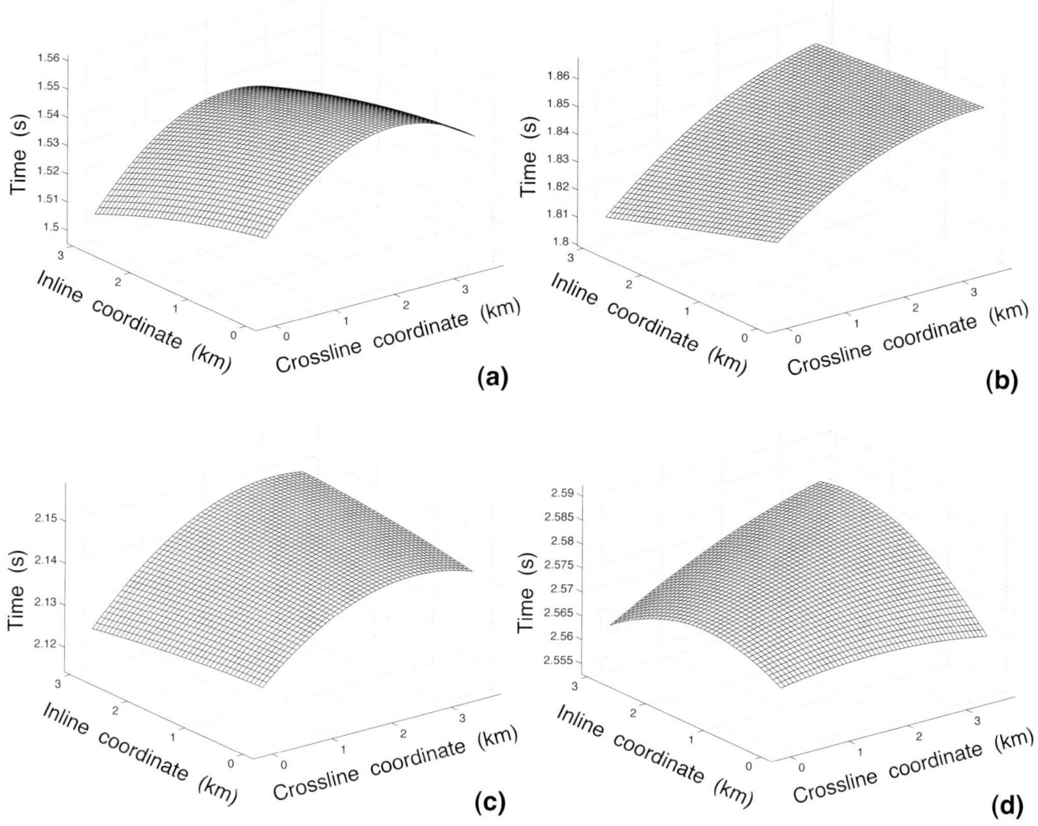

Figure 2.7: Smoothed two-way zero-offset traveltime surfaces for the reflections at (a) 1.54 s, (b) 1.84 s, (c) 2.14 s, and (d) 2.57 s (Grechka and Tsvankin, 1999a).

plicability of the assumption of weak lateral velocity variation and weak anisotropy is justified by the fact that the effective ellipticity is relatively small (Figure 2.6).

Figure 2.7 shows smoothed surfaces of the two-way zero-offset reflection traveltime $t(\mathbf{y}) = 2\tau_0(\mathbf{y})$ for the four reflection events. The smoothing function had the same form as that used for NMO ellipses (equation 2.12). The maximum apparent dip in Figure 2.7 reaches $30-40$ ms over a distance of about 3 km, giving the apparent horizontal slowness $p = 0.5\,(dt/dy) = 0.005$ s/km. Attributing this spatial variation $t(\mathbf{y})$ to reflector dip ϕ yields a relative correction in the NMO velocity that can be roughly estimated (in the isotropic limit) as $1/\cos\phi \approx 1/\sqrt{1-p^2 V_{\text{cir}}^2}$. Substituting $p = 0.005$ s/km and $V_{\text{cir}} = 4.0$ km/s (see Figure 2.8), we find that the distortion caused by reflector dip is close to just 0.02%, which is an order of magnitude smaller than the estimated errors in the effective NMO ellipses. Even though the apparent dip of the zero-offset traveltime surface is also influenced by lateral velocity variation

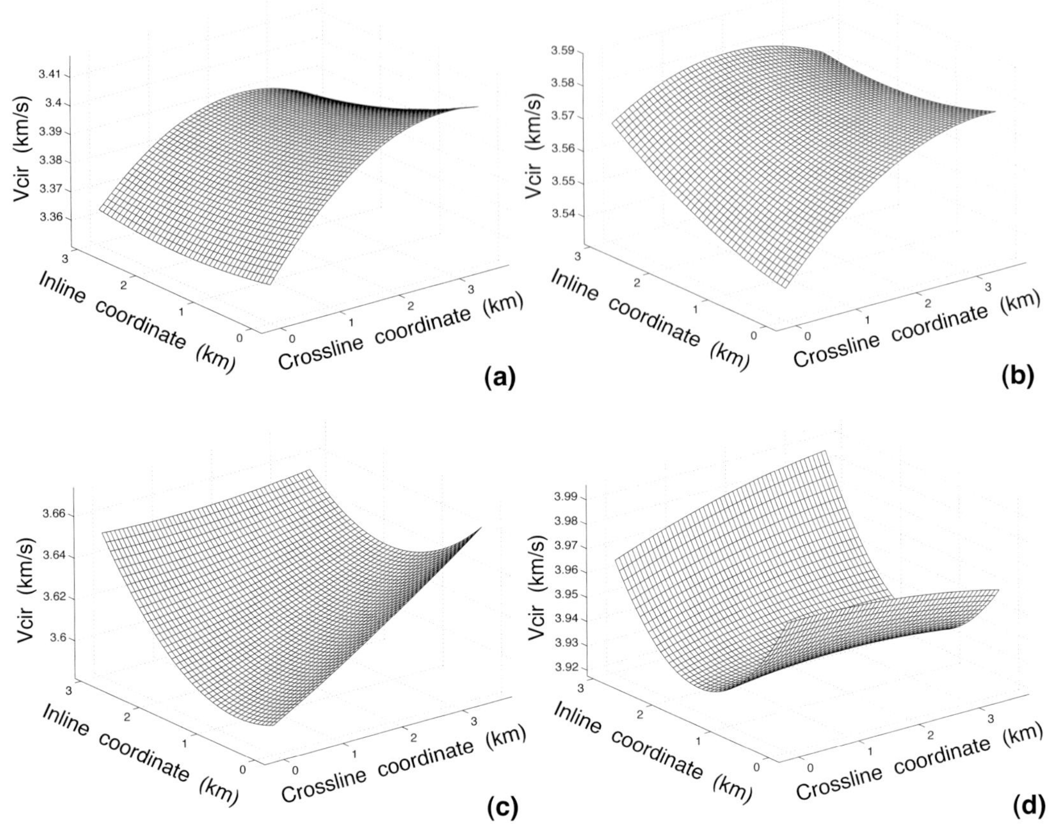

Figure 2.8: Smoothed effective NMO velocities V_{cir} (equation 2.6) for the reflections at (a) 1.54 s, (b) 1.84 s, (c) 2.14 s, and (d) 2.57 s (Grechka and Tsvankin, 1999a).

(Figure 2.8), it is clear that the contribution of dip to NMO velocity can be ignored for all reflection events.

The zero-offset traveltime surfaces $t(\mathbf{y})$ (Figure 2.7), along with the surfaces of the average velocity $V_{\text{cir}}(\mathbf{y})$ (Figure 2.8), can be used to strip the influence of lateral velocity variation from the effective NMO ellipses (Figure 2.9). Note that the curvature of the traveltime surface, responsible for the influence of lateral velocity variation on the NMO ellipse, is determined by the coefficients of our smoothing function (equation 2.12) and remains constant over the survey area. The correction for lateral velocity variation for the shallow reflection (at 1.54 s) was carried out using the single-layer equation 2.3, while for the event at 1.84 s we applied the two-layer equation 2.4. In principle, the reflections at 2.14 s and 2.57 s should be treated by means of a more complicated multiple-layer correction formula (not given here). However, in order to simplify the processing algorithm, the same equation 2.4 was used for

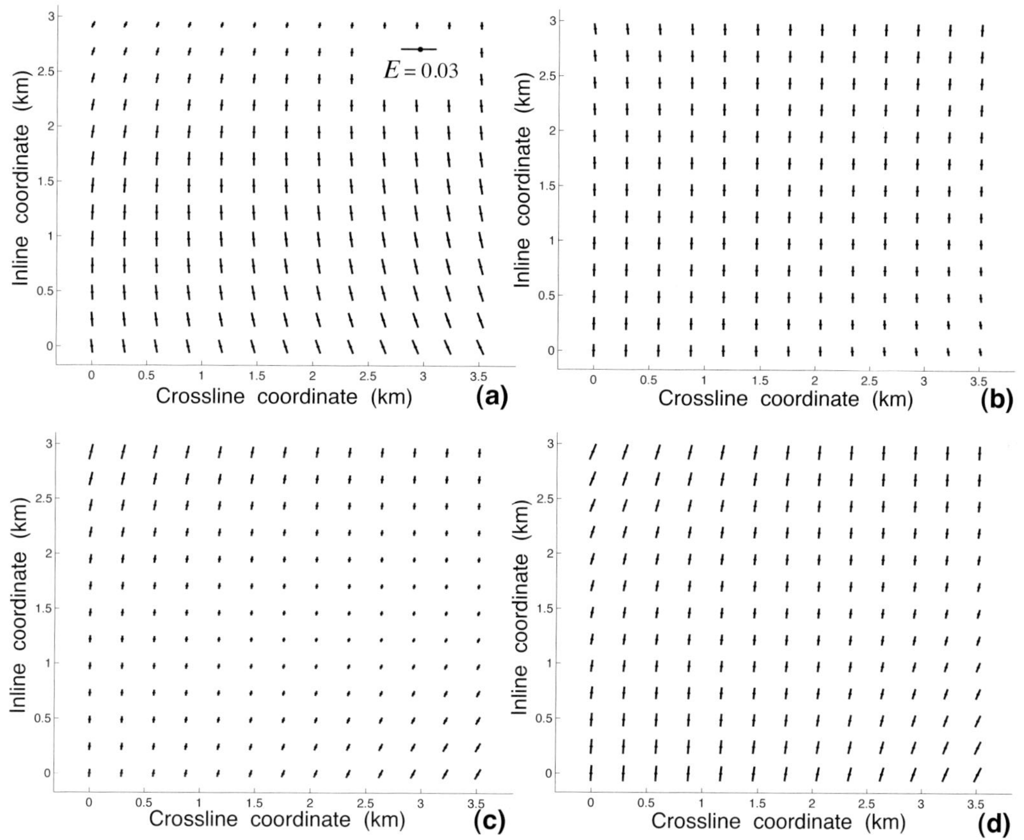

Figure 2.9: Effective NMO ellipses for the reflections at (a) 1.54 s, (b) 1.84 s, (c) 2.14 s, and (d) 2.57 s after correction for lateral velocity variation (Grechka and Tsvankin, 1999a).

both deeper events under the assumption that the entire stratified overburden may be approximated by a single effective layer. For instance, in calculating the correction term for the basement reflection (2.57 s), the first-layer parameters in equation 2.4 (τ_{01} and $V_{\rm cir1}$) were assumed to correspond to the event at 2.14 s.

Comparison of Figures 2.9 and 2.6 illustrates the distortions caused by the lateral velocity variation. As an example, the surface $t(\mathbf{y})$ for the event at 1.54 s has a negative curvature in approximately the east-west (crossline) direction (Figure 2.7a), which leads to an increase in the NMO velocity. Application of equation 2.3 stretches the NMO ellipses in the orthogonal (north-south) direction, thereby restoring the signature related to the azimuthal anisotropy (Figure 2.9a). For the other three reflections, the influence of lateral velocity variation is more complicated because it involves two traveltime surfaces, τ_0 and τ_{01}. Note that the correction did not significantly rotate the NMO ellipses for the deepest event (Figures 2.6d and 2.9d).

Our correction procedure is highly sensitive to the shape of the surfaces $t(\mathbf{y})$. Because these surfaces are differentiated twice, the form of the smoothing function can substantially change the corrected NMO ellipse. We tested two other choices of smoothing: a bicubic polynomial ($k+l \leq 3$ in equation 2.12) and the running average over the first Fresnel zone (see Figure 2.5). For both choices, the corrected effective ellipticity was implausibly high (over 0.3), which is indicative of overfitting the data and amplifying errors in the raw NMO ellipses.

Generalized Dix differentiation

The last step of our processing sequence is computation of interval NMO ellipses using the generalized Dix differentiation (equation 1.32). Presumably, azimuthal variation of the interval NMO velocity in Figure 2.10 is associated exclusively with azimuthal anisotropy and, therefore, carries information about fracturing. If we assume that each layer contains a single set of vertical fractures in an isotropic or VTI matrix, which is consistent with the available geologic information (Corrigan et al., 1996; Withers and Corrigan, 1997), the major axis of the NMO ellipse is parallel to the fracture direction, while the ellipticity E is related to the fracture density (for more details, see Chapter 9).

For the layer between 1.54 s and 1.84 s, the major axis of the NMO ellipse is predominantly oriented east-west, but the ellipticity is rather small, up to 3% (Figure 2.10a). The pattern changes significantly in the producing interval between 1.84 s and 2.14 s. The orientation of the NMO ellipse in that layer varies substantially over the area, with higher ellipticities (up to 8%) in the northwest and southeast corners (Figure 2.10b). In the deepest layer above the basement, the major axis of the NMO ellipse is close to the north-south direction, with the ellipticity somewhat increasing towards the south (Figure 2.10c). The results of the interval moveout analysis here are supported by independent borehole measurements and shear-wave data (see below).

The generalized Dix differentiation, as with its conventional counterpart, suffers from amplification of errors for layers that are thin compared to their depth. The average relative thickness of the intervals used in our analysis is about 15%, which is close to the limit of applicability of the Dix differentiation. The effective NMO ellipses, however, were obtained by two sequential averaging procedures, which increased the stability of the interval moveout estimation. Indeed, the effective NMO ellipse was built by a 3D semblance search for the entire superbin (which amounts to azimuthal averaging of NMO velocity) followed by the spatial smoothing of the NMO ellipses over the survey area.

Interpretation of the interval NMO ellipses

Undoubtedly, the accuracy of the interval NMO ellipses is influenced to some extent by several assumptions and approximations used in the processing sequence. Also, the effective ellipticity is relatively small, which made the recovery of the anisotropic

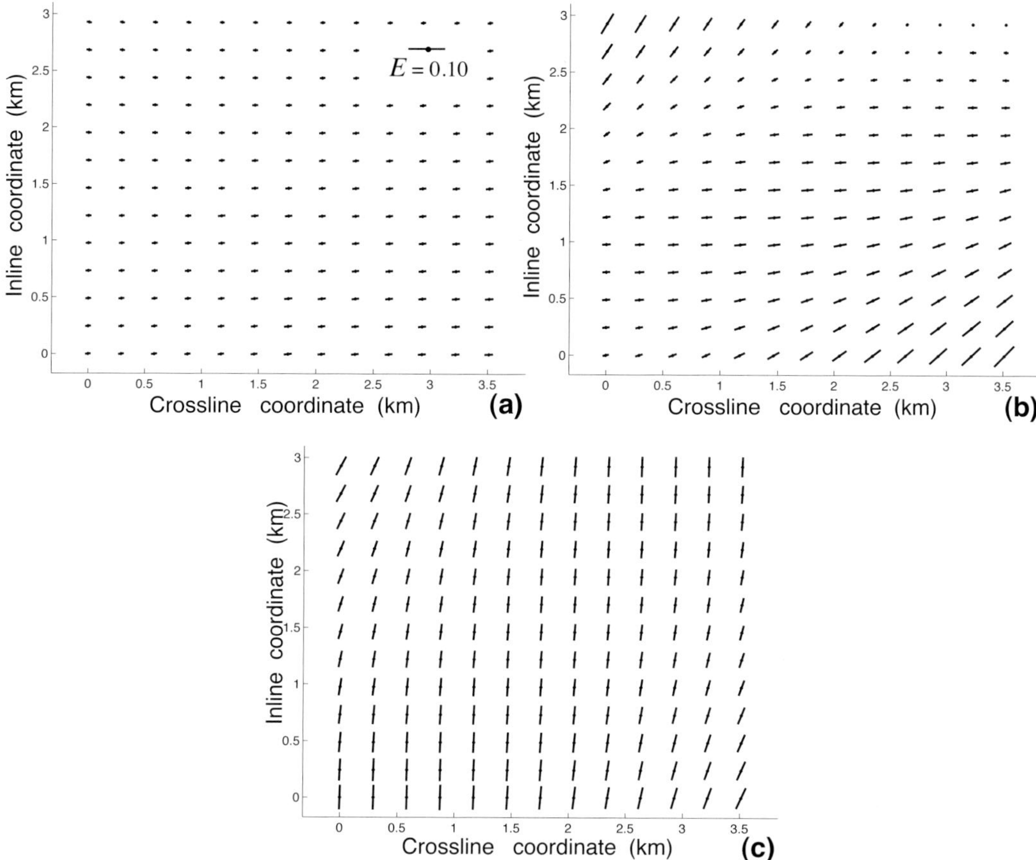

Figure 2.10: Interval NMO ellipses for the horizons between (a) 1.54 s and 1.84 s, (b) 1.84 s and 2.14 s, and (c) 2.14 s and 2.57 s (Grechka and Tsvankin, 1999a).

signature rather difficult; this methodology is more robust in areas with a higher magnitude of azimuthal anisotropy (Jenner et al., 2001). Therefore, it is important to compare our conclusions with information about the fracture orientation in different layers obtained by other methods (Withers and Corrigan, 1997; Corrigan, personal communication):

1. Fractures in the shallow section are oriented approximately N110°E. This is supported by outcrop measurements, FMI/FMS borehole scans, and rotational analysis of four-component shear-wave data. The N110°E orientation is generally close to the major axis of the interval NMO ellipse for the layer immediately above $t = 1.84$ s. As mentioned above, there is no evidence of multiple fracture sets in the study area, so the fast NMO-velocity direction should be aligned with the fracture strike.

2. Outcrop studies and analysis of a nine-component VSP survey indicate that

the fracture azimuth in the interval between $t = 1.84$ s and 2.14 s is about N70°E. The same orientation was obtained by layer-stripping analysis of surface S-wave data. These are producing fractures whose direction and density are of most interest from an exploration standpoint. The average azimuth of the major axis of the interval NMO ellipse (Figure 2.10b) is indeed between N50°E and N70°E, although the orientation of the interval NMO ellipses exhibits substantial spatial variability.

3. Fracture orientation in the interval between $t = 2.14$ s and 2.57 s is not well established. There is, however, outcrop evidence for a direction close to north-south, which agrees with the azimuth of the major axis of the interval NMO ellipse.

An important issue for fracture characterization is to understand which medium parameters (in addition to the fracture direction) can be estimated from the interval P-wave NMO ellipses. The answer is model-dependent and becomes increasingly complex for lower anisotropic symmetries. For the simplest azimuthally anisotropic model, fracture-related horizontal transverse isotropy, the semimajor axis of the P-wave NMO ellipse is equal to the vertical velocity and points in the fracture direction. The difference between the semiaxes is roughly proportional to the anisotropy parameter $\delta^{(V)}$ (see Appendix 1B), which depends on both the fracture density and fluid saturation (Tsvankin, 1997b; Bakulin et al., 2000a). Therefore, estimation of the fracture parameters generally requires supplementing the P-wave NMO ellipse with other data or making restrictive assumptions about the fractures; this problem is addressed in more detail in Chapter 9.

2.2 Examples of depth-domain inversion for TI media with dipping interfaces

The main difficulties in anisotropic velocity analysis and inversion using surface seismic data are caused by the multiparameter nature of the problem and inherent trade-offs between the model parameters. For the most often used anisotropic model, vertical transverse isotropy, P-wave kinematic signatures are controlled by the vertical velocity V_{P0} and the anisotropy parameters ϵ and δ. However, only two combinations of these parameters – the NMO velocity for a horizontal reflector ($V_{\mathrm{nmo},P}$) and the anellipticity parameter η (equations 1.52 and 1.53) – can be estimated from P-wave reflection traveltimes, if the medium above the reflector is *laterally homogeneous* (Alkhalifah and Tsvankin, 1995; Tsvankin, 1996, 2005). While $V_{\mathrm{nmo},P}$ and η are sufficient for time-domain imaging in VTI media, they cannot be used to resolve the vertical velocity and build velocity models needed for depth migration.

As discussed in the previous section and Chapter 1, lateral heterogeneity above the reflector has a strong influence on P-wave normal moveout. Application of the formalism based on NMO-velocity surfaces to piecewise-homogeneous VTI media shows that the effective NMO ellipse in the presence of intermediate dipping interfaces depends on the parameters V_{P0}, ϵ, and δ individually (equations 1.54 – 1.56). In general, however, there is no guarantee that all three parameters can be resolved from surface

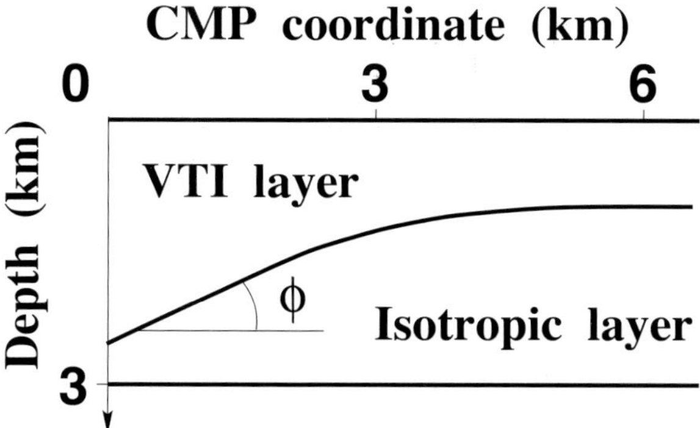

Figure 2.11: Model used for parameter estimation from P-wave reflection traveltimes (Le Stunff et al., 2001). The top layer is VTI with P-wave vertical velocity $V_{P0,t}$=2.5 km/s and anisotropy parameters ϵ_t=0.2 and δ_t=0.05. The bottom layer is isotropic with P-wave velocity $V_{P,b}$=3.5 km/s; the depth of the lower boundary Z_r=3 km. The intermediate interface has the dip ϕ=25° at the zero CMP coordinate and is horizontal near the right edge of the model.

Number of common midpoints	500
CMP spacing	12.5 m
Number of receivers per CMP gather	40
Receiver spacing	100 m
Minimum offset	0 m
Maximum offset	4000 m

Table 2.1: Ray-tracing parameters for the model in Figure 2.11. The number of receivers in a CMP gather and the maximum offset were reduced for CMP locations near the edges of the model.

P-wave data. Still, as illustrated by the synthetic example below, for relatively simple models with dipping interfaces above the reflector it might be possible to build a VTI depth model solely from P-wave reflection traveltimes. In addition, we demonstrate with physical-modeling data that depth-domain parameter estimation may also be feasible for certain types of TI media with a tilted symmetry axis (TTI).

2.2.1 VTI model with a curved interface

Following Le Stunff et al. (2001), we consider P-wave moveout inversion for a model composed of two homogeneous layers (VTI and isotropic) separated by a curved interface (Figure 2.11). P-wave reflection traveltimes for both interfaces were calculated by anisotropic ray tracing; the modeling parameters are listed in Table 2.1.

After adding Gaussian noise with zero mean and a standard deviation of 2.5 ms to the traveltimes to simulate picking errors, the data were processed by a tomographic algorithm designed for VTI media (Le Stunff and Grenié, 1998). The initial model was purely isotropic with parameters estimated from conventional time-to-depth conversion using the NMO velocities and zero-offset traveltimes. The tomographic procedure updates the model using a linearized least-squares algorithm to minimize the misfit between the input and computed traveltimes (Tarantola, 1987). The inversion was constrained by making the following assumptions:

1. The model consists of two homogeneous layers.
2. The bottom layer is isotropic.
3. The bottom reflector is horizontal.

The intermediate interface was parameterized by a B-spline with 12 nodes (one node per every 40 common midpoints). The model parameters include the P-wave vertical velocity $V_{P0,t}$ (the subscript t stands for *top*) and the anisotropy parameters ϵ_t and δ_t in the VTI layer, the velocity $V_{P,b}$ (b stands for *bottom*) in the isotropic layer, the depth Z_r of the horizontal reflector, and the spline coefficients responsible for the shape and position of the intermediate interface. At each iteration all parameters were updated simultaneously, and the algorithm successfully converged towards the correct solution after only a few iterations (Figure 2.12). The velocities $V_{P0,t}$ and $V_{P,b}$ can be used to reconstruct the actual depths and dips of both interfaces. Clearly, in this example P-wave reflection traveltimes provide enough information for building a model in the *depth* domain.

Le Stunff et al. (2001) demonstrated that the success of the tomographic inversion procedure was ensured by the reflection traveltimes for the bottom reflector near the left edge of the model, where the intermediate interface is dipping. Indeed, P-wave moveout in VTI media is controlled by the zero-dip NMO velocity and parameter η only if the medium above the reflector is laterally homogeneous. The dip of the intermediate interface causes the reflection traveltimes for the bottom of the model to depend on all four relevant parameters ($V_{P0,t}$, $V_{P,b}$, ϵ_t, and δ_t). In particular, the time slope (i.e., the horizontal derivative of the zero-offset traveltime) of the bottom reflection (p_b) combined with that of the reflection from the intermediate interface (p_t) constrains the vertical velocity $V_{P0,t}$. Under the assumption of weak anisotropy and mild dip of the intermediate interface, the two slopes are related by (Le Stunff et al., 2001)

$$p_b \, V_{P,b} = p_t \left(V_{P,b} - V_{P0,t} \right) . \tag{2.13}$$

Because the velocity $V_{P,b}$ in the isotropic layer can be estimated from the conventional Dix formula near the right edge of the model, equation 2.13 helps obtain $V_{P0,t}$. Substitution of the exact parameters p_b, p_t, and $V_{P,b}$ into equation 2.13 yields $V_{P0,t} = 2.42$ km/s, which is close to the actual value (2.5 km/s). (If the intermediate interface were horizontal, then $p_b = p_t = 0$, and equation 2.13 would contain no information about the velocities.) The parameter δ_t can be determined by combining $V_{P0,t}$ with the zero-dip NMO velocity in the VTI layer estimated from moveout analysis on the right side of the model.

The NMO velocity of the bottom reflection on the left side of the model can be

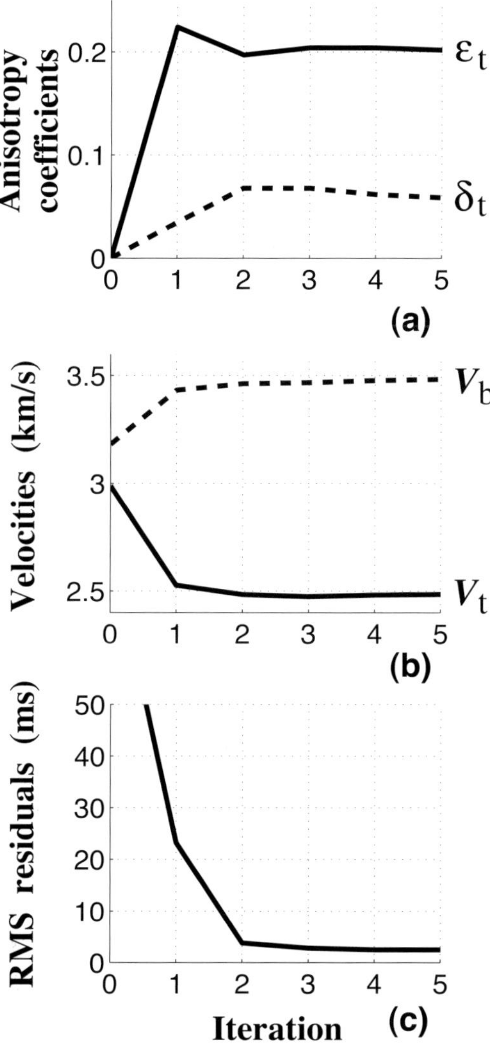

Figure 2.12: Model parameters and time residuals obtained at each iteration of the tomographic algorithm (Le Stunff et al., 2001). (a) The anisotropy parameters ϵ_t (solid curve) and δ_t (dashed) in the VTI layer; (b) the velocities $V_{P0,t}$ (solid) and $V_{P,b}$ (dashed); and (c) the rms traveltime residuals. The residuals at the fourth and fifth iterations do not exceed the standard deviation of the noise (2.5 ms) added to the data.

obtained using the Dix-type averaging procedure described in Chapter 1 (Figure 1.10). Because both layers are homogeneous and the model is 2D, the NMO-velocity surfaces of all reflections are cylinders with axes confined to the vertical incidence plane. Intersections of these cylinders with the interfaces represent ellipses (or circles in special cases) with one axis in the incidence plane. Hence, the in-plane NMO velocity can be obtained by Dix-type averaging of the in-plane axes of these ellipses. This averaging procedure, carried out by Le Stunff et al. (2001), shows that the NMO velocity from the bottom of the model can be inverted for the parameter δ_t, which is sufficient for resolving the rest of the model parameters.

The time slope and NMO velocity of the bottom reflection therefore provide redundant information about the depth scale of the model and the anisotropy parameters in the VTI layer. The greater the dip of the intermediate interface, the tighter are δ_t and the vertical velocity $V_{P0,t}$ constrained by surface P-wave data. Because time slopes and NMO velocities are determined by reflection traveltimes, the tomographic algorithm (using the assumptions listed above) converges toward the correct depth model.

2.2.2 Physical model with a bending TTI layer

The main physical reasons for transverse isotropy (hexagonal symmetry) observed in sedimentary basins are the intrinsic anisotropy of shales and periodic fine layering. Horizontally layered, unfractured sediments often have a near-vertical symmetry axis and are adequately described by VTI models. In active tectonic areas, however, TI layers (or interbedding isotropic sediments) can be dipping, thus giving rise to transverse isotropy with a tilted axis of symmetry (TTI).[2] For example, uptilted shale layers near salt domes can have a large inclination of the symmetry axis (Tsvankin, 2005). The TTI model also is rather typical for overthrust areas, such as the Canadian Foothills, where shale horizons are often bent by tectonic processes and may have dips exceeding 45° (e.g., Leslie and Lawton, 1996).

The presence of TTI formations causes serious imaging problems such as mispositioning of target reflectors and poor focusing of seismic sections. Higher-quality images of both dipping and horizontal features in the overthrust environment can be produced by migration algorithms capable of handling TTI media (Isaac and Lawton, 1999; Vestrum et al., 1999). The main difficulty, however, is in estimating the anisotropy parameters from surface reflection data, especially in structurally complex areas.

P-wave kinematic signatures in TTI media are controlled by the dip and azimuth of the symmetry axis, the symmetry-direction velocity V_{P0}, and the parameters ϵ and δ defined with respect to the symmetry axis (i.e., the anisotropy parameters are defined as though the symmetry axis were vertical). Because depth-domain inversion of P-wave data is generally nonunique for vertical transverse isotropy, it should be

[2]Tilted transverse isotropy can also be produced by a system of dipping fractures embedded in isotropic host rock (e.g., Dewangan and Tsvankin, 2006a, 2006b).

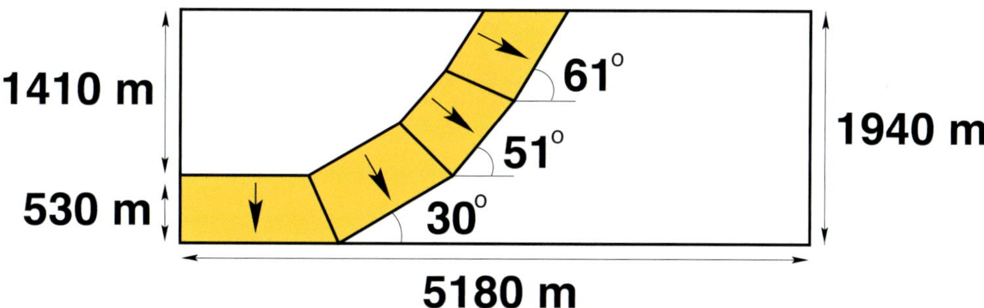

Figure 2.13: 2D physical model of an anisotropic thrust sheet (Leslie and Lawton, 1996). In each block of the bending TI layer, the symmetry axis (marked by the arrows) is perpendicular to the layer boundaries. The distances were scaled by a factor of 10^4.

Number of common midpoints	504
CMP spacing	10 m
Offset spacing	20 m
Minimum offset	200 m
Maximum offset	2000 m

Table 2.2: Acquisition parameters used in the physical modeling.

even more ill-posed when the symmetry axis is tilted in an unknown direction. As demonstrated above, however, some of the trade-offs between the VTI parameters can be resolved in the presence of intermediate dipping interfaces. Grechka et al. (2001), whose main results are reproduced here, arrive at a similar conclusion for a physical model that contains a bending TI layer with the symmetry axis orthogonal to the layer boundaries.

Physical-modeling data set

The model was built at the University of Calgary by Leslie and Lawton (1996) to imitate overthrust structures typical for the Central Alberta Foothills in Canada (Figure 2.13). Four blocks of phenolic laminate were glued together to make up a bending TI layer with the symmetry axis orthogonal to the layer boundaries. The phenolic material, which is known to be orthorhombic, was cut in the direction of one of the symmetry planes to produce a 2D TI sheet with parameters believed to be typical for shales in the Canadian Foothills. The P-wave velocities in the symmetry direction ($V_{P0} = 2925$ m/s) and isotropy plane ($V_{\text{hor}} = 3365$ m/s) yield the anisotropy parameter $\epsilon = 0.16$, while $\delta = 0.08$. The TI layer was embedded in a purely isotropic plexiglass material with the P-wave velocity $V_P = 2740$ m/s.

Leslie and Lawton (1996) acquired an ultrasonic P-wave reflection survey at the top of the model; the acquisition parameters are listed in Table 2.2. A time section for the smallest offset in the data (Figure 2.14) shows a false anticline structure at a time of approximately 1.4 s. This pull-up of the bottom of the model is caused by the combined influence of the anisotropy and high velocity in the phenolic blocks. Leslie and Lawton (1998) presented a series of prestack depth migrations of the data and showed that the bottom interface cannot be flattened without taking anisotropy into account. Their results also suggest that P-wave reflection traveltimes might contain sufficient information for obtaining the model parameters needed for anisotropic imaging.

Parameter estimation

To avoid ambiguities in inverting of the available set of measurements for the model parameters, it is necessary to make several assumptions about the model.

1. All blocks making up the model in Figure 2.13 are assumed to be homogeneous. Allowing for lateral velocity variation leads to trade-offs between the parameters, such as those described for VTI media by Grechka (1998).

2. The four TTI blocks contain the same material with the symmetry axis orthogonal to the block boundaries. Without this assumption, the number of independent parameters is too large to be constrained by the data.

3. The plexiglass blocks (white in Figure 2.13) are isotropic. In principle, the observed P-wave reflection traveltimes for the top of the TTI layer can be reproduced using an elliptically anisotropic overburden with the same value of the zero-dip NMO velocity (Alkhalifah and Tsvankin, 1995).

4. The reflector at the bottom of the model is horizontal, which implies that the anticline structure at approximately 1.4 s in Figure 2.14 is an artifact. This assumption, which might be difficult to use in field-data applications, would not be needed if it were possible to estimate the NMO velocity of the event denoted by **Ra**.

Under these assumptions, Grechka et al. (2001) estimated all model parameters in the depth domain from the zero-offset traveltimes and NMO velocities of the reflection events **R1** − **R5** (Figures 2.14–2.16). The reflections **Ra** and **Rb**, which also carry information about the parameters of the TTI layer, do not produce sufficiently focused semblance maxima.

We begin with the event **R1** reflected from the segment $[D_2 D_3]$ beneath the isotropic and homogeneous plexiglass layer (Figure 2.15; see assumptions 1 and 3). The two-way zero-offset traveltime and NMO velocity of **R1** were picked from semblance panels similar to the one in Figure 2.16a and averaged over several CMP locations, which gave the mean values $\tau_{\rm R1} = 1.437$ s and $V_{\rm nmo,\,R1} = V_{\rm iso} = 2742$ m/s (Tables 2.3 and 2.4). Based on the spatial extent of the semblance maxima, the standard deviations of $\tau_{\rm R1}$ and $V_{\rm nmo,\,R1}$ caused by picking inaccuracies were estimated at 0.004 s and 30 m/s, respectively (Figure 2.16a). The zero time slope of **R1** in Figure 2.14 indicates that the segment $[D_2 D_3]$ is horizontal. The computed depth of $[D_2 D_3]$, which, according to assumption 4, is supposed to be constant under the whole model, is $z = z_{\rm R1} = \tau_{\rm R1}\, V_{\rm iso}/2 = 1969 \pm 22$ m. Note that the estimated depth z is somewhat

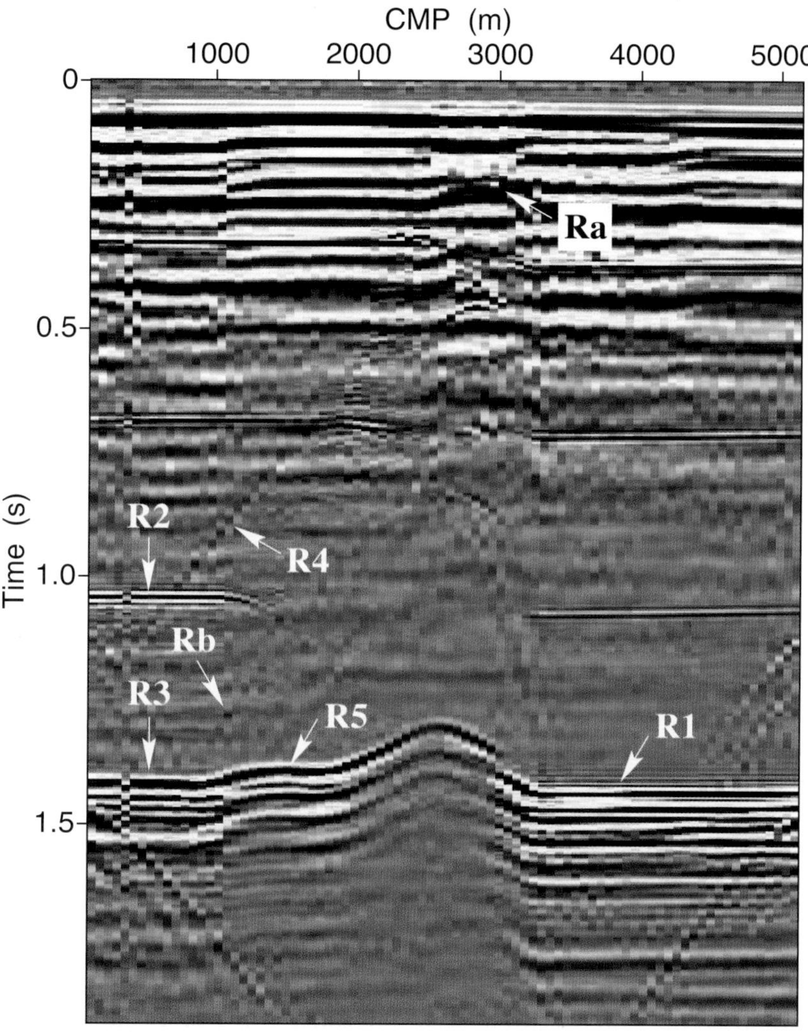

Figure 2.14: Common-offset (200 m) time section of reflection data acquired over the model from Figure 2.13 (Grechka et al., 2001). The arrows mark the reflection events (**R1** – **R5**, **Ra**, and **Rb**) discussed in the text.

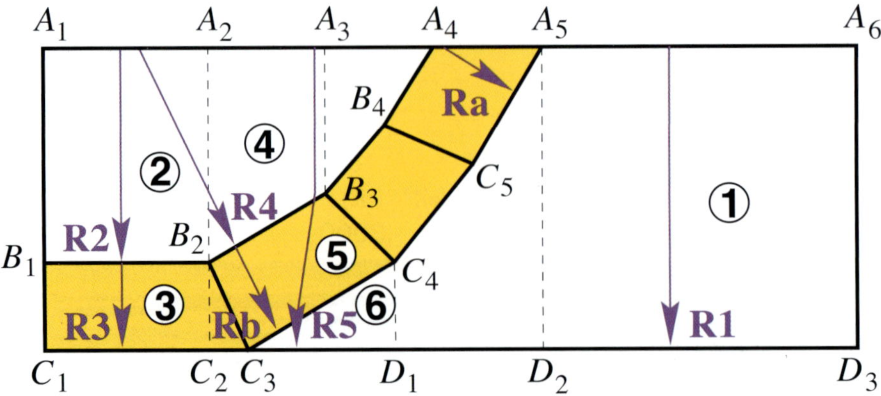

Figure 2.15: Cartoon of the zero-offset rays for the events **R1**–**R5**, **Ra**, and **Rb** (Grechka et al., 2001). The circled numbers denote model blocks.

Reflection event	Attributes	Inverted parameters
R1	τ_{R1}, $V_{\mathrm{nmo,R1}}$	$z(z_{\mathrm{R1}})$, V_{iso}
R2	τ_{R2}, $V_{\mathrm{nmo,R2}}$	z_{R2}
R3	τ_{R3}, $V_{\mathrm{nmo,R3}}$	V_{P0}, δ
R4	p_{R4}	ϕ
R5	τ_{R5}, $V_{\mathrm{nmo,R5}}$	$z_{\perp}$, ϵ

Table 2.3: For each reflection event in the left column, the table lists the attributes used in the inversion and the estimated model parameters (Grechka et al., 2001).

greater than the actual value (1940 m) because it was determined using the traveltimes picked from the semblance maxima rather than from the actual first breaks. Hereafter, the notation $a = b \pm c$ means that the quantity a is equal to b with the error bar or standard deviation c. The error bars calculated for all intermediate quantities (not given in the text) were used to evaluate the accuracy of the inverted model parameters.

The event **R2** is similar to **R1** because it also represents a reflection from the bottom of an isotropic homogeneous layer. Because the NMO velocity $V_{\mathrm{nmo,R2}}$ (Figure 2.16b; $V_{\mathrm{nmo,R2}} = V_{\mathrm{iso}} = 2750$ m/s) practically coincides with $V_{\mathrm{nmo,R1}}$ and the plexiglass is supposed to be homogeneous and isotropic, one would correctly conclude that blocks 1 and 2 (Figure 2.15) are made of the same material. On the time section from Figure 2.14, reflection **R2** can be seen as a horizontal event at the traveltime $\tau_{\mathrm{R2}} = 1.05$ s, implying that the segment $[B_1 B_2]$ is horizontal. The estimated thickness of block 2 is $z_{\mathrm{R2}} = \tau_{\mathrm{R2}} V_{\mathrm{nmo,R2}}/2 = 1444$ m.

The reflection **R3**, generated at the bottom of the horizontal block 3 (Figures 2.14 and 2.15), contains information about the zero-dip NMO velocity in the TI layer. The symmetry axis in block 3 is vertical (i.e., the block is VTI), and the interval V_{nmo} is

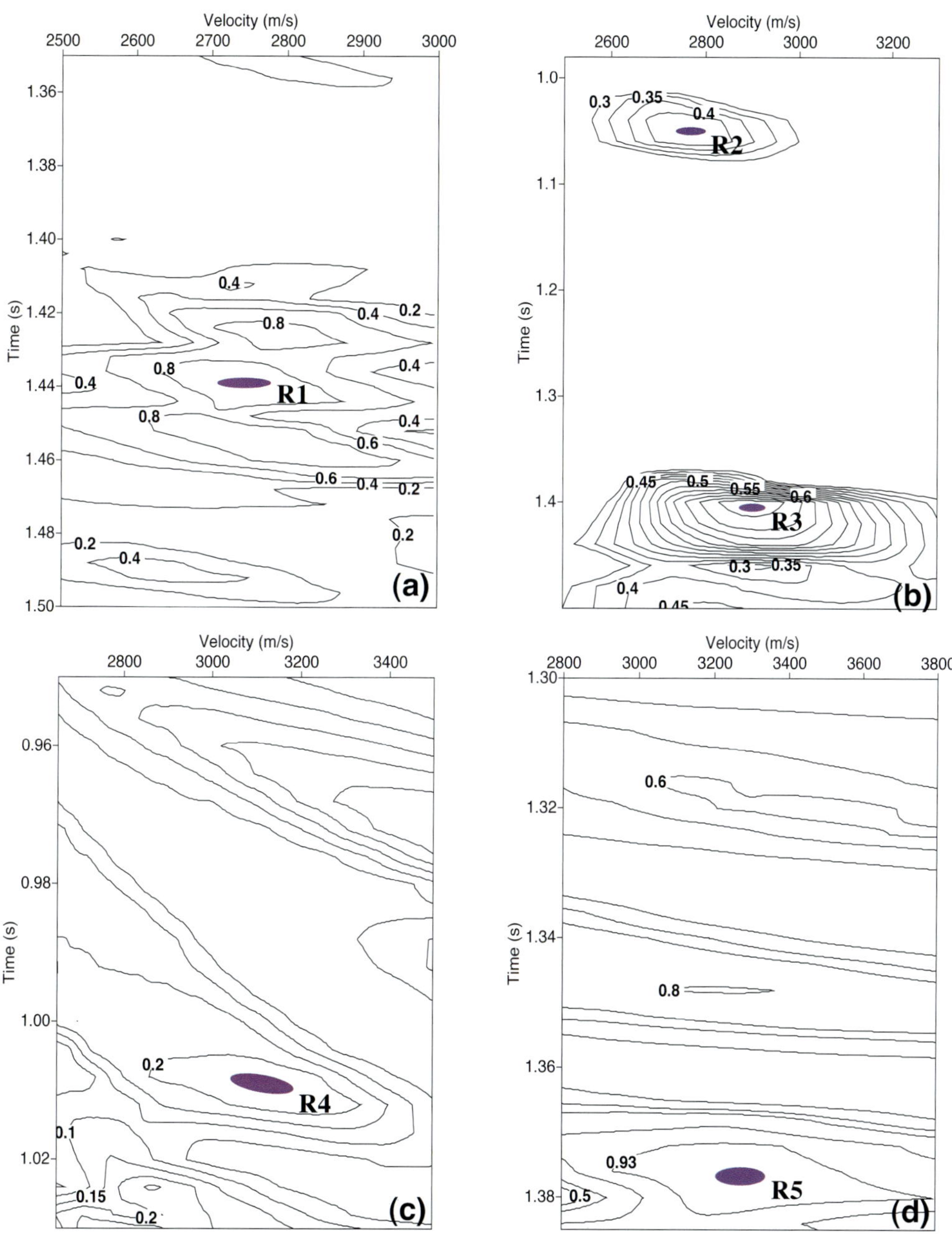

Figure 2.16: Semblance contours at CMP locations (a) 3780 m, (b) 630 m, (c) 680 m, and (d) 1480 m (Grechka et al., 2001). The magenta areas mark the semblance maxima used to pick the zero-offset traveltimes and NMO velocities of the reflection events **R1**–**R5**.

	Actual	Inverted
V_{iso} (m/s)	2740	2742 ± 30
V_{P0} (m/s)	2925	2963 ± 157
ϵ	0.16	0.16 ± 0.06
δ	0.08	0.09 ± 0.06

Table 2.4: Comparison of the actual and inverted parameters (Grechka et al., 2001).

related to the symmetry-direction velocity V_{P0} as

$$V_{\text{nmo}} = V_{P0}\sqrt{1+2\delta}\,. \tag{2.14}$$

Because the ray **R3** propagates through a laterally homogeneous medium (assumption 1), the interval NMO velocity in the TI block can be found using conventional Dix differentiation:

$$V_{\text{nmo}} = \sqrt{\frac{\tau_{\text{R3}}\,V^2_{\text{nmo, R3}} - \tau_{\text{R2}}\,V^2_{\text{nmo, R2}}}{\tau_{\text{R3}} - \tau_{\text{R2}}}} = 3192\,\text{m/s}\,, \tag{2.15}$$

where $\tau_{\text{R3}} = 1.405$ s and $V_{\text{nmo, R3}} = 2870$ m/s are the measured zero-offset traveltime and NMO velocity for the reflection **R3** (Figure 2.16b). Under assumption 4, the depth of the segment $[C_1C_2]$ is equal to that of segment $[D_2D_3]$: $z_{\text{R3}} = z_{\text{R1}} = 1969$ m. Therefore, we can compute the vertical velocity in block 3 as

$$V_{P0} = 2\,\frac{z_{\text{R3}} - z_{\text{R2}}}{\tau_{\text{R3}} - \tau_{\text{R2}}} = 2963\,\text{m/s}\,. \tag{2.16}$$

Substituting V_{P0} into equation 2.14 with $V_{\text{nmo}} = 3192$ m/s gives $\delta = 0.09$. As a result, conventional moveout analysis near the left edge of the model helps constrain the symmetry-direction velocity V_{P0} and the parameter δ in the TTI layer. The other relevant anisotropy parameter ϵ still has to be found.

The reflections **R4** and **Rb** (the latter is barely visible) form parallel straight lines on the time section in Figure 2.14, which indicates that the segments $[B_2B_3]$ and $[C_3C_4]$ of TTI block 5 (Figure 2.15) are parallel to each other. Approximation of the picked zero-offset traveltimes of the event **R4** with a straight line gives the half-slope $p_{\text{R4}} = 1.82\times10^{-4}$ s/m. The NMO velocity $V_{\text{nmo, R4}} = 3120$ m/s corresponding to the semblance maximum in Figure 2.16c satisfies the isotropic relationship (within the standard deviation)

$$V_{\text{nmo, R4}} = \frac{V_{\text{iso}}}{\sqrt{1 - p^2_{\text{R4}}\,V^2_{\text{iso}}}}\,. \tag{2.17}$$

Although equation 2.17 is also valid for elliptically anisotropic media with the zero-dip NMO velocity V_{iso}, we use assumption 3 to identify blocks 2 and 4 as part of the homogeneous, isotropic section of the model. Then the dip ϕ of $[B_2B_3]$ is given by

$$\phi = \sin^{-1}(p_{\text{R4}}\,V_{\text{iso}}) = 29.8^\circ\,. \tag{2.18}$$

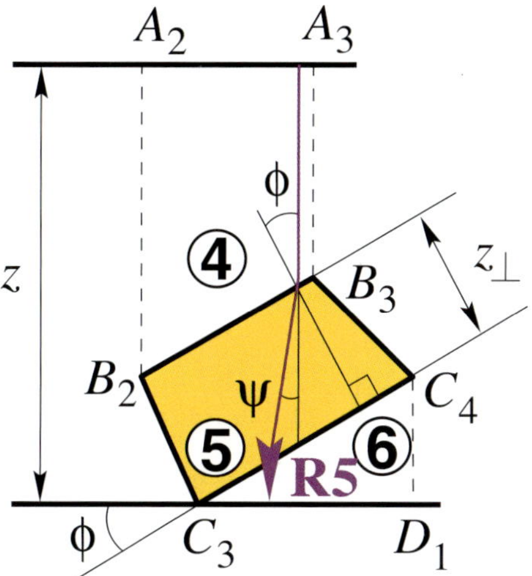

Figure 2.17: Notation associated with the zero-offset ray of the reflection **R5** from the horizontal interface $[C_3 D_1]$ (Figure 2.15) below TTI block 5 (Grechka et al., 2001).

Neither the zero-offset traveltime τ_{Rb} nor the NMO velocity $V_{\text{nmo, Rb}}$ of the reflection **Rb** could be picked on semblance panels because of the event's low amplitude (Figure 2.14). This, however, does not hamper parameter estimation because both τ_{Rb} and $V_{\text{nmo, Rb}}$ are controlled by model parameters that can be estimated from other reflections.

The thickness $z_\perp$ of the TTI layer and the anisotropy parameter ϵ can be found from the traveltime $\tau_{\text{R5}} = 1.378$ s and the NMO velocity $V_{\text{nmo, R5}} = 3250$ m/s of the reflection **R5** (Figures 2.16d and 2.17). Note than neither τ_{R5} nor $V_{\text{nmo, R5}}$ depends on the CMP coordinate (Figure 2.14). This holds only if the segments $[B_2 B_3]$ and $[C_3 C_4]$ are parallel to each other, which has already been established by the analysis of **R4** and **Rb**. Consequently, the zero-offset ray **R5** inside the isotropic blocks 4 and 6 has to be vertical (Figures 2.15 and 2.17). Then the zero-offset traveltime τ_{R5} of the event **R5** can be written as

$$\tau_{\text{R5}} = \tau_{\text{tti, R5}} + \tau_{\text{iso, R5}} \,, \tag{2.19}$$

where $\tau_{\text{tti, R5}}$ and $\tau_{\text{iso, R5}}$ are the traveltimes inside the TTI and isotropic blocks, respectively. Using the relationships between the group-velocity vector and the horizontal (p) and vertical $[q = q(p)]$ components of the slowness vector in TTI block 5 (Appendix 1A), we obtain

$$\tau_{\text{tti, R5}} = 2\, z_\perp \,(q - p\, q')\, \frac{\cos\psi}{\cos(\phi + \psi)} \tag{2.20}$$

and

$$\tau_{\text{iso, R5}} = \frac{2}{V_{\text{iso}}} \left[z - z_\perp \, \frac{\cos\psi}{\cos(\phi + \psi)} \right] , \tag{2.21}$$

where $q' = dq/dp$ and ψ is the angle between the zero-offset ray and vertical inside block 5 (Figure 2.17). Equation 1.71 helps express ψ through the slownesses p and q (the sign of the right-hand side was changed because of the choice of the coordinate directions):

$$\psi = \tan^{-1}\left(\frac{dq}{dp}\right) \equiv \tan^{-1}(q')\,. \tag{2.22}$$

The slownesses p and q satisfy Snell's law at the top and bottom boundaries of block 5:

$$p = \left(q - \frac{1}{V_{\rm iso}}\right)\tan\phi\,. \tag{2.23}$$

In combination with the Christoffel equation in the TTI layer, equation 2.23 can be used to obtain both slowness components and then $\tau_{\rm R5}$ (equation 2.19) for a given set of the TTI parameters. Therefore, the traveltime of the event **R5** yields one equation for estimating the parameters $z_\perp$ and ϵ.

A second equation containing $z_\perp$ and ϵ is provided by the NMO velocity $V_{\rm nmo,\,R5}$, which can be derived using the Dix-type averaging of NMO-velocity surfaces (Figure 1.10) discussed in Chapter 1 (Grechka et al., 2001):

$$V^2_{\rm nmo,\,R5} = \frac{1}{\tau_{\rm R5}}\left[\tau_{\rm iso,\,R5}V^2_{\rm iso} + \tau_{\rm tti,\,R5}\left(\frac{V_{\rm tti,\,nmo}(p)}{1 - q'\tan\phi}\right)^2\right], \tag{2.24}$$

where

$$V^2_{\rm tti,\,nmo}(p) = \frac{d^2q/dp^2}{pq' - q} \tag{2.25}$$

is the interval NMO velocity within the TTI block evaluated for the horizontal slowness component p of the zero-offset ray.

Joint nonlinear inversion of equations 2.19 and 2.24 using the values of $\tau_{\rm R5}$ and $V_{\rm nmo,\,R5}$ picked from the semblance panel in Figure 2.16d yielded accurate estimates of $z_\perp = 594$ m and $\epsilon = 0.16$ (Table 2.4). Still, it should be emphasized that the model was relatively simple and inversion was performed under several restrictive assumptions.

2.3 P-wave stacking-velocity tomography for VTI media

The two examples in the previous section indicate that the presence of dipping interfaces in the overburden might help resolve the parameters V_{P0}, ϵ, and δ from conventional-spread P-wave reflection traveltimes. Here, we describe the tomography-style algorithm introduced by Grechka et al. (2002a) for P-wave moveout inversion in layered VTI media and study the influence of interface dip and curvature on the parameter-estimation results. First, we examine models with plane dipping interfaces and then discuss a more general tomographic algorithm that accounts for interface curvature.

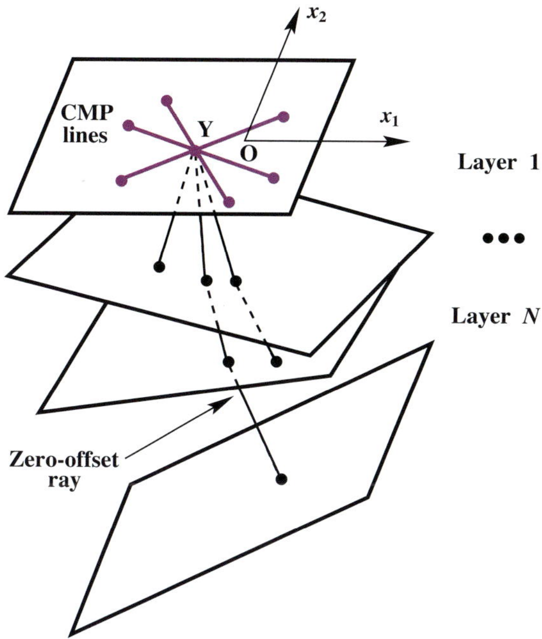

Figure 2.18: Zero-offset rays in a model containing a stack of homogeneous VTI layers separated by plane dipping interfaces (Grechka et al., 2002a).

2.3.1 Models with plane dipping interfaces

Methodology of tomographic inversion

We consider a model composed of N homogeneous VTI layers (some of them may be isotropic) separated by plane, dipping, nonintersecting interfaces (Figure 2.18). The algorithm operates with conventional-spread P-wave data acquired in wide-azimuth 3D surveys. P-wave kinematics is controlled by the interval vertical velocities $V_{P0,n}$, anisotropy parameters ϵ_n and δ_n, and the spatial positions of the layer boundaries. Each interface is defined by its depth z_n measured under the coordinate origin and the dip ϕ_n and azimuth ψ_n of the interface normal. Thus, the model vector

$$\mathbf{m} \equiv \{V_{P0,n},\, \epsilon_n,\, \delta_n,\, \phi_n,\, \psi_n,\, z_n\}\,, \quad (n = 1,\, \ldots,\, N) \tag{2.26}$$

consists of $6N$ independent quantities. It is convenient to split $\mathbf{m}$ into two vectors $\mathbf{l}$ and $\mathbf{i}$, where $\mathbf{l}$ contains the interval Thomsen parameters,

$$\mathbf{l} \equiv \{V_{P0,n},\, \epsilon_n,\, \delta_n\}\,, \quad (n = 1,\, \ldots,\, N)\,, \tag{2.27}$$

and $\mathbf{i}$ describes the interfaces,

$$\mathbf{i} \equiv \{\phi_n,\, \psi_n,\, z_n\}\,, \quad (n = 1,\, \ldots,\, N)\,. \tag{2.28}$$

Velocity analysis of 3D wide-azimuth P-wave data recorded at common midpoints with coordinates $\mathbf{y} = [y_1, y_2]$ can provide the one-way zero-offset reflection traveltimes $\tau_0(\mathbf{y}, n)$ for all interfaces and the corresponding NMO ellipses $V_{\text{nmo}}(\alpha)$ expressed in terms of the 2×2 symmetric matrices $\mathbf{W}$ (equations 1.3 and 1.4). The matrices $\mathbf{W}(\mathbf{y}, n)$ can be obtained from azimuthal velocity analysis based on the hyperbolic moveout equation parameterized by the NMO ellipse (see Chapter 1). On the zero-offset (or stacked) time sections of reflection events, we can estimate the reflection time slopes and, therefore, the ray parameters $\mathbf{p}(\mathbf{y}, n) = [p_1(\mathbf{y}, n),\, p_2(\mathbf{y}, n)]$ of the zero-offset rays at the surface.

Because the layers in our model are homogeneous and the interfaces are planar (Figure 2.18), the slownesses $p_1(\mathbf{y}, n)$ and $p_2(\mathbf{y}, n)$ are independent of the CMP coordinate $\mathbf{y}$. Taking into account that

$$\frac{\partial \tau_0(\mathbf{y}, n)}{\partial y_j} = p_j(n)\,, \quad (j = 1, 2)\,, \tag{2.29}$$

the traveltimes $\tau_0(\mathbf{y}, n)$ can be expressed as linear functions of y_j:

$$\tau_0(\mathbf{y}, n) = \tau_0(\mathbf{O}, n) + p_1(n)\, y_1 + p_2(n)\, y_2\,; \tag{2.30}$$

$\tau_0(\mathbf{O}, n)$ are the traveltimes recorded at the coordinate origin $\mathbf{O}$. Using equation 2.30, the input data $\mathbf{d}(\mathbf{y}, n)$ can be represented in the following form:

$$\mathbf{d}(\mathbf{y}, n) \equiv \{\tau_0(\mathbf{O}, n),\, p_1(n),\, p_2(n),\, W_{11}(\mathbf{y}, n),\, W_{12}(\mathbf{y}, n),\, W_{22}(\mathbf{y}, n)\}\,, \tag{2.31}$$

where $n = 1, \ldots, N$.

To build the effective NMO ellipse at the surface, we employ the Dix-type averaging procedure that operates with the NMO-velocity surfaces computed along the zero-offset ray (Chapter 1). Because each layer is homogeneous, the averaging formula involves the intersections $\mathbf{W}_n(\mathbf{y})$ of the interval NMO-velocity cylinders with the medium interfaces (Figure 1.10). As discussed in Chapter 1, the NMO-velocity cylinder is computed for a fictitious reflector orthogonal to the slowness vector of the interval zero-offset ray. For example, suppose the model contains two layers separated by a dipping interface (Figure 2.19). Then the intersection $\bar{\mathbf{W}}(\mathbf{y})$ of the effective NMO-velocity cylinder (measured at the surface) with a plane parallel to the intermediate interface is given by

$$\tau_0(\mathbf{y}) \left[\bar{\mathbf{W}}(\mathbf{y})\right]^{-1} = \tau_{0,1}(\mathbf{y}) \left[\mathbf{W}_1(\mathbf{y})\right]^{-1} + \tau_{0,2}(\mathbf{y}) \left[\mathbf{W}_2(\mathbf{y})\right]^{-1}\,, \tag{2.32}$$

where $\tau_{0,1}$ and $\tau_{0,2}$ are the interval zero-offset traveltimes, $\mathbf{W}_1(\mathbf{y})$ (shaded) and $\mathbf{W}_2(\mathbf{y})$ (dashed) are the intersections of the interval NMO-velocity cylinders in the first and second layer with the intermediate interface. The ellipse described by $\bar{\mathbf{W}}(\mathbf{y})$ is then projected along the effective NMO-velocity cylinder in the first layer onto the horizontal plane to find the ellipse $\mathbf{W}(\mathbf{y})$ estimated from surface data.

In models with plane interfaces (Figure 2.18), zero-offset rays from the same interface are parallel to one another at different CMP locations $\mathbf{y}$. As a result, the interval

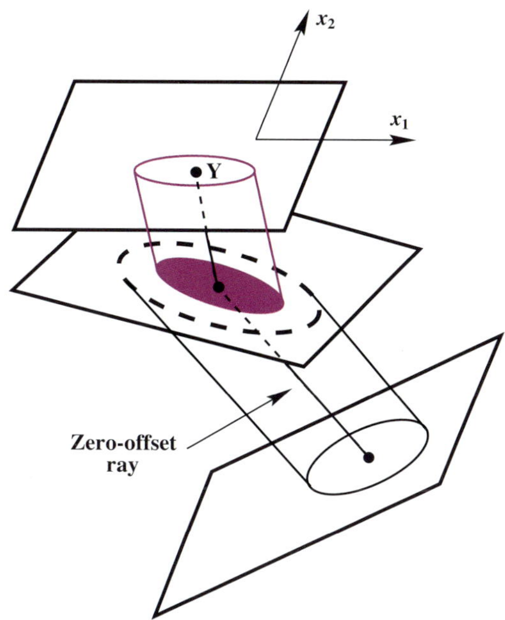

Figure 2.19: Effective NMO ellipse is obtained by averaging the intersections of the interval NMO-velocity cylinders with the model interfaces (see Figure 1.10) along the zero-offset ray (Grechka et al., 2002a).

slownesses, group-velocity vectors, and matrices $\mathbf{W}_n$ (equation 2.32) are independent of the CMP coordinate $\mathbf{y}$, whereas the interval traveltimes $\tau_{0,n}(\mathbf{y})$ are linear functions of $\mathbf{y}$. Therefore, the NMO ellipses $\mathbf{W}(\mathbf{O}, n)$ and $\mathbf{W}(\mathbf{y}, n)$ measured at points $\mathbf{O}$ and $\mathbf{y}$ differ only because of variations in the length of the interval ray segments. This suggests that the dependence of $\mathbf{W}(\mathbf{y}, n)$ on the location $\mathbf{y}$ does not provide any new information about the model parameters. The conclusion that it is sufficient to measure the NMO ellipses at a single common midpoint is supported by synthetic test results below.

Feasibility of parameter estimation

As follows from the above discussion, although traveltime data at different common midpoints may help suppress noise, they give the same information about the medium parameters as does the data vector

$$\mathbf{d}(\mathbf{O}, n) = \{\tau_0(\mathbf{O}, n),\ \mathbf{p}(n),\ \mathbf{W}(\mathbf{O}, n)\} \tag{2.33}$$

at a single CMP. Thus, we evaluate the feasibility of the inversion using the dependence of $\mathbf{d}(\mathbf{O}, n)$ on the vector $\mathbf{m}$ of model parameters (equation 2.26). For brevity, henceforth the CMP coordinate will be omitted.

For an N-layered VTI model, the vectors $\mathbf{d}$ and $\mathbf{m}$ contain $6N$ components each. Therefore, the vector $\mathbf{m}$ can be obtained uniquely only if all components of $\mathbf{d}$ are independent. Unfortunately, this is not the case for the top (subsurface) layer. The P-wave NMO ellipse $\mathbf{W}(1)$ for a dipping reflector overlaid by a homogeneous VTI medium provides only two equations for the medium parameters because its orientation is fixed by the reflector azimuth ψ_1, which can be found from the reflection time slopes:

$$\tan\psi_1 = \frac{p_2(1)}{p_1(1)} = \frac{W_{22}(1) - W_{11}(1) + \sqrt{[W_{22}(1) - W_{11}(1)]^2 + 4\,W_{12}^2(1)}}{2\,W_{12}(1)}\,; \tag{2.34}$$

$W_{12}(1) \neq 0$. The semiaxes of the NMO ellipse in the top layer constrain the zero-dip NMO velocity and the parameter η, so the data vector in that layer contains only five independent components. The singular value decomposition (SVD) performed below shows that if all interfaces have distinctly different dip azimuths, equation 2.34 is the *only* relationship between the components of $\mathbf{m}$.

For inversion purposes, it is convenient to split the vector $\mathbf{d}(\mathbf{O}, n)$ into two parts. For a given ("trial") set of the interval VTI parameters $\tilde{\mathbf{l}}_n$ (equation 2.27), the traveltime $\tau_0(n)$ and the horizontal slownesses $p_1(n)$ and $p_2(n)$ can be used to estimate the depths, dips, and azimuths of the interfaces $\tilde{\mathbf{i}}_n$ (equation 2.28). Indeed, knowledge of the parameters of the first layer $\tilde{\mathbf{l}}_1$ is sufficient for computing the vertical slowness component from the Christoffel equation and obtaining the slowness vector $[p_1(1),\, p_2(1),\, p_3(1)]$. Because the slowness vector of the zero-offset ray is orthogonal to the reflector, it defines the reflector normal. Then we can find the group-velocity vector (ray) in the first layer and use the traveltime $\tau_0(1)$ to determine the depth z_1 of the first reflector.

Once the first interface has been reconstructed, the zero-offset ray from the second interface can be "traced back" into the medium using its horizontal slownesses $p_1(2)$ and $p_2(2)$ at the surface. The slowness vector of the zero-offset ray in the second layer (obtained from Snell's law) and the zero-offset time $\tau_0(2)$ yield the orientation and depth of the second reflector. By continuing this procedure downward, we compute the dips, strikes (or azimuths), and depths of all interfaces in the trial model. Clearly, any errors in the input data or trial layer parameters $\tilde{\mathbf{l}}_n$ would distort the estimates of the interfaces $\tilde{\mathbf{i}}_n$.

The best-fit vector of the layer parameters $\mathbf{l}_n$ has to be found by inverting the NMO ellipses because the rest of the input data has already been used to reconstruct the interfaces. Parameter estimation is performed by applying the least-squares method to minimize the difference between the measured and computed matrices $\mathbf{W}(\mathbf{O}, n)$.

To study the feasibility of the inversion procedure, it is sufficient to carry out SVD analysis of the $3N \times 3N$ matrix of Fréchet derivatives

$$\boldsymbol{F} = \frac{\partial\,\mathbf{W}(n)}{\partial\,\mathbf{l}_k}\,, \quad (k, n = 1, \ldots, N)\,. \tag{2.35}$$

The vector $\mathbf{i}$, which specifies the interfaces, is assumed to be such that the zero-offset traveltimes and horizontal slowness components for each trial model exactly match those in the data. As discussed above, the matrices $\mathbf{W}(n)$ are computed using the formalism based on NMO-velocity surfaces.

A typical result of SVD analysis of NMO ellipses in a two-layer model is shown in Figure 2.20. Whereas the last singular value is always equal to zero, the other five do not vanish if the azimuths of the interfaces differ (see the curves marked by squares, diamonds and triangles). The presence of two vanishing singular values when the strikes of both interfaces coincide (circles in Figure 2.20) is not surprising, because in this case the axes of the NMO ellipse for the bottom of the model are aligned with the dip and strike directions (i.e., the model becomes 2D), and there is one less independent data component.

SVD analysis repeated for a number of different common midpoints produced the same singular values as those in Figure 2.20. Moreover, the singular values do not change when the NMO ellipses computed at several CMP locations are used in the Fréchet matrix 2.35 simultaneously. Therefore, other than for noise-suppression purposes, it is indeed sufficient to carry out the inversion using the data vector measured at a single CMP.

To get a more quantitative assessment of the feasibility of the inversion for the model from Figure 2.20, the singular values were computed for different orientations of the first interface. The results indicate that for noise-free data it should be possible to estimate the five parameter combinations for any angles ϕ_1 and ψ_1, except for $\phi_1 = 0°$ and $\psi_1 = 0°$ or $180°$ (Figure 2.21). When $\phi_1 = 0°$, the first layer is horizontal and the NMO ellipse $W(1)$ degenerates into a circle that constrains just one combination of the medium parameters. For $\psi_1 = 0°$ or $180°$, the two reflectors are co-oriented, which reduces the number of equations provided by the NMO ellipses to four.

The observations drawn from Figures 2.20 and 2.21 can be extended to an arbitrary number of VTI layers. At a maximum, P-wave NMO ellipses constrain $3N$–1 combinations of the $3N$ interval parameters $\{V_{P0,n}, \epsilon_n, \delta_n\}$, if the model interfaces have distinctly different azimuths. Otherwise, P-wave traveltimes contain less information about the medium. For instance, if the azimuths of all interfaces are identical, the NMO ellipses for different reflectors are co-oriented. Thus, only their semiaxes constrain the layer parameters, and the number of independent equations reduces to $2N$. In the limiting case of horizontal layers, the NMO ellipses become circles defined by the N interval zero-dip NMO velocities. Hence, unambiguous inversion is impossible without additional information about the model; some practical possibilities are discussed below.

Parameter estimation using a priori information

The results of the previous section suggest that in some cases a priori knowledge of a single layer parameter may be sufficient to overcome the nonuniqueness. For example, if the vertical velocity $V_{P0,1}$ in the top (subsurface) layer is known, the anisotropy parameters ϵ_1 and δ_1 can be obtained from the NMO ellipse $\mathbf{W}(1)$. According to the SVD results, this should be sufficient for estimating the remaining medium parameters, if the interfaces have different orientations.

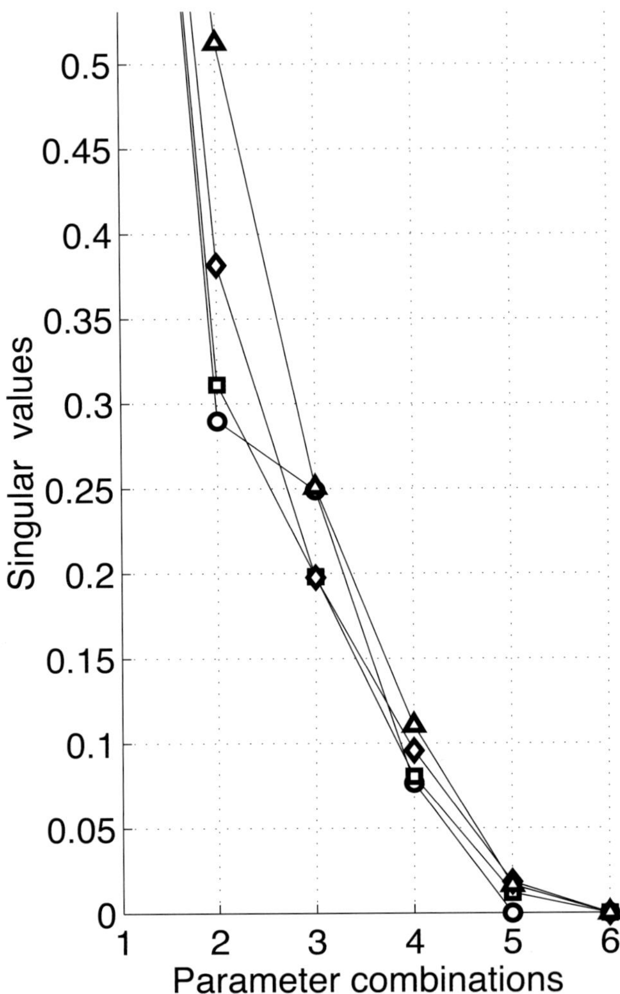

Figure 2.20: SVD analysis for a two-layer VTI model with $V_{P0,1}$=2 km/s, ϵ_1=0.15, δ_1=0.05, $V_{P0,2}$=3 km/s, ϵ_2=0.25, and δ_2=0.1 (Grechka et al., 2002a). The singular values are normalized by the largest one. The curves correspond to different azimuths of the intermediate (first) interface: ψ_1=0° (○), ψ_1=30° (□), ψ_1=60° (◇), and ψ_1=90° (△); the azimuth of the bottom interface is ψ_2=0°. The interface depths beneath the CMP location are z_1=1 km and z_2=3 km, and the dips are ϕ_1=40° and ϕ_2=20°.

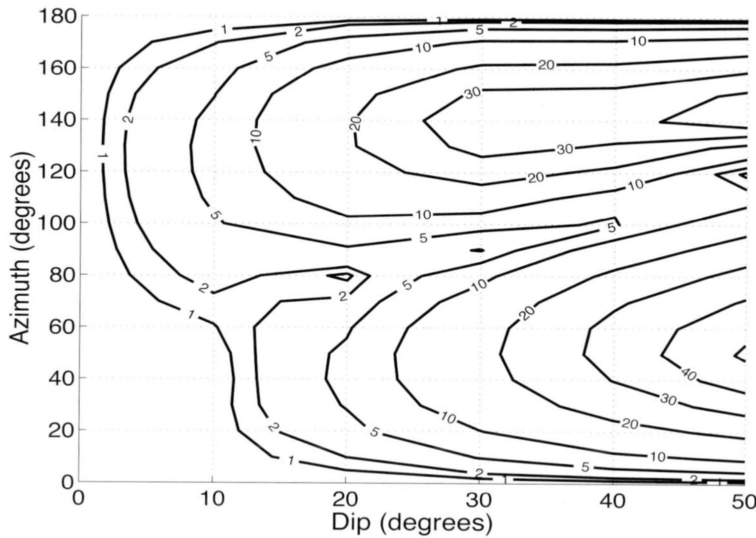

Figure 2.21: Contours of the fifth eigenvalue (multiplied by 1000) as a function of the dip ϕ_1 and azimuth ψ_1 of the intermediate interface in a two-layer VTI model (Grechka et al., 2002a). The parameters $V_{P0,1}$, ϵ_1, δ_1, z_1, $V_{P0,2}$, ϵ_2, δ_2, z_2, ϕ_2, and ψ_2 are the same as those in Figure 2.20.

This conclusion was verified by several numerical tests, with a typical example displayed in Figure 2.22 and Table 2.5. We traced zero-offset reflected rays through a three-layer VTI model for nine CMP locations, added Gaussian noise to the computed NMO ellipses and zero-offset traveltimes, and estimated the layer parameters using the algorithm described above. Although multiple common midpoints do not provide new information for the inversion, they help mitigate the influence of random noise.

To constrain the inversion, the parameter δ_1 was assumed to be known, which made it possible to obtain close estimates of most other parameters (Table 2.5). As with any other technique that operates with reflection traveltimes, the accuracy of the inverted parameters is lower in the deeper layers. First, inversion errors generally accumulate with depth because of the influence of distortions in the shallow part of the model. Second, the sensitivity of reflection traveltime to the interval parameters of a layer with fixed thickness is inversely proportional to its depth. The low accuracy in the parameter ϵ_3 is also caused by insufficient angle coverage of the reflected rays in the bottom layer.

The choice of δ_1 as the known parameter was arbitrary. Provided the interfaces have distinctly different azimuths, it is possible to reconstruct the whole model if the vertical velocity $V_{P0,n}$ or one of the anisotropy parameters (ϵ_n or δ_n) in any layer is fixed at the correct value. The inversion procedure also works well when one or more layers are isotropic with both ϵ_n and δ_n set to zero.

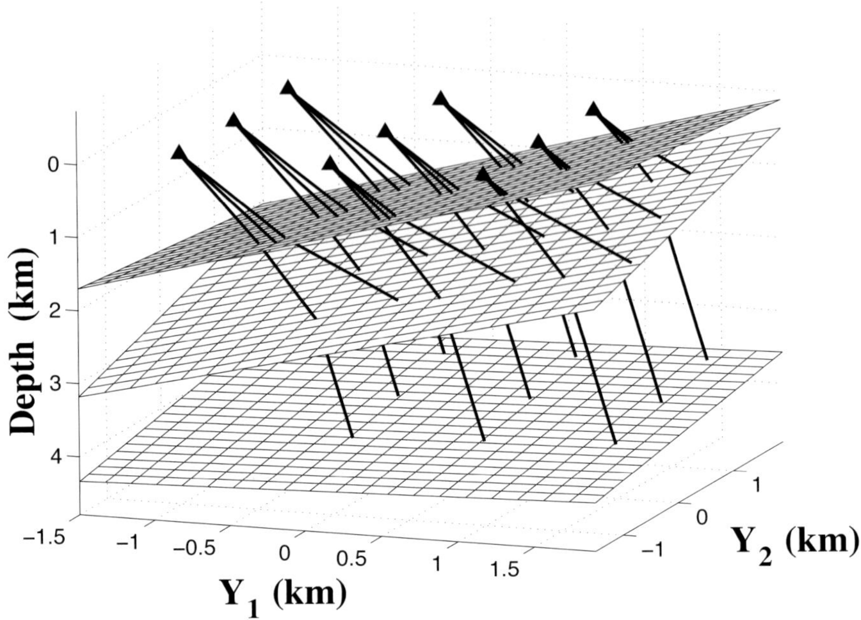

Figure 2.22: Zero-offset P-wave rays in the three-layer VTI model with the parameters listed in Table 2.5 (Grechka et al., 2002a).

	$V_{P0,1}$ (km/s)	ϵ_1	δ_1	$V_{P0,2}$ (km/s)	ϵ_2	δ_2	$V_{P0,3}$ (km/s)	ϵ_3	δ_3
Actual	1.00	0.08	0.04	2.00	0.20	0.10	3.00	0.10	0.05
Inverted	0.99	0.09	–	2.02	0.18	0.09	2.96	0.18	0.07

	z_1 (km)	ϕ_1 (deg)	ψ_1 (deg)	z_2 (km)	ϕ_2 (deg)	ψ_2 (deg)	z_3 (km)	ϕ_3 (deg)	ψ_3 (deg)
Actual	1.00	30.0	−10.0	2.00	30.0	30.0	4.00	10.0	70.0
Inverted	1.00	29.8	−10.0	1.98	29.9	30.4	3.98	10.3	70.5

Table 2.5: Comparison of the actual and inverted parameters for the three-layer VTI model in Figure 2.22 (Grechka et al., 2002a). The parameter δ_1 is assumed to be known. The standard deviations of Gaussian noise added to the NMO velocities and zero-offset traveltimes are 2.0% and 0.5%, respectively.

If all interfaces in the three-layer medium discussed above have the same azimuth, the axes of the NMO ellipses W(1), W(2), and W(3) are parallel to the dip and strike directions, and the moveout data provide only six equations (two semiaxes of each ellipse). The model has a total of nine interval parameters, so three of them have to be specified in advance.

One practical possibility is to assume that the vertical velocities are known, for example, from check shots (Figure 2.23 and Table 2.6). We modeled the NMO ellipses and zero-offset traveltimes for several CMP locations distributed along the dip direction, fixed the interval vertical velocities at the actual values, and performed the inversion of noise-contaminated data. The interval parameters ϵ_n and δ_n in all three layers were estimated with high accuracy (Table 2.6).

Another way to reduce the number of unknowns is to impose an empirical relationship (e.g., based on rock-physics information for a specific region) between ϵ and δ in at least one layer. As discussed by Grechka et al. (2002a), one special case when this approach does not help is elliptical anisotropy. Even if all layers are known to be elliptically anisotropic (i.e., $\epsilon_n = \delta_n$), and the number of the relevant VTI parameters reduces to $2N$, the $3N - 1$ equations for the NMO ellipses do not have a unique solution. This conclusion agrees with the results of Dellinger and Muir (1988) that were obtained using linear transformations (stretching) of the isotropic wave equation.

Information from intersecting boundaries

A priori information might not be needed at all for models with intersecting/multiple boundaries in some of the layers. The presence of different reflector dips or azimuths in the same depth interval causes reflected rays to span more spatial directions, which helps constrain the interval anisotropy parameters. Suppose, for example, that the intermediate interface in the two-layer VTI model from Figure 2.20 is bent into two plane portions with the same dip $\phi_1 = 40°$ but different azimuths $\psi_1 = 30°$ and $90°$. Recording reflections from the bottom of the model that cross both portions of the intermediate interface yields an additional NMO ellipse (i.e., three more equations). The absence of vanishing singular values for this problem (Figure 2.24; the smallest singular value is 0.02) indicates that all parameters can be resolved uniquely.

Parameter estimation can also become feasible if the model contains a fault plane, and the data include reflections from both the fault and layer boundaries (Figure 2.25). A similar model was used by Alkhalifah and Tsvankin (1995), who developed a dip-moveout (DMO) inversion method to estimate the interval values of η from surface P-wave data (also see Tsvankin, 2005). That DMO-inversion method, however, assumes that each zero-offset reflected ray crosses only horizontal interfaces on its way to the surface. In our models intermediate interfaces are dipping, which creates the dependence of NMO ellipses from both horizontal and dipping reflectors on the interval parameters ϵ and δ. Even for co-oriented interfaces, fault planes in some of the layers can help resolve all relevant VTI parameters. For example, the semiaxes of the three NMO ellipses corresponding to the zero-offset rays marked in Figure 2.25 provide six independent equations for the six unknowns.

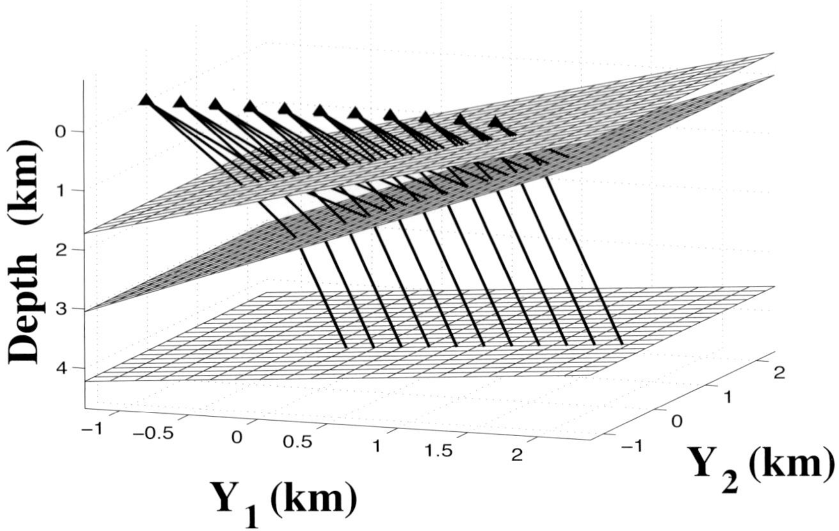

Figure 2.23: Zero-offset rays in a three-layer VTI model with co-oriented interfaces; $\psi_1=\psi_2=\psi_3=0°$ (Grechka et al., 2002a). The other parameters are the same as those in Table 2.5.

	$V_{P0,1}$ (km/s)	ϵ_1	δ_1	$V_{P0,2}$ (km/s)	ϵ_2	δ_2	$V_{P0,3}$ (km/s)	ϵ_3	δ_3
Actual	1.00	0.08	0.04	2.00	0.20	0.10	3.00	0.10	0.05
Inverted	–	0.08	0.04	–	0.19	0.10	–	0.08	0.05

Table 2.6: Comparison of the actual and inverted parameters for the three-layer VTI model from Figure 2.23 (Grechka et al., 2002a). The interval vertical velocities are assumed to be known. The standard deviations of Gaussian noise added to the NMO velocities and zero-offset traveltimes are 2.0% and 1.0%, respectively.

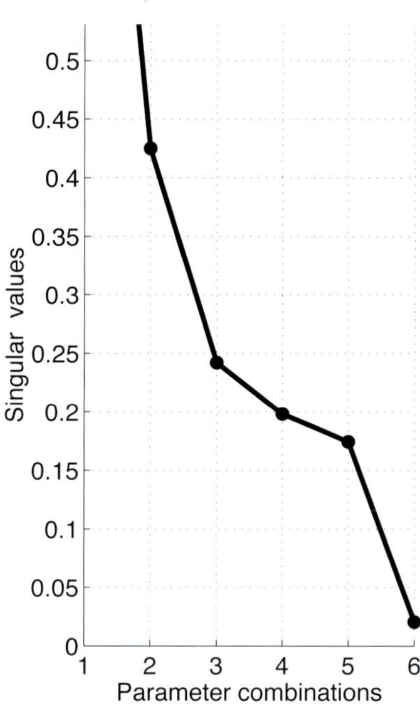

Figure 2.24: SVD analysis for a two-layer VTI model similar to the one in Figure 2.20 (Grechka et al., 2002a). This time, however, the intermediate interface contains two plane segments with the same dip ϕ_1=40° but different azimuths, ψ_1=30° and 90°. The azimuth of the bottom interface is ψ_2=0°.

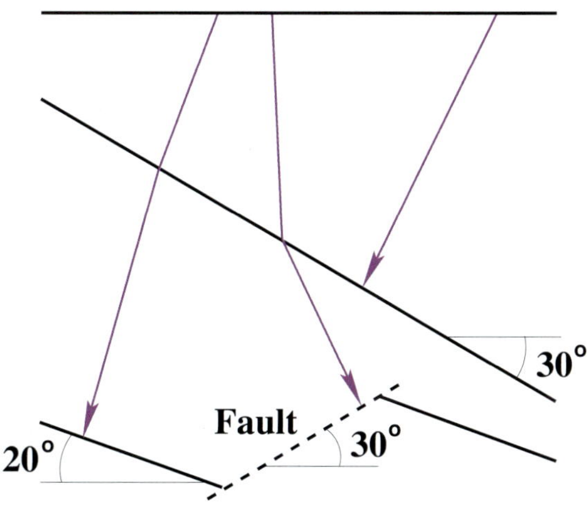

Figure 2.25: Presence of fault-plane reflections in layered VTI media might be sufficient to reconstruct the model in depth using P-wave reflection data (Grechka et al., 2002a).

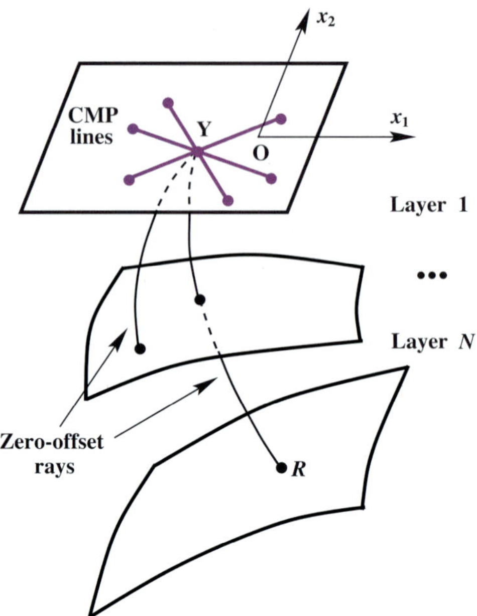

Figure 2.26: Multiazimuth CMP recording over a layered model with curved interfaces (Grechka et al., 2002a).

2.3.2 Models with curved interfaces

Stacking-velocity inversion described above can be extended in a straightforward way to wide-azimuth P-wave data acquired over layered VTI media with curved interfaces (Figure 2.26). As long as each layer is homogeneous, variable dip of the reflector and intermediate interfaces provides additional information for the inversion procedure. The forward-modeling algorithm, however, needs to account for the interface curvature.

NMO ellipse in the presence of curved interfaces

Reflector curvature has no influence on the NMO ellipse because V_{nmo} is independent of reflection-point dispersal and can be obtained by tracing rays through the zero-offset reflection point (e.g., Hubral and Krey, 1980). The Dix-type averaging equations for the effective NMO ellipse, however, have to be corrected for the curvature of intermediate interfaces.

If reflection data are acquired at a curved surface described by the radius vector $\mathbf{s}(x_1, x_2, x_3) = 0$, the NMO velocity $V_{\text{nmo}}(\mathbf{s}, \ell)$ in the direction ℓ on $\mathbf{s}$ is given by (Grechka et al., 2002a)

$$V_{\text{nmo}}^{-2}(\mathbf{s}, \ell) = \ell \, (\mathbf{W} + \tau_0 \, \mathbf{p} \cdot \boldsymbol{\kappa}) \, \ell^{\mathbf{T}} \, , \tag{2.36}$$

where $\mathbf{W}$ is the NMO ellipse obtained as the intersection of the NMO-velocity surface

$\mathbf{U}$ with the plane $\mathcal{P}$ tangent to $\mathbf{s}$ at the common midpoint, $\mathbf{T}$ denotes transposition, and $\boldsymbol{\kappa}$ is the $3\times2\times2$ "curvature" tensor with components

$$\kappa_{mij} = \left.\frac{\partial^2 s_m(h_1, h_2)}{\partial h_i \partial h_j}\right|_{h_1=h_2=0}, \quad (m = 1, 2, 3;\ i, j = 1, 2)\,; \tag{2.37}$$

h_1 and h_2 are the horizontal coordinates with respect to the CMP location.

If surface $\mathbf{s}$ represents the boundary between two layers, the NMO ellipse is discontinuous across the tangent plane $\mathcal{P}$, with the jump depending on the difference between the slowness vectors above ($\mathbf{p}^{(+)}$) and below ($\mathbf{p}^{(-)}$) the surface. As shown by Grechka et al. (2002a), the NMO ellipse $\mathbf{W}^{(+)}$ on the "positive" side of the interface can be found as

$$\mathbf{W}^{(+)} = \mathbf{W}^{(-)} - \tau_0 \left(\mathbf{p}^{(+)} - \mathbf{p}^{(-)}\right) \cdot \boldsymbol{\kappa}\,, \tag{2.38}$$

where the term $\tau_0 \left(\mathbf{p}^{(+)} - \mathbf{p}^{(-)}\right) \cdot \boldsymbol{\kappa}$ represents a correction for the interface curvature.

Equation 2.38 can be used to generalize the Dix-type averaging procedure described in Chapter 1 (Figure 1.10) to media with curved interfaces. The effective NMO ellipse at the earth's surface can still be computed from equation 1.60, but the intersections of the effective NMO-velocity surfaces with curved interfaces have to include the correction term in equation 2.38. This formalism allows us to compute the hyperbolic portion of pure-mode reflection moveout for layered models with curved interfaces by tracing just the zero-offset ray for each reflection event.

Feasibility of parameter estimation

The vector $\mathbf{d}$ of input data remains the same as that for models with plane interfaces:

$$\mathbf{d}(\mathbf{y}, n) \equiv \{\tau_0(\mathbf{y}, n),\ \mathbf{p}(\mathbf{y}, n),\ \mathbf{W}(\mathbf{y}, n)\}\,. \tag{2.39}$$

The zero-offset traveltimes $\tau_0(n)$, reflection slopes $\mathbf{p}(n) = [p_1(n),\ p_2(n)]$ (horizontal slownesses), and matrices $\mathbf{W}(n)$ describing the NMO ellipses are acquired for reflectors ($n = 1, \ldots, N$) at a number of CMP locations $\mathbf{y} = [y_1,\ y_2]$.

Our goal is to determine whether the data can be inverted for all relevant interval parameters

$$\mathbf{m} \equiv \{V_{P0,n},\ \epsilon_n,\ \delta_n,\ \zeta_{j_1 j_2, n}\}\,, \quad (n = 1, \ldots, N;\ j_1 = 1, \ldots, J_1;\ j_2 = 1, \ldots, J_2)\,, \tag{2.40}$$

where the matrices $\boldsymbol{\zeta}_n$ contain the coefficients of the basis functions describing the depths $z_n(y_1, y_2)$ of the model interfaces. The inversion procedure can be simplified by adopting a linear relationship between $z_n(y_1, y_2)$ and $\zeta_{j_1 j_2, n}$, such as the polynomial representation

$$z_n(y_1, y_2) = \sum_{j_1=1}^{J_1} \sum_{j_2=1}^{J_2} \zeta_{j_1 j_2, n}\, y_1^{j_1 - 1}\, y_2^{j_2 - 1}\,. \tag{2.41}$$

Equation 2.41 provides sufficient accuracy and flexibility in describing smooth interfaces for purposes of tomographic inversion.

The input parameters in definition 2.39 do not constrain the interval values of V_{P0}, ϵ, and δ if the layers are separated by plane nonintersecting boundaries (see above). In the presence of interface curvature, however, the dependence of the data vector $\mathbf{d}$ (equation 2.39) on the CMP coordinate $\mathbf{y}$ provides additional information about the model parameters because zero-offset rays change direction with CMP location, even though each layer is still homogeneous. Therefore, the spatial variation of the vector $\mathbf{d}$ might help constrain the inversion and resolve the interval Thomsen parameters and the depth scale of the model.

Similar to the approach outlined above for plane interfaces, the reflection slopes $\mathbf{p}(n)$ and the zero-offset traveltimes $\tau_0(n)$ can be used to build the interfaces $z_n(\mathbf{y})$ for a given estimate of the interval parameters (i.e., for a trial model),

$$\mathbf{l} = \{V_{P0,n},\, \epsilon_n,\, \delta_n\}\,, \quad (n = 1,\, \dots,\, N)\,. \tag{2.42}$$

This is achieved by tracing zero-offset rays downward and computing the coordinates of the reflection points and the corresponding slowness vectors. In contrast to media with plane interfaces, however, the number of common midpoints and their spatial distribution determine our ability to reconstruct the shape of curved interfaces. Because the slowness vector of each zero-offset ray is orthogonal to the reflector at the reflection point R (Figure 2.26), it provides the orientation (i.e., the azimuth and polar angle) of the unit normal $\mathbf{b}_n(R)$ to the reflecting interface. In addition, the zero-offset traveltime yields the depth of the reflection point R. Therefore, the triplet $\{\tau_0(n,\mathbf{y}),\, p_1(n,\mathbf{y}),\, p_2(n,\mathbf{y})\}$ at the CMP location $\mathbf{y}$ provides three constraints on the quantities $\zeta_{j_1 j_2,n}$ specifying the reflector (equation 2.41). This implies that the number M of common midpoints required to obtain $J_1 \times J_2$ coefficients $\boldsymbol{\zeta}_n$ has to satisfy the inequality

$$M \geq \frac{J_1 J_2}{3}\,. \tag{2.43}$$

For example, four appropriately chosen common midpoints ($M = 4$) in Figure 2.27 should be sufficient for reconstructing $J_1 J_2 = 3 \times 3 = 9$ coefficients that define each model interface. We solve for $\boldsymbol{\zeta}_n$ by applying the least-squares method using all available CMP locations.

While the traveltimes $\tau_0(n,\mathbf{y})$ and slopes $\mathbf{p}(n,\mathbf{y})$ are used to obtain $z_n(\mathbf{y})$, the information required to estimate the interval parameters $\mathbf{l}$ (equation 2.42) is provided by the NMO ellipses $\mathbf{W}(n)$. To prove that for a certain class of VTI models the layer parameters can be estimated uniquely, we present an example of SVD analysis for the two-layer VTI model in Figure 2.27. The NMO ellipses for the two reflectors measured at four CMP locations (triangles) provide $2 \times 4 \times 3 = 24$ equations for the six components of the vector $\mathbf{l}$.

Thus, the feasibility of the inversion can be evaluated by applying SVD to the 24×6 matrix of the Fréchet derivatives,

$$\boldsymbol{F} = \frac{\partial\, \mathbf{W}(n,\mathbf{y})}{\partial\, \mathbf{l}}\,, \tag{2.44}$$

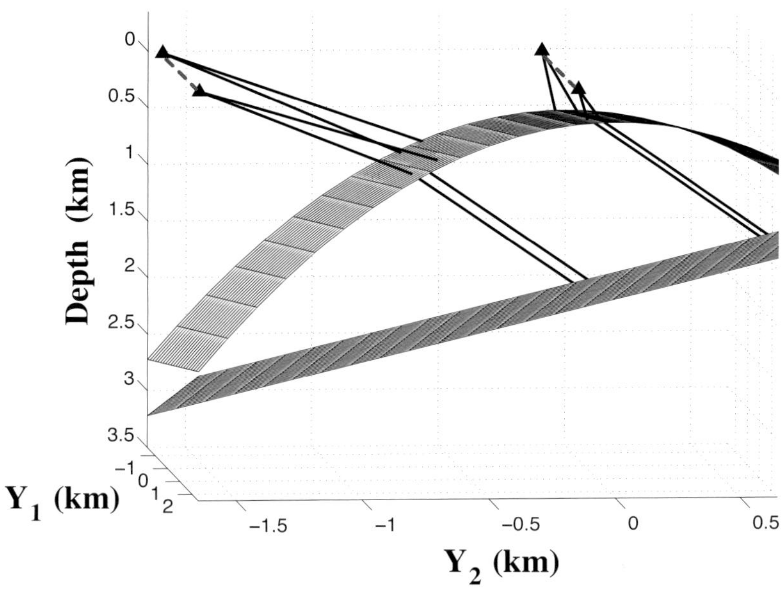

Figure 2.27: Zero-offset rays in a two-layer VTI model with significant curvature of the shallow interface (Grechka et al., 2002a). The SVD analysis in Figure 2.28 is performed for the CMP locations marked by the triangles. The interval parameters are $V_{P0,1}$=1 km/s, ϵ_1=0.2, δ_1=0.1, $V_{P0,2}$=2 km/s, ϵ_2=0.15, and δ_2=0.05. Both interfaces are described by equation 2.41 with $J_1 = J_2 = 3$, so $\boldsymbol{\zeta}_1$ and $\boldsymbol{\zeta}_2$ are 3×3 matrices.

computed for the correct model parameters **m**. Because none of the singular values vanishes (Figure 2.28), the input data provide sufficient information for parameter estimation, and this VTI model can be fully reconstructed in the depth domain from P-wave reflection traveltimes.

Clearly, it is not always possible to resolve the VTI parameters from P-wave data. For example, as the curvature of the intermediate interface in Figure 2.27 decreases and both interfaces become plane, at least one singular value goes to zero (see above). The inversion also becomes nonunique if both layers are elliptically anisotropic with $\epsilon_n = \delta_n$ (Dellinger and Muir, 1988).

Inversion examples

The SVD results were verified by performing inversion of P-wave data computed for 240 common midpoints placed every 25 m along the two dashed lines between the triangles in Figure 2.27. This provided an overdetermined system of $240 \times 3 = 720$ equations for reconstructing the matrix $\boldsymbol{\zeta}$ describing each interface. We added Gaussian noise to the modeled NMO velocities and zero-offset traveltimes, and estimated the parameter vector **m** (equation 2.40) using the two-step inversion procedure discussed above. The inversion results for both layers are satisfactory, with the maximum

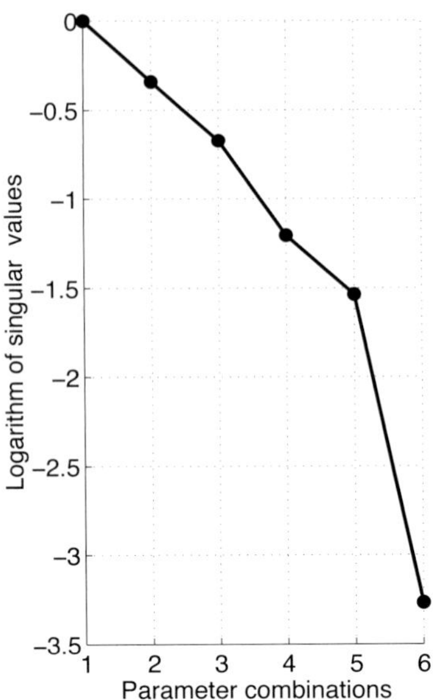

Figure 2.28: Normalized singular values for the model from Figure 2.27 (Grechka et al., 2002a).

	$V_{P0,1}$ (km/s)	ϵ_1	δ_1	$V_{P0,2}$ (km/s)	ϵ_2	δ_2
Actual	1.00	0.20	0.10	2.00	0.15	0.05
Inverted	1.02	0.17	0.07	2.05	0.11	0.03

Table 2.7: Inversion results for the two-layer VTI model in Figure 2.27 (Grechka et al., 2002a). The standard deviation of Gaussian noise added to the NMO velocities and zero-offset traveltimes is 2% and 1%, respectively.

error in the anisotropy parameters not exceeding 0.04 (Table 2.7). The relative errors in the interval vertical velocities $V_{P0,n}$ and absolute errors in ϵ_n and δ_n are comparable to the standard deviation of the noise added to the NMO velocities. The inverted parameters do not depend on the initial guess (the initial model was isotropic), which suggests that the least-squares objective function in a certain vicinity of the correct solution is unimodal. Knowledge of the interval parameters $V_{P0,n}$, ϵ_n and δ_n is sufficient to reconstruct the depth and shape of the interfaces and to build the entire VTI model in depth.

Another test was carried out for a three-layer VTI model with a more complicated shape of the interfaces (Figure 2.29). The data vector $\mathbf{d}(\mathbf{y}, n)$ was computed at 600 common midpoints located with a spacing of 15 m along two straight lines. The accuracy of parameter estimation for noise-contaminated data (Table 2.8) is comparable to that for the simpler model in Table 2.7. Therefore, for a subset of layered

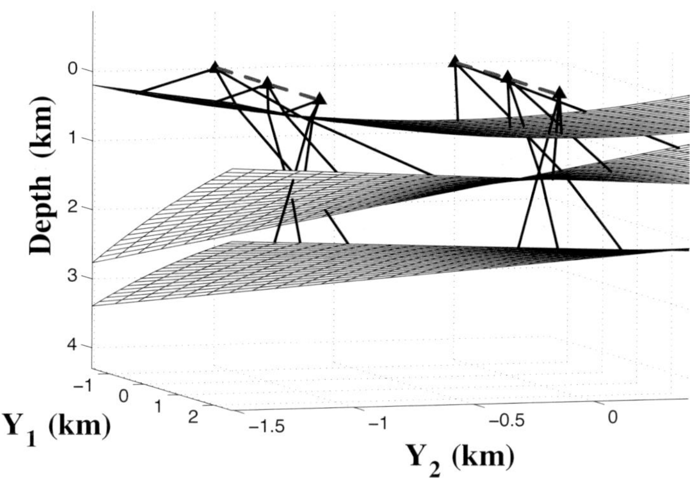

Figure 2.29: Three-layer VTI model with several zero-offset rays (Grechka et al., 2002a). Inversion was carried out for 600 common midpoints placed on the two dashed lines.

	$V_{P0,1}$ (km/s)	ϵ_1	δ_1	$V_{P0,2}$ (km/s)	ϵ_2	δ_2	$V_{P0,3}$ (km/s)	ϵ_3	δ_3
Actual	1.80	0.20	0.10	2.00	0.15	0.05	2.30	0.10	0.03
Inverted	1.77	0.23	0.12	1.97	0.17	0.07	2.26	0.12	0.05

Table 2.8: Inversion results for the three-layer VTI model from Figure 2.29 (Grechka et al., 2002a). The standard deviation of Gaussian noise is 2% for the NMO velocities and 1% for the zero-offset traveltimes.

VTI models with curved interfaces, P-wave reflection traveltimes provide sufficient information for velocity analysis in the depth domain.

2.4 Parameter estimation for layered TTI media

For laterally homogeneous VTI media above a horizontal or dipping reflector, the P-wave NMO ellipse is controlled by the zero-dip NMO velocity and parameter η. When the symmetry axis is tilted, the anisotropic contribution to the NMO ellipse is no longer fully governed by η (Tsvankin, 1997b, 2005). Also, reflection traveltimes in tilted TI media vary with the tilt ν and azimuth β of the symmetry axis. Therefore, even in the absence of intermediate dipping interfaces, P-wave moveout for TTI models depends on the symmetry-direction velocity V_{P0}, the anisotropy parameters ϵ and δ defined with respect to the symmetry axis, and the angles ν and β.

In the special case of an HTI medium ($\nu = 90°$), the azimuth β of the symmetry axis is constrained by the orientation of the P-wave NMO ellipse for a horizontal reflector. The semiaxes of the NMO ellipse yield the velocity V_{P0} and the parameter $\delta^{(V)}$ ($\delta^{(V)} \approx 2\epsilon - \delta$; see Appendix 1B) defined in the symmetry-axis plane (Tsvankin, 2005). Additional information needed to resolve the parameters $\epsilon^{(V)}$ or ϵ can be obtained from the NMO ellipse of a dipping event (Contreras et al., 1999) or from nonhyperbolic moveout. Therefore, P-wave moveout inversion in HTI media provides enough information to perform depth processing.

The TTI model can be considered as intermediate between VTI and HTI. This does not mean, however, that moveout inversion in TTI media can be understood by simply examining the results for the two extreme orientations of the symmetry axis. Indeed, the tilt of the symmetry axis represents an extra parameter to be recovered from moveout data. Clearly, the NMO ellipse for a single reflector does not provide enough information to constrain all five TTI parameters. Below, following the results of Grechka and Tsvankin (2000), we examine the possibility of combining the P-wave NMO ellipses for two reflectors with different orientations.

2.4.1 Special cases of homogeneous TTI media

For a homogeneous anisotropic layer, the matrix $\mathbf{W}$ responsible for the NMO ellipse can be obtained from the horizontal slowness components p_i of the zero-offset ray (equation 1.7):

$$\mathbf{W} = \frac{p_1 q_{,1} + p_2 q_{,2} - q}{q_{,11} q_{,22} - q_{,12}^2} \begin{pmatrix} q_{,22} & -q_{,12} \\ -q_{,12} & q_{,11} \end{pmatrix}, \tag{2.45}$$

where $q \equiv q(p_1, p_2) \equiv p_3$ is the vertical component of the slowness vector, $q_{,i} \equiv \partial q / \partial p_i$, and $q_{,ij} \equiv \partial^2 q / (\partial p_i \partial p_j)$; p_1, p_2, and the derivatives of q are evaluated for the zero-offset ray. The vertical slowness $q = q(p_1, p_2)$ for a given anisotropic model can be found by solving the Christoffel equation. Then, as discussed in Chapter 1, differentiation of the Christoffel equation yields the derivatives $q_{,i}$ and $q_{,ij}$. The NMO ellipse defined by $\mathbf{W}$ from equation 2.45 has the form (equation 1.8)

$$\begin{aligned} V_{\text{nmo}}^{-2}(\alpha) &\equiv V_{\text{nmo}}^{-2}(\alpha, p_1, p_2) \\ &= \frac{p_1 q_{,1} + p_2 q_{,2} - q}{q_{,11} q_{,22} - q_{,12}^2} \left[q_{,22} \cos^2\alpha - 2 q_{,12} \sin\alpha \cos\alpha + q_{,11} \sin^2\alpha \right]. \end{aligned} \tag{2.46}$$

Normal moveout for two special types of homogeneous TTI media (a layer above a steep reflector and an elliptically anisotropic layer) is discussed in Appendix 2B. Below, we analyze the P-wave NMO velocity in a TI layer with the symmetry axis orthogonal to the reflector. For these three special cases the P-wave NMO ellipse and normal moveout as a whole possess certain distinct features that need to be taken into account in the inversion procedure. Also, to gain insight into the influence of ϵ, δ, and the symmetry-axis orientation on the P-wave NMO ellipse, we simplify equations 2.45 and 2.46 for horizontal and dipping reflectors under the assumption of weak anisotropy.

Symmetry axis perpendicular to the reflector

The symmetry axis in dipping TI formations (e.g., shale layers in overthrust tectonic areas) often is perpendicular to the layering. The relative simplicity of this model makes it possible to obtain a concise exact representation of the NMO ellipse.

Because the dip plane of the reflector contains the symmetry axis, it becomes a vertical symmetry plane for the whole model and determines the orientation of the NMO ellipse. The semiaxis of the NMO ellipse that points in the dip direction is expressed through the reflector dip ϕ as (Tsvankin, 1995a, 2005)

$$V_{\rm nmo}^{(1)}(\phi) = \frac{V_{\rm nmo}(0)}{\cos\phi}\,. \tag{2.47}$$

The NMO velocity from a horizontal interface $[V_{\rm nmo}(0)]$ is obtained for a vertical symmetry axis (because the symmetry axis stays perpendicular to the reflector). Equation 2.47 is identical to the well-known cosine-of-dip dependence of NMO velocity in isotropic media (Levin, 1971), with $V_{\rm nmo}(0)$ replacing the medium velocity.

Substituting the ray parameter $p = \sqrt{p_1^2 + p_2^2}$ of the zero-offset ray into equation 2.47, we obtain

$$V_{\rm nmo}^{(1)}(p) = \frac{V_{\rm nmo}(0)}{\sqrt{1 - p^2 V_0^2}}\,, \tag{2.48}$$

where $V_0 = \sin\phi/p$ is the symmetry-direction velocity of the mode under consideration (it may be either a P- or an S-wave). In anisotropic media $V_{\rm nmo}(0)$ and V_0 generally differ, and equation 2.48 does not coincide with the corresponding isotropic expression. Straightforward but tedious algebraic transformations of equation 2.46 show that the second (strike-line) semiaxis of the NMO ellipse is simply given by

$$V_{\rm nmo}^{(2)} = V_{\rm nmo}(0)\,. \tag{2.49}$$

For P-waves, this result can be more easily obtained in the weak-anisotropy approximation (Appendix 2B) by substituting the relation $\tan\nu = \tan\phi = p_1/p_3$ into equation 2.100 for W_{22}. Therefore, as in isotropic media (Levin, 1971), $V_{\rm nmo}^{(2)}$ is equal to the zero-dip NMO velocity and does not vary with dip.

Combining the two semiaxes of the NMO ellipse (equations 2.48 and 2.49) and estimating p from the zero-offset section, we can find $V_{\rm nmo}(0)$, the symmetry-direction velocity V_0, and the tilt $\nu = \phi$ of the symmetry axis. Because for P-waves

$$V_{\rm nmo}(0) = V_{P0}\sqrt{1 + 2\delta}\,, \tag{2.50}$$

the NMO ellipse for a P-wave dipping event yields the parameter δ in addition to the velocity V_{P0}. The anisotropy parameter ϵ in this model, however, cannot be constrained by conventional-spread P-wave moveout. Given a typical level of noise in reflection traveltimes, this approach should give stable estimates of V_{P0} and δ, if the dip reaches at least 20°.

The velocity V_{P0} and parameter δ can be determined even from 2D P-wave data acquired in the dip plane if the velocity $V_{\rm nmo}(0)$ was found from a horizontal event.

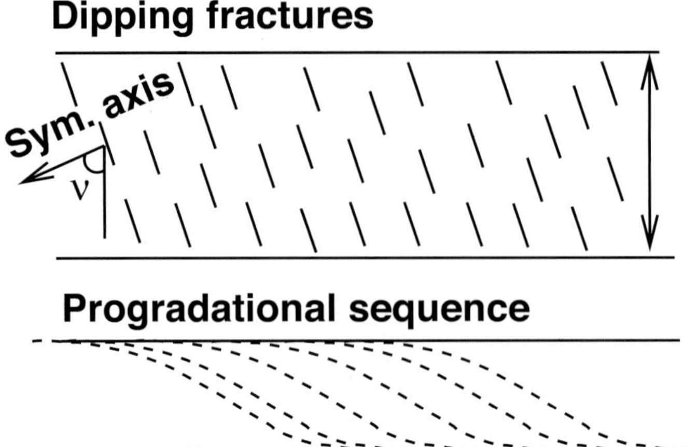

Figure 2.30: Schematic depth section illustrating that a horizontal TI layer with a tilted symmetry axis can describe dipping fracture systems and progradational sequences (Dewangan and Tsvankin, 2006a).

Also, NMO velocity can be inverted for V_{P0} and δ if the reflecting interface has a gradually changing slope and the symmetry axis remains orthogonal to the reflector (e.g., see the physical model in Figure 2.13).

Weakly anisotropic horizontal layer

A horizontal TI layer with a tilted symmetry axis may represent an adequate effective model for obliquely dipping, penny-shaped fractures embedded in isotropic host rock (see a field-data example in Angerer et al., 2002) and for thin-bedded progradational clastic or carbonate sequences (Figure 2.30). To gain insight into the dependence of the P-wave NMO ellipse on the TTI parameters, in Appendix 2B we linearize the matrix $\mathbf{W}$ (equation 2.45) in ϵ and δ (equations 2.98 – 2.100). For a horizontal reflector ($p_1 = p_2 = 0$) and the symmetry axis confined to the $[x_1, x_3]$-plane, this approximation simplifies to

$$W_{11} = \frac{1}{V_{P0}^2}\left[1 - 2\delta + 2\epsilon \sin^2\nu - 14(\epsilon - \delta)\sin^2\nu\cos^2\nu\right], \tag{2.51}$$

$$W_{12} = 0\,, \tag{2.52}$$

$$W_{22} = \frac{1}{V_{P0}^2}\left[1 - 2\delta - 2(\epsilon - \delta)\sin^2\nu\,(1 + \cos^2\nu)\right]. \tag{2.53}$$

Because $W_{12} = 0$, the semiaxes of the NMO ellipse are aligned with the coordinate axes x_1 and x_2. This is an expected result because the $[x_1, x_3]$-plane contains the symmetry axis (it is often called the *symmetry-axis plane*) and, therefore, represents a symmetry plane of the whole model. For vanishing W_{12}, the matrix elements W_{11}

and W_{22} are reciprocal to the squared semiaxes of the NMO ellipse (i.e., to the squared NMO velocities along the axes of the ellipse):

$$W_{ii} = \frac{1}{[V_{\rm nmo}^{(i)}(0)]^2}\,, \qquad (i = 1, 2)\,. \tag{2.54}$$

Equations 2.51 and 2.54 for the velocity $V_{\rm nmo}^{(1)}(0)$ in the symmetry-axis plane are equivalent to the linearized 2D NMO equation given by Tsvankin (2005; equation 3.45). The orientation of the NMO ellipse depends on the azimuth β of the symmetry axis, while the values of the semiaxes give two more equations for the remaining four parameters. Note that the angle β cannot be found unambiguously because the symmetry axis can be parallel to either the major or minor axis of the NMO ellipse. For horizontal transverse isotropy ($\nu = 90°$) arising from parallel penny-shaped cracks, the symmetry axis points in the direction of the minor axis (e.g., Tsvankin, 1997b).

If the symmetry axis is vertical (VTI; $\nu = 0$), NMO velocity is independent of azimuth and is equal to $V_{P0}\sqrt{1+2\delta}$. It is interesting that the NMO ellipse can also degenerate into a circle for an *azimuthally anisotropic* layer with a tilted symmetry axis. Indeed, $W_{11} = W_{22}$ (equations 2.51 and 2.53) for $\nu \neq 0$ when

$$2\epsilon - \delta - 6(\epsilon - \delta)\cos^2\nu = 0\,. \tag{2.55}$$

For instance, if the medium is HTI ($\cos\nu = 0$), equation 2.55 gives

$$2\epsilon = \delta\,. \tag{2.56}$$

As mentioned above, the ellipticity of P-wave NMO velocity in HTI media is governed by the parameter $\delta^{(\rm V)} \approx 2\epsilon - \delta$, which goes to zero if equation 2.56 is satisfied.

An example of extremely weak azimuthal variation of NMO velocity for a TTI layer with a tilt angle of 45° is shown in Figure 2.31. Equation 2.55 with $\nu = 45°$ reduces to $\epsilon = 2\delta$, so for the model from Figure 2.31 the semiaxes of the NMO ellipse should equal one another. Although equation 2.55 is an approximation valid only for weakly anisotropic media, the exact NMO ellipse is almost circular.

Another important point illustrated by Figure 2.31 is that the theoretical NMO ellipse is close to the ellipse reconstructed from ray-traced traveltimes for spreadlength equal to the reflector depth. The maximum difference between the two ellipses, caused by the deviation of the moveout curve from the analytic hyperbola, is only about 1%. This example provides another confirmation that for typical TI models nonhyperbolic moveout influences P-wave traveltimes only when the offset-to-depth ratio exceeds unity.

Weakly anisotropic dipping layer

To estimate all five relevant layer parameters (V_{P0}, ϵ, δ, β, and ν), we suggest combining the NMO ellipses for a horizontal and a dipping reflector in the inversion procedure. Although two NMO ellipses seem to provide enough equations for the inversion,

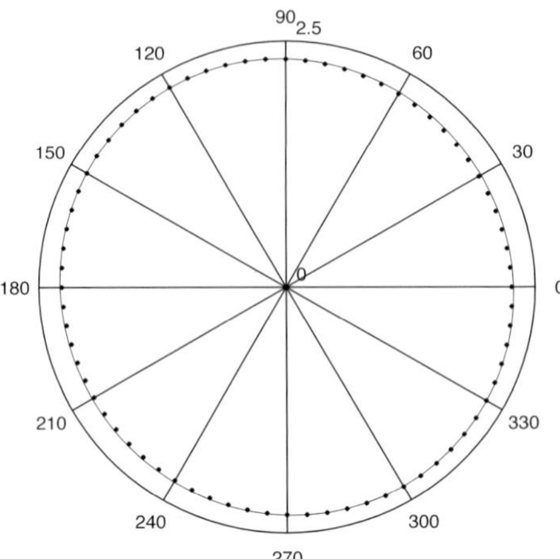

Figure 2.31: Quasi-circular P-wave NMO ellipse in a horizontal TTI layer (Grechka and Tsvankin, 2000). The solid line is the exact NMO ellipse; the dotted line is the best-fit ellipse reconstructed from ray-traced traveltimes computed in four azimuths (0°, 45°, 90°, and 135°; the azimuth is marked on the perimeter) for spreadlength equal to the layer thickness. The relevant medium parameters are V_{P0}=2.0 km/s, ϵ=0.2, δ=0.1, β=30°, and ν=45°. The value "2.5" on top of the plot indicates the range of V_{nmo} in km/s.

in some cases the parameters cannot be resolved individually. The weak-anisotropy approximation for the NMO ellipse in Appendix 2B (equations 2.98 – 2.100) helps identify the trade-offs between the model parameters and study the stability of the inversion procedure.

For instance, the anisotropy parameters in all dip-dependent terms appear only as the difference $\epsilon - \delta$; indeed, for elliptical anisotropy ($\epsilon = \delta$) the dip-dependence of NMO velocity is purely isotropic (Appendix 2B). Still, ϵ and δ can be found individually using the NMO ellipse for horizontal events (equations 2.51 and 2.53), unless the tilt ν is relatively small (estimates are given in the next section). For small ν the medium approaches VTI, and the NMO velocity for a horizontal reflector yields a single combination of V_{P0} and δ, while the dip-dependence of the NMO ellipse is controlled by $\eta \approx \epsilon - \delta$ (Alkhalifah and Tsvankin, 1995; Tsvankin, 2005). Thus, ϵ and δ can be estimated separately only for a sufficiently large tilt of the symmetry axis.

A similar constraint applies to the reflector dip, which should not be too mild to ensure sufficient sensitivity of the NMO ellipse of the dipping event to the anisotropy parameters. (The anisotropic dip terms $\hat{W}_{ij}$ in the weak-anisotropy approximations 2.98 – 2.100 can be used to make appropriate estimates.) On the whole, accurate estimation of the five relevant parameters of TTI media from the P-wave NMO ellipses for a horizontal and a dipping reflector should be feasible if both the reflector dip and the tilt of the symmetry axis are not too small. This statement is quantified below by performing inversion of noise-contaminated data.

2.4.2 Inversion for a single TTI layer

Parameter estimation for a homogeneous TTI medium is based on equation 2.46 for the NMO ellipse. Note that the horizontal slownesses p_1 and p_2 control the reflection slope on the zero-offset (stacked) section and, therefore, can be obtained directly from reflection data. The algorithm combines the P-wave NMO ellipses for a horizontal [$\tilde{\mathbf{W}}^{\rm hor}$ or $\tilde{V}_{\rm nmo}^{\rm hor}(\alpha)$] and a dipping [$\tilde{\mathbf{W}}^{\rm dip}$ or $\tilde{V}_{\rm nmo}^{\rm dip}(\alpha)$] event to invert for the five relevant medium parameters (V_{P0}, ϵ, δ, ν, and β). The least-squares objective function is defined as

$$\mathcal{F} = \int_0^{2\pi} \left[1 - \frac{V_{\rm nmo}^{\rm hor}(\alpha)}{\tilde{V}_{\rm nmo}^{\rm hor}(\alpha)}\right]^2 d\alpha + \int_0^{2\pi} \left[1 - \frac{V_{\rm nmo}^{\rm dip}(\alpha)}{\tilde{V}_{\rm nmo}^{\rm dip}(\alpha)}\right]^2 d\alpha \,, \qquad (2.57)$$

where the modeled NMO ellipses $V_{\rm nmo}^{\rm hor}(\alpha)$ and $V_{\rm nmo}^{\rm dip}(\alpha)$ are calculated from equation 2.46. The minimization of the objective function is carried out using the simplex method. With a realistic starting model (i.e., $|\epsilon| < 1$, $|\delta| < 1$) and noise-free data, the inversion algorithm converges towards the correct set of parameters for a range of TTI models. The influence of errors in the input data is studied in detail below.

Influence of tilt

Figure 2.32 illustrates the influence of the tilt of the symmetry axis on parameter estimation. The exact NMO velocities were computed for a homogeneous TTI layer in four azimuths (0°, 45°, 90° and 135°) and distorted by Gaussian noise with a standard deviation of 2% to simulate errors in velocity picking. Then we obtained the ellipses that provide the best approximation for the noise-contaminated NMO velocities and carried out the inversion using equation 2.57.

Each dot in Figure 2.32 corresponds to the inversion result for a particular realization of errors in the input NMO velocities; the inversion procedure for each model was repeated 200 times. The mean values of all parameters were recovered accurately despite the fact that we intentionally used an incorrect shear-wave velocity $V_{S0} = V_{P0}/2 = 1$ km/s (the actual value is 1.2 km/s). As has been well established in the literature (Tsvankin and Thomsen, 1994; Tsvankin, 2005), V_{S0} has negligible influence on P-wave moveout.

Figure 2.32 confirms that the inversion becomes more stable as the symmetry axis deviates from the vertical. While for $\nu = 60°$ and $\nu = 80°$ the inverted ϵ and δ are tightly clustered near the actual values, the deviations visibly increase for $\nu = 40°$ and, especially, for $\nu = 20°$. The inversion results for $\nu < 30 - 40°$ are so sensitive to errors in the input parameters that P-wave NMO velocity cannot be used to estimate ϵ and δ individually. Nonetheless, the inverted ϵ and δ show a linear trend corresponding to the correct value of η. For vertical transverse isotropy ($\nu = 0°$), η is the only combination of the anisotropy parameters constrained by the P-wave NMO velocity.

The accuracy in the P-wave symmetry-direction velocity V_{P0} (not shown in Figure 2.32) exhibits a similar dependence on the tilt, with the scatter in the inversion results monotonically decreasing with ν. The standard deviation in V_{P0} changes from

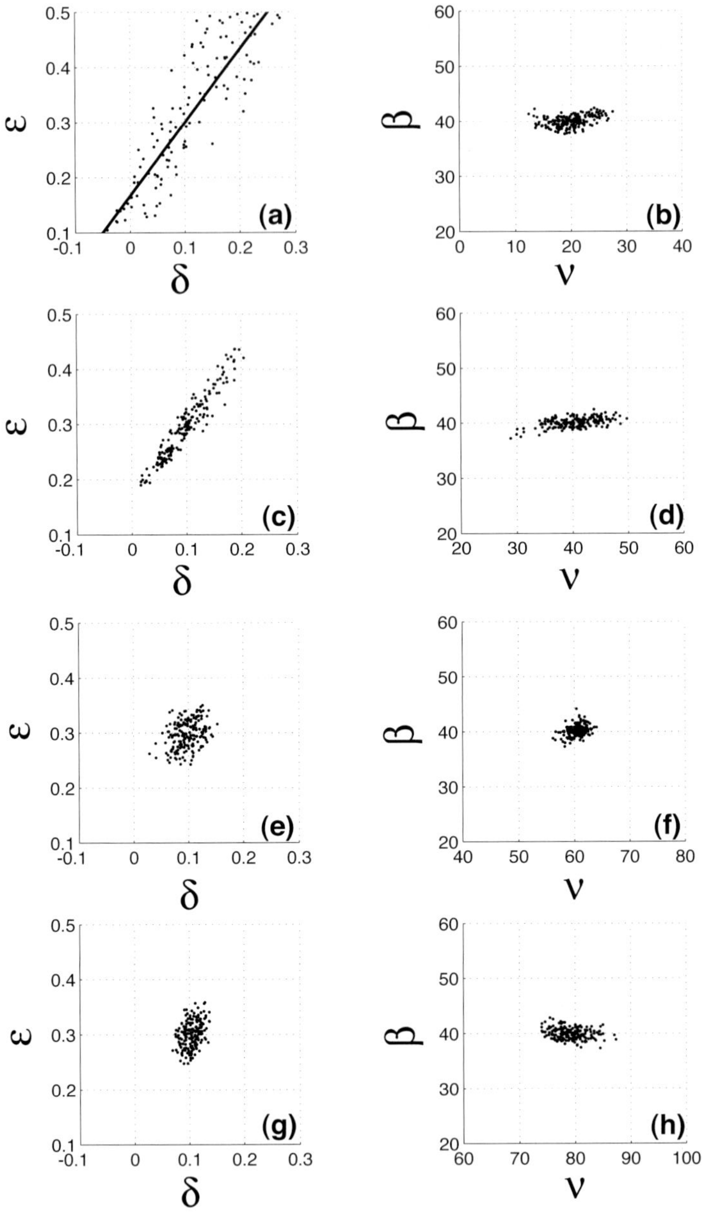

Figure 2.32: Inverted parameters ϵ, δ, β, and ν (the angles are in degrees) for a homogeneous TTI layer with different inclinations of the symmetry axis (Grechka and Tsvankin, 2000). The input data include the P-wave NMO ellipses for a horizontal interface and a reflector with the dip ϕ=60° and azimuth 50°. Both ellipses were distorted by Gaussian noise with a standard deviation of 2%. The tilt ν is equal to 20° in (a) and (b), 40° in (c) and (d), 60° in (e) and (f), and 80° in (g) and (h). The other parameters are V_{P0}=2 km/s, ϵ=0.3, δ=0.1, and β=40°. The solid line on plot (a) indicates ϵ and δ values corresponding to the correct value of η=0.167.

0.01 km/s (just 0.5% of the correct V_{P0}) for a near-horizontal symmetry axis ($\nu = 80°$) to 0.26 km/s (13%) for a tilt of 20°.

An interesting observation from the right-column plots in Figure 2.32 is that the azimuth β of the symmetry axis is well-constrained for all four tilts ν, while the scatter in the tilt itself is more significant. Indeed, the angle β is responsible for the orientation of the NMO ellipse for the horizontal reflector. Overall, the accuracy in both β and ν is satisfactory for a wide range of tilts. The only exception is quasi-VTI models with small $\nu < 10°$ (not shown), for which the orientation of the symmetry axis no longer has much influence on normal moveout. Also, the scatter in the orientation of the symmetry axis increases as the medium approaches isotropic ($|\epsilon| < 0.05$, $|\delta| < 0.05$), and kinematic signatures lose their sensitivity to the symmetry-axis direction.

Time processing for mild tilts

The stability analysis indicates that the parameters responsible for P-wave kinematics cannot be resolved unambiguously for small and moderate tilts of the symmetry axis ($\nu < 30° - 40°$). Hence, an important practical question is whether moveout inversion for mild ν provides any useful information for seismic processing.

P-wave moveout in VTI media ($\nu = 0°$) constrains two parameter combinations [$V_{\rm nmo}(0)$ and η], which are fully responsible for all time-processing steps (NMO, DMO, and time migration). For TTI models with $\nu < 30° - 40°$, our algorithm produces a family of equivalent models, all of which have close NMO ellipses for the two reflectors used in the inversion. To verify whether these equivalent models can replace the actual model in time processing, we computed nonhyperbolic moveout for a horizontal reflector and NMO ellipses for dipping reflectors for a representative subset of models obtained in Figure 2.32a,b. A typical result, displayed in Figure 2.33, demonstrates that long-spread (nonhyperbolic) moveout curves for the actual model and a substantially different set of the inverted parameters are indeed close to each other in all azimuthal directions. We conclude that these two models have similar impulse responses for poststack time migration.

The NMO ellipses for equivalent models (not shown here) somewhat diverge from the corresponding exact ellipses with increasing dip if the reflector has an azimuth distinctly different from the one used in the inversion. Still, for practical purposes of dip-moveout processing it should be acceptable to use a model found from the parameter-estimation procedure described here, especially if the subsurface structure does not contain a wide range of reflector azimuths. Therefore, although our algorithm cannot resolve the medium parameters for mild tilts of the symmetry axis, it provides an approximate TTI model for time-domain processing.

Influence of reflector dip

The results in Figure 2.32 were obtained for a rather favorable, large dip $\phi = 60°$ of one of the reflectors. As expected, the inversion becomes less stable with decreasing

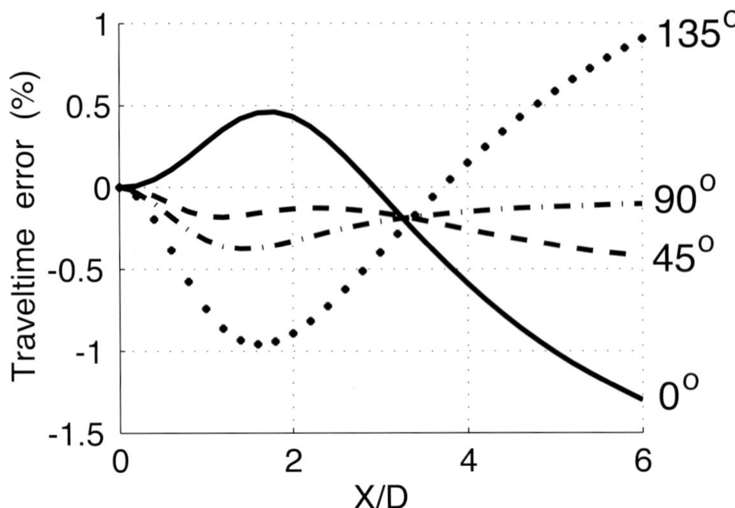

Figure 2.33: Comparison of long-spread reflection moveout from a horizontal interface for the actual TTI model and an equivalent model obtained by the moveout inversion (Grechka and Tsvankin, 2000). The curves, computed for several azimuths marked on the plot, show the percentage difference between the reflection traveltimes for the two models as a function of the offset-to-depth ratio. The parameters of the actual model are V_{P0}=2 km/s, ϵ=0.3, δ=0.1, ν=20°, and β=40°; for the equivalent model (taken from Figure 2.32a,b), V_{P0}=2.2 km/s, ϵ=0.14, δ=0.03, ν=21.5°, and β=40.8° .

ϕ (Figure 2.34). Each row of plots in Figure 2.34 should be compared with Figures 2.32e,f, which were generated for the same tilt $\nu = 60°$ of the symmetry axis. While the scatter in the inverted parameters is comparable for the dips $\phi = 40°$ and $\phi = 60°$, the results deteriorate for the smaller $\phi = 30°$ (Figures 2.34c,d) and become quite unstable for $\phi = 20°$ (Figures 2.34a,b). The inverted ϵ and δ in Figure 2.34a still cluster around a straight line, but they no longer yield an accurate estimate of η. Note that the scatter in the parameters responsible for the symmetry-axis orientation is much more sensitive to the reflector dip (the right column in Figure 2.34) than to the tilt of the symmetry axis (Figure 2.32). The standard deviation in the symmetry-direction velocity V_{P0} (not shown) also increases from 1.2% for $\phi = 40°$ to 3.1% for $\phi = 30°$ to 6.8% for $\phi = 20°$.

Therefore, for the inversion to be reasonably stable, the dip should reach at least 30°. Similar values of the minimum dip are required in the dip-moveout inversion for the parameter η in VTI media (Alkhalifah and Tsvankin, 1995; Tsvankin, 2005). On the other hand, for steep dips approaching 90°, specular reflections (even at zero offset) and the NMO ellipse may not exist at all (see Appendix 2B).

We conclude that the moveout inversion based on the P-wave NMO ellipses for a horizontal and a dipping reflector yields sufficiently stable results if both the reflector dip ϕ and tilt ν of the symmetry axis exceed $30° - 40°$.

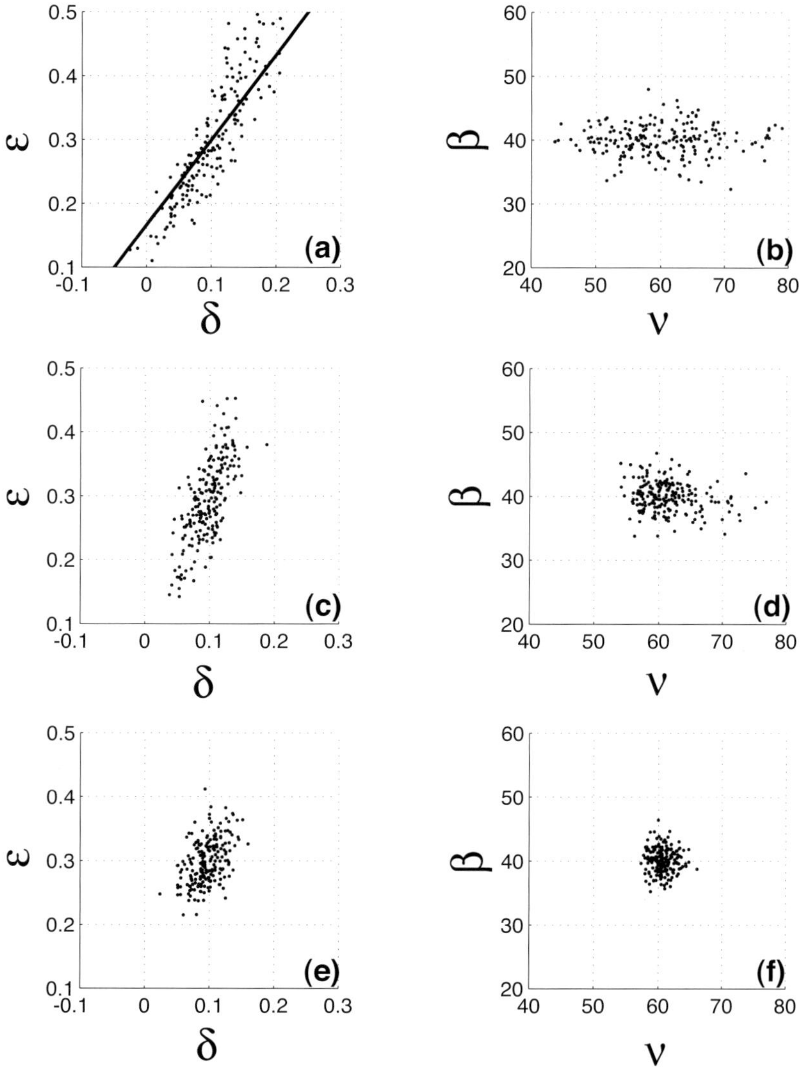

Figure 2.34: Same as Figure 2.32, but for variable reflector dip; the second reflector is horizontal (Grechka and Tsvankin, 2000). The reflector azimuth and the parameters V_{P0}, ϵ, δ, and β are the same as those in Figure 2.32; the tilt of the symmetry axis ν=60°. The reflector dip ϕ is equal to 20° in (a) and (b), 30° in (c) and (d), and 40° in (e) and (f). The solid line on plot (a) corresponds to the correct η=0.167.

Inversion for elliptical anisotropy

In addition to the constraints on the angles ν and ϕ, stable moveout inversion requires that the parameters ϵ and δ differ sufficiently from one another. If $\epsilon = \delta$ (the medium is elliptical), the dip-dependence of P-wave NMO velocity is purely isotropic (equation 2.91), and the only information contained in the NMO ellipse for the dipping event is that $\eta = 0$. This implies that the single P-wave NMO ellipse for a horizontal event (three equations) has to be inverted for the four remaining parameters (V_{P0}, ϵ, ν, and β). Because this inversion is ambiguous, it is possible to find a family of elliptical models with identical NMO ellipses for horizontal and dipping reflectors. To overcome the ambiguity, one of the parameters has to be known a priori. For instance, if the symmetry axis is assumed to be horizontal ($\nu = 90°$; HTI medium), the remaining three parameters can be obtained from the NMO ellipse of a horizontal event.

Grechka and Tsvankin (2000) show that the parameters of elliptical media can be resolved for arbitrary orientation of the symmetry axis if P-wave moveout data are supplemented with the SV-wave NMO velocity for a horizontal reflector (the velocity of SV-waves in elliptical media is independent of angle). The addition of shear data also helps increase the stability of the inversion procedure and evaluate the S-wave symmetry-direction velocity V_{S0}.

2.4.3 Inversion for multiple TTI layers

The parameter-estimation algorithm operating with dip-dependent NMO ellipses can be extended to a stack of horizontal TTI layers with arbitrary symmetry-axis orientation above a dipping reflector (a model of this type is shown in Figure 2.35). The exact interval NMO ellipse $\mathbf{W}_\ell$ in layer ℓ can be found by means of the generalized Dix differentiation (equation 1.32):

$$\mathbf{W}_\ell^{-1} = \frac{\tau(\ell)\,\mathbf{W}^{-1}(\ell) - \tau(\ell-1)\,\mathbf{W}^{-1}(\ell-1)}{\tau(\ell) - \tau(\ell-1)}, \tag{2.58}$$

where $\mathbf{W}(\ell-1)$ and $\mathbf{W}(\ell)$ describe the NMO ellipses for the reflections from the top and bottom of the layer, and $\tau(\ell-1)$ and $\tau(\ell)$ are the corresponding zero-offset traveltimes. The effective and interval NMO ellipses in equation 2.58 are evaluated for the slowness components of the zero-offset ray. For the horizontal events, the zero-offset slowness vector is vertical for all reflections, and the interval ellipse can be found directly from the effective ellipses for the reflections from the top and bottom of any layer.

For dipping events, however, the ray parameters of the zero-offset ray vary depending on the reflector dip and medium properties immediately above the reflecting interface. Therefore, parameter estimation not only requires the presence of a dipping reflector in each layer (Figure 2.35), but has to be performed from top to bottom in the layer-stripping mode. After carrying out the inversion in the top layer using the single-layer algorithm, it is necessary to calculate the interval matrix $\mathbf{W}$ in this layer

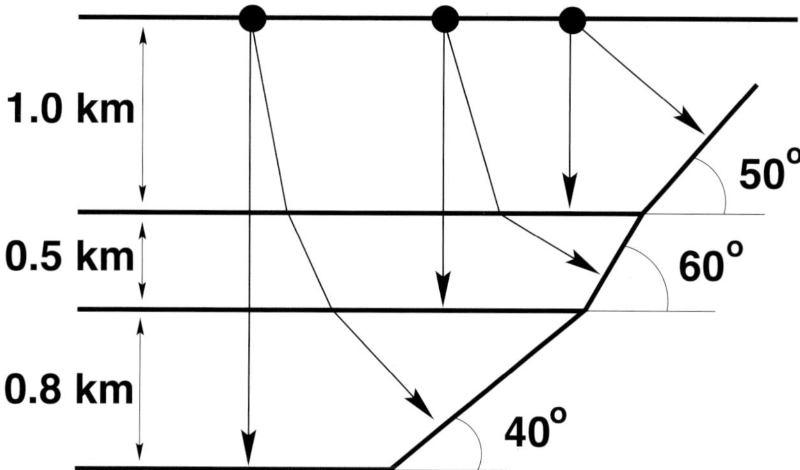

Figure 2.35: Schematic 2D depth section of the TTI model with layer parameters (from top to bottom) listed in Table 2.9 (Grechka and Tsvankin, 2000). The azimuth of the dipping reflector is 0°; the layer thicknesses and reflector dips ϕ are shown on the plot. The symmetry axis in each layer deviates from the dip plane.

for the slownesses p_1 and p_2 of the zero-offset reflection from the dipping interface in the second layer. Then we obtain the interval NMO ellipse for the dipping event in the second layer from equation 2.58, combine it with the corresponding interval ellipse for the horizontal reflector, perform the inversion in the second layer, and continue the layer-stripping procedure downward.

This algorithm was applied to synthetic data generated for the three-layer TTI model with a dipping reflector (e.g., a fault plane) in Figure 2.35. Whereas the reflector azimuth (i.e., the azimuth of the dip plane) is fixed, the azimuth β of the symmetry axis varies from layer to layer, so the model does not have a throughgoing vertical symmetry plane (Table 2.9). We performed 3D anisotropic ray tracing and computed P-wave reflection traveltimes for the horizontal and dipping reflectors along CMP lines in four azimuths (0°, 45°, 90°, and 135°) with respect to the dip plane. Then we fit hyperbolas to the computed traveltimes on conventional-length spreads and obtained the best-fit effective NMO ellipses needed for the inversion procedure.

The errors in the interval parameters for all layers are rather small (Table 2.10) and mostly caused by the influence of nonhyperbolic moveout (associated with both vertical heterogeneity and anisotropy) on the effective NMO ellipses. Despite the error amplification in the generalized Dix differentiation, the maximum deviations of the interval NMO velocities from the exact values are quite moderate [0.5%, 3%, and 1.9% for the first (top), second (middle), and third (bottom) layer, respectively] and correspond to the horizontal events. While the largest error in the interval NMO ellipse is observed in the second layer, the inversion results are least accurate for the

Layer	V_{P0} (km/s)	V_{S0} (km/s)	ϵ	δ	ν (degrees)	β (degrees)
1	2.00	1.20	0.15	0.10	50.0	20.0
2	2.50	1.60	0.20	0.10	60.0	40.0
3	3.00	1.80	0.30	0.15	40.0	60.0

Table 2.9: Interval parameters of the TTI model from Figure 2.35.

Layer	V_{P0} (km/s)	ϵ	δ	ν (degrees)	β (degrees)
1	2.02	0.14	0.09	53.6	19.6
2	2.52	0.19	0.10	60.1	38.9
3	2.91	0.38	0.14	43.7	64.5

Table 2.10: Inverted interval parameters of the model from Figure 2.35 and Table 2.9 (Grechka and Tsvankin, 2000). The maximum source-receiver offset used in estimating moveout velocities is equal to the distance between the CMP and the corresponding reflector. SV-wave reflections were not used in this test, and the shear-wave symmetry-direction velocity V_{S0} could not be constrained.

bottom layer (see Table 2.10). This is explained not just by error accumulation with depth and the reduced contribution of the deeper layers to the effective traveltime, but also by the smaller tilt of the symmetry axis in the third layer (see Figure 2.32).

Overall, the accuracy of the parameter estimation in vertically heterogeneous TTI media can be predicted on the basis of the single-layer error analysis with a correction that accounts for the error amplification inherent in Dix-type layer-stripping algorithms. As in conventional Dix velocity analysis, the accuracy in the interval values is inversely proportional to the thickness of the interval. Let us denote the relative time thickness of the ℓth layer (for which the differentiation is performed) as

$$\sigma_\ell = \frac{\tau_\ell}{\tau(\ell)} \,. \tag{2.59}$$

Then equation 2.58 can be rewritten in the form

$$\mathbf{W}_\ell^{-1} = \frac{1}{\sigma_\ell}\Big[\mathbf{W}^{-1}(\ell) \;-\; (1-\sigma_\ell)\,\mathbf{W}^{-1}(\ell-1)\Big], \tag{2.60}$$

which explicitly shows that errors in the effective NMO ellipses will propagate into the interval ellipse with amplification factor $1/\sigma_\ell$ (for small σ_ℓ).

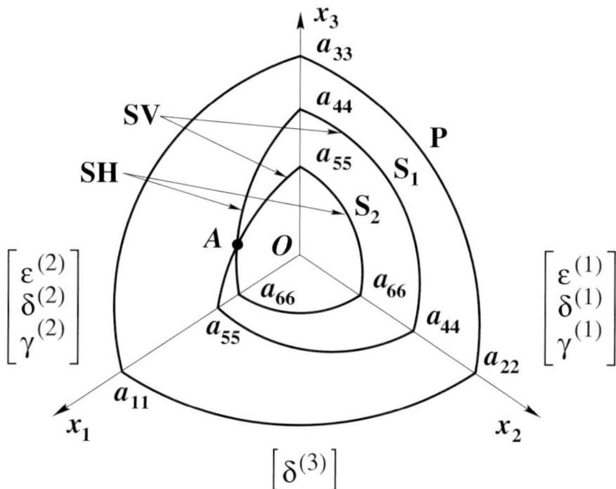

Figure 2.36: Sketch of the phase-velocity surfaces in orthorhombic media (Tsvankin, 2005). The parameters $\epsilon^{(1)}$, $\delta^{(1)}$, and $\gamma^{(1)}$ are defined in the symmetry plane $[x_2, x_3]$, while $\epsilon^{(2)}$, $\delta^{(2)}$, and $\gamma^{(2)}$ in the plane $[x_1, x_3]$ (see Appendix 1B); $a_{ij} \equiv \sqrt{c_{ij}/\rho}$ are the density-normalized stiffness coefficients. If the Cartesian coordinate planes coincide with the symmetry planes, the stiffness matrix c_{ij} has nine independent elements. Point A marks a shear-wave singularity where the phase velocities of the split S-waves coincide.

2.5 Time-domain velocity analysis for orthorhombic media

The moveout-inversion methodology described above for transverse isotropy can be extended to lower-symmetry orthorhombic (also referred to as orthotropic) models, which are typical for naturally fractured formations (e.g., Crampin, 1991; Schoenberg and Helbig, 1997; Bakulin et al., 2000b; Tsvankin, 2005). For instance, orthorhombic symmetry can be caused by a set of parallel vertical fractures embedded in a background VTI medium or by two orthogonal (or nonorthogonal but identical) vertical fracture systems in a purely isotropic or VTI matrix (e.g., Wild and Crampin, 1991). In principle, the presence of multiple fracture sets with different directions lowers the medium symmetry to monoclinic or even triclinic. If, however, the host rock is isotropic, the symmetry of the fracture-induced anisotropy remains close to orthorhombic for any fracture orientations, shapes, and types of infill (Grechka and Kachanov, 2006a,b; for more details, see Chapter 9).

Orthorhombic media have three mutually orthogonal planes of mirror symmetry, one of which is assumed here to be horizontal (Figure 2.36). We do not, however, restrict ourselves to any particular physical model and consider a general orthorhombic stiffness matrix with nine independent elements c_{ij}. Inverting reflection moveout for the azimuth of the vertical symmetry planes and nine stiffness coefficients (which can vary in space) is an extremely difficult problem, especially with the limited angle coverage of reflection data. Significant progress, however, can be achieved by combining the stiffnesses in a way that simplifies analytic description of seismic velocities and amplitudes.

Tsvankin (1997a, 2005) took advantage of the identical form of the Christoffel equation in the symmetry planes of orthorhombic and TI media to introduce a notation based on the same principle as the widely used Thomsen's (1986) parameters for vertical transverse isotropy. The definitions and descriptions of Tsvankin's parameters are given in Appendix 1B. The vertical velocities of the P-wave (V_{P0}) and one of the split S-waves (V_{S0}) are chosen as the reference isotropic quantities, while the anisotropy is controlled by the dimensionless parameters $\epsilon^{(1,2)}$, $\delta^{(1,2,3)}$, and $\gamma^{(1,2)}$ defined in the symmetry planes of the model.

This notation preserves the attractive features of Thomsen's parameters in describing symmetry-plane velocities, traveltimes, and plane-wave reflection coefficients (Tsvankin, 2005). It also provides a unified framework for treating orthorhombic and TI models, which is particularly important for inversion algorithms. As discussed in Appendix 1B, both VTI and HTI media represent special cases of the more general orthorhombic symmetry.

The ϵ- and δ-parameters capture the combinations of the stiffnesses responsible for P-wave kinematic signatures both within and outside symmetry planes, even for strongly anisotropic media. P-wave velocities and traveltimes (including reflection moveout) in orthorhombic models with a fixed orientation of the symmetry planes depend on just six parameters (V_{P0}, $\epsilon^{(1,2)}$, and $\delta^{(1,2,3)}$). For example, the parameters $\delta^{(1)}$ and $\delta^{(2)}$ govern near-vertical P-wave velocity variations and concisely describe the P-wave NMO ellipse for horizontal reflectors (see equation 2.62).

As is the case for TI media, the dimensionless anisotropy parameters are well-suited for developing weak-anisotropy approximations for various seismic signatures. Below, we linearize the NMO ellipse in $\epsilon^{(1,2)}$ and $\delta^{(1,2,3)}$ to evaluate the contribution of anisotropy to P-wave moveout for arbitrary reflector dip. Then we develop a parameter-estimation procedure to build an orthorhombic model suitable for P-wave time imaging. This section is based on the results of Grechka and Tsvankin (1999b), who gave a detailed analysis of P-wave moveout inversion for vertically heterogeneous orthorhombic media.

2.5.1 NMO ellipse in a homogeneous orthorhombic layer

Horizontal layer

As mentioned above, we consider an orthorhombic medium with a horizontal symmetry plane. Then the azimuths of the vertical symmetry planes determine the orientation of the NMO ellipse in a horizontal orthorhombic layer (equation 1.20):

$$V_{\text{nmo}}^{-2}(\alpha) = W_{11}\cos^2\alpha + W_{22}\sin^2\alpha = \frac{\cos^2\alpha}{[V_{\text{nmo}}^{(2)}]^2} + \frac{\sin^2\alpha}{[V_{\text{nmo}}^{(1)}]^2}\,, \tag{2.61}$$

where α is the azimuth with respect to the x_1-axis, and $V_{\rm nmo}^{(2)}$ and $V_{\rm nmo}^{(1)}$ are the NMO velocities in the symmetry planes $[x_1, x_3]$ and $[x_2, x_3]$ (respectively), which can be found by analogy with vertical transverse isotropy (Tsvankin, 1997a). For P-waves,

$$V_{\rm nmo}^{(i)} = V_{P0}\sqrt{1+2\delta^{(i)}}\,, \qquad (i=1,2)\,. \tag{2.62}$$

Therefore, the eccentricity of the P-wave NMO ellipse is determined by the difference between the parameters $\delta^{(1)}$ and $\delta^{(2)}$. We should mention that the linearized δ-coefficients introduced by Mensch and Rasolofosaon (1997) within the framework of the weak-anisotropy approximation are not appropriate for the exact equations 2.61 and 2.62.

By reconstructing the NMO ellipse from wide-azimuth data, we can find the orientation of the vertical symmetry planes and the velocities $V_{\rm nmo}^{(1,2)}$ within them. The only case for which the symmetry-plane azimuths cannot be resolved is when the ellipse degenerates into a circle (i.e., $\delta^{(1)} = \delta^{(2)}$). Likewise, the symmetry-plane directions can be estimated from the azimuthally-dependent NMO velocities of either split shear wave.

Layer above a dipping reflector

If the bottom of a homogeneous orthorhombic layer is dipping, the azimuthal variation of NMO velocity is influenced by both the reflector orientation and azimuthal anisotropy. Unless the dip plane of the reflector coincides with one of the symmetry planes of the medium, the axes of the NMO ellipse deviate from the symmetry-plane directions. According to Snell's law, the slowness vector of the zero-offset ray (but not necessarily the ray itself) is normal to the reflector at the zero-offset reflection point. Hence, the horizontal components (p_1 and p_2) of the zero-offset slowness vector, which can be obtained from the time slopes of the zero-offset reflection, constrain the reflector azimuth. The symmetry-plane directions of the layer, however, have to be estimated from the NMO ellipse.

The starting point for our analysis of the P-wave NMO ellipse is the exact equation 2.46, valid for a homogeneous layer of arbitrary symmetry. Although numerical modeling of the NMO ellipse is relatively straightforward, its dependence on the medium parameters is hidden in the slowness components and their derivatives. Therefore, it is convenient to simplify equation 2.46 in the weak-anisotropy approximation. To express the vertical slowness $q = p_3$ through p_1 and p_2, we use the P-wave phase-velocity function linearized in the parameters $\epsilon^{(1,2)}$ and $\delta^{(1,2,3)}$ (Tsvankin, 1997a, 2005):

$$\begin{aligned} V^2 &= V_{P0}^2\,[\,1 + 2n_1^4\,\epsilon^{(2)} + 2n_2^4\,\epsilon^{(1)} + 2n_1^2\,n_3^2\,\delta^{(2)} + 2n_2^2\,n_3^2\,\delta^{(1)} \\ &\quad + 2n_1^2\,n_2^2\,(2\epsilon^{(2)} + \delta^{(3)})\,]\,; \end{aligned} \tag{2.63}$$

$\mathbf{n} = \mathbf{p}V$ is the unit vector in the slowness direction. Substituting the vector $\mathbf{p}$ into equation (2.63) yields an explicit expression for the vertical slowness q:

$$q^2 = \left(\frac{1}{V_{P0}^2} - p_1^2 - p_2^2\right) - 2V_{P0}^2 \Big\{(p_1^2 + p_2^2)\left[p_1^2\left(\epsilon^{(2)} - \delta^{(2)}\right) + p_2^2\left(\epsilon^{(1)} - \delta^{(1)}\right)\right] + \frac{p_1^2\,\delta^{(2)} + p_2^2\,\delta^{(1)}}{V_{P0}^2} + p_1^2 p_2^2\left(\epsilon^{(1)} - \epsilon^{(2)} - \delta^{(3)}\right)\Big\}. \tag{2.64}$$

Note that equation 2.64 contains the combinations $(\epsilon^{(2)} - \delta^{(2)})$ and $(\epsilon^{(1)} - \delta^{(1)})$, which represent the linearized versions of the anellipticity parameters $\eta^{(2)}$ and $\eta^{(1)}$ defined in Appendix 1B. The kinematic equivalence between the symmetry planes of orthorhombic and VTI media implies that $\eta^{(2)}$ and $\eta^{(1)}$ are responsible for the 2D P-wave NMO-velocity function in the $[x_1, x_3]$- and $[x_2, x_3]$-planes, respectively. This equivalence, however, is valid only if one of the symmetry planes coincides with the dip plane of the reflector; otherwise, reflected rays propagate outside the vertical incidence plane and are influenced by azimuthal velocity variations. Another combination of the anisotropy parameters in equation 2.64, $(\epsilon^{(1)} - \epsilon^{(2)} - \delta^{(3)})$, is the linearized anellipticity parameter $\eta^{(3)}$ defined in the horizontal symmetry plane (Appendix 1B).

Differentiating equation 2.64 with respect to p_1 and p_2, substituting q and its derivatives into equation 2.46 and further linearizing the NMO ellipse in the anisotropy parameters leads to (Grechka and Tsvankin, 1999b):

$$\begin{aligned} V_{\rm nmo}^{-2}(\alpha, p_1, p_2) &= \cos^2\alpha \left\{[V_{\rm nmo}^{(2)}]^{-2} - p_1^2 + \sum_{i=1}^{3} d_{1i}\,\eta^{(i)}\right\} \\ &+ 2\sin\alpha\cos\alpha \left\{-p_1\,p_2 + \sum_{i=1}^{3} d_{2i}\,\eta^{(i)}\right\} \\ &+ \sin^2\alpha \left\{[V_{\rm nmo}^{(1)}]^{-2} - p_2^2 + \sum_{i=1}^{3} d_{3i}\,\eta^{(i)}\right\}, \end{aligned} \tag{2.65}$$

where $V_{\rm nmo}^{(1,2)}$ are the linearized symmetry-plane NMO velocities for a horizontal reflector. The quantities d_{ki} are defined as

$$\begin{aligned} d_{11} &= -2p_2^2\,(1 - 4p_1^2\,\tilde V^2)\,(1 - p_1^2\,\tilde V^2 - p_2^2\,\tilde V^2)\,, \\ d_{12} &= -2(6p_1^2 + p_2^2 - 9p_1^4\,\tilde V^2 - 5p_1^2\,p_2^2\,\tilde V^2 + 4p_1^6\,\tilde V^4 + 4p_1^4\,p_2^2\,\tilde V^4)\,, \\ d_{13} &= 2p_2^2\,(1 - 5p_1^2\,\tilde V^2 + 4p_1^4\,\tilde V^4)\,, \\ d_{21} &= -4p_1 p_2\,(1 - 2p_2^2\,\tilde V^2)\,(1 - p_1^2\,\tilde V^2 - p_2^2\,\tilde V^2)\,, \\ d_{22} &= -4p_1 p_2\,(1 - 2p_1^2\,\tilde V^2)\,(1 - p_1^2\,\tilde V^2 - p_2^2\,\tilde V^2)\,, \\ d_{23} &= 4p_1 p_2\,(1 - p_1^2\,\tilde V^2 - p_2^2\,\tilde V^2 + 2p_1^2\,p_2^2\,\tilde V^4)\,, \\ d_{31} &= -2(6p_2^2 + p_1^2 - 9p_2^4\,\tilde V^2 - 5p_1^2\,p_2^2\,\tilde V^2 + 4p_2^6\,\tilde V^4 + 4p_1^2\,p_2^4\,\tilde V^4)\,, \\ d_{32} &= -2p_1^2\,(1 - 4p_2^2\,\tilde V^2)\,(1 - p_1^2\,\tilde V^2 - p_2^2\,\tilde V^2)\,, \\ d_{33} &= 2p_1^2\,(1 - 5p_2^2\,\tilde V^2 + 4p_2^4\,\tilde V^4)\,; \end{aligned}$$

$$\tilde{V} = \frac{1}{2}\left(V_{\rm nmo}^{(1)} + V_{\rm nmo}^{(2)}\right). \tag{2.66}$$

For a horizontal reflector, $p_1 = p_2 = d_{ki} = 0$, and equation 2.65 reduces to the NMO ellipse for a horizontal layer (equation 2.61).

Although P-wave phase velocity depends on six parameters, only five combinations of them ($V_{\rm nmo}^{(1)}$, $V_{\rm nmo}^{(2)}$, $\eta^{(1)}$, $\eta^{(2)}$, and $\eta^{(3)}$) determine the P-wave NMO ellipse in the weak-anisotropy limit. This indicates that the NMO ellipse for a dipping reflector ($p_1 \neq 0$ and/or $p_2 \neq 0$) could be inverted for $\eta^{(1,2,3)}$ provided $V_{\rm nmo}^{(1)}$, $V_{\rm nmo}^{(2)}$, and the orientation of the vertical symmetry planes have been estimated from wide-azimuth horizontal events.

To assess the possibility of recovering all three η-parameters from a single dipping event, let us assume that the dip plane of the reflector coincides with the $[x_1, x_3]$ symmetry plane. Then the zero-offset ray and its slowness vector are confined to the dip plane, and the horizontal slowness p_2 goes to zero. Evaluating the coefficients d_{ki} for $p_2 = 0$ and substituting the results into equation 2.65 gives the following expression for the NMO ellipse:

$$\begin{aligned} V_{\rm nmo}^{-2}(\alpha, p_1, p_2 = 0) &= \cos^2\alpha \left\{[V_{\rm nmo}^{(2)}]^{-2} - p_1^2 - 2p_1^2\,\eta^{(2)}\left(4p_1^4\,\tilde{V}^4 - 9p_1^2\,\tilde{V}^2 + 6\right)\right\} \\ &+ \sin^2\alpha \left\{[V_{\rm nmo}^{(1)}]^{-2} - 2p_1^2\,[\eta^{(1)} - \eta^{(3)} + \eta^{(2)}\,(1 - p_1^2\,\tilde{V}^2)]\right\}. \end{aligned} \tag{2.67}$$

As expected, the axes of the NMO ellipse are aligned with the symmetry planes of the medium. Because the symmetry-plane azimuths can be found from horizontal events, the orientation of the ellipse in equation 2.67 does not carry new information for the inversion. The elliptical semiaxes can constrain only two combinations of the medium parameters, which is insufficient to resolve all three η-parameters individually.

The NMO velocity in the dip plane can be obtained from equation 2.67 by setting $\alpha = 0°$:

$$V_{\rm nmo}^{-2}(0, p_1) = [V_{\rm nmo}^{(2)}]^{-2} - p_1^2 - 2p_1^2\,\eta^{(2)}\left(4p_1^4\,\tilde{V}^4 - 9p_1^2\,\tilde{V}^2 + 6\right), \tag{2.68}$$

which reduces to the weak-anisotropy approximation for the dip-line P-wave NMO velocity in VTI media (equation 6.7 in Tsvankin, 2005), if we choose $\tilde{V}$ as $V_{\rm nmo}^{(2)}$ (it is acceptable in the weak-anisotropy approximation) and recall that $\eta^{(2)}$ and $V_{\rm nmo}^{(2)}$ are equivalent to the VTI parameters η and $V_{\rm nmo}(0)$, respectively. Because reflected rays excited on the dip line cannot deviate from the incidence (symmetry) plane, $V_{\rm nmo}(0, p_1)$ should indeed be given by the VTI equation for any strength of anisotropy because of the kinematic equivalence between the symmetry planes of orthorhombic and VTI media. According to the results of Alkhalifah and Tsvankin (1995) for vertical transverse isotropy, the NMO velocity in the $[x_1, x_3]$ plane can be inverted in a stable fashion for the coefficient $\eta^{(2)}$, if the dip (represented by the horizontal slowness p_1) reaches at least 25°.

The NMO velocity $V_{\rm nmo}(\pi/2, p_1)$ measured in the strike direction (the second semiaxis of the NMO ellipse) depends on $\eta^{(2)}$ and the difference between the parameters $\eta^{(1)}$ and $\eta^{(3)}$ (equation 2.67). Therefore, if $V_{\rm nmo}^{(1)}$ and $V_{\rm nmo}^{(2)}$ have been obtained from horizontal events, the semiaxes of the NMO ellipse for a dipping reflector can be inverted for the parameter $\eta^{(2)}$ and the difference $\eta^{(1)} - \eta^{(3)}$. Likewise, if $[x_2, x_3]$ represents the dip plane, the NMO ellipse constrains $\eta^{(1)}$ and the difference between $\eta^{(2)}$ and $\eta^{(3)}$.[3] Therefore, if the strike of a dipping reflector is close to either vertical symmetry plane, the P-wave NMO ellipse does not constrain all three η-parameters.

The VTI equation 2.68 can be used on the dip line even outside the symmetry planes, but with the azimuthally-dependent values of the zero-dip NMO velocity and parameter η (Grechka and Tsvankin, 1999b). This result, valid only for weak anisotropy, follows from the identical form of the phase-velocity equation in any vertical plane of orthorhombic and VTI media (Tsvankin, 1997a, 2005). A more detailed discussion of the equivalence between VTI and orthorhombic media for out-of-plane propagation can be found in Rommel and Tsvankin (2000).

P-wave NMO ellipse for strong anisotropy

In the inversion procedure described below, we use the exact NMO equation 2.46 valid for any magnitude of anisotropy. Therefore, it is necessary to find out whether the parameters $V_{\rm nmo}^{(1)}$, $V_{\rm nmo}^{(2)}$, $\eta^{(1)}$, $\eta^{(2)}$, and $\eta^{(3)}$ fully control the P-wave NMO ellipse for homogeneous orthorhombic models with strong velocity variations. It should be emphasized that here and below we use the definitions of the η-parameters given in equations 1.88 – 1.90 rather than their linearized approximations.

If the dip plane of the reflector coincides with one of the vertical symmetry planes, the dip-line NMO velocity is given by the VTI equations and, therefore, is governed by the zero-dip NMO velocity and the corresponding η-parameter. Even in this special case, however, the strike-line NMO velocity is influenced by azimuthal velocity variations and has to be studied separately.

To demonstrate that the five parameters listed above indeed describe the P-wave NMO ellipse for strong anisotropy and arbitrary reflector dip, we performed a series of numerical tests. A typical result for a reflector with dip-plane azimuth diverging by 30° from the $[x_1, x_3]$ symmetry plane is displayed in Figure 2.37. The NMO ellipse was characterized by its semiaxes $V_{e\ell 1}$ and $V_{e\ell 2}$ (Figures 2.37a,b) and the angle β between the major axis and the $[x_1, x_3]$-plane (Figure 2.37c). All three parameters were calculated for a wide range of reflector dips governed by the horizontal slowness $p = \sqrt{p_1^2 + p_2^2}$. In the first test, we varied the ϵ- and δ-parameters keeping $V_{\rm nmo}^{(1,2)}$ and $\eta^{(1,2,3)}$ constant (compare the solid and dashed lines). Clearly, the NMO ellipses for these vastly different, strongly anisotropic models practically coincide for all dips. In contrast, variation in the relevant parameters $\eta^{(1)}$ and $\eta^{(2)}$ (dotted line) leads to noticeable changes in the semiaxes and orientation of the NMO ellipse as the dip increases. This indicates that it may be possible to estimate the parameters $\eta^{(1,2,3)}$ from the NMO ellipse of a dipping event.

[3]In the weak-anisotropy approximation, the definition of $\eta^{(3)}$ is symmetric in the sense that it does not change if x_2 is used as the symmetry axis of the equivalent VTI medium in the $[x_1, x_2]$-plane (Rommel and Tsvankin, 2000).

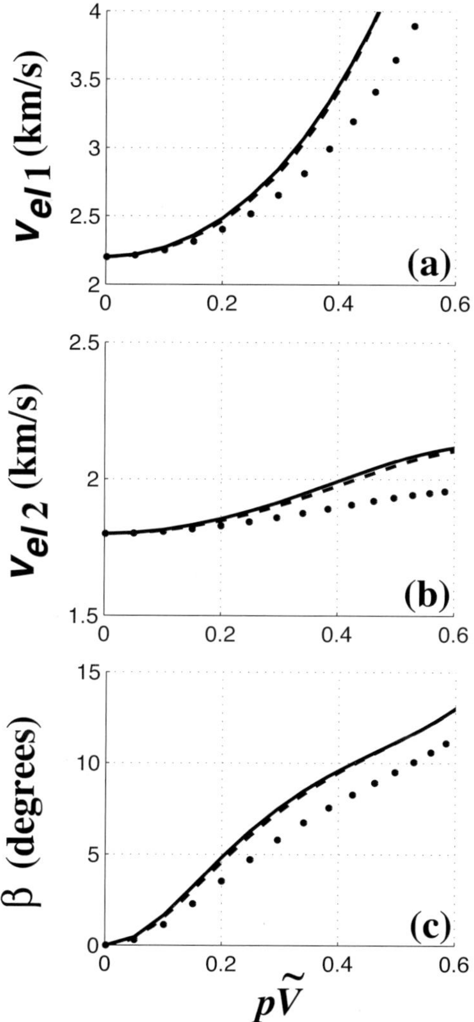

Figure 2.37: P-wave NMO ellipse of a dipping event for three orthorhombic models (Grechka and Tsvankin, 1999b). $V_{e\ell 1}$ and $V_{e\ell 2}$ [plots (a) and (b)] are the semiaxes of the ellipse, and β [plot (c)] is the rotation angle of the major axis with respect to the $[x_1, x_3]$-plane. The dip plane of the reflector makes an angle of 30° with the $[x_1, x_3]$ symmetry plane; the dip changes in accordance with the horizontal slowness $p=\sqrt{p_1^2+p_2^2}$. The velocity $\tilde{V}$ is defined by equation 2.66. The solid line (model 1): $V_{\text{nmo}}^{(1)}$=1.8 km/s, $V_{\text{nmo}}^{(2)}$=2.2 km/s, $\eta^{(1)}=0.2$, $\eta^{(2)}=0.3$, $\eta^{(3)}=0.15$, $\epsilon^{(1)}=0.2$, $\epsilon^{(2)}=0.695$, $\delta^{(1)}=0$, $\delta^{(2)}=0.25$, and $\delta^{(3)}=-0.27$. The dashed line (model 2): all $V_{\text{nmo}}^{(i)}$ and $\eta^{(i)}$ are the same as those for model 1, but $\epsilon^{(1)}=-0.031$, $\epsilon^{(2)}=0.3$, $\delta^{(1)}=-0.165$, $\delta^{(2)}=0$, and $\delta^{(3)}=-0.27$. The dotted line (model 3): $V_{\text{nmo}}^{(1)}$=1.8 km/s, $V_{\text{nmo}}^{(2)}$=2.2 km/s, $\eta^{(1)}=0.1$, $\eta^{(2)}=0.2$, and $\eta^{(3)}=0.15$.

Similar results have been obtained for different reflector azimuths and a plausible range of the anisotropy parameters (Grechka and Tsvankin, 1999b). Thus, for a fixed orientation of the vertical symmetry planes, the exact NMO velocity $V_{\text{nmo}}(\alpha, p_1, p_2)$ is a function of only five medium parameters – $V_{\text{nmo}}^{(1,2)}$ and $\eta^{(1,2,3)}$. In principle, $\eta^{(3)}$ can be replaced by Tsvankin's "generic" parameter $\delta^{(3)}$ (see Appendix 1B), but we prefer to use $\eta^{(3)}$ to maintain uniformity in the description of the NMO ellipse.

2.5.2 Inversion for a single layer

To resolve the five relevant parameters along with the symmetry-plane azimuths, we need at least two NMO ellipses corresponding to distinct reflector dips and/or azimuths. Although in principle the inverse problem can be posed for two arbitrary dips, we consider only the practically important special case when one of the reflectors is horizontal. The NMO ellipse of a horizontal event (equations 2.61 and 2.62) constrains the azimuths of the symmetry planes (as long as the ellipse does not degenerate into a circle) and the NMO velocities $V_{\text{nmo}}^{(1,2)}$ within them. Then the zero-offset traveltimes of the dipping event, measured for a wide range of azimuths, can be used to estimate the horizontal slowness components p_1 and p_2 of the zero-offset ray, while the corresponding NMO ellipse can be inverted for the parameters $\eta^{(1,2,3)}$. We carry out this inversion by applying the simplex method to the following least-squares problem:

$$\mathcal{F} \equiv \sum_{i,j=1}^{2} \left[\hat{W}_{ij} - W_{ij}(\eta^{(1)}, \eta^{(2)}, \eta^{(3)})\right]^2 = \min\,, \tag{2.69}$$

where the matrix $\hat{\mathbf{W}}$ corresponds to the NMO ellipse estimated from azimuthally varying traveltimes, and $\mathbf{W}$ is computed from equation 2.45. Because $W_{12} = W_{21}$, the sum in equation 2.69 includes three terms.

The feasibility of reconstructing NMO ellipses from wide-azimuth reflection traveltimes and inverting them for the η-parameters is confirmed by the test in Figure 2.38. The NMO ellipses for horizontal and dipping reflectors were estimated on spreadlength equal to the CMP-reflector distance and substituted into objective function 2.69. All three anellipticity parameters were found with high accuracy: $\eta^{(1)} = 0.084$, $\eta^{(2)} = 0.041$, and $\eta^{(3)} = 0.123$. The small inversion errors are caused by the influence of nonhyperbolic moveout on the finite-spread moveout velocities. The larger distortion in $\eta^{(3)}$ is indicative of a lower sensitivity of the input data to this parameter (see the discussion below).

The results for the model in Figure 2.38 indicate that a certain percentage error in the NMO velocities (less than 1.5%) translates into a similar absolute error in $\eta^{(1)}$ and $\eta^{(2)}$. The same conclusion was drawn for η estimation in VTI media by Alkhalifah and Tsvankin (1995) and Grechka and Tsvankin (1998b).The magnitude of nonhyperbolic moveout for horizontal reflectors increases with $\eta^{(1)}$ and $\eta^{(2)}$, but the associated NMO-velocity error on spreads close to the reflector depth seldom exceeds 2 – 3%. If the data after hyperbolic moveout correction exhibit residual moveout, a

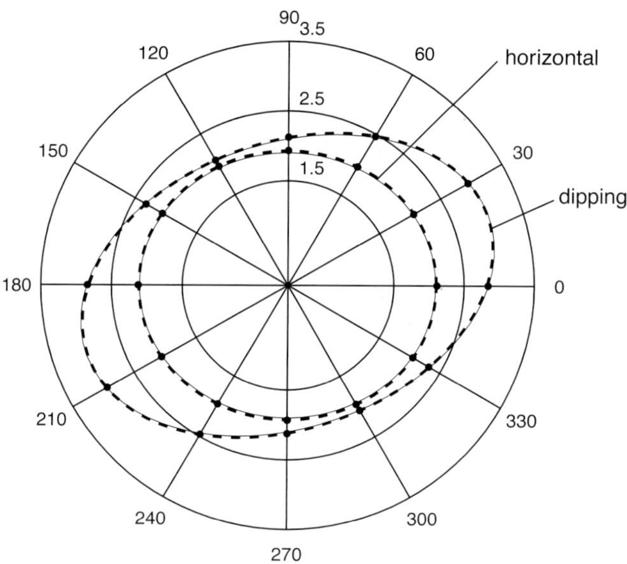

Figure 2.38: Comparison of the exact P-wave NMO ellipses for horizontal and dipping reflectors computed from equation 2.46 (solid lines) with the moveout (stacking) velocities (dots) obtained by least-squares fitting of a hyperbola to ray-traced traveltimes; the spreadlength is equal to the distance between the CMP and the reflector (Grechka and Tsvankin, 1999b). The dashed lines mark the best elliptical fit to the estimated moveout velocities. For the dipping reflector, the azimuth of the dip plane with respect to the $[x_1, x_3]$ symmetry plane is 30°, and the dip is 40°. The medium parameters are $V_{\rm nmo}^{(1)}$=1.90 km/s, $V_{\rm nmo}^{(2)}$=2.10 km/s, $\eta^{(1)}$=0.1, $\eta^{(2)}$=0.05, and $\eta^{(3)}$=0.15 ($\epsilon^{(1)}$=0.1, $\epsilon^{(2)}$=0.17, $\delta^{(1)}$=0, $\delta^{(2)}$=0.11, and $\delta^{(3)}$=−0.16).

more accurate NMO ellipse can be obtained by applying a nonhyperbolic moveout equation (see Chapter 3).

An analysis of the influence of errors in the input data on the inversion results is presented in Figure 2.39. If the elements of the matrices $\mathbf{W}$ are taken from the two exact NMO ellipses, the estimated $\eta^{(1,2,3)}$ deviate from the actual values by no more than 0.005. These minor errors are explained by the fact that we intentionally used inaccurate values of $V_{P0} = \tilde{V} = 2.0$ km/s (the actual $V_{P0} = 1.9$ km/s) and $V_{S0} = 0.95$ km/s (the actual $V_{S0} = 0.6$ km/s). This is another illustration of the negligible influence of the vertical velocities (which are generally unknown) on the NMO ellipses; hence, it is justified to use reasonable estimates for V_{P0} and V_{S0} in the inversion procedure.

It is clear from Figure 2.39 that the different η-parameters are most sensitive to errors in different measured quantities. For instance, $\eta^{(2)}$ (the solid line in Figures 2.39a,b) strongly depends on the larger semiaxes of both NMO ellipses (which are close to the $[x_1, x_3]$ plane) and is almost insensitive to errors in the smaller semiaxes (Figures 2.39c,d). These conclusions are in good agreement with the analytic results obtained in the weak-anisotropy approximation. According to equation 2.68,

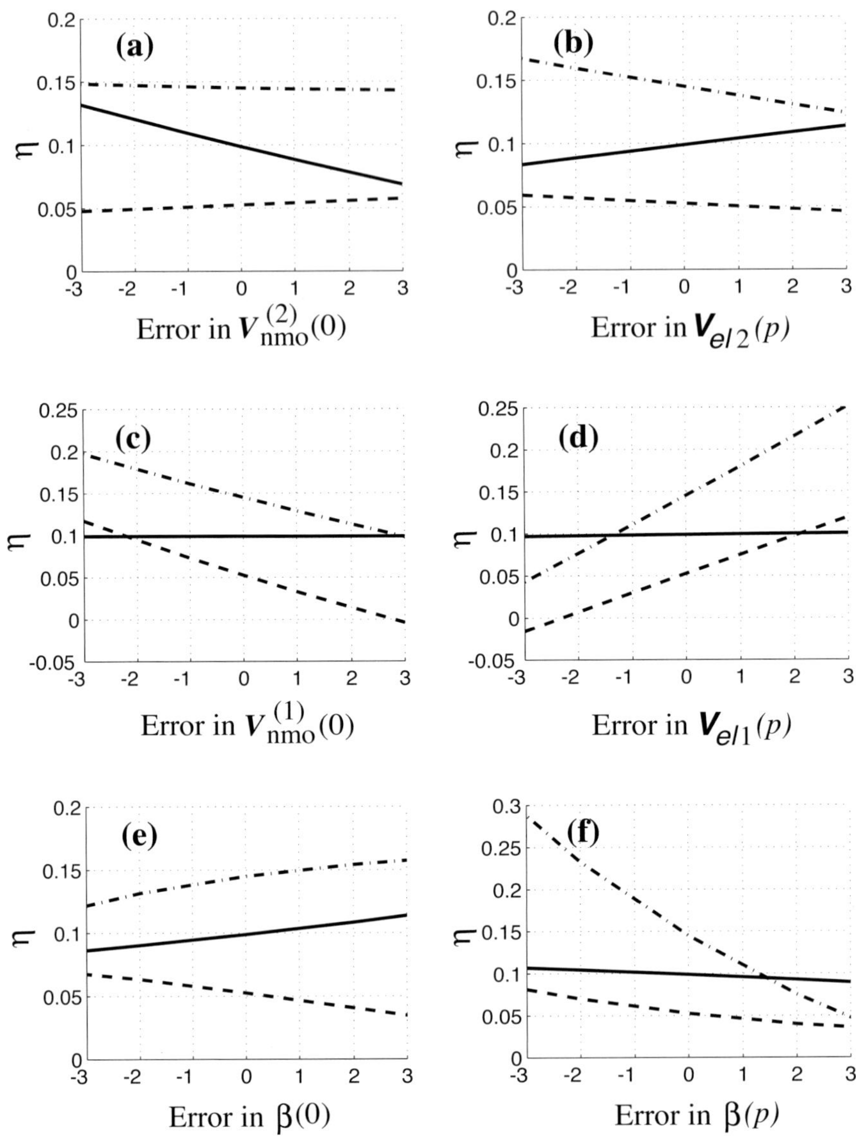

Figure 2.39: Inversion of noise-contaminated P-wave NMO ellipses of horizontal and dipping events for $\eta^{(1)}$ (dashed line), $\eta^{(2)}$ (solid), and $\eta^{(3)}$ (dash-dotted) (Grechka and Tsvankin, 1999b). The dip plane makes an angle of 30° with the $[x_1, x_3]$ symmetry plane, and the dip is 40°. The horizontal axis is the error in the semiaxes of the NMO ellipse (in percent) or in the orientation angle β (in degrees). The left column [plots (a), (c), and (e)] corresponds to the horizontal event, and the right column [(b), (d), and (f)] to the dipping event. The actual parameters are $\eta^{(1)}$=0.05, $\eta^{(2)}$=0.10, and $\eta^{(3)}$=0.15.

$\eta^{(2)}$ controls the major semiaxis of the NMO ellipse (i.e., the dip-line NMO velocity) for a reflector with the dip plane parallel to the x_1-axis. The inversion of the dip-line NMO velocity for $\eta^{(2)}$ in this case involves just $V_{\rm nmo}^{(2)}$ – the zero-dip NMO velocity in the same (x_1) direction. Because for the model in Figure 2.39 the azimuth of the dip plane is relatively close to the x_1-axis, $\eta^{(2)}$ remains primarily responsible for the larger semiaxis $V_{\rm el2}$ of the ellipse for the dipping event and is sensitive to errors in $V_{\rm nmo}^{(2)}$.

If $[x_1, x_3]$ represents the dip plane of the reflector, the parameters $\eta^{(1)}$ and $\eta^{(3)}$ influence just the strike-line NMO velocity (i.e., the smaller semiaxis of the NMO ellipse), and only as the difference $\eta^{(1)} - \eta^{(3)}$ (equation 2.67). In addition, note that the NMO velocity in the strike direction depends on $V_{\rm nmo}^{(1)}$ but does not contain $V_{\rm nmo}^{(2)}$. This explains the results of the inversion for $\eta^{(1)}$ and $\eta^{(3)}$ in Figure 2.39. First, both $\eta^{(1)}$ and $\eta^{(3)}$ are sensitive mostly to errors in the smaller semiaxis $V_{\rm el1}$ of the NMO ellipse for the dipping event, as well as in $V_{\rm nmo}^{(1)}$. Second, the difference $\eta^{(1)} - \eta^{(3)}$ in Figures 2.39c,d is constrained much tighter than either parameter individually.

The influence of errors in the orientation angle β (Figures 2.39e,f) is more difficult to interpret, but it can also be understood in terms of the weak-anisotropy approximation 2.65. Note that the parameter $\eta^{(3)}$ is much more sensitive than are $\eta^{(1)}$ and $\eta^{(2)}$ to errors in the angle β for the dipping event. Because $\eta^{(3)}$ is defined in the horizontal symmetry plane, its influence on NMO velocity for moderate reflector dips is smaller than that of $\eta^{(1)}$ and $\eta^{(2)}$. A more stable estimate of $\eta^{(3)}$ requires the presence of a steeper reflector with the dip plane sufficiently deviating from the vertical symmetry planes. Alternatively, $\eta^{(3)}$ can be obtained from nonhyperbolic moveout inversion of wide-azimuth P-wave data (see Chapter 3).

Overall, if the dip plane of the reflector is closer to the x_1 direction, the parameter most tightly constrained by P-wave moveout data is $\eta^{(2)}$. This is not surprising because $\eta^{(2)}$ is fully responsible for the dip-dependent NMO velocity in the $[x_1, x_3]$-plane. Extensive numerical testing shows that estimation of all three η-parameters requires a minimum reflector dip of about 25°, which agrees with the results of Alkhalifah and Tsvankin (1995) for η-inversion in VTI media.

2.5.3 Inversion for layered media

Layer-stripping methodology

Inversion of NMO ellipses for a horizontally layered orthorhombic medium above a dipping reflector (see Figure 1.3) is based on the generalized Dix differentiation (equation 2.58). The layer-stripping algorithm is similar to that described in the previous section for layered TTI media. It is assumed that dipping reflectors (e.g., a fault plane; the dip does not have to be constant) are present in each depth interval (Figure 2.40).

First, Dix-type differentiation of the effective NMO ellipses for horizontal interfaces yields the interval matrices $\mathbf{W}_\ell$, which are used to estimate the orientation of the symmetry planes [described by the rotation angles $\beta_\ell(0)$] and the interval symmetry-plane NMO velocities $V^{(1)}_{\text{nmo},\ell}$ and $V^{(2)}_{\text{nmo},\ell}$. Note that the generalized Dix equation allows the orientation of the vertical symmetry planes to vary from layer to layer in arbitrary fashion. The results of the single-layer inversion show that resolving all three interval η-parameters requires at least one dipping reflector per layer, with the dip plane deviating sufficiently from the symmetry planes of the layer.

The parameters $\eta_1^{(1)}$, $\eta_1^{(2)}$, and $\eta_1^{(3)}$ in the top layer are estimated by the single-layer algorithm and used to compute the interval matrix $\mathbf{W}_1$ for the horizontal slownesses (time slopes) p_1 and p_2 corresponding to the dipping reflector in the second layer. Then $\mathbf{W}_1$ is combined with the effective matrix $\mathbf{W}(2)$ for that dipping event to obtain the interval matrix $\mathbf{W}_2$ in the second layer from equation 2.58. Using the interval velocities $V^{(1)}_{\text{nmo},2}$ and $V^{(2)}_{\text{nmo},2}$ and the angle β_2 found from the horizontal events, the matrix $\mathbf{W}_2$ is inverted for the parameters $\eta_2^{(1,2,3)}$. The layer-stripping procedure is continued downward, as long as both horizontal and dipping events are available.

Implementation of this algorithm requires computing not only the interval matrices $\mathbf{W}_\ell$, but also the interval traveltimes τ_ℓ as functions of p_1 and p_2. Although P-wave traveltimes in general depend on the velocity V_{P0} and parameters $\epsilon^{(1,2)}$ and $\delta^{(1,2,3)}$, Grechka and Tsvankin (1999b) show that the moveout parameters $V^{(1,2)}_{\text{nmo}}$ and $\eta^{(1,2,3)}$ are sufficient for converting the vertical zero-offset traveltime $\tau_{0,\ell}$ in a horizontal orthorhombic layer (known from horizontal events) into the traveltime along an arbitrary oblique ray.

The principle of this layer-stripping procedure is close to that of the 2D dip-moveout (DMO) inversion for η in VTI media (Alkhalifah and Tsvankin, 1995; Tsvankin, 2005). Here, however, we operate with NMO ellipses influenced by the 3D character of wave propagation in orthorhombic media, rather than with NMO velocities measured in a vertical symmetry plane of the model.

Synthetic example

The algorithm was applied to exact (ray-traced) reflection traveltimes generated for the three-layer orthorhombic model with a throughgoing dipping reflector (e.g., a fault plane) in Figure 2.40. The NMO ellipses were obtained by hyperbolic semblance analysis for spreadlengths close to the distance between the CMP and the reflector.

Even for vertically heterogeneous orthorhombic media, the deviation of the exact traveltimes from the hyperbola parameterized by the analytic NMO velocity become pronounced only for offsets that exceed the CMP-reflector distance (Figure 2.41). The reflected rays used in Figure 2.41 cross three orthorhombic layers with different orientation of the vertical symmetry planes and, therefore, bear the full impact of azimuthal anisotropy and vertical heterogeneity. The magnitude of nonhyperbolic moveout is more significant for the reflection from the horizontal interface, while the traveltimes of the dipping event stay close to the analytic hyperbola up to relatively large offsets. Indeed, if both the spreadlength and CMP-to-reflector distance are fixed, the range of take-off angles for reflected rays decreases with dip, which mitigates the influence of anisotropy and reduces nonhyperbolic moveout (see Chapter 3).

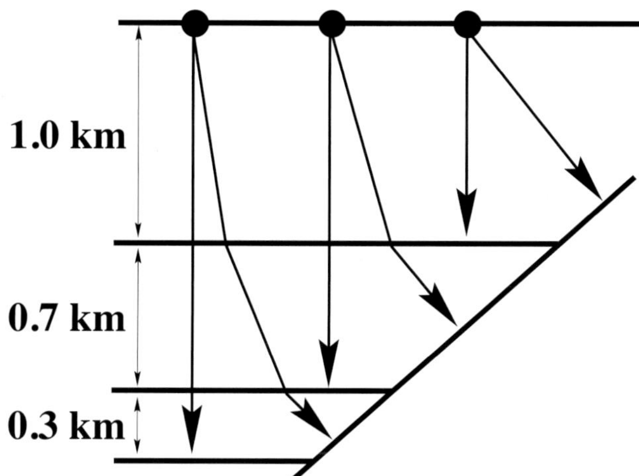

Figure 2.40: Schematic 2D section of a layered orthorhombic model used to test the inversion algorithm (Grechka and Tsvankin, 1999b). The azimuth β of the $[x_1, x_3]$ symmetry plane changes from layer to layer (Table 2.11). The azimuth of the dip plane of the reflector is 30°, and the dip is 40°.

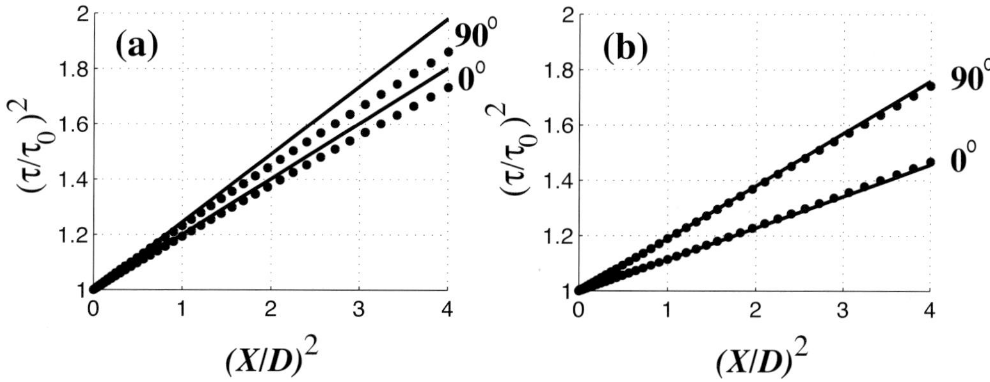

Figure 2.41: Accuracy of the hyperbolic moveout equation for reflections from (a) the deepest horizontal boundary in Figure 2.40 (bottom of layer 3) and (b) the dipping interface in layer 3 (Grechka and Tsvankin, 1999b). P-wave traveltimes computed for two azimuths (0° and 90°) by 3D anisotropic ray tracing (dots) are compared with the hyperbolic moveout curves parameterized by the exact NMO velocity (solid lines). The traveltime τ is normalized by the zero-offset time τ_0; D is the distance between the CMP and the reflector. The model parameters are listed in Table 2.11.

Layer	Actual					
	$V_{\rm nmo}^{(1)}$ (km/s)	$V_{\rm nmo}^{(2)}$ (km/s)	β	$\eta^{(1)}$	$\eta^{(2)}$	$\eta^{(3)}$
1	2.50	2.76	0.0	0.10	0.05	0.15
2	2.90	3.18	−10.0	0.15	0.10	0.15
3	3.20	3.77	−20.0	0.20	0.15	0.20
Layer	Inverted					
	$V_{\rm nmo}^{(1)}$ (km/s)	$V_{\rm nmo}^{(2)}$ (km/s)	β	$\eta^{(1)}$	$\eta^{(2)}$	$\eta^{(3)}$
1	2.54	2.79	0.0	0.07	0.04	0.13
2	2.97	3.23	−8.0	0.12	0.08	0.13
3	3.25	3.83	−18.0	0.18	0.13	0.18

Table 2.11: Comparison of the actual and inverted interval parameters for the three-layer orthorhombic model in Figure 2.40 (Grechka and Tsvankin, 1999b). The NMO ellipses of both horizontal and dipping events were estimated from ray-traced traveltimes; the spreadlength $x_{\rm max}$ for layer 1 is 1.2 km (the spreadlength-to-depth ratio $x_{\rm max}/D$=1.2); for layer 2, $x_{\rm max}$=1.8 km ($x_{\rm max}/D$=1.06); and for layer 3, $x_{\rm max}$=1.8 km ($x_{\rm max}/D$=0.9).

The errors in the effective NMO ellipses (mostly for horizontal events) are marginally higher than those for the single-layer model (Figure 2.38), still allowing the interval parameters to be recovered with sufficiently high accuracy (Table 2.11). The maximum error in the parameters $\eta^{(1,2,3)}$ is slightly over 0.03, with the error in the relatively thin deepest layer limited by just 0.02.

2.6 Summary

In this chapter, the elliptical NMO equation was applied to the inversion of P-wave conventional-spread data from layered, azimuthally anisotropic media. One of the main issues in moveout-based anisotropic parameter estimation is separating the influence of anisotropy and lateral heterogeneity (e.g., in the form of velocity gradients or dipping interfaces) on reflection traveltimes. In the first section we discussed the inversion of NMO ellipses for a stack of horizontal layers with mild lateral velocity variation. The processing sequence included 3D "global" semblance analysis based on the equation of the NMO ellipse, spatial smoothing of the effective NMO ellipses and zero-offset traveltimes, correction for lateral velocity variation, and generalized Dix differentiation. The resulting interval NMO ellipses quantify the magnitude of azimuthal anisotropy (measured by P-wave NMO velocity) within the layer of interest. We demonstrated the effectiveness of this methodology on a fracture-characterization case study from the Powder River Basin, Wyoming. The estimates of depth-varying fracture trends in the survey area are in good agreement with borehole measurements and rotational analysis of four-component shear-wave data.

Another type of lateral heterogeneity, which has a strong influence on normal moveout, is the dip or curvature of layer boundaries. Although structural complexity in general may seriously hamper anisotropic parameter estimation, we presented two examples of successful 2D depth-domain inversion of P-wave reflection traveltimes for TI models with nonhorizontal intermediate interfaces. The Dix-type formalism based on the concept of NMO-velocity surfaces shows that interface dip makes reflection traveltimes in VTI media dependent on the individual values of Thomsen parameters. Synthetic data for a model that includes a VTI layer with a curved lower boundary above an isotropic layer were successfully inverted for the parameters V_{P0}, ϵ, and δ. Information about the vertical velocity and, therefore, the depth scale of the model was contained in reflected rays that crossed the dipping segment of the intermediate interface. It was essential, however, to constrain the inversion procedure by assuming that the bottom layer is isotropic.

We also used a 2D physical-modeling data set to estimate the parameters of a bending TI layer with a tilted symmetry axis (TTI) from P-wave traveltimes. The model represents a simplified version of a typical section from the Alberta Foothills, with the horizontal target reflector beneath a bending TTI thrust sheet. The parameter ϵ was obtained using the traveltimes of a wave that crosses one of the dipping TTI blocks and reflects from the bottom of the model. The moveout inversion also gave an accurate estimate of the thickness of the TTI layer, which helped reconstruct the depth scale of the section. As was the case for the VTI model discussed above, we had to make several assumptions about the model. In particular, the symmetry axis was set orthogonal to the bedding and the TTI layer was assumed to be homogeneous and embedded in isotropic host rock.

Then we presented a 3D method (*stacking-velocity tomography*) designed to invert wide-azimuth, conventional-spread P-wave moveout for the parameters of homogeneous VTI layers separated by plane dipping or curved interfaces. The objective function at each iteration of the algorithm is obtained in two steps. First, for a certain trial set of the interval VTI parameters, P-wave zero-offset traveltimes and reflection slopes (horizontal slownesses) are used to reconstruct the reflectors and build the trial model in depth. Second, we compute effective NMO ellipses for the trial model and define the objective function through the misfit between the modeled ellipses and those obtained from the data. For plane nonintersecting layer boundaries, the parameters V_{P0}, ϵ, and δ cannot be recovered from P-wave moveout alone. Nonetheless, if the reflectors have distinctly different azimuths, a priori knowledge of any single interval parameter is sufficient to overcome the ambiguity and reconstruct the whole model in depth. For example, parameter estimation becomes unique if the uppermost layer is known to be isotropic. To carry out 2D inversion on the dip line of co-oriented reflectors, it is necessary to specify one parameter (e.g., the vertical velocity) of each layer. The inversion is better posed if some of the layers contain additional reflecting boundaries, such as fault planes.

Despite the higher complexity of models with curved interfaces, the increased angle coverage of reflected rays helps resolve the trade-offs between the medium parameters. Singular value decomposition (SVD) shows that in the presence of sufficient interface curvature all parameters needed for anisotropic depth processing can be obtained solely from conventional-spread P-wave moveout. By performing tests on noise-contaminated data, we demonstrated that for a range of VTI models stacking-velocity tomography reconstructs the interfaces and estimates the interval parameters with high accuracy. It should be emphasized, however, that all layers have to be homogeneous with nonelliptical anisotropy (i.e., $\epsilon \neq \delta$). Both SVD analysis and moveout inversion are implemented using an efficient modeling technique based on the theory of NMO-velocity surfaces (see section 1.4) generalized for wave propagation through curved interfaces.

3D moveout inversion for tilted TI media can be performed by combining the P-wave NMO ellipses for two reflectors with different orientation. The parameters V_{P0}, ϵ, and δ of a homogeneous TTI layer, along with the tilt ν and azimuth β of the symmetry axis, can be obtained from the NMO ellipses for a horizontal and a dipping reflector as long as $\nu > 30°$ and the dip varies between 30° and 80°. Another condition required for stable parameter estimation is that the medium not be elliptical (i.e., ϵ cannot be close to δ). Although for mild tilts ν the medium parameters cannot be resolved separately, the NMO-velocity inversion typically provides enough information for building TTI models suitable for time processing (NMO, DMO, time migration). A strong assumption used in the inversion procedure is that the symmetry-axis orientation is the same for horizontal and dipping reflectors. If the symmetry axis is known to be perpendicular to the reflector, the NMO ellipse of a single dipping event can be inverted for V_{P0} and δ; the parameter ϵ, however, can be constrained only by nonhyperbolic moveout. The algorithm was also extended to vertically heterogeneous TTI models with a throughgoing dipping reflector (e.g., a fault plane) using the generalized Dix equation.

For the lower-symmetry orthorhombic medium above a reflector with arbitrary dip, the P-wave NMO ellipse is controlled by the azimuths of the vertical symmetry planes, the symmetry-plane NMO velocities $V_{\rm nmo}^{(1,2)}$ for a horizontal reflector, and the anellipticity parameters $\eta^{(1,2,3)}$. The symmetry-plane azimuths and velocities $V_{\rm nmo}^{(1,2)}$ can be estimated from the NMO ellipse of a horizontal event, while the inversion for $\eta^{(1,2,3)}$ requires the NMO ellipse for a reflector with a dip of at least 25° and dip plane that deviates from both vertical symmetry planes. The interval parameters $V_{\rm nmo}^{(1,2)}$ and $\eta^{(1,2,3)}$ in layered orthorhombic media can be obtained by applying the generalized Dix equation to effective NMO ellipses of horizontal and dipping events. In Chapter 3 we discuss an alternative way of estimating the parameters $V_{\rm nmo}^{(1,2)}$ and $\eta^{(1,2,3)}$ based on nonhyperbolic moveout inversion of long-spread, wide-azimuth data.

The moveout-inversion algorithms in this chapter are developed for models composed of vertically homogeneous layers. Grechka and Tsvankin (2002b) examined the distortions in parameter estimation caused by the intralayer vertical heterogeneity unaccounted for in velocity analysis. To describe P-wave moveout for a horizontal or a dipping reflector overlaid by a *vertically heterogeneous* isotropic medium, the assumed effective *homogeneous* overburden has to be anisotropic.[4] For such a model, combining the P-wave NMO velocity of a horizontal event with the vertical velocity always results in nonnegative values of the effective parameter δ. Also, the inversion of the P-wave NMO ellipse for a dipping reflector yields a nonnegative effective parameter η, which increases with dip. As shown by Grechka and Tsvankin (2002b), for truly anisotropic (VTI) media, the influence of vertical heterogeneity between medium interfaces can lead to a bias toward larger values of the parameters δ and η estimated from P-wave moveout.

[4]This apparent anisotropy is caused not only by typical velocity gradients, but also by rapid velocity variations routinely observed in well logs.

Appendices for Chapter 2

2A NMO ellipse in a layer with weak lateral velocity variation

Here, we follow Grechka and Tsvankin (1999a) in developing a correction for the influence of weak lateral velocity variation (for brevity, we will call it lateral heterogeneity, or LH) on the NMO ellipse in a single horizontal anisotropic layer. The general 3D NMO equation 1.3 parameterized by the matrix $\mathbf{W}$ (equation 1.4) accounts for both anisotropy and lateral heterogeneity, unless deviations from hyperbolic moveout make the very notion of normal-moveout velocity meaningless. Our goal is to find the relationship between the matrix $\mathbf{W}^{\text{het}}$, responsible for the NMO ellipse in the LH layer, and the matrix $\mathbf{W}^{\text{hom}}$ for a reference homogeneous medium.

Assuming that lateral velocity variation is weak and following the approach developed by Grechka (1998), we apply first-order perturbation theory to express the matrix $\mathbf{W}^{\text{het}}$ as the sum of $\mathbf{W}^{\text{hom}}$ and the contribution (assumed to be small) of LH. We restrict ourselves to anisotropic models with a horizontal symmetry plane (i.e., the medium can be transversely isotropic, orthorhombic, or monoclinic), which imposes additional constraints on the character of lateral velocity variation: While each elastic constant can vary laterally in a different fashion, it is assumed that these spatial variations do not destroy the symmetry of the phase- and group-velocity surfaces with respect to the horizontal plane.

Using equation 1.4 for a common midpoint (CMP) located at $\mathbf{y} = \{y_1, y_2\} = 0$ (Figure 2.42), we can write

$$W_{ij}^{\text{het}} = \tau_0^{\text{het}} \left. \frac{\partial^2 \tau^{\text{het}}}{\partial x_i \partial x_j} \right|_{\mathbf{x}=0}, \qquad (i, j = 1, 2) \qquad (2.70)$$

and

$$W_{ij}^{\text{hom}} = \tau_0^{\text{hom}} \left. \frac{\partial^2 \tau^{\text{hom}}}{\partial x_i \partial x_j} \right|_{\mathbf{x}=0}, \qquad (i, j = 1, 2), \qquad (2.71)$$

where τ^{het} and τ^{hom} are the one-way traveltimes from the zero-offset reflection point to the surface location $\mathbf{x}$, and $\tau_0^{\text{het}} = \tau^{\text{het}}(\mathbf{x}=0)$ and $\tau_0^{\text{hom}} = \tau^{\text{hom}}(\mathbf{x}=0)$ are the zero-offset traveltimes; τ^{hom} is obtained for a reference homogeneous layer with the parameters corresponding to those at the CMP location. The one-way traveltimes from the zero-offset reflection point appear in equations 2.70 and 2.71 because reflection-point dispersal has no influence on normal-moveout velocity, even if the medium is anisotropic and heterogeneous (Hubral and Krey, 1980; Tsvankin, 2005).

Since the layer is horizontal and has a horizontal symmetry plane, in the absence of lateral heterogeneity the zero-offset reflection point $\mathbf{R}$ coincides with the projection of the CMP onto the boundary and has the coordinates $\{0, 0, x_3\}$ (Figure 2.42). The

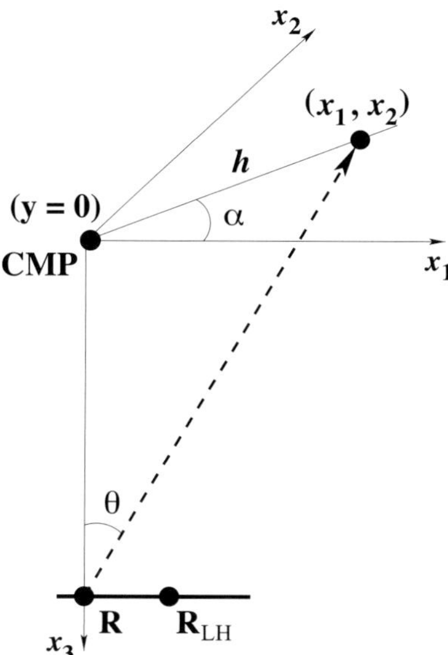

Figure 2.42: In the derivation of the NMO velocity for a layer with a weak lateral velocity variation, the actual reflection raypath in CMP geometry can be replaced with a nonspecular raypath (dashed line) traced through the zero-offset reflection point **R** in the reference homogeneous medium (Grechka and Tsvankin, 1999a). The zero-offset reflection point in the laterally heterogeneous layer is denoted by $\mathbf{R}_{\mathrm{LH}}$.

traveltime τ^{hom} in the homogeneous layer is given by

$$\tau^{\mathrm{hom}}(x_1, x_2) = \tau^{\mathrm{hom}}(h, \alpha) = \frac{\sqrt{h^2 + x_3^2}}{\mathcal{G}(\alpha, \theta)}, \tag{2.72}$$

where x_1 and x_2 are the receiver coordinates, $h = \sqrt{x_1^2 + x_2^2}$ is half the source-receiver offset, and $\mathcal{G}(\alpha, \theta)$ is the group velocity as a function of the polar angle $\theta = \tan^{-1}(h/x_3)$ and of the azimuth of the CMP line $\alpha = \tan^{-1}(x_2/x_1)$.

In general, lateral velocity variation leads to a shift of the zero-offset reflection point from **R** to a new location $\mathbf{R}_{\mathrm{LH}}$ (Figure 2.42). Nonetheless, since LH is assumed to be weak (i.e., quadratic and higher-order terms in the spatial derivatives of the elastic constants can be ignored), the traveltime changes caused by lateral velocity variation can be computed along the *unperturbed* ray propagating in the homogeneous model (Grechka and McMechan, 1997). Hence, we obtain the traveltime τ^{het} as an integral along the nonspecular raypath originated at point **R**:

$$\tau^{\mathrm{het}}(x_1, x_2) = \int_0^{\sqrt{h^2+x_3^2}} \frac{d\zeta}{\mathcal{G}(\alpha, \theta, \zeta)}, \tag{2.73}$$

where the group velocity $\mathcal{G}$ now depends on the distance ζ along the ray. Introducing the horizontal displacement ξ along the ray, the traveltime can be rewritten as

$$\tau^{\text{het}}(x_1, x_2) = \frac{\sqrt{h^2 + x_3^2}}{h} \int_0^h \frac{d\xi}{\mathcal{G}(\alpha, \theta, y_1(\xi), y_2(\xi))}; \tag{2.74}$$

$$y_1 = \xi \cos\alpha\,, \quad y_2 = \xi \sin\alpha\,. \tag{2.75}$$

Assuming that the lateral variation in group velocity is sufficiently smooth, $\mathcal{G}$ can be expanded in a double Taylor series in the vicinity of the common midpoint $y_1 = y_2 = 0$:

$$\mathcal{G}(\alpha, \theta, \xi) = \mathcal{G}(\alpha, \theta, y_1, y_2) = \mathcal{G}_0 \left[1 + \frac{1}{\mathcal{G}_0} \sum_{i=1}^{2} \mathcal{G}_{,i}\, y_i + \frac{1}{2\mathcal{G}_0} \sum_{i,j=1}^{2} \mathcal{G}_{,ij}\, y_i\, y_j + \ldots \right], \tag{2.76}$$

where

$$\mathcal{G}_0 \equiv \mathcal{G}(\alpha, \theta, 0, 0)\,, \tag{2.77}$$

$$\mathcal{G}_{,i} \equiv \left. \frac{\partial \mathcal{G}(\alpha, \theta, y_1, y_2)}{\partial y_i} \right|_{y_1 = y_2 = 0}, \tag{2.78}$$

$$\mathcal{G}_{,ij} \equiv \left. \frac{\partial^2 \mathcal{G}(\alpha, \theta, y_1, y_2)}{\partial y_i \partial y_j} \right|_{y_1 = y_2 = 0}. \tag{2.79}$$

Because lateral velocity variation is weak, all terms involving y_i in equation 2.76 are small compared to unity. Keeping only linear terms in the spatial derivatives $\mathcal{G}_{,i}$ and $\mathcal{G}_{,ij}$, we find $1/\mathcal{G}$ as

$$\frac{1}{\mathcal{G}(\alpha, \theta, y_1, y_2)} = \frac{1}{\mathcal{G}_0} \left[1 - \frac{1}{\mathcal{G}_0} \sum_{i=1}^{2} \mathcal{G}_{,i}\, y_i - \frac{1}{2\mathcal{G}_0} \sum_{i,j=1}^{2} \mathcal{G}_{,ij}\, y_i y_j \right]. \tag{2.80}$$

Substituting equations 2.75 and 2.80 into equation 2.74 and evaluating the integral yields

$$\begin{aligned} \tau^{\text{het}}(x_1, x_2) = {} & \frac{\sqrt{h^2 + x_3^2}}{\mathcal{G}_0} \left[1 - \frac{h}{2\mathcal{G}_0} \Big(\mathcal{G}_{,1} \cos\alpha + \mathcal{G}_{,2} \sin\alpha \Big) \right. \\ & \left. - \frac{h^2}{6\mathcal{G}_0} \Big(\mathcal{G}_{,11} \cos^2\alpha + 2\mathcal{G}_{,12} \sin\alpha \cos\alpha + \mathcal{G}_{,22} \sin^2\alpha \Big) \right]. \end{aligned} \tag{2.81}$$

The term in front of the brackets in equation 2.81 is the traveltime $\tau^{\text{hom}}(x_1, x_2)$ (see equation 2.72). Using the relations $x_1 = h \cos\alpha$ and $x_2 = h \sin\alpha$ (Figure 2.42), equation 2.81 can be rewritten in the form

$$\begin{aligned} \tau^{\text{het}}(x_1, x_2) = {} & \tau^{\text{hom}}(x_1, x_2) \left[1 - \frac{1}{2\mathcal{G}_0} \Big(\mathcal{G}_{,1} x_1 + \mathcal{G}_{,2} x_2 \Big) \right. \\ & \left. - \frac{1}{6\mathcal{G}_0} \Big(\mathcal{G}_{,11} x_1^2 + 2\mathcal{G}_{,12} x_1 x_2 + \mathcal{G}_{,22} x_2^2 \Big) \right]. \end{aligned} \tag{2.82}$$

Equation 2.82 expresses the contribution of the lateral velocity variation to the one-way traveltime in terms of the spatial derivatives of the group-velocity function. Note that in our approximation the zero-offset ray in the LH layer remains vertical, and

$$\tau_0^{\rm het} = \tau_0^{\rm hom}\,. \tag{2.83}$$

The matrix $\mathbf{W}^{\rm het}$ defined by equation 2.70 depends on the second-order derivatives of the traveltime $\tau^{\rm het}$ (equation 2.82) with respect to the coordinates (x_1, x_2) at zero offset:

$$\left.\frac{\partial^2 \tau^{\rm het}}{\partial x_i\, \partial x_j}\right|_{\mathbf{x}=0} = \left.\frac{\partial^2 \tau^{\rm hom}}{\partial x_i\, \partial x_j}\right|_{\mathbf{x}=0} - \left.\frac{\partial \tau^{\rm hom}}{\partial x_i}\right|_{\mathbf{x}=0} \frac{\mathcal{G}_{,j}}{2\mathcal{G}_0} - \left.\frac{\partial \tau^{\rm hom}}{\partial x_j}\right|_{\mathbf{x}=0} \frac{\mathcal{G}_{,i}}{2\mathcal{G}_0}$$

$$- \tau_0^{\rm hom} \left.\left[\frac{\partial}{\partial x_i}\left(\frac{\mathcal{G}_{,j}}{2\mathcal{G}_0}\right) + \frac{\partial}{\partial x_j}\left(\frac{\mathcal{G}_{,i}}{2\mathcal{G}_0}\right) + \frac{\mathcal{G}_{,ij}}{3\mathcal{G}_0}\right]\right|_{\mathbf{x}=0}. \tag{2.84}$$

The second and third terms on the right-hand side of equation 2.84 depend on the horizontal components of the slowness vector $\mathbf{p}$ in the homogeneous layer:

$$\frac{\partial \tau^{\rm hom}}{\partial x_i} \equiv p_i\,. \tag{2.85}$$

At zero offset, $p_1|_{\mathbf{x}=0} = p_2|_{\mathbf{x}=0} = 0$ because the slowness vector of the zero-offset ray is vertical (orthogonal to the horizontal reflector), so the terms containing $\partial\tau^{\rm hom}/\partial x_i$ and $\partial\tau^{\rm hom}/\partial x_j$ go to zero. The spatial derivatives of $\mathcal{G}_{,1}/(2\mathcal{G}_0)$ and $\mathcal{G}_{,2}/(2\mathcal{G}_0)$ evaluated at $\mathbf{x} = 0$ in equation 2.84 vanish as well. Indeed, $\mathcal{G}_0$ is an even function of x_1 (for $x_2 = 0$) and x_2 (for $x_1 = 0$) because group velocity in a medium with a horizontal symmetry plane is symmetric with respect to the vertical direction in any vertical plane. Although the group-velocity derivatives with respect to y_i ($\mathcal{G}_{,1}$ and $\mathcal{G}_{,2}$) vary with the spatial position, they remain even functions of the polar angle θ and, therefore, of x_1 and x_2 (equation 2.78). Thus, equation 2.84 reduces to

$$\left.\frac{\partial^2 \tau^{\rm het}}{\partial x_i \partial x_j}\right|_{\mathbf{x}=0} = \left.\frac{\partial^2 \tau^{\rm hom}}{\partial x_i \partial x_j}\right|_{\mathbf{x}=0} - \tau_0^{\rm hom} \left.\frac{\mathcal{G}_{,ij}}{3\mathcal{G}_0}\right|_{\mathbf{x}=0}. \tag{2.86}$$

The group velocity $\mathcal{G}_0$ and its spatial derivatives $\mathcal{G}_{,ij}$ in equation 2.86 are computed for the vertical ray traced from the zero-offset reflection point. Since in a medium with a horizontal symmetry plane the vertical phase and group velocities are equal to each other, we can replace $\mathcal{G}_0$ and $\mathcal{G}_{,ij}$ with the vertical phase velocity V_0 and its derivatives with respect to y_i. Finally, multiplying equation 2.86 with $\tau_0^{\rm het} = \tau_0^{\rm hom} = \tau_0$ (see equation 2.83), we obtain the following relationship between the matrices $\mathbf{W}^{\rm hom}$ and $\mathbf{W}^{\rm het}$:

$$W_{ij}^{\rm het} = W_{ij}^{\rm hom} - \left.\frac{\tau_0^2}{3V_0} \frac{\partial^2 V_0}{\partial y_i \partial y_j}\right|_{\mathbf{y}=0}, \qquad (i, j = 1, 2)\,. \tag{2.87}$$

It should be mentioned that equation 2.76 can be replaced with a direct expansion of the "group slowness" $1/\mathcal{G}$ in y_1 and y_2. Then the final result of the appendix has a form similar to equation 2.87 but the correction term involves the vertical slowness $1/V_0$, which simplifies the transition to equation 2.3 in the main text. It is more physically intuitive, however, to express the contribution of lateral heterogeneity through the spatial velocity derivatives.

2B NMO velocity in homogeneous TTI media: Special cases

2B.1 Normal moveout for steep reflectors

Generation of a specular zero-offset reflection requires that a segment of the incident be parallel to the reflecting interface (i.e., the phase-velocity vector for this segment has to be orthogonal to the reflector). If the medium above the reflector is homogeneous and has a horizontal plane of mirror symmetry, the wavefront of the downgoing wave contains the full range of phase angles with the vertical. As a result, specular reflections in a homogeneous layer with a horizontal symmetry plane (such as VTI or HTI) exist for the whole range of dips from 0° to 90° (Tsvankin, 1997c, 2005).

A tilt of the symmetry axis, however, makes the wavefront asymmetric with respect to the horizontal plane, and the portion of the wavefront propagating toward the reflector may contain only a limited range of phase angles. Tsvankin (1997c, 2005) studied the existence of specular reflections for 2D TTI models with the symmetry axis confined to the dip plane of the reflector. As illustrated in Figure 2.43, if the symmetry axis is tilted towards the reflector and $\epsilon > 0$, the maximum phase angle in the segment of the wavefront approaching the reflector ($\theta_{\max}$) is smaller than 90°. Steep interfaces with the dip $\phi > \theta_{\max}$ reflect all incident rays downward and, therefore, become invisible on surface reflection data. If the medium is heterogeneous and velocity increases with depth, the "missing" dips may produce reflected arrivals at the surface, but these reflections would represent *turning* rays. In contrast, for the symmetry axis tilted away from the reflector, it is possible to record reflections in homogeneous media even from overhang structures with dips exceeding 90°.

Grechka and Tsvankin (2000) treated a more general situation of the symmetry axis making an arbitrary azimuthal angle with the dip plane of the reflector. Figure 2.44 shows the maximum dip that generates a zero-offset reflected ray as a function of the azimuth of the symmetry axis for several TTI models. For the maximum dip, the zero-offset ray is horizontal, and the NMO velocity is infinite, which means that the NMO ellipse degenerates into two parallel straight lines (Grechka and Tsvankin, 1998a).

If the symmetry axis is confined to the dip plane and points towards the reflector ($\beta = 0°$), the maximum dip for the model with $\nu = 25°$, $\epsilon = 0.25$, and $\delta = 0.05$ is only 76° (Figure 2.44b), which agrees with the result of Tsvankin (1997c, his Figure 3). As

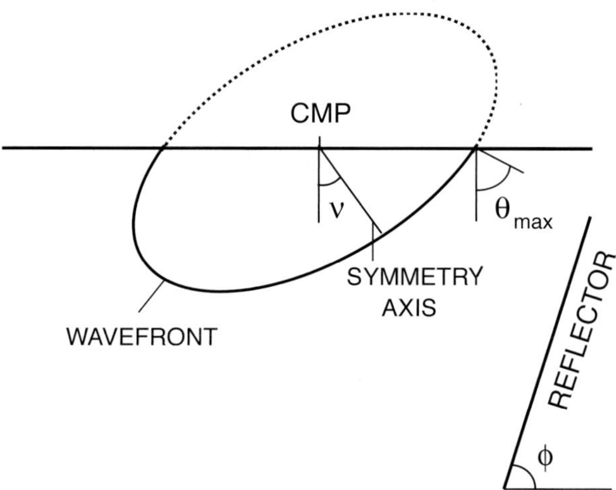

Figure 2.43: P-wavefront for a TI medium with the symmetry axis tilted towards the reflector and $\epsilon > 0$ (Tsvankin, 1997c). The increase in phase and group velocity away from the symmetry axis reduces the angular range of the wavefront normals in the segment of the wavefront propagating towards the reflector. The maximum phase (wavefront) angle in the lower right quadrant is $\theta_{\text{max}} < 90°$. Note that the maximum phase angle in the lower left quadrant (not marked) exceeds 90°.

the symmetry axis deviates from the dip plane, the range of "missing" dips becomes more narrow and vanishes altogether for $\beta = 90°$, when the horizontal projection of the symmetry axis coincides with the reflector strike. For $\beta > 90°$, the maximum dip exceeds 90°, so it is possible to record surface reflections from overhang structures. When the symmetry axis is in the dip plane but tilted *away* from the reflector ($\beta = 180°$), the maximum dip may be larger than 100° (Figures 2.44b,c).

As discussed by Tsvankin (1997c, 2005), it does not take a large tilt of the symmetry axis for the maximum dip to deviate considerably from 90°. Due to the asymmetric shape of the P-wavefront with respect to 45°, the smallest θ_{max} usually corresponds to tilt angles of $20° - 35°$. Figure 2.44 also shows that for typical TI models the maximum dip is more sensitive to the parameter ϵ than it is to δ.

2B.2 Elliptical anisotropy

A TI model with any orientation of the symmetry axis becomes elliptically anisotropic if $\epsilon = \delta$. Then the P-wave slowness surface and wavefront (group-velocity surface) have an ellipsoidal shape, while the slowness surface and wavefront of SV-waves are spherical because the SV-wave velocity is equal to V_{S0} in all directions. Although the condition $\epsilon = \delta$ is uncommon for subsurface rocks (Thomsen, 1986; Tsvankin, 2005), moveout inversion is sensitive to the proximity of the model to elliptical.

If the symmetry axis of an elliptical medium is confined to the dip plane of the reflector, the dip-line NMO velocity represents the same function of the ray parameter

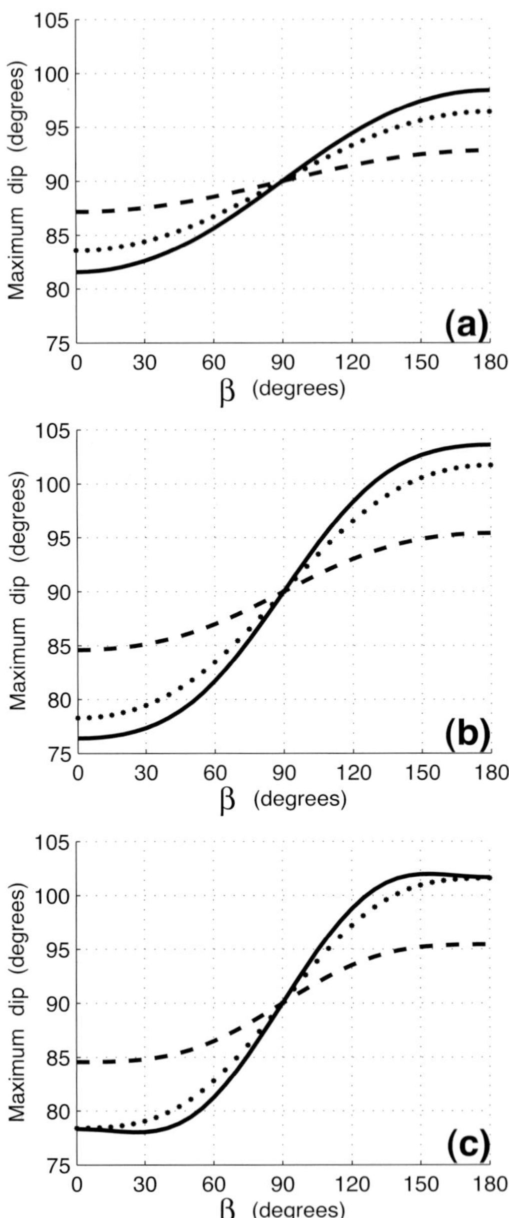

Figure 2.44: Dip of the steepest reflector that generates a zero-offset P-wave reflection in a homogeneous TTI layer (Grechka and Tsvankin, 2000). The angle β is the azimuth of the symmetry axis with respect to the dip plane of the reflector. For β=0°, the symmetry axis points towards the reflector; for β=180° it is tilted away from the reflector. The tilt of the symmetry axis is (a) ν=10°, (b) ν=25°, and (c) ν=40°. The solid curves are computed for ϵ=0.25 and δ=0.05; dotted for ϵ=0.25 and δ=0.15; and dashed for ϵ=0.10 and δ=0.05.

p of the zero-offset ray as in isotropic media (Tsvankin, 2005):

$$V_{\text{nmo}}(p) = \frac{V_{\text{nmo}}(0)}{\sqrt{1 - p^2 V_{\text{nmo}}^2(0)}} \,. \tag{2.88}$$

Analysis of equation 2.45 shows that if $\epsilon = \delta$, the dependence of the P-wave NMO ellipse on p_1 and p_2 remains isotropic for arbitrary orientations of the symmetry axis and reflector normal. The P-wave phase-velocity function for elliptical anisotropy is given by

$$V_P(\theta) = V_{P0} \sqrt{1 + 2\delta \sin^2 \theta} \,, \tag{2.89}$$

where θ is the angle between the phase-velocity (slowness) vector and the symmetry axis. Representing θ through the slowness vector $\mathbf{p} = [p_1, p_2, q]$ and the symmetry-axis orientation (defined by the unit vector $\mathbf{a}$ or the angles ν and β) yields

$$\cos\theta = V_P\,(\mathbf{p} \cdot \mathbf{a}) = V_P\,(p_1 \sin\nu \cos\beta + p_2 \sin\nu \sin\beta + q \cos\nu) \,. \tag{2.90}$$

Combining equations 2.89 and 2.90 and taking into account that $V_P^2 = (p_1^2 + p_2^2 + q^2)^{-1}$ leads to a quadratic equation for q as a function of p_1 and p_2. Substituting $q(p_1, p_2)$ into equation 2.45, we find the following equation for the matrix $\mathbf{W}$:

$$W_{ij}(p_1, p_2) = W_{ij}(0, 0) - p_i\, p_j \,. \tag{2.91}$$

Equation 2.91 shows that the dependence of W_{ij} on p_1 and p_2 is identical to that for isotropic media. The influence of elliptical anisotropy on the NMO velocity in equation 2.91 is hidden in the NMO ellipse $W_{ij}(0, 0)$ for a horizontal event. Implications of this result for the inversion procedure are discussed in the main text.

The isotropic form of the function $W_{ij}(p_1, p_2)$ also means that conventional dip-moveout algorithms developed for isotropic media are valid for elliptical anisotropy with any symmetry-axis orientation. Since reflection moveout in homogeneous elliptical media is purely hyperbolic (Uren et al., 1990), the diffraction curve on the zero-offset section (i.e., in the poststack domain) also has the same shape as in isotropic media. Hence, the poststack time-migration impulse response for elliptical anisotropy can be computed with isotropic algorithms using the correct (anisotropic) NMO velocity in any given azimuth. Nevertheless, if the symmetry axis is tilted, both the NMO velocity and migration impulse response in elliptical media vary with azimuth, which should be taken into account in the moveout correction and time migration. Dellinger and Muir (1988) describe a sequence of stretching operations that can be applied to velocity surfaces to correct for elliptical anisotropy in seismic imaging (provided that the medium parameters are known).

The above conclusions, as well as equation 2.91, are entirely valid for SH-waves because their anisotropy in any TI medium is elliptical.

2B.3 Weakly anisotropic TTI layer

The matrix $\mathbf{W}$ (equation 2.45) that determines the NMO ellipse can be linearized with respect to the parameters ϵ and δ under the assumption of weak anisotropy

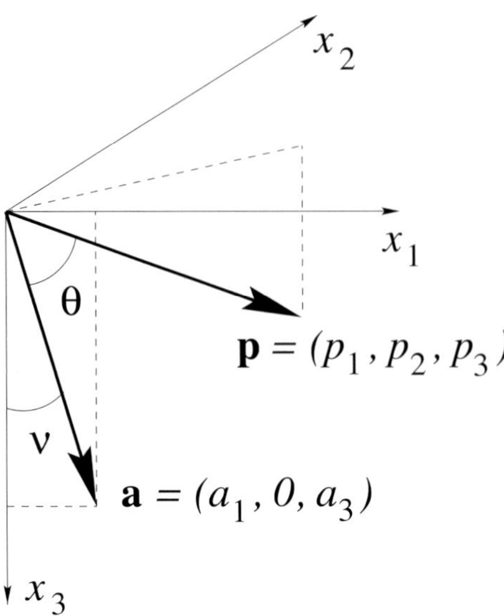

Figure 2.45: In the derivation of the weak-anisotropy approximation for the NMO ellipse in a TTI layer, the symmetry axis (described by the unit vector **a**) is assumed to be confined to the $[x_1, x_3]$-plane (Grechka and Tsvankin, 2000). The slowness vector **p** of the zero-offset ray is orthogonal to the reflector and makes the angle θ with the symmetry axis.

($|\epsilon| \ll 1$ and $|\delta| \ll 1$). Without losing generality, we assume that the symmetry axis described by the unit vector **a** lies in the coordinate plane $[x_1, x_3]$ (Figure 2.45). Then

$$\mathbf{a} = [a_1, 0, a_3] = [\sin\nu, 0, \cos\nu] . \tag{2.92}$$

To obtain the elements of the matrix **W**, we need to solve the Christoffel equation for the vertical slowness $q \equiv p_3$ as a function of the horizontal slowness components p_1 and p_2 of the zero-offset ray. The Christoffel equation for a TI medium with the symmetry axis in the x_3-direction has the form

$$\Gamma \equiv (c_{11}s^2 + c_{44}c^2 - 1)(c_{44}s^2 + c_{33}c^2 - 1) - (c_{13} + c_{44})^2 s^2 c^2 = 0 , \tag{2.93}$$

where c_{ij} is the stiffness matrix, $s = |\mathbf{p}|\sin\theta$, and $c = |\mathbf{p}|\cos\theta$; θ is the angle between the slowness vector **p** and the symmetry axis **a**. Expressing s and c through the components of the vectors **p** and **a**, we find

$$s^2 \equiv [\mathbf{a}\times\mathbf{p}]\cdot[\mathbf{a}\times\mathbf{p}] = p_2^2 + (a_3 p_1 - a_1 q)^2 ,$$

and

$$c^2 \equiv (\mathbf{a}\cdot\mathbf{p})^2 = (a_1 p_1 + a_3 q)^2 .$$

Next, we replace the stiffnesses with the Thomsen parameters defined with respect to the symmetry axis:

$$c_{33} = \rho V_{P0}^2 , \quad c_{44} = \rho V_{S0}^2 , \quad c_{11} = \rho V_{P0}^2 (1 + 2\epsilon) ,$$

$$c_{13} = \rho \sqrt{(V_{P0}^2 - V_{S0}^2)\left[V_{P0}^2(1+2\delta) - V_{S0}^2\right]} - \rho V_{S0}^2 \,; \tag{2.94}$$

ρ is the density.

The linearized solution of the Christoffel equation 2.93 for the vertical slowness q can be represented as the sum of its value $\tilde{q}$ for isotropic media and the anisotropic correction term Δq:

$$q \equiv p_3 = \tilde{q} + \Delta q \,. \tag{2.95}$$

For P-waves, the vertical slowness in an isotropic medium with velocity V_{P0} is given by

$$\tilde{q} = \sqrt{\frac{1}{V_{P0}^2} - p_1^2 - p_2^2} \,. \tag{2.96}$$

Δq can be considered as the linear term in a Taylor series expansion of q in ϵ and δ for fixed horizontal slownesses p_1 and p_2:

$$\Delta q = -\frac{1}{\partial\Gamma/\partial p_3}\left(\frac{\partial\Gamma}{\partial\epsilon}\epsilon + \frac{\partial\Gamma}{\partial\delta}\delta\right), \tag{2.97}$$

with the partial derivatives obtained by differentiating the Christoffel equation 2.93. Combining equations 2.96 and 2.97 yields $q = \tilde{q} + \Delta q$ (equation 2.95) as a function of p_1 and p_2. Then the vertical slowness $q(p_1, p_2)$, along with its derivatives $q_{,i} = \partial q/\partial p_i$ and $q_{,ij} = \partial^2 q/(\partial p_i \partial p_j)$, is substituted into equation 2.45 for the matrix $\mathbf{W}$.

Further linearization of $\mathbf{W}$ in ϵ and δ using symbolic software Mathematica™ leads to the following final result:

$$W_{11} = \frac{1}{V_{P0}^2}\left[1 - 2\delta + 2\epsilon a_1^2 - 14\,(\epsilon-\delta)\,a_1^2\,a_3^2\right] - p_1^2 + \hat{W}_{11}(\epsilon-\delta)\,, \tag{2.98}$$

$$W_{12} = -\,p_1\,p_2 + \hat{W}_{12}(\epsilon-\delta)\,, \tag{2.99}$$

and

$$W_{22} = \frac{1}{V_{P0}^2}\left[1 - 2\delta - 2\,(\epsilon-\delta)\,a_1^2\,(1+a_3^2)\right] - p_2^2 + \hat{W}_{22}(\epsilon-\delta)\,. \tag{2.100}$$

Here, $y_1 = p_1^2 V_{P0}^2$, $y_2 = p_2^2 V_{P0}^2$, and

$$\begin{aligned}
\hat{W}_{11} &= 2p_1^2(-6 + 9y_1 - 4y_1^2)\,(1 - 8a_1^2a_3^2) \\
&+ 8a_1a_3p_1\tilde{q}\,(1-2a_1^2)\,(3 - 7y_1 + 4y_1^2) + 2p_2^4a_3^4V_{P0}^2(1-4y_1) \\
&+ 4p_2^2[a_3^2\,(1-4a_1^2)\,(-1 + 5y_1 + 4y_1^2) + 2a_1a_3^3p_1\tilde{q}\,V_{P0}^2(-3+4y_1)]\,,
\end{aligned} \tag{2.101}$$

$$\begin{aligned}
\hat{W}_{12} &= 8p_1p_2\,\{-y_2^2a_3^4 + y_2a_3^2[2 - 5a_1^2 - 2y_1(1-4a_1^2)] \\
&\qquad -a_3^2(1-4a_1^2) + y_1\,(2 - a_1^2 - 12a_1^2a_3^2) - y_1^2\,(1-8a_1^2a_3^2)\} \\
&+ 8a_1a_3p_2\tilde{q}\,\{a_3^2[1 - y_2(1-4y_1)] - y_1(5-8a_1^2) + 4y_1^2(1-2a_1^2)\}\,,
\end{aligned} \tag{2.102}$$

$$\begin{aligned}
\hat{W}_{22} &= -\,4p_1^2a_3^2(1-4a_1^2) + 2p_1^2\,y_1 - 16p_1^2y_1a_1^2a_3^2 \\
&+ 8a_1a_3p_1\tilde{q}\,\{a_3^2 - y_1(1-2a_1^2) - y_2[5a_1^2 - 4y_1(1-2a_1^2)]\,\} \\
&+ 4p_2^2[-3a_3^4 + 5y_1a_3^2(1-4a_1^2) - 2y_1^2(1-8a_1^2a_3^2)\,] \\
&+ 2p_2^2\,y_2a_3^2\,[9a_3^2 - 8y_1(1-4a_1^2) + 16a_1a_3p_1\tilde{q}\,V_{P0}^2\,] - 8p_2^2y_2^2a_3^4\,.
\end{aligned} \tag{2.103}$$

Chapter 3

Nonhyperbolic moveout analysis of wide-azimuth P-wave data

In conventional seismic data processing, reflection traveltime is often assumed to be described by a hyperbolic equation with the quadratic term determined by NMO velocity. However, the influence of heterogeneity (either lateral or vertical) and non-elliptical anisotropy causes deviations from hyperbolic moveout, which typically become substantial for offset-to-depth ratios exceeding unity. Practical difficulties in working with long-spread data and insufficient understanding of nonhyperbolic moveout often force seismic processors to mute out the large-offset portion of the moveout curve. Nonetheless, advances in acquisition technology have significantly increased the range of source-receiver offsets in 3D surveys, and nonhyperbolic moveout has proved useful in such applications as anisotropic parameter estimation, suppression of multiples, and large-angle AVO (amplitude-variation-with-offset) analysis. Among the first to recognize the benefits of employing nonhyperbolic moveout in interval anisotropic velocity parameter estimation was Sena (1991), who developed traveltime approximations for multilayered VTI and HTI media based upon the so-called "skewed" hyperbolic moveout formulation of Byun et al. (1989). A detailed overview of nonhyperbolic moveout analysis for pure (PP and SS) modes and PS-waves in layered VTI models can be found in Tsvankin (2005).

This chapter discusses the influence of azimuthal anisotropy on nonhyperbolic moveout and introduces an efficient moveout-inversion algorithm for wide-azimuth P-wave data from orthorhombic and HTI media. We start with a brief description of nonhyperbolic moveout equations for layered anisotropic models. Deviation from hyperbolic moveout for pure reflected waves is largely governed by the quartic coefficient A_4 of the traveltime series $t^2(x^2)$. We present an exact expression for A_4 that takes into account reflection-point dispersal in CMP gathers and is valid for arbitrarily anisotropic, heterogeneous media.

First, this equation for the quartic moveout coefficient is applied to P-waves reflected from the bottom of a dipping transversely isotropic layer with the symmetry axis confined to the dip plane. Using the weak-anisotropy approximation, we simplify the coefficient A_4 and examine its dependence on the anisotropy parameters, tilt of the symmetry axis, and reflector dip. In particular, A_4 is proportional to the parameter η and vanishes for elliptical models with any symmetry-axis direction.

For an orthorhombic layer with a horizontal symmetry plane, the P-wave quartic moveout coefficient is controlled by the anellipticity parameters $\eta^{(1)}$, $\eta^{(2)}$, and $\eta^{(3)}$. The analytic solution, supported by numerical examples, describes the complicated, multilobe shape of the azimuthally varying coefficient A_4 for both horizontal and dipping reflectors. Whereas the magnitude of A_4 on dip lines typically becomes small for dips exceeding 45°, nonhyperbolic moveout near the strike direction can remain significant even for subvertical reflectors.

To carry out nonhyperbolic moveout analysis of wide-azimuth P-wave data from horizontally layered orthorhombic and HTI media, we develop a semblance operator based on a generalized version of the Alkhalifah-Tsvankin moveout equation. By decoupling the orientation of the NMO ellipse from the principal azimuths of the quartic moveout coefficient, this equation can be adapted even for models with depth-varying azimuths of the vertical symmetry planes. Application of the algorithm to wide-azimuth P-wave data acquired above a fractured reservoir at Weyburn field in Canada illustrates the potential of nonhyperbolic moveout analysis in fracture characterization.

The moveout-inversion algorithm produces the effective parameters for the section above the reflector. To avoid the stability problems of Dix-type layer stripping in estimating the interval parameters, we suggest obtaining the interval moveout function from the velocity-independent layer-stripping (VILS) method, which operates directly with reflection traveltimes. Synthetic and field-data examples demonstrate that, in contrast to Dix-type techniques, both 2D and 3D versions of VILS remain robust in the presence of typical levels of correlated traveltime noise.

Finally, we show that nonhyperbolic moveout inversion provides sufficient information for P-wave time-domain processing in orthorhombic media. The symmetry-plane NMO velocities and the anellipticity parameters control all 3D P-wave time-processing steps in orthorhombic media including NMO and DMO corrections, and prestack and poststack time migration.

3.1 Nonhyperbolic moveout equations

Reflection traveltime of pure (nonconverted) modes is conventionally approximated by a Taylor series expansion of the squared traveltime t^2 (e.g., Taner and Koehler, 1969):

$$t^2 = A_0 + A_2 x^2 + A_4 x^4 + \dots , \tag{3.1}$$

where x is the source-receiver offset. The moveout coefficients are given by

$$A_0 = t_0^2 , \quad A_2 = \left. \frac{d(t^2)}{d(x^2)} \right|_{x=0} , \quad A_4 = \frac{1}{2} \frac{d}{d(x^2)} \left[\frac{d(t^2)}{d(x^2)} \right] \Bigg|_{x=0} ; \tag{3.2}$$

$t_0 = t(0)$ is the zero-offset traveltime, and A_2 is related to the NMO velocity as $A_2 = V_{\rm nmo}^{-2}$. The first two terms on the right-hand side of equation 3.1 represent the hyperbolic part of the moveout curve (see equation 1.1) while the quartic coefficient A_4 describes nonhyperbolic moveout. The addition of higher-order terms to equation 3.1 is generally considered impractical.

Although the three-term series in equation 3.1 provides a better approximation for long-spread moveout than does the conventional hyperbolic equation, it loses accuracy for offset-to-depth ratios reaching $1.5-2$. Tsvankin and Thomsen (1994) modified equation 3.1 by adding a denominator to the quartic moveout term with the goal of making $t(x)$ convergent at $x \to \infty$:

$$t^2 = t_0^2 + \frac{x^2}{V_{\text{nmo}}^2} + \frac{A_4 x^4}{1 + A x^2}, \tag{3.3}$$

where the coefficient A depends on the horizontal group velocity V_{hor},

$$A = \frac{A_4}{V_{\text{hor}}^{-2} - V_{\text{nmo}}^{-2}}. \tag{3.4}$$

The definition of A in equation 3.4 ensures that $t^2|_{x \to \infty}$ approaches a hyperbola parameterized by the velocity V_{hor}. Equation 3.3 represents an accurate and efficient tool for modeling of reflection traveltimes of P-waves and, in some cases, converted PS-waves (Tsvankin, 2005).

The Tsvankin-Thomsen equation was originally derived for vertical transverse isotropy, but its generic form makes it suitable for anisotropic models of any symmetry. For wide-azimuth data, the velocity V_{nmo} is obtained from the equation of the NMO ellipse described in detail in Chapters 1 and 2. The influence of azimuthal anisotropy, however, substantially complicates analysis of the coefficient A_4, as discussed below.

Alkhalifah and Tsvankin (1995) suggested a particularly convenient, simplified form of equation 3.3 for P-waves in VTI media that includes just two moveout parameters, the NMO velocity of horizontal events and the anellipticity parameter η. The Alkhalifah-Tsvankin equation has been widely used for estimating η from nonhyperbolic semblance analysis and for building vertically heterogeneous VTI models in the time domain (e.g., Alkhalifah, 1997; Toldi et al., 1999). Below, this equation is extended to orthorhombic and HTI media by making the NMO velocity and parameter η azimuthally dependent.

A common alternative representation of long-spread moveout in horizontally layered models is based on the equation of shifted hyperbola (de Bazelaire and Viallix, 1994; Castle, 1994). Siliqi and Bousqué (2000) apply the shifted-hyperbola formalism to nonhyperbolic moveout inversion and time-domain processing in VTI media. Also, Sayers and Ebrom (1997) propose to describe nonhyperbolic moveout in azimuthally anisotropic media by expanding group velocity in spherical harmonics. For a more detailed discussion of these methods we refer the reader to the original papers.

3.2 3D description of the quartic moveout coefficient

The key issue in applying equation 3.3 to modeling and inversion of reflection moveout in azimuthally anisotropic media is derivation of the azimuthally varying quartic moveout coefficient A_4. The dependence of A_4 on the medium parameters also provides valuable analytic insight into the properties of nonhyperbolic moveout. Here, we give an analytic representation of the coefficient A_4 and discuss its behavior for tilted TI and orthorhombic media.

3.2.1 General equation for the quartic coefficient A_4

An exact expression for the quartic moveout coefficient in arbitrarily anisotropic, heterogeneous media (Figure 3.1) was presented by Pech et al. (2003). They expanded the two-way traveltime in a Taylor series in half-offset and applied the normal-incidence-point (NIP) theorem (Chernjak and Gritsenko, 1979; Hubral and Krey, 1980). This theorem helps relate the Taylor-series coefficients to the spatial derivatives of the traveltime between the surface and the reflector. The quartic moveout coefficient A_4 in its most general form can be written as (Pech et al., 2003)

$$A_4(\mathbf{L}) = \frac{1}{16}\left[\frac{\partial^2\tau}{\partial y_k \partial y_l}\frac{\partial^2\tau}{\partial y_m \partial y_n} + \frac{\tau_0}{3}\frac{\partial^4\tau}{\partial y_k \partial y_l \partial y_m \partial y_n} - \tau_0 \frac{\partial^3\tau}{\partial y_k \partial y_l \partial x_i}\left(\frac{\partial^2\tau}{\partial x_i \partial x_j}\right)^{-1}\frac{\partial^3\tau}{\partial x_j \partial y_m \partial y_n}\right] L_k L_l L_m L_n\,, \tag{3.5}$$

where $\mathbf{L} = [\cos\alpha, \sin\alpha, 0]$ is a unit vector parallel to the CMP line (Figure 3.1), $\mathbf{y}$ defines the CMP location, τ is the one-way traveltime between the CMP and point $\mathbf{x}$ on the reflector, and τ_0 is the one-way zero-offset time. The traveltime derivatives are evaluated at the zero-offset reflection point $\mathbf{x}^{(0)}$ corresponding to the CMP location $\mathbf{y}$. Summation over repeated indices from one to two is implied.

Equation 3.5 is obtained without making specific assumptions about the anisotropy or heterogeneity of the model; also, it is generally valid for reflectors of irregular shape. The traveltime, however, is supposed to be differentiable with respect to the spatial coordinates near the common midpoint, which is not the case, for example, in shadow zones. The Taylor series expansion for traveltime can break down as a whole for models with strong lateral velocity variations and in the vicinity of caustics (Grechka and Tsvankin, 1998a). Nonetheless, for sufficiently smooth subsurface models commonly used in seismology, equation 3.5 gives an accurate representation of the quartic moveout coefficient and the magnitude of nonhyperbolic moveout.

The form of the azimuthal dependence of the coefficient A_4 in equation 3.5 is governed by the derivatives of the traveltime τ with respect to the coordinates of the common midpoint $\mathbf{y}$ and point $\mathbf{x}$ on the reflector. For relatively simple models, τ can be expressed explicitly as a function of $\mathbf{y}$ and $\mathbf{x}$, and the derivatives in equation 3.5

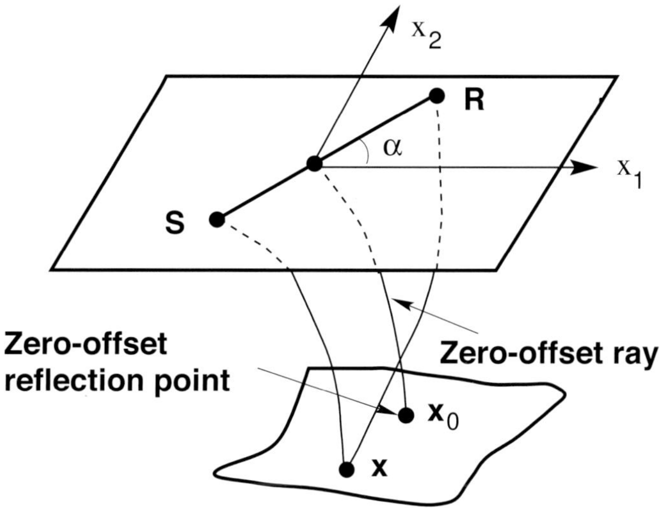

Figure 3.1: Reflection traveltimes from a curved interface beneath an anisotropic, heterogeneous medium are recorded in a CMP gather (Pech et al., 2003). The quartic moveout coefficient A_4 varies with the azimuth α of the CMP line. The derivation of A_4 in Pech et al. (2003) accounts for reflection-point dispersal.

can be found in closed form (e.g., see Appendix 3A). If the medium is laterally heterogeneous and/or has a low anisotropic symmetry, it is convenient to express A_4 in terms of the horizontal slowness component of the zero-offset ray (Grechka et al., 1999b).

Most important, all derivatives in equation 3.5 can be evaluated using quantities computed during the tracing of the zero-offset ray. Note, however, that A_4 depends on fourth-order traveltime derivatives, which implies that the smoothness of the model should be higher than that required in conventional second-order dynamic ray tracing.

3.2.2 Quartic coefficient in a tilted TI layer

Equation 3.5 was applied by Pech et al. (2003) to a homogeneous TI layer with a tilted symmetry axis above a plane dipping reflector (Figure 3.2). We assume that the symmetry axis is confined to the dip plane of the reflector, which commonly holds for dipping shale layers.

As before, the TTI model is described by the velocities of P-waves (V_{P0}) and S-waves (V_{S0}) in the symmetry-axis direction and the anisotropy parameters ϵ, δ, and γ defined with respect to the symmetry axis. Our convention is that the tilt ν of the symmetry axis is considered positive if the axis points towards the reflector (i.e., if the symmetry axis and the reflector normal deviate from the vertical in the same direction). Since the dip plane of the reflector contains the symmetry axis, it represents a vertical symmetry plane for the whole model. Therefore, the dip

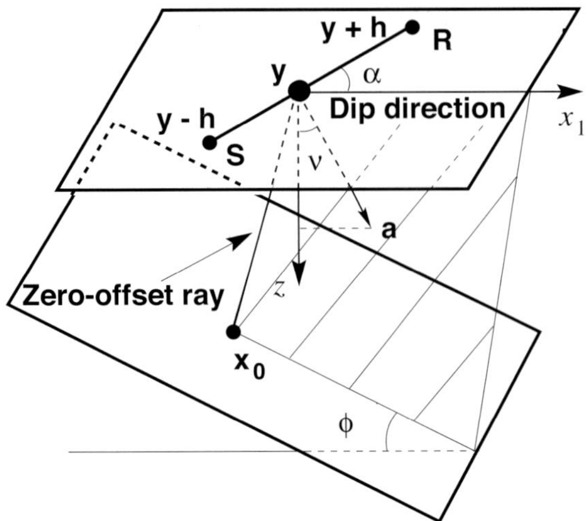

Figure 3.2: Reflected wave is recorded above a homogeneous TI layer with a plane dipping lower boundary (Pech et al., 2003). The symmetry axis (vector **a**) is confined to the dip plane $[x_1, z]$ and rotated from the vertical by angle ν.

and strike directions of the reflector determine the principal axes of the azimuthally varying coefficient A_4.

Weak-anisotropy approximation

While the exact equation 3.5 is suitable for computational purposes, it is too complex to explain the behavior of the quartic moveout coefficient in TTI media. In Appendix 3A we linearize equation 3.5 for P-waves in ϵ and δ and show that the approximate coefficient A_4 is proportional to the anellipticity parameter η:

$$A_4^{\mathrm{TTI}} = -\frac{2\eta}{t_{P0}^2 V_{P0}^4} F(\alpha, \phi, \nu), \tag{3.6}$$

where the function F is defined in equation 3.65, t_{P0} is the two-way zero-offset P-wave traveltime, α is the azimuth of the CMP line with respect to the dip plane, and ϕ is the reflector dip.

Note that any kinematic signature of SV-waves (i.e., of the mode polarized in the plane formed by the slowness vector and the symmetry axis) for weak transverse isotropy can be obtained from the corresponding P-wave signature by making the following substitutions (Tsvankin, 2005): $V_{P0} \to V_{S0}$, $\delta \to \sigma$, and $\epsilon \to 0$. The parameter $\sigma \equiv (V_{P0}/V_{S0})^2 (\epsilon - \delta)$ is largely responsible for SV-wave anisotropy in TI media. This substitution rule is fully applicable to equation 3.6, with t_{P0} replaced by the SV-wave zero-offset traveltime t_{S0}.

The linearized solution for A_4 is sufficiently close to the exact quartic coefficient for relatively small values of the anisotropy parameters (Figure 3.3). Equation 3.6

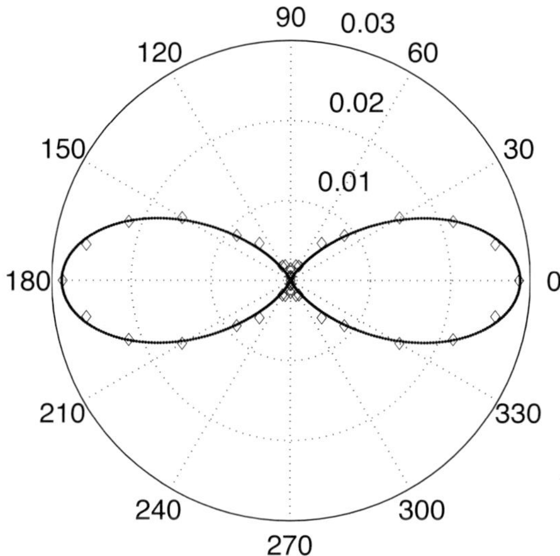

Figure 3.3: Accuracy of the linearized approximation for the coefficient A_4 in a horizontal TTI layer (Pech et al., 2003). The solid line is the magnitude of A_4 computed from equation 3.6, and diamonds mark $|A_4|$ obtained for a fixed azimuth (numbers on the perimeter) by fitting a quartic polynomial to ray-traced $t^2(x^2)$-curve; the spreadlength-to-depth ratio is 1.2. The model parameters are V_{P0}=1 km/s, ϵ=0.1, δ=0.025, and ν=80°.

(solid curve) provides a good approximation to the best-fit values of $|A_4|$ (diamonds) obtained from ray-traced traveltimes for a TTI model with a large tilt of the symmetry axis. As follows from the results of Tsvankin and Thomsen (1994) and Tsvankin (2005), the weak-anisotropy approximation for A_4 can rapidly lose its accuracy with increasing ϵ and δ. Equation 3.6, however, can still be used to study the azimuthal signature of nonhyperbolic moveout in tilted TI media.

It is clear from equation 3.6 that regardless of the tilt of the symmetry axis and reflector dip, nonhyperbolic moveout of P-waves for weak transverse isotropy is controlled by the parameter η, rather than by ϵ and δ individually. If the medium is elliptically anisotropic ($\eta = 0$), A_4 vanishes and reflection moveout becomes purely hyperbolic. This is a general result valid for an elliptically anisotropic layer with any strength of anisotropy (Uren et al., 1990).

Dip component

Equations 3.6 and 3.65 can be used to find the coefficient A_4 in the dip and strike directions. On the dip line ($\alpha = 0°$),

$$A_{4,\text{dip}}^{\text{TTI}}(\phi) = -\frac{2\eta}{t_{P0}^2 V_{P0}^4} \cos^3\phi \cos(3\phi - 4\nu) \, . \tag{3.7}$$

Since the coefficient A_4 is proportional to $\cos^3\phi$, the magnitude of nonhyperbolic moveout has a rapidly decreasing trend with dip. Equation 3.7, however, becomes less accurate for steep reflectors because when ϕ approaches 90°, several terms involving the anisotropy parameters can no longer be treated as small. Evaluation of the exact equation 3.5 shows that A_4 for a vertical reflector ($\phi = 90°$) is relatively small but usually does not go to zero, unless the symmetry axis is vertical or horizontal. Also, for large dips and certain relative orientations of the symmetry axis and the reflector normal (typically, if the symmetry axis is tilted toward the reflector), reflection events might not be recorded at the surface at all (Tsvankin, 1997c, 2005).

According to equation 3.7, the dip-line quartic moveout coefficient (and, therefore, nonhyperbolic moveout as a whole) vanishes if $\cos(3\phi - 4\nu) = 0$, or $(3\phi - 4\nu) = n\pi/2$ ($n = \pm 1, \pm 3, \pm 5, ...$). In the special case of VTI media ($\nu = 0°$), the dip-line coefficient A_4 goes to zero for a dip of 30° (see a more detailed discussion of the VTI model below). When $\phi = 30°$, A_4 also vanishes for $\nu = 90°$ (HTI medium) and $\nu = \pm 45°$. Hence, the absence or low magnitude of dip-line nonhyperbolic moveout in nonelliptical ($\eta \neq 0$) TTI media can be used to constrain the relationship between the reflector dip and the tilt of the symmetry axis.

Equation 3.7 is written in terms of dip, which cannot be estimated from surface reflection data unless the velocity model is known. For purposes of anisotropic parameter estimation, it is more convenient to rewrite the quartic coefficient as a function of the horizontal slowness p of the zero-offset ray (i.e., of the time slope on the zero-offset section). The dip can be conveniently expressed through $p = \sin\phi / V(\phi)$ using the weak-anisotropy approximation for the phase velocity V (see Appendix 6A in Tsvankin, 2005). Substituting $\phi(p)$ into equation 3.7 and linearizing the result in the anisotropy parameters yields

$$A_{4,\mathrm{dip}}^{\mathrm{TTI}}(p) = \frac{8\eta\,(1-y)^3}{t_{P0}^2\,[V_{\mathrm{nmo}}^{\mathrm{TTI}}(0)]^4}\left[\left(y - \frac{1}{4}\right)\sqrt{1-y}\,\cos 4\nu + \left(y - \frac{3}{4}\right)\sqrt{y}\,\sin 4\nu\right], \qquad (3.8)$$

where $V_{\mathrm{nmo}}^{\mathrm{TTI}}(0)$ is the NMO velocity from a horizontal reflector, and $y \equiv p^2\,[V_{\mathrm{nmo}}^{\mathrm{TTI}}(0)]^2$. Hence, $A_{4,\mathrm{dip}}^{\mathrm{TTI}}(p)$ depends on the vertical traveltime t_{P0} and three parameters: $V_{\mathrm{nmo}}^{\mathrm{TTI}}(0)$, η, and ν [or $V_{\mathrm{nmo}}^{\mathrm{TTI}}(0)$, $(\eta\cos 4\nu)$ and $(\eta \sin 4\nu)$].

In anisotropic parameter estimation, nonhyperbolic moveout is used in combination with NMO velocity. The dip-line P-wave NMO velocity for weakly anisotropic TTI media is given by Tsvankin (1997c, 2005) as a function of the dip ϕ. Rewriting his result through the ray parameter p, we find

$$V_{\mathrm{nmo}}^{\mathrm{TTI}}(p) = \frac{V_{\mathrm{nmo}}^{\mathrm{TTI}}(0)}{\sqrt{1-y}}\left[1 + \eta b \cos 4\nu - \eta d \sin 4\nu\right], \qquad (3.9)$$

where

$$b \equiv \frac{y}{1-y}\left(4y^2 - 9y + 6\right), \qquad (3.10)$$

$$d \equiv \sqrt{\frac{y}{1-y}}\left(4y^2 - 7y + 3\right). \qquad (3.11)$$

For vertical transverse isotropy ($\nu = 0°$), equation 3.9 reduces to the expression obtained by Alkhalifah and Tsvankin (1995), which is given in Chapter 1 as one of the elements of the NMO-velocity cylinder (equation 1.54).

Thus, both the dip-line NMO velocity and the quartic moveout coefficient are controlled by the same parameter combinations: $V_{\rm nmo}^{\rm TTI}(0)$, $(\eta \cos 4\nu)$, and $(\eta \sin 4\nu)$. In principle, it might be possible to estimate these three combinations from P-wave moveout if $V_{\rm nmo}$ and A_4 are measured on the dip line for two different dips (e.g., in overthrust areas or near salt domes). However, the trade-off between $V_{\rm nmo}$ and A_4 typically leads to substantial uncertainty in the quartic coefficient. Also, when the magnitude of the anisotropy parameters reaches $0.15 - 0.2$, the weak-anisotropy approximation 3.9 loses its accuracy and the exact $V_{\rm nmo}$ becomes dependent on the individual values of ϵ and δ (Tsvankin, 1997c, 2005).

Strike component

The strike component of the quartic moveout coefficient can be obtained by substituting $\alpha = 90°$ into equation 3.6:

$$A_{4,\rm strike}^{\rm TTI}(\phi) = -\frac{2\eta}{t_{P0}^2 V_{P0}^4} \cos^4(\phi - \nu) \, . \tag{3.12}$$

Both the dip and strike components of A_4 are proportional to η, but their dependencies on the reflector dip ϕ and the symmetry-axis tilt ν differ. Equation 3.12 shows that $A_{4,\rm strike}^{\rm TTI}$ goes to zero only if the symmetry axis and the reflector normal are orthogonal (i.e., the symmetry axis is confined to the reflecting plane). For example, if the reflector is vertical ($\phi = 90°$), the strike-line quartic coefficient vanishes for VTI media ($\nu = 0°$). Indeed, for such a model reflected rays are confined to the horizontal (isotropy) plane, where velocity is independent of angle, which makes reflection moveout for any azimuth purely hyperbolic.

In general, the fact that the dip and strike components of the coefficient A_4 vanish for different combinations of ν and ϕ is favorable for a potential 3D inversion procedure that involves nonhyperbolic moveout. The behavior of A_4 for some special cases of TI media is discussed below.

Azimuthal dependence

Unlike NMO velocity, which typically has elliptical azimuthal dependence, the variation of the quartic moveout coefficient with azimuth has a much more complicated character. The nonlinear relationship between A_4 and the angles ν, ϕ, and α in equation 3.6 can lead to multiple zeros of the function $A_4(\alpha)$, whose positions are sensitive to both tilt and dip.

A typical azimuthal variation of the coefficient A_4 in TTI media is displayed in Figure 3.4. Since the model has a vertical symmetry plane, $A_4(\alpha)$ is symmetric with respect to both $\alpha = 0°$ and $\alpha = 90°$, and the moveout signature as a whole is repeated in each quadrant. Indeed, since $A_4(\alpha)$ and $A_4(\alpha \pm \pi)$ correspond to the same CMP

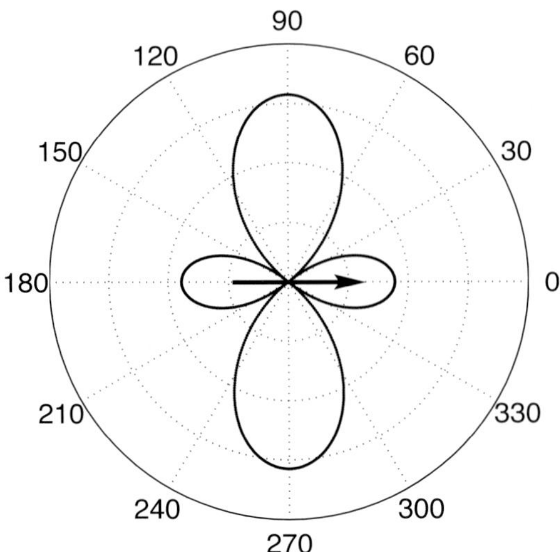

Figure 3.4: Magnitude of the azimuthally varying quartic moveout coefficient A_4 in a dipping TI layer computed from the weak-anisotropy approximation 3.6 (Pech et al., 2003). The tilt of the symmetry axis is ν=40°, and the reflector dip is ϕ=15°; the other parameters change only the scale of the plot (intentionally undefined here). The azimuth α (numbers on the perimeter) is measured from the dip plane marked by the arrow.

line and $A_4(\alpha) = A_4(-\alpha)$, it follows that $A_4(\alpha) = A_4(\pi - \alpha)$, which means that the quartic coefficient is symmetric with respect to the strike direction, $\alpha = \pm\pi/2$.

Clearly, the quartic coefficient exhibits a more complex behavior than does the NMO ellipse and vanishes at an azimuth of 38°. If $\eta > 0$ (A_4 is proportional to η), the sign of A_4 changes from positive for the lobe centered at the dip direction to negative near the strike direction.

These results indicate that the azimuthal dependence of the quartic coefficient can provide useful information for anisotropic parameter estimation. In particular, the azimuthal directions of CMP lines with vanishing A_4 depend on certain combinations of ν and ϕ defined by equation 3.6 and can be used to constrain the orientation of the symmetry axis.

Symmetry axis orthogonal to reflector

As discussed in Chapter 2, models with the symmetry axis orthogonal to the reflector adequately describe, for example, dipping shale layers in fold-and-thrust belts (e.g., the Canadian Foothills) and near salt domes. Then $\nu = \phi$, and both the zero-offset ray and its slowness vector are parallel to the symmetry axis, so some isotropic relationships remain valid. For example, the dip-line NMO velocity satisfies the conventional (isotropic) cosine-of-dip equation 2.47, while V_{nmo} on the strike line is independent of dip (equation 2.49).

Substituting $\nu = \phi$ into equation 3.6, we obtain

$$A_4^{\mathrm{TTI}}(\phi = \nu) = -\frac{2\eta}{t_{P0}^2 V_{P0}^4}\left(1 - \sin^2\phi\,\cos^2\alpha\right)^2. \tag{3.13}$$

According to equation 3.13, the quartic coefficient goes to zero when

$$|\cos\alpha| = \frac{1}{\sin\phi}. \tag{3.14}$$

Condition 3.14 can be satisfied only on the dip line ($\alpha = 0°$) of a vertical reflector ($\phi = 90°$, which implies a horizontal symmetry axis). Away from the dip line, the coefficient A_4 for a vertical reflector in HTI media varies as

$$A_4^{\mathrm{TTI}}(\phi = \nu = 90°) = -\frac{2\eta}{t_{P0}^2 V_{P0}^4}\,\sin^4\alpha\,. \tag{3.15}$$

The dip ($\alpha = 0°$) and strike ($\alpha = 90°$) components of A_4 in equation 3.13 are:

$$A_{4,\mathrm{dip}}^{\mathrm{TTI}}(\phi = \nu) = -\frac{2\eta}{t_{P0}^2 V_{P0}^4}\,\cos^4\phi\,, \tag{3.16}$$

$$A_{4,\mathrm{strike}}^{\mathrm{TTI}}(\phi = \nu) = -\frac{2\eta}{t_{P0}^2 V_{P0}^4}\,. \tag{3.17}$$

Equation 3.17, which shows that the strike-line component of A_4 is independent of dip (or tilt), is well known for a weakly anisotropic VTI medium above a horizontal reflector, i.e., when $\nu = \phi = 0°$ (Tsvankin and Thomsen, 1994; Tsvankin, 2005). $A_{4,\mathrm{strike}}^{\mathrm{TTI}}$ has the same value for an HTI layer ($\nu = 90°$) bounded by a vertical reflector ($\phi = 90°$). Since the strike line for this HTI model is perpendicular to the symmetry axis and reflected rays are confined to the horizontal plane, reflection moveout in the strike direction is identical to that for a horizontal VTI layer.

The dip-line component of A_4 (equation 3.16) decays with dip as $\cos^4\phi$. Therefore, the magnitude of nonhyperbolic moveout for $\nu = \phi$ rapidly decreases away from the strike direction, even if the dip is relatively mild.

Dipping VTI layer

Setting the tilt ν of the symmetry axis in equation 3.6 to zero yields the weak-anisotropy approximation for the quartic coefficient in VTI media:

$$A_4^{\mathrm{VTI}} = -\frac{2\eta\cos^4\phi}{t_{P0}^2 V_{P0}^4}\left(1 - 4\sin^2\phi\,\cos^2\alpha\right). \tag{3.18}$$

For a vertical reflector ($\phi = 90°$), A_4 vanishes regardless of the azimuth of the CMP line because reflected rays are confined to the horizontal isotropy plane where the velocity is constant and the moveout is purely hyperbolic. If the reflector is horizontal ($\phi = 0°$), the model as a whole is azimuthally isotropic, and the approximate A_4 is

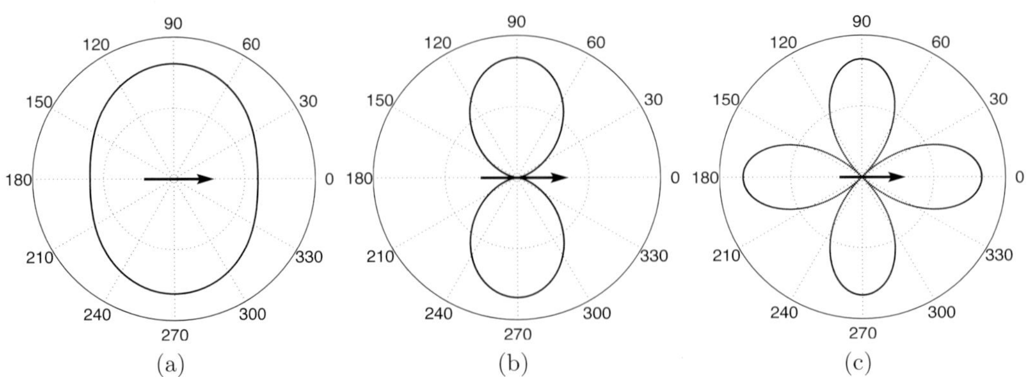

Figure 3.5: Magnitude of the coefficient A_4 computed from equation 3.18 as a function of azimuth (numbers on the perimeter) for a dipping VTI layer (Pech et al., 2003). The reflector dip is (a) 15°, (b) 30°, and (c) 45°. The arrows mark the dip direction.

determined by equation 3.17. A discussion of the exact quartic moveout coefficient of both P- and S-waves in horizontally layered VTI media can be found in Tsvankin (2005).

For dips between 0° and 90°, the coefficient A_4 goes to zero in the azimuthal directions satisfying

$$|\cos\alpha| = \frac{1}{2\sin\phi}\,. \tag{3.19}$$

If the dip is smaller than 30°, equation 3.19 does not have a solution, and A_4 has the same sign for the entire range of azimuths (Figure 3.5a). For a dip of 30°, A_4 goes to zero only in the dip plane ($\alpha = 0°$; Figure 3.5b). This analytic result is in good agreement with the numerical study of P-wave NMO velocity in Tsvankin (1995a, 2005), which shows that the dip-line moveout approaches a hyperbola for reflector dips close to 30°. The influence of the factor $\cos^4\phi$ rapidly reduces the magnitude of A_4^{VTI} for large dips.

For any dip in the range $30° < \phi < 90°$, equation 3.19 yields one azimuth of vanishing A_4 between the dip and strike directions. If the dip is equal to 45°, the quartic coefficient goes to zero for $\alpha = 45°$ (Figure 3.5c). The sign of A_4 for $\phi = 45°$ and $\eta > 0$ is positive near the dip plane and negative near the strike direction.

Horizontal HTI layer

For a horizontal HTI layer ($\phi = 0°$ and $\nu = 90°$), equation 3.6 reduces to

$$A_4^{\mathrm{HTI}} = -\frac{2\eta}{t_{P0}^2 V_{P0}^4}\cos^4\alpha\,. \tag{3.20}$$

Equation 3.20 has the same azimuthal dependence ($\cos^4\alpha$) as does the exact expression for A_4 obtained by Al-Dajani and Tsvankin (1998). The quartic coefficient

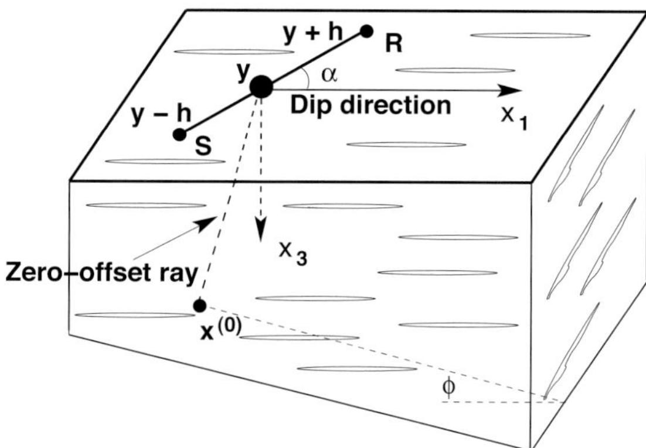

Figure 3.6: Reflected wave is recorded above a homogeneous orthorhombic layer with a dipping lower boundary (Pech and Tsvankin, 2004). The vertical symmetry plane $[x_1, x_3]$ coincides with the dip plane of the reflector.

vanishes in the isotropy plane orthogonal to the symmetry axis ($\alpha = 90°$); reflection moveout in that plane is purely hyperbolic. In the symmetry-axis plane ($\alpha = 0°$), A_4 has the same form as that in VTI media (equation 3.17).

3.2.3 Quartic coefficient in an orthorhombic layer

Next, we analyze the coefficient A_4 for an orthorhombic layer with a horizontal symmetry plane above a horizontal or a dipping reflector (Figure 3.6). The vertical symmetry planes of the layer coincide with the Cartesian coordinate planes $[x_1, x_3]$ and $[x_2, x_3]$. A basic description of orthorhombic symmetry can be found in section 2.5 (in particular, see Figure 2.36). The definitions of Tsvankin's anisotropy parameters used here are given in Appendix 1B.

Horizontal reflector

In principle, the quartic moveout coefficient in a horizontal orthorhombic layer can be obtained from the general equation 3.5. Al-Dajani et al. (1998), who used a different formalism, presented the following exact expression for the P-wave coefficient A_4:

$$A_4(\alpha) = A_4^{(1)} \sin^4 \alpha + A_4^{(x)} \sin^2 \alpha \cos^2 \alpha + A_4^{(2)} \cos^4 \alpha , \qquad (3.21)$$

where α is the azimuth with respect to the $[x_1, x_3]$ symmetry plane, $A_4^{(1)}$ and $A_4^{(2)}$ are the A_4-values in the planes $[x_2, x_3]$ and $[x_1, x_3]$, respectively, and $A_4^{(x)}$ controls the cross-term that contributes only in off-symmetry directions and reaches its maximum magnitude at $\alpha = 45°$.

Because of the kinematic equivalence between the symmetry planes of orthorhombic and transversely isotropic media, the exact coefficients $A_4^{(1)}$ and $A_4^{(2)}$ can be adapted from the corresponding VTI equation (Tsvankin and Thomsen, 1994; Tsvankin, 2005). Alkhalifah and Tsvankin (1995) obtain a concise and accurate approximation for the P-wave coefficient A_4 in VTI media by setting the shear-wave vertical velocity V_{S0} to zero. Their result remains entirely valid for the coefficients $A_4^{(1)}$ and $A_4^{(2)}$, which can be represented as

$$A_4^{(1)} = -\frac{2\eta^{(1)}}{t_{P0}^2 \left[V_{\rm nmo}^{(1)}\right]^4}, \tag{3.22}$$

$$A_4^{(2)} = -\frac{2\eta^{(2)}}{t_{P0}^2 \left[V_{\rm nmo}^{(2)}\right]^4}, \tag{3.23}$$

where t_{P0} is the two-way zero-offset P-wave traveltime, $V_{\rm nmo}^{(1)}$ and $V_{\rm nmo}^{(2)}$ are the symmetry-plane NMO velocities (equation 2.62), and $\eta^{(1)}$ and $\eta^{(2)}$ are the anellipticity parameters defined in the vertical symmetry planes (equations 1.88 and 1.89).

The exact coefficient $A_4^{(x)}$ has an extremely complicated form, which hinders its potential application in moveout analysis. This complexity, however, is mostly caused by the shear-wave vertical velocities, which have negligible influence on P-wave kinematics (Tsvankin, 1997a, 2005). Setting the vertical velocities of S-waves to zero yields (Al-Dajani et al., 1998):

$$A_4^{(x)} = \frac{2}{t_{P0}^2 \left[V_{\rm nmo}^{(1)}\right]^2 \left[V_{\rm nmo}^{(2)}\right]^2} \left[1 - \sqrt{\frac{(1+2\eta^{(1)})\,(1+2\eta^{(2)})}{1+2\eta^{(3)}}}\right]; \tag{3.24}$$

$\eta^{(3)}$ is the anellipticity parameter defined in the horizontal symmetry plane (equation 1.90).

Al-Dajani et al. (1998) show that substitution of equations 3.22–3.24 into equation 3.21 provides an accurate description of the azimuthally varying quartic moveout coefficient. Linearization of $A_4^{(1)}$, $A_4^{(2)}$, and $A_4^{(x)}$ in the anisotropy parameters results in the weak-anisotropy approximation for A_4:

$$A_4(\alpha) = -\frac{2}{t_{P0}^2 V_{P0}^4}\left(\eta^{(1)}\sin^2\alpha \,-\, \eta^{(3)}\sin^2\alpha\,\cos^2\alpha \,+\, \eta^{(2)}\cos^2\alpha\right), \tag{3.25}$$

where V_{P0} is the P-wave vertical velocity. Despite the similarity of equation 3.25 and equation 1.3 of the NMO ellipse, the azimuthal variation of the coefficient A_4 can be far different from elliptical. First, the trigonometric functions in the cross-term are squared [the NMO ellipse contains the product $(\sin\alpha\,\cos\alpha)$], which influences the shape of the curve $A_4(\alpha)$. Second, if the parameters $\eta^{(1)}$ and $\eta^{(1)}$ have different signs, the coefficient A_4 goes to zero and changes sign between the symmetry planes.

Equation 3.25 has the same form as the linearized quartic coefficient in a VTI layer (equation 3.17), but the parameter η now varies with azimuth:

$$\eta(\alpha) = \eta^{(1)}\sin^2\alpha \,-\, \eta^{(3)}\sin^2\alpha\,\cos^2\alpha \,+\, \eta^{(2)}\cos^2\alpha\,. \tag{3.26}$$

This result is not surprising because any vertical plane in weakly anisotropic orthorhombic models is equivalent to VTI in terms of P-wave kinematic signatures. Indeed, Tsvankin (1997a, 2005) showed that the linearized P-wave phase-velocity function for a fixed azimuth has the same form in orthorhombic and VTI media. However, the anisotropy parameters of the "equivalent" VTI medium vary with azimuth, and the anellipticity parameter in equation 3.26 can be obtained as $\eta(\alpha) = \epsilon(\alpha) - \delta(\alpha)$, where $\epsilon(\alpha)$ and $\delta(\alpha)$ are given by equations 1.108 and 1.109 in Tsvankin (2005).

Weak-anisotropy approximation for a dipping layer

For simplicity, we assume that the symmetry plane $[x_1, x_3]$ coincides with the dip plane of the reflector, which makes $[x_1, x_3]$ the only symmetry plane for the model as a whole. Therefore, as in isotropic or VTI media, the zero-offset reflected ray is confined to the dip plane (Figure 3.6), which would not be the case for arbitrary orientation of the vertical symmetry planes.

The discussion below is based on the results of Pech and Tsvankin (2004) who obtained the weak-anisotropy approximation for the P-wave coefficient A_4 in an orthorhombic layer by linearizing the exact equation 3.5:

$$A_4 = -\frac{1}{2t_{P0}^2 V_{P0}^4} \Big\{ \eta^{(2)} \cos^2\phi \left[2\cos 2\phi \,(1 + \cos 2\alpha \cos 2\phi) + \cos 4\phi - 1 \right] + 4\eta^{(1)} \cos^2\phi \sin^2\alpha - 2\eta^{(3)} \sin^2\alpha \,(\cos 2\alpha + \cos 2\phi) \Big\}, \tag{3.27}$$

where ϕ, as before, is the dip, and α is the azimuth of the CMP line with respect to the $[x_1, x_3]$-plane. The derivation of equation 3.27 in Pech and Tsvankin (2004) is similar to that in Appendix 3A for TTI media and is not reproduced here.

Equation 3.27 shows that the azimuthal variation $A_4(\alpha)$ is controlled just by the parameters $\eta^{(1,2,3)}$ and dip ϕ. As is the case for the TTI model discussed above, the quartic moveout coefficient is symmetric with respect to both the dip and strike directions of the reflector, so the azimuthal signature of A_4 is repeated in each quadrant.

Figure 3.7 confirms that equation 3.27 is suitable for a qualitative description of the azimuthal dependence of the quartic moveout coefficient for small and moderate values of the anisotropy parameters (the magnitude of the η-parameters in Figure 3.7 is quite substantial). Nonhyperbolic moveout analysis, however, should not be based on the weak-anisotropy approximation (Tsvankin and Thomsen, 1994; Tsvankin, 2005).

If the reflector is horizontal ($\phi = 0°$), equation 3.27 reduces to equation 3.25 analyzed above. Another special case is that of a dipping VTI layer since transversely isotropic models can be treated as a subset of the more general orthorhombic media. The coefficient A_4 for VTI media can be found from equation 3.27 by setting $\eta^{(1)} = \eta^{(2)} = \eta$ and $\eta^{(3)} = 0$:

$$A_4^{\text{VTI}} = -\frac{2\eta \cos^4\phi}{t_{P0}^2 V_{P0}^4} \left(1 - 4\sin^2\phi \cos^2\alpha \right), \tag{3.28}$$

which coincides with equation 3.18.

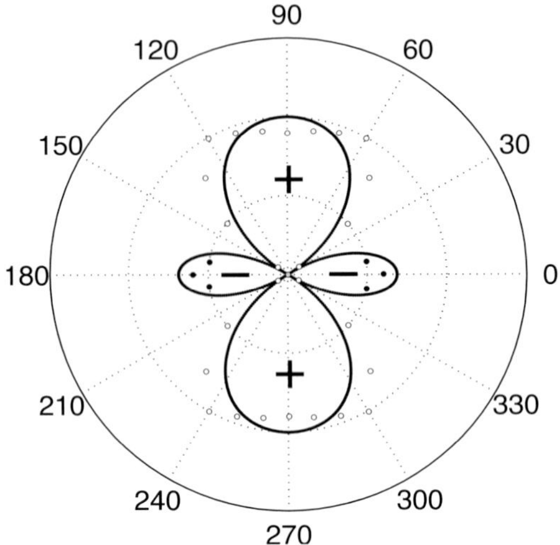

Figure 3.7: Accuracy of the weak-anisotropy approximation for the coefficient A_4 in an orthorhombic layer with the lower boundary dipping at ϕ=15° (Pech and Tsvankin, 2004). The circles mark the relative magnitude of A_4 obtained for each azimuth α (the numbers on the perimeter; α=0° corresponds to the dip plane) by fitting a quartic polynomial to the ray-traced $t^2(x^2)$-curve; the spreadlength-to-depth ratio is 1.5. The empty circles mark positive values of A_4, and the filled circles mark negative A_4. The solid line is the weak-anisotropy approximation 3.27; the sign of A_4 is marked inside each lobe. Both the ray-tracing and weak-anisotropy A_4-functions are normalized by their respective maximum values of $|A_4|$. The anellipticity parameters are $\eta^{(1)}=-0.2$, $\eta^{(2)}=0.2$, and $\eta^{(3)}=0.2$.

Dip-line and strike-line components

On the dip line ($\alpha = 0°$), the coefficient A_4 from equation 3.27 takes the form

$$A_{4,\text{dip}} = -\frac{2\,\eta^{(2)}\cos^4\phi}{t_{P0}^2\,V_{P0}^4}\left(1 - 4\sin^2\phi\right). \tag{3.29}$$

Equation 3.29 becomes identical to the corresponding VTI equation 3.28 for $\alpha = 0°$, if the parameter $\eta^{(2)}$ is replaced with η. Indeed, reflected rays and their phase-velocity vectors on the dip line are confined to the symmetry plane $[x_1, x_3]$ where the kinematics of wave propagation is the same as that in the VTI model with vertical velocity V_{P0} and anisotropy parameters $\epsilon = \epsilon^{(2)}$ and $\delta = \delta^{(2)}$ (so $\eta = \eta^{(2)}$).

The properties of the coefficient $A_{4,\text{dip}}$ for VTI media were discussed above. In particular, the combined influence of the factors $\cos^4\phi$ and $(1 - 4\sin^2\phi)$ in equation 3.29 leads to a rapid decrease in the magnitude of the dip-line quartic coefficient with dip, and $A_{4,\text{dip}} = 0$ for $\phi = 30°$. If $\eta^{(2)} > 0$, $A_{4,\text{dip}}$ is negative for $\phi < 30°$ and positive for $\phi > 30°$.

Substitution of $\alpha = 90°$ into equation 3.27 yields the strike-line coefficient:

$$A_{4,\text{strike}} = -\frac{2}{t_{P0}^2 V_{P0}^4} \left(\eta^{(1)} \cos^2 \phi \, - \, \eta^{(2)} \cos^2 \phi \, \sin^2 \phi \, + \, \eta^{(3)} \sin^2 \phi \right) . \tag{3.30}$$

In contrast to the dip-line coefficient, $A_{4,\text{strike}}$ depends on all three parameters $\eta^{(1,2,3)}$ because reflected rays recorded on the strike line x_2 deviate from the vertical symmetry plane $[x_2, x_3]$. For the same reason, $A_{4,\text{strike}}$ differs from the strike-line coefficient in VTI media with $\eta = \eta^{(1)}$ (the parameter $\eta^{(1)}$ is defined in the $[x_2, x_3]$-plane). Still, equation 3.30 indicates that the parameter $\eta^{(1)}$ does control the strike-line A_4 for mild dips ϕ.

Unlike the quartic coefficient on the dip line, $A_{4,\text{strike}}$ does not always decrease in absolute value with reflector dip. The function $A_{4,\text{strike}}(\phi)$ is governed by the relative magnitudes of the three anellipticity parameters and becomes primarily dependent on $\eta^{(3)}$ for large dips. If the interface is vertical ($\phi = 90°$), reflected rays travel in the horizontal symmetry plane, and the strike-line quartic coefficient reduces to

$$A_{4,\text{strike}}(\phi = 90°) = -\frac{2\eta^{(3)}}{t_{P0}^2 V_{P0}^4} . \tag{3.31}$$

Equation 3.31 coincides with the weak-anisotropy approximation for A_4 in a horizontal VTI layer with $\eta = \eta^{(3)}$. Therefore, if $|\eta^{(3)}| > |\eta^{(1)}|$, the magnitude of $A_{4,\text{strike}}$ for a vertical reflector is higher than that for a horizontal reflector.

Azimuthal dependence

The azimuthal variation of the quartic coefficient from equation 3.27 strongly depends on the dip as well as the magnitudes and signs of the parameters $\eta^{(1,2,3)}$. For mild dips, $A_4(\alpha)$ is largely controlled by the parameters $\eta^{(1)}$ (near the $[x_2, x_3]$-plane) and $\eta^{(2)}$ (near the $[x_1, x_3]$-plane), with the influence of $\eta^{(3)}$ increasing toward $\alpha = 45°$ but typically remaining relatively weak (see equation 3.25). As illustrated by Figure 3.8, if both $\eta^{(1)}$ and $\eta^{(2)}$ are positive and much greater by absolute value than $\eta^{(3)}$, A_4 typically stays negative for dips smaller than 30° (the same result is valid for a dipping VTI layer). Also, as in VTI media, for a dip of 30° A_4 goes to zero only in the single (dip) direction (Figure 3.8b), although it remains small for a range of azimuths ($|\alpha| < 35 - 40°$) near the dip plane.

For the η-parameters used in Figure 3.8 and dips between 30° and 90°, equation 3.27 yields one azimuth in the range $0° < \alpha < 90°$ for which $A_4 = 0$. If the dip $\phi = 45°$, the quartic coefficient is positive near the dip plane, vanishes at $\alpha \approx 60°$, and becomes negative close to the strike direction (Figure 3.8c). Although typically we expect the quartic coefficient for dips $\phi > 45°$ to be larger in the strike direction than in the dip plane, for the model in Figure 3.8 the strike-line A_4 is small because the terms involving $\eta^{(1)}$ and $\eta^{(2)}$ in equation 3.30 almost cancel each other.

When $\eta^{(1)}$ and $\eta^{(2)}$ have opposite signs, the azimuthal variation of the quartic coefficient has a different character. Then for mild dips A_4 changes sign between the dip

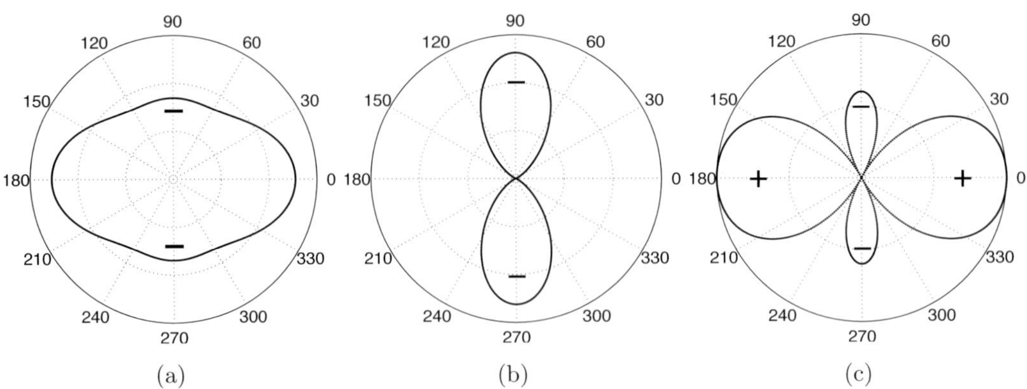

Figure 3.8: Normalized magnitude of the azimuthally varying quartic moveout coefficient A_4 computed from equation 3.27 for a dipping orthorhombic layer (Pech and Tsvankin, 2004). The azimuth α is marked on the perimeter; α=0° corresponds to the dip direction. The anellipticity parameters are $\eta^{(1)}$=0.05, $\eta^{(2)}$=0.1, and $\eta^{(3)}$=0.03. The parameters t_{P0} and V_{P0} change only the scale of the plot (intentionally undefined here). The sign of A_4 for each lobe is marked on the plots. The reflector dip is (a) 15°, (b) 30°, and (c) 45°.

and strike directions (see equations 3.29 and 3.30), with the azimuth of vanishing A_4 dependent on the dip and the relative magnitudes of $\eta^{(1)}$ and $\eta^{(2)}$ (Figure 3.9). Since the quartic coefficient decreases with ϕ most rapidly in the dip plane, the azimuth of $A_4 = 0$ rotates towards the dip direction as ϕ increases. For a dip of 15° (Figure 3.9a), A_4 goes to zero at the angle $\alpha \approx 35°$. When the dip reaches 30° (Figure 3.9b), the quartic coefficient goes to zero in the dip direction, and the azimuthal variation of A_4 is similar to that in Figure 3.8b. However, the sign of A_4 away from the dip plane in Figure 3.9b is positive because $\eta^{(1)} < 0$. Finally, for $\phi > 30°$, the quartic coefficient is positive for all azimuths (Figure 3.9c).

To analyze the influence of the parameter $\eta^{(3)}$ on the quartic coefficient, in the next example we set $\eta^{(1)}$ and $\eta^{(2)}$ to zero (Figure 3.10). The term involving $\eta^{(3)}$ creates a rather complicated azimuthal signature of A_4. The coefficient $A_{4,\text{dip}}$ (equation 3.29) vanishes for both dips because it is proportional to $\eta^{(2)}$. In addition, A_4 goes to zero for another azimuth, which is close to the strike direction for a dip of 15° (Figure 3.10a). The azimuthal variation of A_4 for $\phi = 30°$ has a similar general character, but with completely different relative magnitudes of the lobes (Figure 3.10b).

For nonzero values of $\eta^{(1)}$ and $\eta^{(2)}$, the contribution of the term proportional to $\eta^{(3)}$ generally increases with reflector dip (Figure 3.11). While the dip-line coefficient A_4 depends just on $\eta^{(2)}$, $A_{4,\text{strike}}$ is significantly influenced by $\eta^{(3)}$, in particular for dips exceeding 30°. If $\eta^{(3)}$ is larger in absolute value than $\eta^{(1)}$ and $\eta^{(2)}$, the quartic coefficient typically goes to zero at one azimuth between the symmetry planes for a wide range of dips. For example, if $\phi = 30°$, the influence of $\eta^{(3)}$ produces an additional direction of vanishing A_4 near the reflector strike (Figure 3.11b).

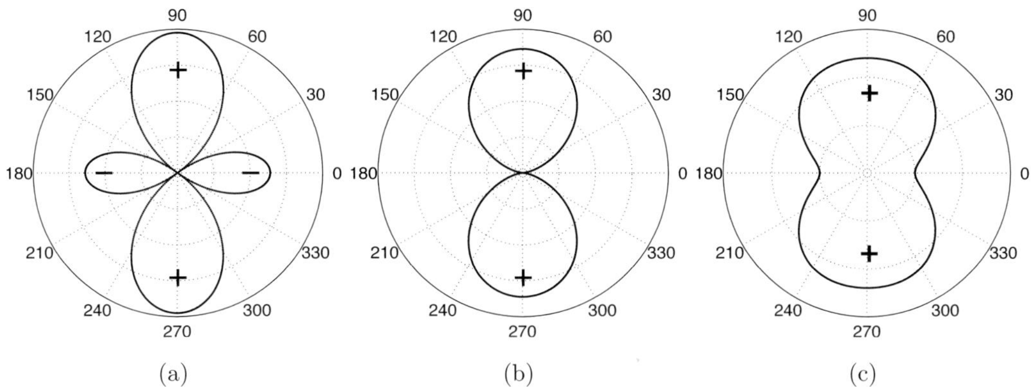

Figure 3.9: Magnitude of A_4 computed from equation 3.27 for a dipping orthorhombic layer (Pech and Tsvankin, 2004). The anellipticity parameters are $\eta^{(1)}=-0.1$, $\eta^{(2)}=0.1$, and $\eta^{(3)}=0.03$. The reflector dip is (a) 15°, (b) 30°, and (c) 45°.

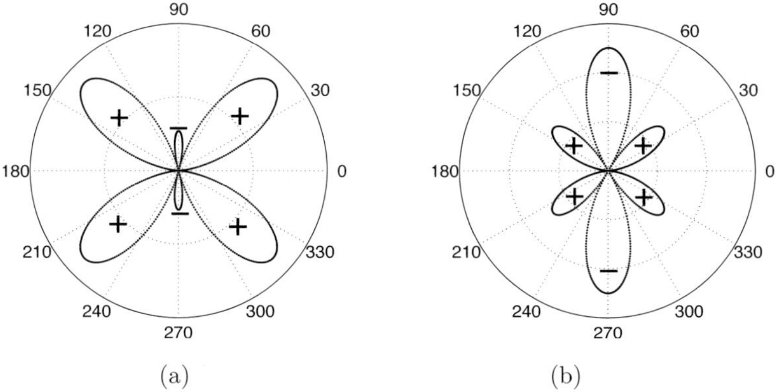

Figure 3.10: Magnitude of A_4 computed from equation 3.27 for a dipping orthorhombic layer (Pech and Tsvankin, 2004). The anellipticity parameters are $\eta^{(1)}=\eta^{(2)}=0$ and $\eta^{(3)}=0.1$. The reflector dip is (a) 15°, and (b) 30°.

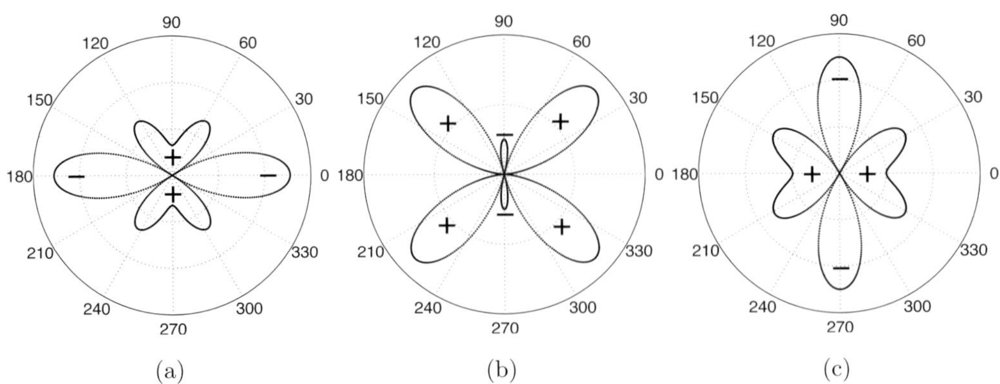

Figure 3.11: Magnitude of A_4 computed from equation 3.27 for a dipping orthorhombic layer (Pech and Tsvankin, 2004). The anellipticity parameters are $\eta^{(1)}=-0.025$, $\eta^{(2)}=0.1$, and $\eta^{(3)}=0.2$. The reflector dip is (a) 15°, (b) 30°, and (c) 45°.

3.3 Nonhyperbolic moveout inversion for orthorhombic and HTI media

Orthorhombic symmetry adequately represents most naturally fractured sedimentary formations, even those with multiple fracture sets (see section 2.5 and Chapter 9). In principle, useful information for anisotropic velocity analysis and fracture characterization can be obtained from NMO ellipses alone, without employing reflection traveltime at long offsets (Chapters 2,6, and 9). The inversion of NMO ellipses, however, is often ambiguous and requires a priori constraints on the medium symmetry and some of the parameters, in particular when only P-wave data are available.

Nonhyperbolic moveout inversion for orthorhombic media helps estimate the anellipticity parameters $\eta^{(1,2,3)}$, which are needed for time-domain P-wave processing and can provide information for fracture characterization. Some case studies (see section 3.4) also indicate that the nonhyperbolic portion of the moveout function is more sensitive to azimuthal anisotropy than is the NMO ellipse. Application of nonhyperbolic moveout analysis, however, is hampered by the trade-off between the NMO velocities and the quartic coefficient A_4, which can cause significant errors in the parameters $\eta^{(1,2,3)}$. Processing of long-spread data also has to deal with such issues as a higher level of noise at large offsets, distortions due to lateral heterogeneity, and NMO stretch.

3.3.1 Moveout equations for orthorhombic media

The description of the quartic moveout coefficient in the previous section makes it possible to extend the nonhyperbolic moveout equations of Tsvankin and Thomsen (1994) and Alkhalifah and Tsvankin (1995) to layered orthorhombic and HTI media. Using the results of Vasconcelos and Tsvankin (2006) and Xu and Tsvankin (2006a), we give an analytic description of nonhyperbolic moveout for a single orthorhombic layer followed by a more general formalism for vertically heterogeneous, azimuthally anisotropic models.

Single horizontal layer

To make the Tsvankin-Thomsen equation 3.3 suitable for an orthorhombic layer, the moveout coefficients V_{nmo}, A_4, and A must be obtained as functions of azimuth:

$$t^2(x,\alpha) = t_{P0}^2 + \frac{x^2}{V_{\text{nmo}}^2(\alpha)} + \frac{A_4(\alpha)\,x^4}{1+A(\alpha)\,x^2}\,; \tag{3.32}$$

as before, α is the angle with the $[x_1, x_3]$ symmetry plane. Because of the kinematic equivalence between the symmetry planes of orthorhombic and VTI media, the VTI version of equation 3.3 is entirely valid in the $[x_1, x_3]$- and $[x_2, x_3]$-planes (i.e., for $\alpha = 0°$ and $\alpha = 90°$). The symmetry-plane moveout coefficients V_{nmo}, A_4, and A can be adapted in a straightforward way from the corresponding VTI equations.

The azimuthally varying P-wave NMO velocity satisfies the elliptical equation 2.61:

$$V_{\text{nmo}}^{-2}(\alpha) = \frac{\sin^2\alpha}{[V_{\text{nmo}}^{(1)}]^2} + \frac{\cos^2\alpha}{V_{\text{nmo}}^{(2)}]^2}\,, \tag{3.33}$$

$$V_{\text{nmo}}^{(i)} = V_{P0}\sqrt{1+2\delta^{(i)}}\,, \quad (i=1,2)\,. \tag{3.34}$$

The azimuthally varying quartic moveout coefficient can be obtained from equation 3.21:

$$A_4(\alpha) = A_4^{(1)}\sin^4\alpha + A_4^{(x)}\sin^2\alpha\cos^2\alpha + A_4^{(2)}\cos^4\alpha\,, \tag{3.35}$$

with $A_4^{(1)}$, $A_4^{(2)}$, and $A_4^{(x)}$ given by equations 3.22, 3.23, and 3.24. The coefficient $A(\alpha)$ (equation 3.4) also depends on the horizontal group velocity, which can be computed from the velocity equations in the $[x_1, x_2]$-plane.

Substitution of $V_{\text{nmo}}(\alpha)$ and $A_4(\alpha)$ (equations 3.33 and 3.35) into the general nonhyperbolic moveout equation 3.32 yields a close approximation for P-wave reflection traveltimes up to offset-to-depth ratios of about three (Al-Dajani et al., 1998). Application of equation 3.32 in moveout analysis, however, involves estimating three independent, azimuthally varying moveout coefficients.

For vertical transverse isotropy, the coefficients A_4 and A can be expressed through the velocity V_{nmo} and the anellipticity parameter η, which leads to the two-parameter Alkhalifah-Tsvankin (1995) moveout equation widely used in seismic processing:

$$t^2(x) = t_{P0}^2 + \frac{x^2}{V_{\text{nmo}}^2} - \frac{2\eta\,x^4}{V_{\text{nmo}}^2\left[t_{P0}^2 V_{\text{nmo}}^2 + (1+2\eta)x^2\right]}\,. \tag{3.36}$$

Equation 3.36 can be directly applied in both vertical symmetry planes of the orthorhombic layer with the appropriate pair of the parameters $V_{\text{nmo}}^{(i)}$ and $\eta^{(i)}$ $(i=1,2)$.

Furthermore, the approximate kinematic equivalence between vertical planes of orthorhombic and VTI media (see equation 3.25) justifies extending the Alkhalifah-Tsvankin equation to off-symmetry directions by taking into account the azimuthal variation of $V_{\rm nmo}$ and η:

$$t^2(x,\alpha) = t_{P0}^2 + \frac{x^2}{V_{\rm nmo}^2(\alpha)} - \frac{2\eta(\alpha)\, x^4}{V_{\rm nmo}^2(\alpha)\left[t_{P0}^2\, V_{\rm nmo}^2(\alpha) + (1+2\eta(\alpha))\, x^2\right]}\,, \tag{3.37}$$

where $V_{\rm nmo}(\alpha)$ is the NMO ellipse (equation 3.33), and the azimuthally varying parameter η can be found from equation 3.26:

$$\eta(\alpha) = \eta^{(1)} \sin^2\alpha \;-\; \eta^{(3)} \sin^2\alpha \cos^2\alpha \;+\; \eta^{(2)} \cos^2\alpha\,. \tag{3.38}$$

Equation 3.37 depends on a smaller number of moveout parameters than equation 3.32 and is easier to implement in data processing and inversion (see below). In the symmetry plane $[x_1, x_3]$ $(\alpha = 0°)$, equation 3.37 reduces to the VTI moveout equation parameterized by $V_{\rm nmo}^{(2)}$ and $\eta^{(2)}$, while for the $[x_2, x_3]$-plane $(\alpha = 90°)$, the parameters are $V_{\rm nmo}^{(1)}$ and $\eta^{(1)}$. The accuracy of equation 3.37 in both symmetry planes is exactly the same as that in the corresponding VTI medium and can be assessed using the results of Grechka and Tsvankin (1998b) and Tsvankin (2005).

For off-symmetry directions, use of the equivalent VTI model and of equation 3.38 for $\eta(\alpha)$ is strictly valid only in the weak-anisotropy approximation. For instance, equation 3.37 does not preserve the correct asymptotic behavior of reflection traveltime for $x \to \infty$ outside the symmetry planes. Still, it should be emphasized that the original Alkhalifah-Tsvankin equation adapted here to orthorhombic media is not limited to weak anisotropy and does not involve linearization in the parameters ϵ, δ, or η. The numerical examples below demonstrate that equation 3.37 (combined with equation 3.33 for $V_{\rm nmo}$ and equation 3.38 for η) is sufficiently accurate even for orthorhombic layers with strong azimuthal anisotropy.

In principle, it may be possible to extend the formalism based on shifted hyperbolas (Siliqi and Bousqué, 2000) and other existing VTI moveout equations (e.g., Fomel, 2004; van der Baan 2004; Douma and Calvert, 2006) to orthorhombic media by making the pertinent moveout parameters azimuthally dependent. However, such generalized equations for wide-azimuth data still have to be developed.

Layered orthorhombic/TI media

The moveout equations described above can be extended to a horizontally stratified medium composed of orthorhombic, VTI, and HTI layers. First, we assume that the vertical symmetry planes in each orthorhombic and HTI layer have the same azimuth and, therefore, represent planes of symmetry for the whole model. Because of the kinematic equivalence between the symmetry planes of orthorhombic, VTI, and HTI media, P-wave nonhyperbolic moveout in the symmetry-plane directions can be modeled by equations 3.36 or 3.37 with the effective parameters $V_{\rm nmo}$ and η computed from the VTI averaging expressions (Appendix 4B in Tsvankin, 2005).

To apply equation 3.37 in off-symmetry directions of layered media, the effective NMO ellipse $V_{\rm nmo}(\alpha)$ can be found from generalized Dix equation 1.31 introduced in Chapter 1. The averaging equations for the effective quartic coefficient A_4 and parameter $\eta(\alpha)$ are more complicated and, in principle, can be derived using the exact equation 3.5. Al-Dajani and Tsvankin (1998) show that if azimuthal anisotropy is not severe, a close approximation for $\eta(\alpha)$ is provided by applying the VTI averaging equation for each azimuth (equation 4.47 in Tsvankin, 2005):

$$\eta(\alpha) = \frac{1}{8}\left\{\frac{1}{V_{\rm nmo}^4(\alpha)\, t_{P0}}\left[\sum_{i=1}^{N}\left(V_{\rm nmo}^{(i)}(\alpha)\right)^4\left(1+8\eta^{(i)}(\alpha)\right)t_{P0}^{(i)}\right]-1\right\}, \tag{3.39}$$

where $V_{\rm nmo}(\alpha)$ is the effective NMO ellipse, and $V_{\rm nmo}^{(i)}(\alpha)$ and $\eta^{(i)}(\alpha)$ are the interval parameters in layer i. According to the numerical analysis in Xu and Tsvankin (2006a), the azimuthal variation of the parameter η from equation 3.39 is accurately described by the single-layer equation 3.38. The parameters $\eta^{(1,2,3)}$, however, then represent effective quantities for the section above the reflector.

Since the symmetry-plane orientation prior to moveout analysis is generally unknown, the equations for the NMO ellipse and the effective parameter $\eta(\alpha)$ have to include the azimuth φ of the symmetry plane $[x_1, x_3]$:

$$V_{\rm nmo}^{-2}(\alpha) = \frac{\sin^2(\alpha-\varphi)}{[V_{\rm nmo}^{(1)}]^2} + \frac{\cos^2(\alpha-\varphi)}{[V_{\rm nmo}^{(2)}]^2}, \tag{3.40}$$

$$\eta(\alpha) = \eta^{(1)}\sin^2(\alpha-\varphi) - \eta^{(3)}\sin^2(\alpha-\varphi)\cos^2(\alpha-\varphi) + \eta^{(2)}\cos^2(\alpha-\varphi)\,. \tag{3.41}$$

For models with depth-varying azimuths of the vertical symmetry planes, the NMO ellipse can still be computed from the generalized Dix equation. The absence of throughgoing symmetry planes, however, causes misalignment of the NMO ellipse with the principal axes of the azimuthally dependent parameter $\eta(\alpha)$ (Xu and Tsvankin, 2006a). The simplest way to account for this misalignment is to introduce a separate azimuth φ_1 responsible for the azimuthal variation of the effective parameter $\eta(\alpha)$ in equation 3.41:

$$\eta(\alpha) = \eta^{(1)}\sin^2(\alpha-\varphi_1) - \eta^{(3)}\sin^2(\alpha-\varphi_1)\cos^2(\alpha-\varphi_1) + \eta^{(2)}\cos^2(\alpha-\varphi_1)\,. \tag{3.42}$$

Below, we study the accuracy of the nonhyperbolic moveout equation 3.37 for layered azimuthally anisotropic media using equation 3.40 for the NMO ellipse and two different forms of the effective parameter $\eta(\alpha)$ (equations 3.41 and 3.42).

3.3.2 Moveout-inversion algorithm

To estimate the azimuth φ (and possibly φ_1) and the effective moveout parameters $V_{\rm nmo}^{(1)}$, $V_{\rm nmo}^{(2)}$, $\eta^{(1)}$, $\eta^{(2)}$, and $\eta^{(3)}$, Vasconcelos and Tsvankin (2006) devised nonhyperbolic semblance search based on equations 3.37 and 3.40–3.42. If the model is known to have aligned vertical symmetry planes, the parameter $\eta(\alpha)$ is described by equation 3.41; otherwise, the algorithm employs equation 3.42. Nonhyperbolic moveout inversion in orthorhombic media requires long-spread P-wave data to be recorded for a wide azimuthal range. Whereas the NMO ellipse can be reconstructed using a minimum of three differently oriented lines, the inversion for the parameters $\eta^{(1,2,3)}$ is less stable and cannot be accomplished without a much more complete azimuthal coverage.

One common way of processing wide-azimuth data is to divide the traces in each common-midpoint gather into azimuthal sectors with a chosen increment in azimuth (sometimes as large as 45°). Then moveout inversion for $V_{\text{nmo}}(\alpha)$ and $\eta(\alpha)$ can be performed in each sector separately, and the parameters obtained for all sectors can be combined into the "best-fit" orthorhombic model. Azimuthal sectoring, however, can introduce bias into velocity analysis because data sampling in offset and azimuth is never uniform, and the inversion results are sensitive to the number and size of the sectors (Grechka and Tsvankin, 1999a; Lynn, 2007). Moreover, combining traces with different azimuths into the same sector can be particularly harmful for nonhyperbolic moveout inversion because of pronounced traveltime variations with azimuth at long offsets.

More robust results can be achieved by processing all traces for a wide-azimuth CMP gather or "superbin" (a composite CMP gather that includes source-receiver pairs with nearby midpoints) simultaneously using a semblance operator. In Chapter 2, this approach was efficiently applied to estimation of NMO ellipses. Nonhyperbolic moveout inversion operates with long-spread data acquired for a wide range of offsets (from $x_{\min}$ to $x_{\max}$) and azimuths (from $\alpha_{\min}$ to $\alpha_{\max}$). The 3D semblance function to be maximized for each zero-offset reflection time t_{P0} is defined as

$$S(t_{P0}, V_{\text{nmo}}^{(1)}, V_{\text{nmo}}^{(2)}, \eta^{(1)}, \eta^{(2)}, \eta^{(3)}, \varphi, \varphi_1) = \frac{\sum_{t_{P0}-\Delta T/2}^{t_{P0}+\Delta T/2} \left[\sum_{x_{\min}}^{x_{\max}} \sum_{\alpha_{\min}}^{\alpha_{\max}} U(x, \alpha, t)\right]^2}{\sum_{t_{P0}-\Delta T/2}^{t_{P0}+\Delta T/2} \sum_{x_{\min}}^{x_{\max}} \sum_{\alpha_{\min}}^{\alpha_{\max}} U^2(x, \alpha, t)}, \tag{3.43}$$

where $U(x, \alpha, t)$ is the seismic trace in a wide-azimuth CMP gather, and the semblance is calculated for the time window ΔT around the reflection time t obtained from equation 3.37.

To increase the efficiency of the algorithm, the full 3D semblance search is preceded by two preliminary parameter-estimation steps designed to build an initial model. First, the algorithm described in Chapter 2 is applied to reconstruct the best-fit NMO ellipse for each reflection event using the offset range limited by the reflector depth (see equation 2.9). The orientation and semiaxes of the NMO ellipse yield the initial values of the azimuth φ and velocities $V_{\text{nmo}}^{(1)}$ and $V_{\text{nmo}}^{(2)}$. Second, the parameters $\eta^{(1)}$ and $\eta^{(2)}$ are estimated from 2D nonhyperbolic moveout inversion in the narrow azimuthal sectors aligned with the axes of the NMO ellipse. These two steps typically produce good approximations for the parameters φ and $V_{\text{nmo}}^{(1,2)}$, while the initial values of $\eta^{(1,2)}$ can substantially deviate from the actual effective parameters if the symmetry-plane azimuths vary with depth. Finally, the Powell method (Press et al., 1987) is used to invert for all moveout parameters (φ, φ_1, $V_{\rm nmo}^{(1,2)}$, and $\eta^{(1,2,3)}$) by maximizing semblance for the full range of offsets and azimuths. In most cases, the search does not start far from the maximum of the objective function, and the algorithm converges in less than 20 iterations.

3.3.3 Tests on synthetic data

Synthetic examples are designed to evaluate the robustness of the inversion in the presence of noise and the convergence of the algorithm when the search starts from different initial solutions in the model space. It is particularly important to study the performance of equation 3.37 for models composed of layers with strong azimuthal anisotropy.

Single orthorhombic layer

The first test (Figure 3.12) was carried out for a single horizontal orthorhombic layer using full-azimuth, long-spread P-wave synthetic data generated by anisotropic ray tracing (Gajewski and Pšenčík, 1987). First, following the methodology described above, the NMO ellipse was reconstructed using source-receiver pairs with offset-to-depth ratios up to unity, which gave the initial values for the symmetry-plane azimuth $\varphi = 130°$ and the velocities $V_{\rm nmo}^{(1)} = 2345$ m/s and $V_{\rm nmo}^{(2)} = 2715$ m/s (the model parameters are listed in the caption of Figure 3.12). The high semblance value (0.97) and the accurate estimate of φ confirm that the obtained NMO ellipse provides a good fit to conventional-spread data. The deviations of the symmetry-plane NMO velocities from the actual values are caused by the small influence of nonhyperbolic moveout unaccounted for at this stage of the inversion.

Next, approximate values of $\eta^{(1)}$ and $\eta^{(2)}$ were obtained from 2D nonhyperbolic moveout analysis (i.e., the VTI inversion) in two azimuthal sectors 10°-wide around the identified symmetry-plane directions. The semblance in both sectors is displayed in Figure 3.12 as a function of $V_{\rm nmo}^{(1,2)}$ and the corresponding horizontal velocity $V_{\rm hor}^{(1,2)} = V_{\rm nmo}^{(1,2)}\sqrt{1+2\eta^{(1,2)}}$. The errors in both η-parameters are caused by a small bias in the 2D nonhyperbolic moveout equation and by the azimuthal traveltime variations within both sectors. Application of the 3D nonhyperbolic moveout inversion with these initial values yielded accurate estimates of all model parameters: $\varphi = 130°$, $V_{\rm nmo}^{(1)} = 2277$ m/s, $V_{\rm nmo}^{(2)} = 2703$ m/s, $\eta^{(1)} = 0.18$, $\eta^{(2)} = 0.06$, and $\eta^{(3)} = 0.11$; the semblance for the best-fit model is 0.89.

Although the algorithm in this noise-free test was able to converge toward the correct model, it is important to identify possible trade-offs among the medium parameters by studying the shape of the objective (semblance) function. To display such trade-offs, we held four model parameters at the correct values and computed the semblance near the actual solution as a function of the two remaining parameters. The semblance scan over the parameters $V_{\rm nmo}^{(2)}$ and $\eta^{(2)}$ in Figure 3.13 shows that there exists a family of models with a relatively wide range of $\eta^{(2)}$-values that fit the data, whereas the velocity $V_{\rm nmo}^{(2)}$ is constrained more tightly.

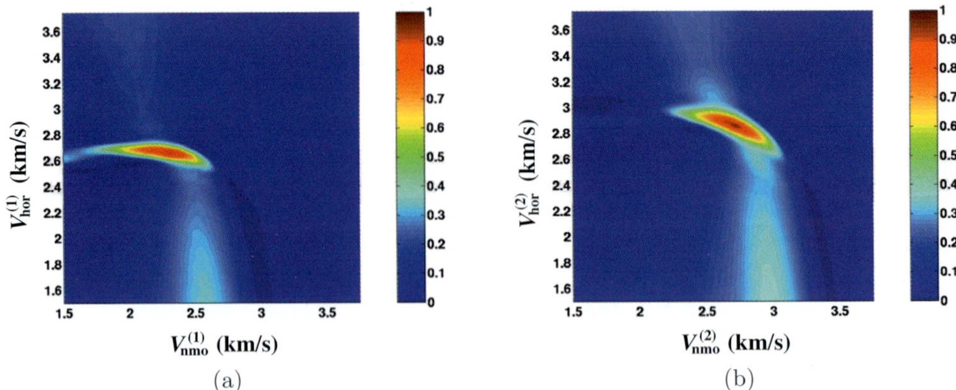

Figure 3.12: Synthetic test for full-azimuth P-wave data from a horizontal orthorhombic layer computed by anisotropic ray tracing (Vasconcelos and Tsvankin, 2006). The plots show semblance contours obtained by scanning over V_{nmo} and the horizontal velocity V_{hor} (i.e., by applying the VTI nonhyperbolic moveout-inversion algorithm) in 10°-wide azimuthal sectors centered at the vertical symmetry planes identified from the NMO ellipse. The spreadlength-to-depth ratio is three. (a) The sector around the $[x_2, x_3]$ symmetry plane at an azimuth of 40° (estimated $\eta^{(1)}$=0.18); (b) the sector around the $[x_1, x_3]$ symmetry plane at an azimuth of 130° (estimated $\eta^{(2)}$=0.05). The model parameters are φ=130°, $V_{nmo}^{(1)}$=2269 m/s, $V_{nmo}^{(2)}$=2699 m/s, $\eta^{(1)}$=0.196, $\eta^{(2)}$=0.065, and $\eta^{(3)}$=0.094.

This trade-off between $V_{nmo}^{(2)}$ and $\eta^{(2)}$ (and between $V_{nmo}^{(1)}$ and $\eta^{(1)}$ as well) has the same character as the interplay between V_{nmo} and η in VTI media (Grechka and Tsvankin 1998b; Tsvankin 2005). Relatively small percentage errors in $V_{nmo}^{(i)}$ can be compensated by (typically larger) absolute errors in $\eta^{(i)}$ ($i = 1,2$) in such a way that the reflection traveltime stays almost the same. Therefore, the inverted parameters $\eta^{(1,2)}$ are sensitive to correlated noise in traveltimes produced, for example, by statics errors. (Because parameter estimation is based on the semblance (coherency) operator, it remains stable in the presence of random noise.) The uncertainty in the parameteres $\eta^{(1,2)}$ reduces with increasing spreadlength-to-depth ratio (Figure 3.13).

Since the parameter $\eta^{(3)}$ has no influence on reflection traveltimes near both symmetry planes, it is constrained more loosely compared to $\eta^{(1,2)}$ (Figure 3.14a). In contrast, the inversion algorithm yields a highly accurate estimate of the azimuth φ (Figure 3.14b) because the orientation of the symmetry planes can be inferred from both the NMO ellipse and azimuthally varying nonhyperbolic moveout.

Layered media

Extensive numerical testing carried out by Xu and Tsvankin (2006a) and Vasconcelos and Tsvankin (2006) shows that the nonhyperbolic equation 3.37 with the best-fit parameters gives a close approximation to P-wave moveout in layered azimuthally

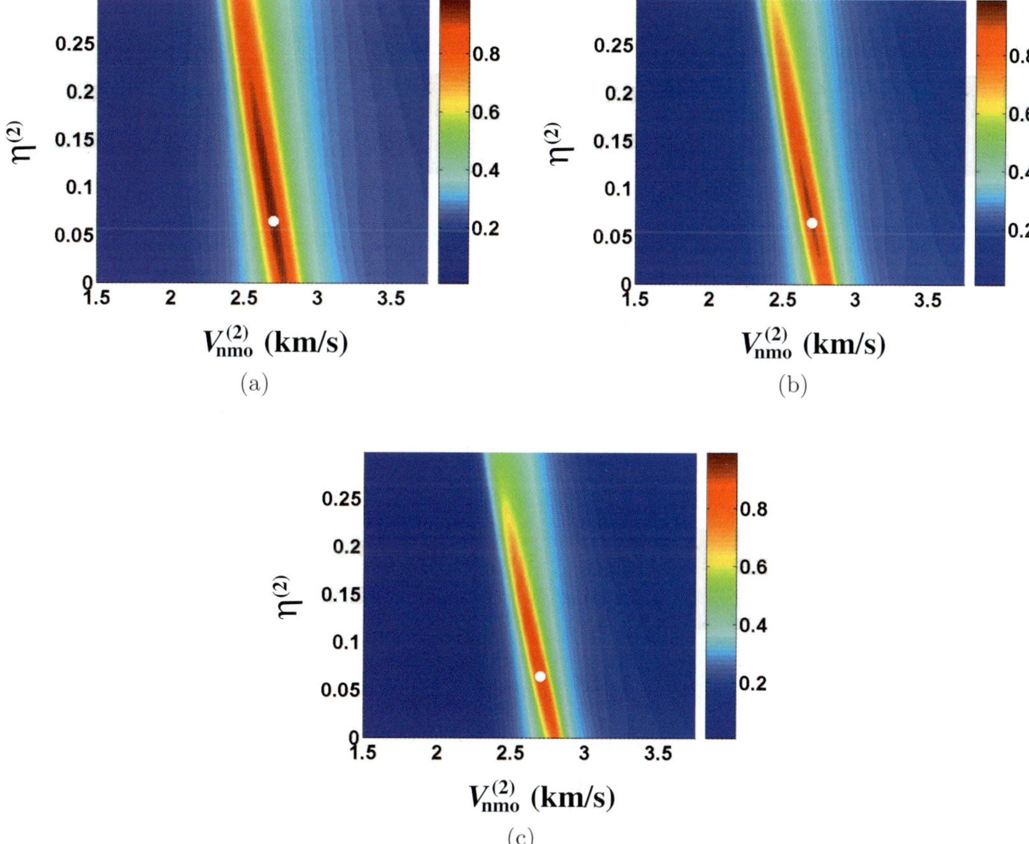

Figure 3.13: Semblance scans over $V_{\rm nmo}^{(2)}$ and $\eta^{(2)}$ computed from equation 3.37 for the following spreadlength-to-depth ratios: (a) $x_{\rm max}/D$=2, (b) $x_{\rm max}/D$=2.5, and (c) $x_{\rm max}/D$=3 (Vasconcelos and Tsvankin, 2006). The model is the same as that in Figure 3.12; the parameters φ, $V_{\rm nmo}^{(1)}$, $\eta^{(1)}$, and $\eta^{(3)}$ are fixed at the correct values. The white dots here and below correspond to the actual model.

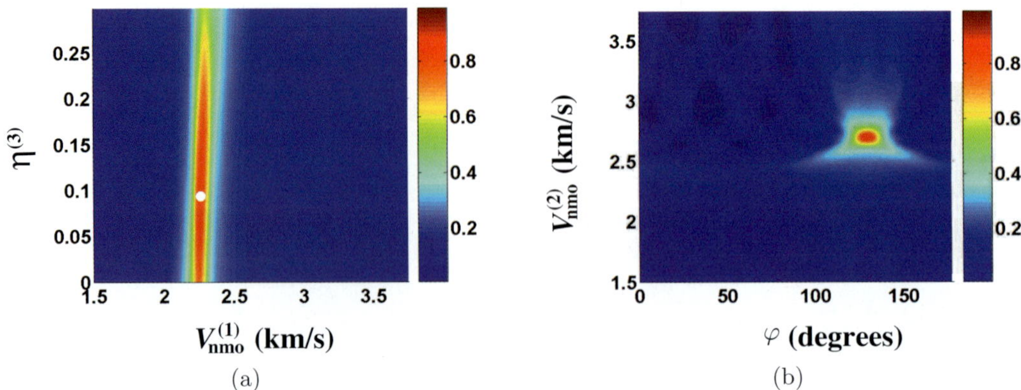

Figure 3.14: Semblance scans over (a) $V_{\text{nmo}}^{(1)}$ and $\eta^{(3)}$; and (b) φ and $V_{\text{nmo}}^{(2)}$ computed from equation 3.37 for the model from Figure 3.12 (Vasconcelos and Tsvankin, 2006). For each scan, the other four parameters are fixed at the correct values; the spreadlength-to-depth ratio is three.

anisotropic media. It should be emphasized that in this section exact traveltimes are not compared with the analytic form of the moveout function. Layer stripping of the effective moveout parameters and interval parameter estimation are discussed in section 3.4.

The model in Figure 3.15 includes an orthorhombic layer overlain by a VTI layer of equal thickness. Application of the moveout inversion resulted in best-fit equation 3.37 that practically coincides with exact traveltimes for the full range of offsets and azimuths. The resolution in the symmetry-plane azimuth φ, however, is lower than that for a homogeneous orthorhombic medium because of the smaller contribution of the azimuthally anisotropic layer to the total traveltime. The error in φ propagates into the rest of the moveout parameters, which represent average values for the two layers. In general, the accuracy in estimating the symmetry-plane orientation depends on the relative thickness (i.e., on the ratio of the thickness and depth) of the orthorhombic layer.

To evaluate the performance of the algorithm for models with misaligned vertical symmetry planes, the VTI layer was replaced with an orthorhombic medium whose $[x_1, x_3]$-plane is rotated by 40° from the x_1-direction in the bottom layer (Figure 3.16). The first test (Figure 3.16a) was carried out with equation 3.41, which ignores the misalignment of $\eta(\alpha)$ and the NMO ellipse. The best-fit azimuth φ is somewhat closer to the $[x_1, x_3]$ symmetry plane in the top layer and deviates from the azimuth φ_1 predicted by equation 3.39 (white dot) by approximately 15°. Evidently, the model with a uniform orientation of the symmetry planes does not adequately describe nonhyperbolic moveout. Another indication of the misalignment of $\eta(\alpha)$ and the NMO ellipse is the relatively low semblance for the best-fit model (0.7; compare with much higher values in Figure 3.15). It is noteworthy that this reduction in semblance

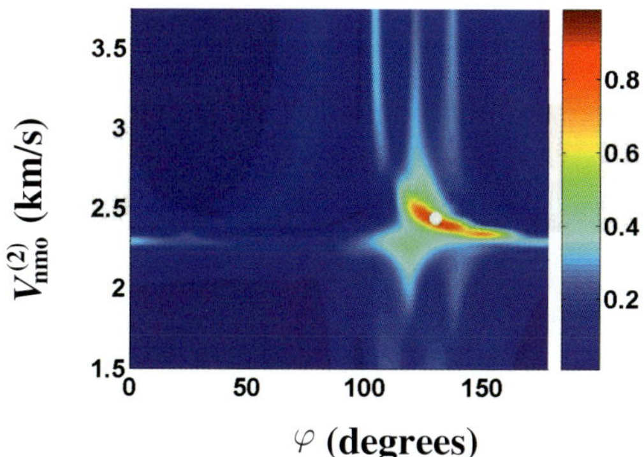

Figure 3.15: Semblance scan over the azimuth φ and the effective NMO velocity $V_{nmo}^{(2)}$ for the reflection from the bottom of a model composed of two layers with equal thickness (Vasconcelos and Tsvankin, 2006). The orthorhombic layer from Figure 3.12 is overlain by a VTI layer with V_{nmo}=2158 m/s and η=0.196. The other moveout parameters are fixed at the best-fit values; the spreadlength-to-depth ratio is three.

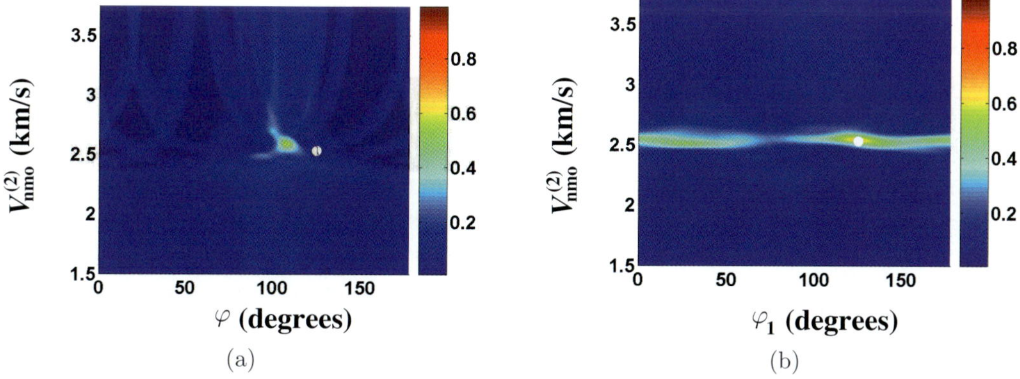

Figure 3.16: Semblance scans for the reflection from the bottom of a model composed of two orthorhombic layers with equal thickness (Vasconcelos and Tsvankin, 2006). The bottom layer is taken from Figure 3.12; its symmetry-plane azimuth φ=130°. The top layer has a different orientation of the symmetry planes (φ=90°), and the parameters $V_{nmo}^{(1)}$=2156 m/s, $V_{nmo}^{(2)}$=2534 m/s, $\eta^{(1)}$=0.398, $\eta^{(2)}$=0.211, and $\eta^{(3)} = 0.193$. The parameter $\eta(\alpha)$ in the inversion is computed from (a) equation 3.41; and (b) equation 3.42. The white dot marks the azimuth φ_1 predicted by equation 3.39. The spreadlength-to-depth ratio is three.

is caused by differences of just 1% or less between equation 3.37 and exact traveltimes.

To account for misaligned symmetry planes, the function $\eta(\alpha)$ should be decoupled from the NMO ellipse by replacing equation 3.41 with equation 3.42. This modification increased the value of semblance by about 20% (Figure 3.16b) because the moveout equation now provides a closer fit to the exact traveltimes. The azimuth φ_1 that corresponds to the maximum semblance value in Figure 3.16b practically coincides with the theoretical prediction (white dot). Still, because φ_1 influences only the nonhyperbolic portion of the moveout curve, it is not as well resolved as is the azimuth φ responsible for the NMO ellipse. In general, the effective value of φ_1 is determined by the relative thicknesses, symmetry-plane orientations, and interval moveout parameters of the constituent layers.

3.3.4 Field-data example

The 3D nonhyperbolic moveout-inversion algorithm was applied by Vasconcelos and Tsvankin (2006) to wide-azimuth P-wave data acquired at Weyburn field located in the Williston basin in Canada. This multicomponent data set was processed and interpreted by the Reservoir Characterization Project at the Colorado School of Mines with the main goal of dynamic monitoring of the CO_2 flood in the fractured reservoir (only the base survey is used here). Jenner (2001) identified laterally varying azimuthal anisotropy at the reservoir level by computing the interval P-wave NMO ellipses and performing azimuthal AVO (amplitude-variation-with-offset) analysis. He found the azimuthal variation of the P-wave signatures in the overburden to be much less pronounced; his conclusions generally agree with the inversion of VSP data conducted by Adam et al. (2003).

A sensitive indicator of azimuthal anisotropy is shear-wave splitting at near-vertical incidence, which was studied at Weyburn field by Cardona (2002). According to his results, the magnitude of S-wave splitting in the overburden is nonnegligible, which likely indicates that much of the section is permeated by natural fractures. Deviation of the shear-wave polarization directions from the axes of the NMO ellipse also allowed Cardona (2002) to discriminate between different anisotropic symmetries at the reservoir level.

Analysis of long-spread moveout could contribute to fracture characterization by identifying the principal azimuthal directions of the quartic moveout coefficient and estimating the parameters $\eta^{(1,2,3)}$. As discussed above, if the azimuths of the symmetry planes (possibly related to fracture orientation) above the reflector vary with depth, the principal directions of $\eta(\alpha)$ may deviate from the axes of the NMO ellipse. Another possible reason for such misalignment is the presence of low anisotropic symmetries (i.e., monoclinic or lower) in the overburden. Also, the parameters $\eta^{(1,2,3)}$ for fracture-induced orthorhombic media are sensitive to the fracture properties (Bakulin et al., 2000b; see Chapter 9).

The distribution of source-to-receiver azimuths and offsets for two 9×9 CMP superbins from the survey is shown in Figures 3.17a,b. NMO-corrected gathers confirm

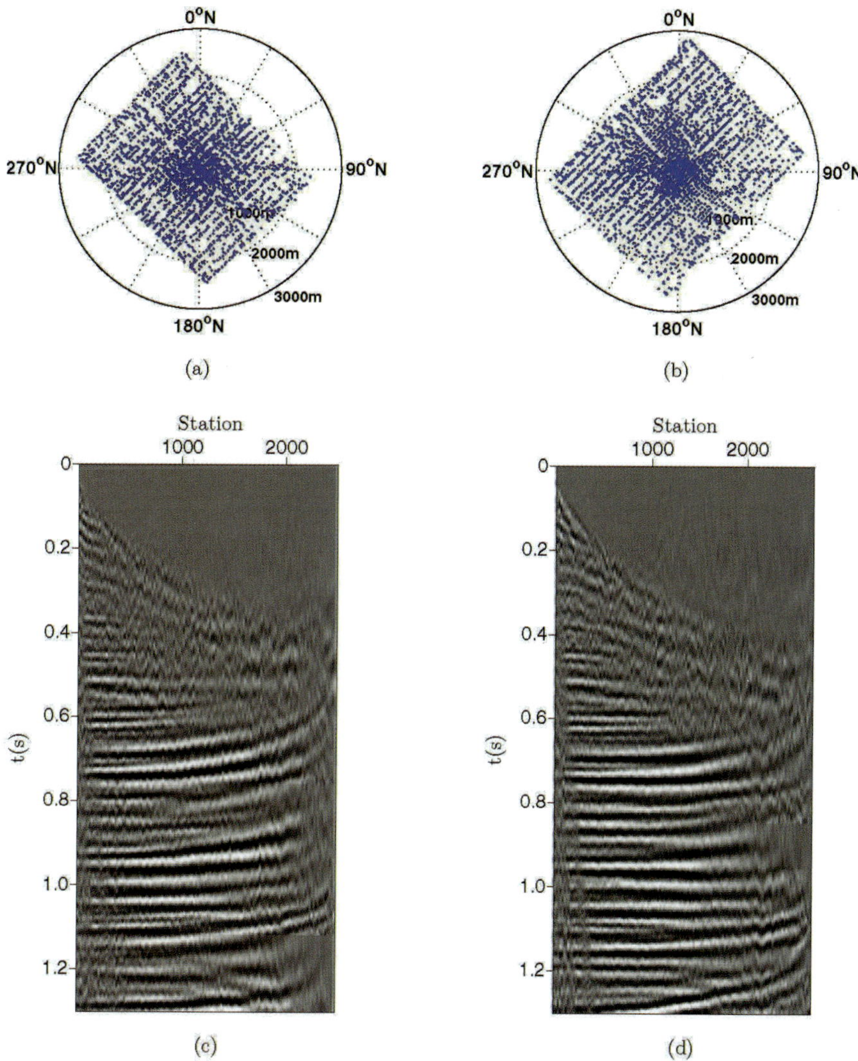

Figure 3.17: Source-to-receiver azimuth and offset coverage for the superbins centered at CMP locations (a) 10103 and (b) 10829 (Vasconcelos and Tsvankin, 2006). Each point marks a source-receiver pair; the polar angle corresponds to the source-to-receiver azimuth in degrees from the north (numbers on the perimeter), whereas the radius is the offset (in meters). The total fold is 2491 for superbin 10103, and 2702 for superbin 10829. On the bottom are gathers for superbins (c) 10103 and (d) 10829 after conventional hyperbolic NMO correction (the moveout velocities are estimated for offset-to-depth ratios of up to unity). The gathers are sorted by increasing offset and contain all azimuths and offsets, which are irregularly sampled [see plots (a) and (b)]. The "jittery" character of the reflections suggests that the traveltimes vary with azimuth. The curvature ("hockey sticks") of the NMO-corrected events at long offsets indicates the influence of nonhyperbolic moveout.

the presence of nonhyperbolic moveout and azimuthal traveltime variations (Figures 3.17c,d). Unfortunately, the range of offsets acquired at Weyburn field is not sufficient for constraining the quartic moveout term for reflections beneath the top of the reservoir. The maximum offset-to-depth ratio for the reservoir top is about 1.8, and far-offset traces for the deep reflections are quite noisy. Also, the reservoir is thinner than one-half of the dominant wavelength, which is below the vertical resolution of any traveltime method. In his estimation of the interval NMO ellipses, Jenner (2001) had to combine the reservoir horizon with some underlying beds into a single coarse layer.

Therefore, nonhyperbolic moveout inversion was performed only for several major horizons in the overburden (Figure 3.18). Describing $\eta(\alpha)$ with equations 3.41 and 3.42 produced similar semblance values, which indicates that the azimuths of the vertical symmetry planes are practically invariant with depth. Clearly, anisotropy is quite substantial through most of the overburden, with the effective η values reaching 0.25 for the reflection from the Mississippian Unconformity. The semblance scans in Figure 3.19 demonstrate that the resolution in $\eta^{(2)}$ (and the other anellipticity parameters) decreases for the deeper horizons because of their smaller offset-to-depth ratios. The relatively large uncertainty in $\eta^{(2)}$ for the Lower Vanguard reflection (Figure 3.19c) is also caused by the high level of noise at the far offsets. In principle, error bars for the η values in Figure 3.18 could be inferred from the size of the semblance contours in Figure 3.19.

In contrast to the anellipticity parameters, the symmetry-plane azimuth φ is not well constrained even for the shallow horizons (Figure 3.20) because of the relatively small magnitude of azimuthal anisotropy. The weak variation of φ with depth and the high accuracy provided by equation 3.41 exclude another possible reason for the instability in estimating φ – misaligned symmetry planes in the overburden layers. To verify whether it is possible to fit the data without taking azimuthal anisotropy into account, semblance analysis was carried out with the VTI moveout equation (i.e., using $V_{\text{nmo}}^{(1)} = V_{\text{nmo}}^{(2)}$, $\eta^{(1)} = \eta^{(2)}$ and $\eta^{(3)} = 0$). The semblance values for the best-fit VTI model were on average 10% smaller than those for orthorhombic media, with the exception of the Lower Vanguard reflection at CMP 10103 (for that event the reduction was only 5%).

The influence of azimuthal anisotropy also manifests itself in the consistency of the estimated symmetry-plane azimuth φ for all four reflection events at a fixed CMP location. If the reflection moveout had no azimuthal signature, we would expect the inverted values of φ to vary significantly from one event to another. Although the estimate of φ for CMP 10103 differs from that for CMP 10829, both inverted azimuths are close to the directions of the off-trend fracture sets measured from borehole data (Figure 3.21).

The substantial magnitude of $\eta^{(1)}$ and $\eta^{(2)}$ for the Viking reflection at CMP 10829 (Figure 3.18) indicates that the upper part of the section is moderately anisotropic. Also, the large difference between $\eta^{(1)}$ and $\eta^{(2)}$ for the Viking is a clear sign of nonnegligible azimuthal anisotropy. Interpretation of the anellipticity parameters for

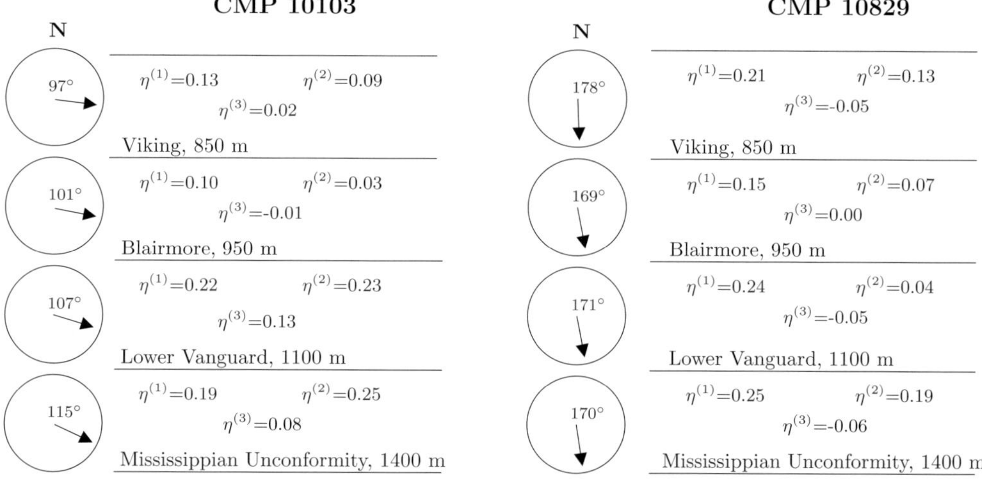

Figure 3.18: Inversion results for the two superbins from Figure 3.17 (Vasconcelos and Tsvankin, 2006). The reflections are from the Viking horizon, the spreadlength-to-depth ratio $x_{\max}/D$=2.5 (the depth is shown next to the event's name); the Blairmore, $x_{\max}/D$=2.0; the Lower Vanguard, $x_{\max}/D$=1.9; and the Mississippian Unconformity, $x_{\max}/D$=1.8. The arrows mark the estimated azimuth of the major axis of the NMO ellipse with respect to north. All η-parameters are the effective values for a given reflection event. The Mississippian Unconformity is close to the top of the reservoir.

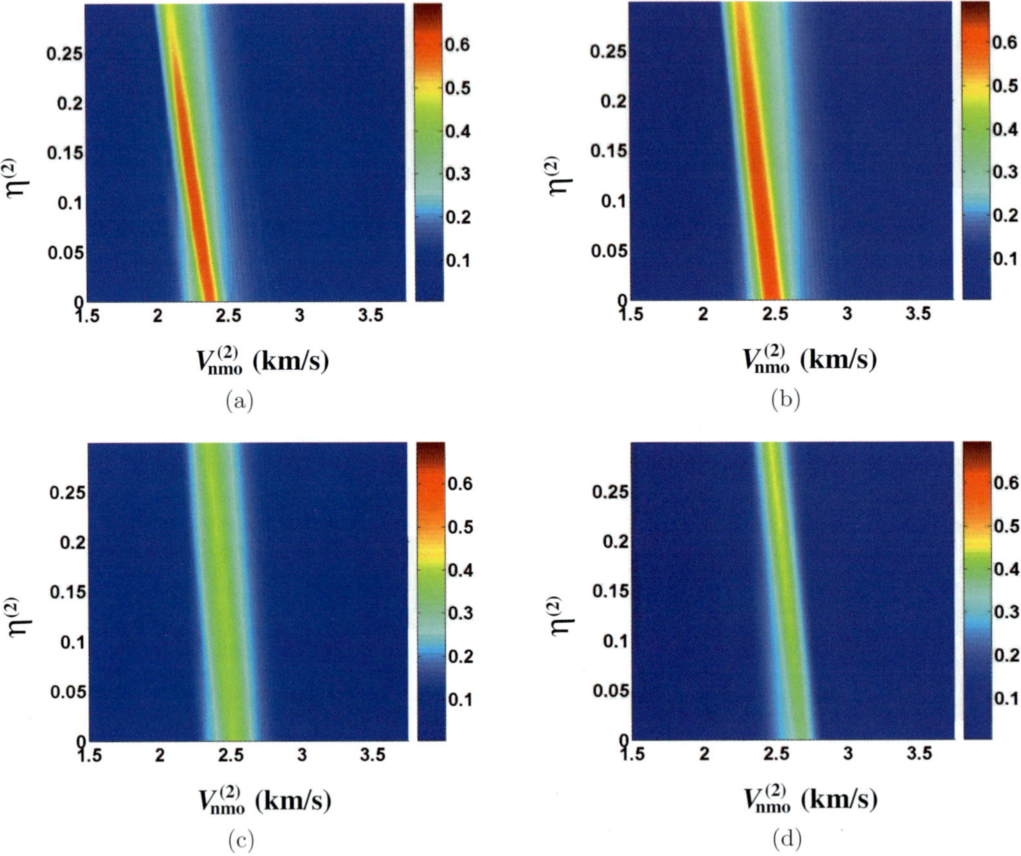

Figure 3.19: Semblance scans over $V_{\mathrm{nmo}}^{(2)}$ and $\eta^{(2)}$ for CMP 10103 (Vasconcelos and Tsvankin, 2006). The reflections are from (a) the Viking horizon, (b) the Blairmore, (c) the Lower Vanguard, and (d) the Mississippian Unconformity. The other moveout parameters are fixed at the best-fit values.

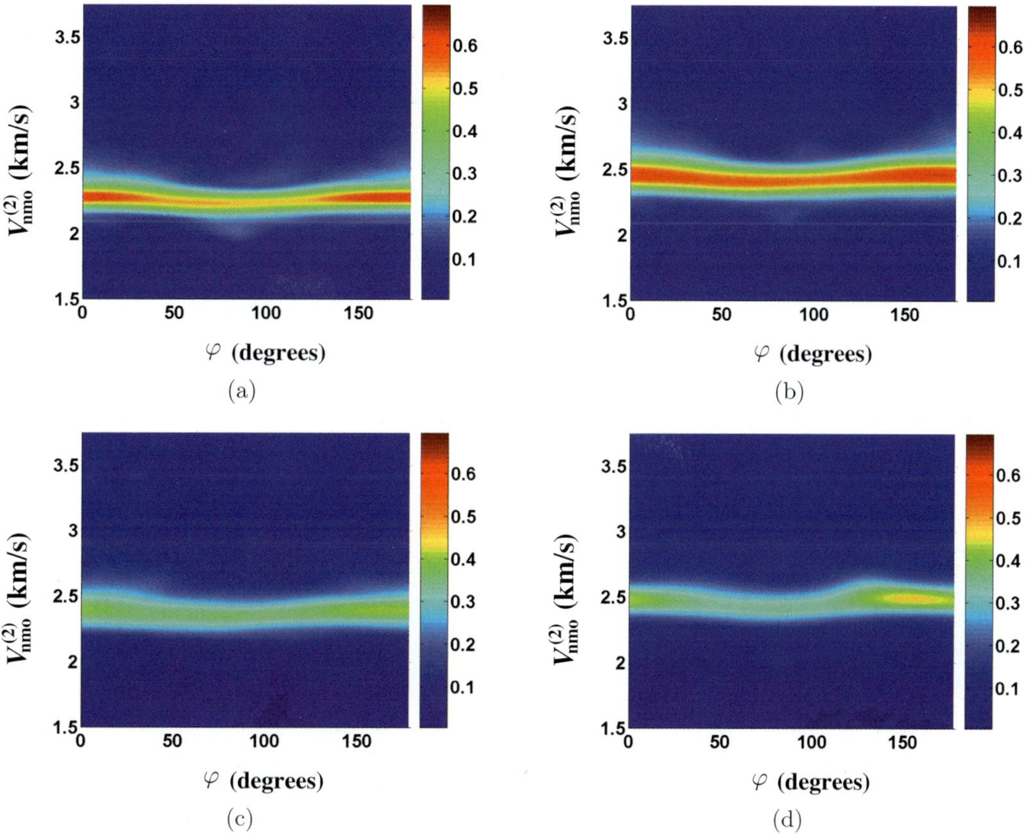

Figure 3.20: Semblance scans over φ and $V_{\rm nmo}^{(2)}$ for CMP 10103 (Vasconcelos and Tsvankin, 2006). The reflections are from (a) the Viking horizon, (b) the Blairmore, (c) the Lower Vanguard, and (d) the Mississippian Unconformity. The azimuth φ is measured counterclockwise from the survey's x-axis, which points eastward.

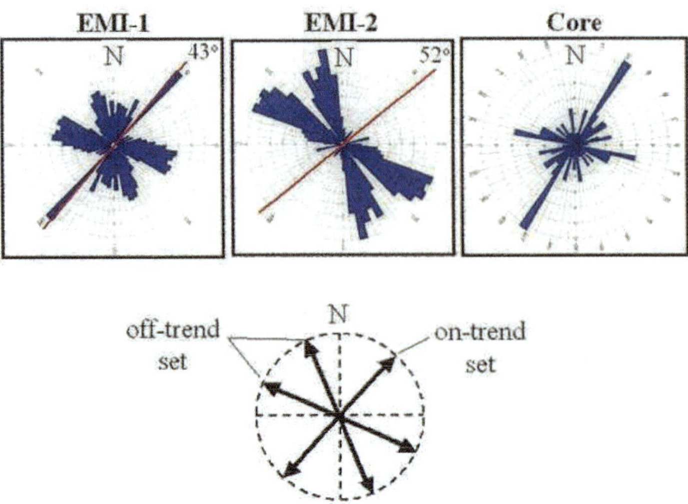

Figure 3.21: Fracture counts for different fracture azimuths from well-log images and core analysis (Cardona, 2002). Well EMI-1 is located close to superbin CMP 10103. The estimated symmetry-plane azimuths for CMP 10103 and CMP 10829 generally agree with the directions of the off-trend fracture sets.

the deeper events is complicated by the fact that they represent effective quantities for all layers above the reflector. The decrease in the effective η-parameters for the Blairmore reflection suggests that the interval between the Viking and Blairmore is weakly anisotropic. The magnitude of nonhyperbolic moveout increases at the Lower Vanguard horizon, in particular for CMP 10103. The Lower Vanguard reflection, however, is quite noisy, and the semblance values for it are about 15% smaller than those for the other horizons. Although the parameter $\eta^{(2)}$ at CMP 10829 increases for the Mississippian Unconformity, the smaller spreadlength-to-depth ratio and the influence of vertical heterogeneity renders the inversion results for that event less reliable. Estimation of the interval moveout parameters from long-spread data is discussed in the next section.

3.4 Interval parameter estimation using nonhyperbolic moveout

The semblance-based moveout inversion described in the previous section produces the effective moveout parameters of layered azimuthally anisotropic media. However, for purposes of anisotropic velocity analysis and seismic fracture characterization it is necessary to estimate the interval parameters of individual layers.

The interval NMO ellipse in horizontally layered media can be obtained in a straightforward way from the generalized Dix equation (see Chapter 2). Layer stripping of the effective parameters $\eta^{(1,2,3)}$ is more complicated, but it can be carried out azimuth-by-azimuth using the approximate kinematic equivalence between vertical planes of VTI and orthorhombic media. This Dix-type approach, however, inherits the well-documented instability problems of interval η-estimation in VTI media (Grechka and Tsvankin, 1998b; Tsvankin, 2005). Here we follow the work of Wang and Tsvankin (2009), who developed a more stable algorithm for interval nonhyperbolic moveout inversion using the velocity-independent layer-stripping method introduced by Dewangan and Tsvankin (2006c).

3.4.1 Velocity-independent layer stripping

In contrast to Dix-type techniques, velocity-independent layer stripping (VILS) is designed to produce the interval *traveltime function* in the horizon of interest. VILS is based on the same idea as the PP+ PS = SS method described in Chapter 4. Similar to the PP+ PS = SS method, VILS does not require knowledge of the velocity field and, therefore, can be applied prior to anisotropic parameter estimation.

2D version of VILS

First, we describe the 2D implementation of VILS designed for vertical symmetry planes of anisotropic media. Figure 3.22 shows ray trajectories of pure-mode (nonconverted) reflections from the top and bottom of the target zone overlain by a laterally homogeneous overburden. The incidence plane is assumed to represent a symmetry plane for the model as a whole, so that wave propagation is two-dimensional; this assumption becomes unnecessary in the 3D extension of the method discussed below. While the target zone can be heterogeneous with intermediate curved interfaces, each layer in the overburden has to be laterally homogeneous with a horizontal symmetry plane. Then the raypath of any reflection from the top of the target zone is symmetric with respect to the reflection point (e.g., points T and R in Figure 3.22).

A detailed discussion of VILS for 2D models can be found in Dewangan and Tsvankin (2006c). The method is based on identifyng target and overburden reflections with shared ray segments by equalizing traveltime slopes (horizontal slownesses) on common-source and common-receiver gathers. First, matching time slopes on common-receiver gathers at the source location $x^{(1)}$ helps identify the receiver location $x^{(3)}$, for which the overburden reflection $x^{(1)}\mathrm{T}x^{(3)}$ has the same horizontal slowness at $x^{(1)}$ as the reflection $x^{(1)}\mathrm{TQR}x^{(2)}$ from the bottom of the target layer. This means that the events $x^{(1)}\mathrm{T}x^{(3)}$ and $x^{(1)}\mathrm{TQR}x^{(2)}$ share the downgoing leg $x^{(1)}\mathrm{T}$. Second, by matching time slopes on common-source gathers at location $x^{(2)}$, one can find the overburden reflection $x^{(4)}\mathrm{R}x^{(2)}$ that has the same upgoing leg $\mathrm{R}x^{(2)}$ as the target event $x^{(1)}\mathrm{TQR}x^{(2)}$. Since any reflection path in the overburden is symmetric with respect to the reflection point, the interval traveltime t^{int} along the raypath TQR in the target zone can be computed as

$$t^{\mathrm{int}}(\mathrm{T},\mathrm{R}) = t^{\mathrm{eff}}(x^{(1)},x^{(2)}) - \frac{1}{2}\left[t^{\mathrm{ovr}}(x^{(1)},x^{(3)}) + t^{\mathrm{ovr}}(x^{(4)},x^{(2)})\right], \qquad (3.44)$$

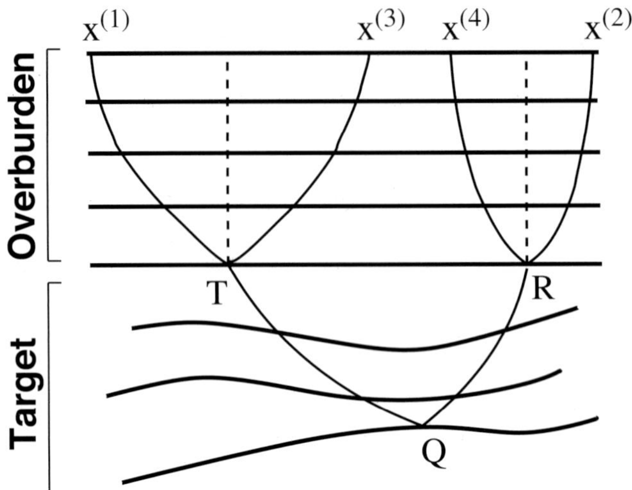

Figure 3.22: 2D diagram of velocity-independent layer stripping (VILS) for pure-mode reflections (Wang and Tsvankin, 2009). The method is applied to the reflection $x^{(1)}\mathrm{TQR}x^{(2)}$ from the bottom of the target layer. Points T and R are located at the bottom of the laterally homogeneous overburden. The leg $x^{(1)}\mathrm{T}$ is shared by the target reflection $x^{(1)}\mathrm{TQR}x^{(2)}$ and the overburden event $x^{(1)}\mathrm{T}x^{(3)}$; the leg $\mathrm{R}x^{(2)}$ is shared by the reflections $x^{(1)}\mathrm{TQR}x^{(2)}$ and $x^{(4)}\mathrm{R}x^{(2)}$.

where the superscripts "eff" and "ovr" refer to the target event $x^{(1)}\mathrm{TQR}x^{(2)}$ and the reflections from the bottom of the overburden, respectively. The corresponding source-receiver pair (T,R) has the following horizontal coordinates:

$$x_{\mathrm{T}} = \frac{x^{(1)} + x^{(3)}}{2}, \qquad x_{\mathrm{R}} = \frac{x^{(2)} + x^{(4)}}{2}. \tag{3.45}$$

Equations 3.44 and 3.45 yield the interval moveout function in the target zone without any information about the velocity model.

2D interval moveout-inversion methods based on ideas similar to those behind VILS were developed by van der Baan and Kendall (2002, 2003) and Fowler et al. (2008). In contrast to VILS, however, these algorithms are implemented only for VTI media and assume the invariance of the horizontal slowness along each ray, which implies that the target zone has to be laterally homogeneous.

3D VILS for wide-azimuth data

The 3D version of the layer-stripping algorithm does not impose any restrictions on the properties (anisotropy, heterogeneity) of the target zone, but each layer in the overburden still has to be laterally homogeneous with a horizontal symmetry plane. For wide-azimuth data (Figure 3.23), identifying the target and overburden reflections with the same ray segments requires estimating two orthogonal horizontal slowness components from reflection time slopes. In Figure 3.23, the horizontal slownesses of

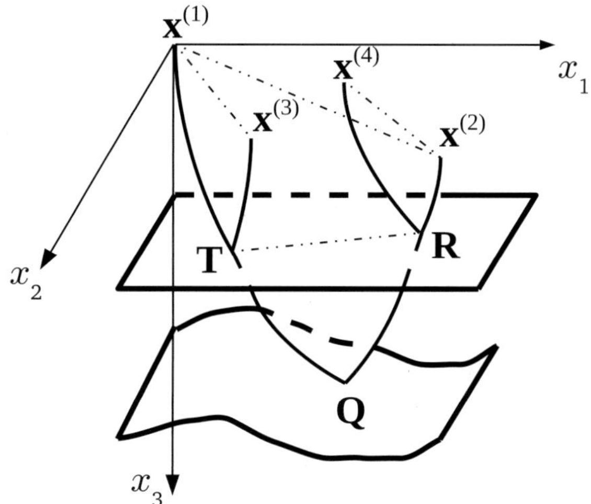

Figure 3.23: 3D diagram of the layer-stripping algorithm (Wang and Tsvankin, 2009). The sources and receivers ($\mathbf{x}^{(1)}$, $\mathbf{x}^{(2)}$, $\mathbf{x}^{(3)}$ and $\mathbf{x}^{(4)}$) are placed at the surface but not necessarily along a straight line. The reflection point $\mathbf{Q}$ is located at the bottom of the target layer, which can be arbitrarily anisotropic and heterogeneous. Points $\mathbf{T}$ and $\mathbf{R}$ are at the bottom of the laterally homogeneous overburden. The leg $\mathbf{x}^{(1)}\mathbf{T}$ is shared by the target event $\mathbf{x}^{(1)}\mathbf{TQR}\mathbf{x}^{(2)}$ and the overburden reflection $\mathbf{x}^{(1)}\mathbf{T}\mathbf{x}^{(3)}$; the leg $\mathbf{R}\mathbf{x}^{(2)}$ is shared by the reflections $\mathbf{x}^{(1)}\mathbf{TQR}\mathbf{x}^{(2)}$ and $\mathbf{x}^{(4)}\mathbf{R}\mathbf{x}^{(2)}$.

the target (eff) and overburden (ovr) reflections at location $\mathbf{x}^{(1)} = [x_1^{(1)}, x_2^{(1)}]$ can be obtained from

$$p_i^{\text{eff}}(\mathbf{x}^{(1)}, \mathbf{x}^{(2)}) = \left. \frac{\partial t^{\text{eff}}(\mathbf{x}, \mathbf{x}^{(2)})}{\partial x_i} \right|_{\mathbf{x}=\mathbf{x}^{(1)}}, \qquad (i = 1, 2) \tag{3.46}$$

and

$$p_i^{\text{ovr}}(\mathbf{x}^{(1)}, \mathbf{x}^{(3)}) = \left. \frac{\partial t^{\text{ovr}}(\mathbf{x}, \mathbf{x}^{(3)})}{\partial x_i} \right|_{\mathbf{x}=\mathbf{x}^{(1)}}, \qquad (i = 1, 2). \tag{3.47}$$

Using equations 3.46 and 3.47, we find the location $\mathbf{x}^{(3)}$, for which the time slopes (horizontal slownesses) of the two events are identical,

$$p_i^{\text{eff}}(\mathbf{x}^{(1)}, \mathbf{x}^{(2)}) = p_i^{\text{ovr}}(\mathbf{x}^{(1)}, \mathbf{x}^{(3)}), \qquad (i = 1, 2). \tag{3.48}$$

Therefore, the reflections $\mathbf{x}^{(1)}\mathbf{TQR}\mathbf{x}^{(2)}$ and $\mathbf{x}^{(1)}\mathbf{T}\mathbf{x}^{(3)}$ have the common leg $\mathbf{x}^{(1)}\mathbf{T}$. The same operation applied at point $\mathbf{x}^{(2)}$ helps identify the overburden reflection $\mathbf{x}^{(4)}\mathbf{R}\mathbf{x}^{(2)}$ that shares the upgoing leg $\mathbf{R}\mathbf{x}^{(2)}$ with the target event $\mathbf{x}^{(1)}\mathbf{TQR}\mathbf{x}^{(2)}$. The interval reflection traveltime can then be obtained from equation 3.44, in which all coordinates become two-component vectors. Since $\mathbf{T}$ and $\mathbf{R}$ represent the midpoints of the corresponding source-receiver pairs, their horizontal coordinates can be easily found from $\mathbf{x}^{(1)}$, $\mathbf{x}^{(2)}$, $\mathbf{x}^{(3)}$, and $\mathbf{x}^{(4)}$.

Thus, VILS makes it possible to construct interval moveout functions in both 2D and 3D without information about the velocity field. Because reflection traveltimes can be estimated with higher accuracy than moveout parameters, VILS helps avoid the stability problems in Dix-type inversion caused by the trade-offs between the effective moveout parameters (see below).

Similar to the original version of the PP+ PS = SS method, this layer-stripping algorithm operates with reflection traveltimes. Grechka and Dewangan (2003) replaced traveltime analysis in the PP+ PS = SS method with a convolution of recorded PP and PS traces (see Chapter 4). Their technique can be adapted to compute interval reflection data using the reflections from the top and bottom of the target layer. Although the convolution of recorded traces cannot produce the correct amplitudes, the constructed arrivals should have the kinematics of primary reflections and, therefore, are suitable for interval moveout analysis.

3.4.2 Synthetic tests for VTI and orthorhombic media

Wang and Tsvankin (2009) applied VILS to interval parameter estimation in layered VTI and orthorhombic media. The tests were conducted on 2D and 3D long-spread P-wave data generated by kinematic ray tracing (Gajewski and Pšenčík, 1987). To remove offset-dependent amplitude variations, reflected arrivals on the synthetic seismograms were obtained by placing the Ricker wavelet at the corresponding reflection traveltime. The interval moveout parameters in the target layer were estimated from both VILS and Dix-type equations using noise-contaminated input traveltimes.

2D inversion for VTI media

For stratified VTI media composed of N horizontal layers, long-spread P-wave traveltime is described by equation 3.36:

$$t^2(x) = t_{P0}^2 + \frac{x^2}{V_{\text{nmo}}^2(N)} - \frac{2\eta(N)\,x^4}{V_{\text{nmo}}^2(N)\left[t_{P0}^2 V_{\text{nmo}}^2(N) + (1+2\eta(N))\,x^2\right]}\,, \tag{3.49}$$

where $V_{\text{nmo}}(N)$ and $\eta(N)$ are the effective parameters usually obtained by applying semblance-based nonhyperbolic moveout inversion. Then the interval NMO velocity in layer i can be computed from the Dix (1955) equation,

$$(V_{\text{nmo}}^{(i)})^2 = \frac{V_{\text{nmo}}^2(i)\,t_{P0}(i) - V_{\text{nmo}}^2(i-1)\,t_{P0}(i-1)}{t_{P0}(i) - t_{P0}(i-1)}\,, \tag{3.50}$$

while equation 3.39 yields the interval η:

$$\eta^{(i)} = \frac{1}{8\,(V_{\text{nmo}}^{(i)})^4}\left[\frac{f(i)\,t_{P0}(i) - f(i-1)\,t_{P0}(i-1)}{t_{P0}(i) - t_{P0}(i-1)} - (V_{\text{nmo}}^{(i)})^4\right]; \tag{3.51}$$

$$f(i) \equiv V_{\text{nmo}}^4(i)\,[1+8\eta(i)]\,.$$

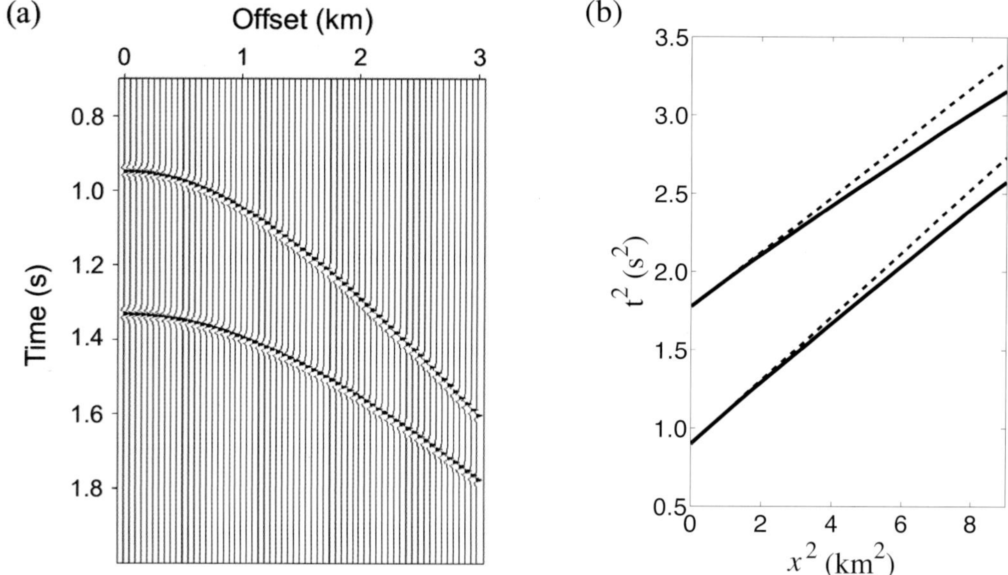

Figure 3.24: (a) Synthetic long-spread reflections from the top and bottom of layer 3 (target) in model 1 (Table 3.1) and (b) the $t^2(x^2)$ function (solid lines) for both events shown on plot (a) (Wang and Tsvankin, 2009). The dashed lines mark the hyperbolic moveout function parameterized by the analytic NMO velocity. The maximum offset-to-depth ratio for the bottom of the model is $x_{\max}/D = 2$.

Implementation of this Dix-type inversion and the accuracy of η-estimation is discussed in detail by Alkhalifah (1997), Grechka and Tsvankin (1998b), and Tsvankin (2005). Although equation 3.49 provides a good approximation for nonhyperbolic moveout in VTI media, this method suffers from instability caused by the trade-off between V_{nmo} and η. Even small correlated traveltime errors, which could be considered as insignificant in data processing, can cause large errors in the effective parameter η. For stratified media, this error is greatly amplified in the layer-stripping equation 3.51, as illustrated by the examples below. The effective η function is often smoothed prior to application of the Dix-type equations, but smoothing does not remove the main source of errors in the interval parameter estimation.

The first numerical test was performed for the three-layer VTI model with the parameters listed in Table 3.1 (Figure 3.24). Both VILS and the Dix-type method were used to estimate the interval parameters V_{nmo} and η in the third (bottom) layer. Although the η values in this model are moderate, the traveltime curves for the top and bottom of the target layer noticeably deviate from the hyperbolic moveout approximation at large offsets (Figure 3.24b).

The reflection traveltimes from the top and bottom of the target (third) layer were reconstructed using 2D semblance search for the effective parameters V_{nmo} and η based

	Layer 1	Layer 2	Layer 3 (target)
Thickness (km)	0.7	0.3	0.5
t_{P0} (s)	0.70	0.25	0.39
V_{nmo} (km/s)	2.10	2.52	2.78
η	0.00	0.10	0.20

Table 3.1: Interval parameters of a three-layer VTI model (model 1).

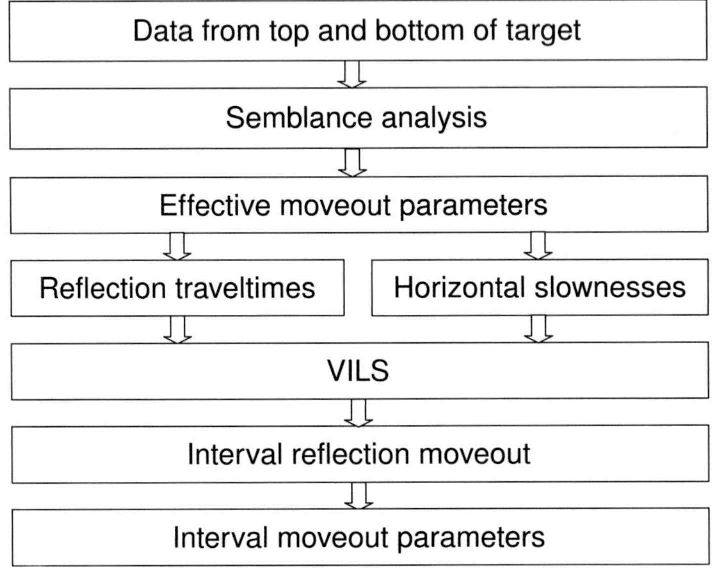

Figure 3.25: Workflow for interval parameter estimation using VILS (Wang and Tsvankin, 2009).

on equation 3.49. Then VILS was applied to compute the interval traveltime, which was inverted for the interval parameters using least-squares fitting of the single-layer equation 3.36 (see the flow chart in Figure 3.25). The interval value of η estimated by VILS is quite accurate, with the error (just 0.02) caused mostly by the slight bias of the moveout equation.

The semblance analysis for the top and bottom of the target layer also provided input data for the Dix-type differentiation described above. In contrast to VILS, the Dix-type algorithm operates with the effective moveout parameters, not traveltimes. As a result, small distortions in the effective η estimates get amplified in the layer-stripping procedure, which leads to an error of 0.06 in the interval η value.

The influence of realistic noise on the interval parameter estimation was studied by adding random, linear, and sinusoidal time errors to the reflection moveout from the bottom of the target layer. The traveltimes from the top of the target were left unchanged. As before, the input data for both VILS and the Dix-type method were

Parameters of error function	$A = 3\,\text{ms},\ n = 3$		$A = 3\,\text{ms},\ n = 2$		$A = 8\,\text{ms},\ n = 3$	
Inversion error	V_{nmo} (%)	η	V_{nmo} (%)	η	V_{nmo} (%)	η
VILS	0.6	0.01	0.0	0.00	2.1	0.08
Dix	11	0.19	8.2	0.13	21	0.41

Table 3.2: Influence of correlated noise on the interval parameter estimation for the third layer in model 1 (Wang and Tsvankin, 2009). A sinusoidal error function $[t = A\sin(n\pi x/x_{\text{max}})]$ was added to the traveltimes from the bottom of the layer. The percentage error in the interval velocity V_{nmo} and the absolute error in the interval η are estimated by VILS and the Dix-type method for different A and n.

obtained from 2D semblance search using equation 3.49.

Since both methods employ semblance analysis, they remain reasonably stable in the presence of random noise. For random errors with the magnitude close to 10 ms, the interval η estimated by VILS is distorted by 0.02 or less, while the Dix-type method produces η errors in the range of $0.05-0.08$.

The second type of tested noise is linear, which can simulate long-period statics errors. For a relatively large error that changes from 6 ms at zero offset to -6 ms at the maximum offset, VILS estimates the interval V_{nmo} and η with distortions of 4% and 0.07, respectively. The errors in V_{nmo} and η after the Dix-type layer stripping are much larger (15% and 0.34, respectively), which makes the inversion practically useless. These results are consistent with the analysis in Grechka and Tsvankin (1998b) and Tsvankin (2005) who demonstrate that linear time noise of a somewhat smaller magnitude can cause errors in the effective η close to 0.1. The Dix-type procedure amplifies such errors by a factor that depends on the relative thickness of the target layer (i.e., on the ratio of its thickness and depth).

Next, the data were contaminated with a sinusoidal time function designed to emulate short-period statics errors: $t = A\sin(n\pi x/x_{\text{max}})$. The interval parameter-estimation results for different values of A and n are listed in Table 3.2. The error in the interval η produced by VILS reaches only 0.08 even for $A = 8$ ms, while the Dix-type method breaks down for $A \geq 3$ ms.

These tests clearly demonstrate the superior stability of VILS in the presence of typical correlated noise in reflection traveltimes. Even relatively small time errors may cause nonnegligible distortions in the effective parameter η, which propagate with amplification into the interval η-values during Dix-type differentiation. In contrast, VILS computes the interval moveout directly from well constrained reflection traveltimes.

The layer-stripping procedure in VILS is based on the assumption that each layer in the overburden is laterally homogeneous and has a horizontal symmetry plane. Lateral velocity gradients or dipping interfaces make the raypaths of overburden events asymmetric with respect to the reflection point, which will cause errors in estimating

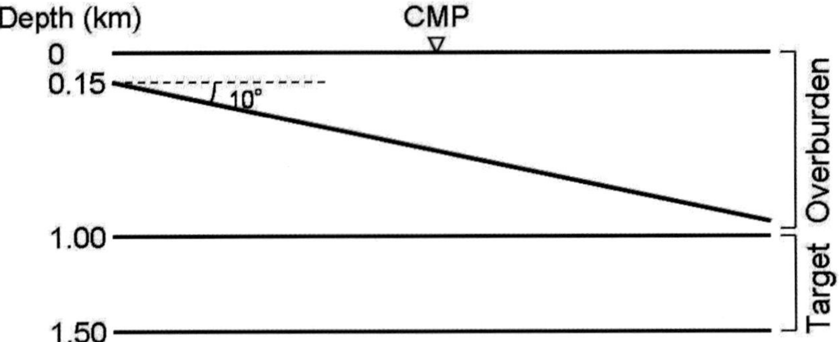

Figure 3.26: Three-layer VTI model used to evaluate the influence of mild dip (10°) in the overburden on the inversion results (Wang and Tsvankin, 2009). Except for the dip of the uppermost interface, all medium parameters are the same as those for model 1 (Table 3.1). The lateral extent of the model is 4 km; the CMP used for parameter estimation is located in the middle.

the interval traveltime in the target layer. Lateral heterogeneity above the reflector also violates the assumptions behind the version of the Dix-type method used here.

To evaluate the influence of mild dips in the overburden on interval parameter estimation, the uppermost reflector in model 1 was tilted by 10° (Figure 3.26). Then VILS and the Dix-type method were applied to synthetic data from the laterally heterogeneous model without taking the dip into account. The interval parameters V_{nmo} and η in the third layer estimated by VILS are distorted by only 2% and 0.05, respectively. In contrast, the Dix-type method produces more significant errors in the interval values, reaching 6% in V_{nmo} and 0.15 in η. Hence, VILS is less sensitive to mild lateral heterogeneity than is Dix differentiation.

3D inversion for orthorhombic media

The azimuthally dependent P-wave reflection moveout in layered orthorhombic media can be described using the nonhyperbolic moveout equation 3.37 parameterized by the effective NMO ellipse (equation 3.40) and the effective parameter $\eta(\alpha)$ (equation 3.42 in the general case or equation 3.41 for aligned vertical symmetry planes). Application of VILS and Dix-type layer stripping to wide-azimuth, long-spread P-wave data is based on the 3D moveout-inversion algorithm discussed in the previous section. The best-fit effective moveout parameters φ, φ_1, $V_{\text{nmo}}^{(1,2)}$, and $\eta^{(1,2,3)}$ for the top and bottom of the target layer are found by multidimensional "global" semblance search that includes all available source-receiver pairs in a CMP gather.

For purposes of Dix-type differentiation, the interval NMO ellipse is obtained from the generalized Dix equation 1.32, and the interval parameter $\eta(\alpha)$ is computed azimuth-by-azimuth from the VTI equation 3.51. Then the interval values of $\eta^{(1,2,3)}$ are estimated by fitting equation 3.41 to the function $\eta(\alpha)$.

	Layer 1	Layer 2	Layer 3 (target)
Symmetry type	ISO	VTI	Orthorhombic
Thickness (km)	0.5	0.5	0.5
t_{P0} (s)	0.50	0.41	0.39
$V_{\rm nmo}^{(1)}$ (km/s)	2	2.49	3.18
$V_{\rm nmo}^{(2)}$ (km/s)	2	2.49	2.64
$\eta^{(1)}$	0	0.05	0.2
$\eta^{(2)}$	0	0.05	0.06
$\eta^{(3)}$	0	0	0.13
φ (°)	–	–	30

Table 3.3: Interval parameters of a model (model 2) used to test the 3D layer-stripping algorithm (Wang and Tsvankin, 2009).

The long-spread, wide-azimuth reflection traveltimes reconstructed by the nonhyperbolic semblance analysis serve as the input data for VILS (see the flow chart in Figure 3.25). To apply VILS to 3D wide-azimuth data (Figure 3.23), it is also necessary to estimate the horizontal slowness components at the source and receiver locations. In principle, the horizontal projection of the slowness vector (equations 3.46 and 3.47) can be computed from reflection traveltimes on common-shot and common-receiver gathers. A more stable and efficient option, however, is to express the horizontal slownesses as functions of offset and azimuth through the best-fit moveout parameters using equation 3.37. Despite the parameter trade-offs, equation 3.37 provides sufficient accuracy for long-spread P-wave moveout and, therefore, for the spatial traveltime derivatives. After the interval traveltime has been computed by VILS, the interval parameters φ, $V_{\rm nmo}^{(1,2)}$, and $\eta^{(1,2,3)}$ are obtained from the single-layer moveout inversion.

The 3D parameter-estimation algorithm was first tested on an orthorhombic target layer overlain by VTI and isotropic layers (model 2; Table 3.3). VILS and the Dix-type method were applied to long-spread ($x_{\rm max}/D = 2$ for the bottom of the model), wide-azimuth data from the top and bottom of the target orthorhombic layer. Since the model is laterally homogeneous, the data were generated for a single source location and a full (180°) range of source-receiver azimuths; receivers were placed on 19 lines with an azimuthal interval of 10°. The azimuthal anisotropy in the target layer makes the reflection traveltimes from its bottom vary with azimuth (Figure 3.27). Without noise in the traveltimes, both methods give similar accuracy in the interval moveout parameters.

As before, linear and sinusoidal time errors were added to the reflection moveout from the bottom of the target layer. For the azimuthally invariant linear error that changes from 6 ms at zero offset to −6 ms at the maximum offset, the interval parameters $V_{\rm nmo}^{(1,2)}$ and $\eta^{(1,2,3)}$ estimated by VILS are distorted by no more than 3% and

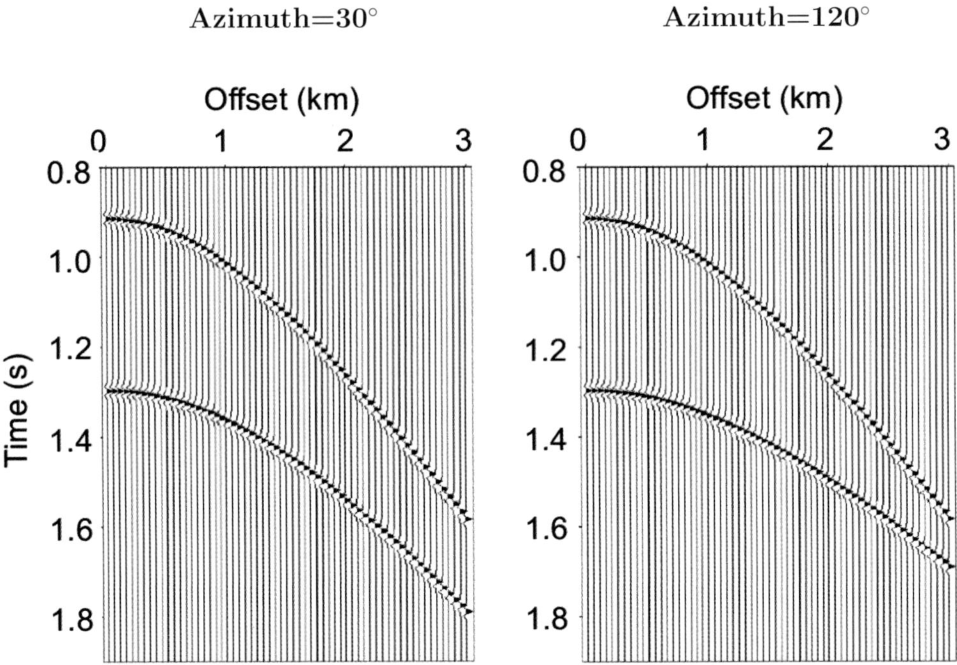

Figure 3.27: Synthetic long-spread P-wave reflections from the top and bottom of layer 3 (target) in model 2 from Table 3.3 (Wang and Tsvankin, 2009). The seismograms are computed in the two orthogonal vertical symmetry planes of the target orthorhombic layer. Note the significant difference between the traveltimes of the deeper event for α=30° and α=120°, especially at offsets approaching 3 km.

Parameters of error function	$A = 3$ ms, $n = 3$, $m = 0$		$A = 3$ ms, $n = 3$, $m = 2$		$A = 10$ ms, $n = 3$, $m = 0$	
Inversion error	V_{nmo} (%)	η	V_{nmo} (%)	η	V_{nmo} (%)	η
VILS	1.2	0.04	1.0	0.02	2.4	0.09
Dix	4.4	0.09	2.3	0.08	10	0.22

Table 3.4: Influence of correlated noise on the interval parameter estimation for the third layer in model 2 (Wang and Tsvankin, 2009). A sinusoidal error function $[A \sin(n\pi x/x_{\text{max}}) \sin m\alpha]$ was added to the traveltimes from the bottom of the layer. The maximum percentage error in the interval velocities $V_{\text{nmo}}^{(1,2)}$ and the maximum absolute error in the interval $\eta^{(1,2,3)}$ are estimated by VILS and the Dix-type method. The errors in the azimuth φ do not exceed 0.5° for either method.

Thickness (km)	0.35		0.15	
Inversion error	$V_{\rm nmo}$ (%)	η	$V_{\rm nmo}$ (%)	η
VILS	2.2	0.05	8.3	0.15
Dix	7.5	0.13	20	0.38

Table 3.5: Maximum errors in the interval parameters $V_{\rm nmo}^{(1,2)}$ and $\eta^{(1,2,3)}$ for two different thicknesses of the target layer in model 2 from Table 3.3 (Wang and Tsvankin, 2009). The traveltimes from the bottom of the target were contaminated by the sinusoidal noise function (see the caption of Table 3.4) with A=3 ms, n=3, and m=0.

0.06, respectively. In comparison, the Dix-type method produces much larger errors reaching 9% in $V_{\rm nmo}^{(1,2)}$ and 0.16 in $\eta^{(1,2,3)}$. The errors in the symmetry-plane azimuth φ for both methods are negligible.

The inversion results for a more complicated, azimuthally varying error function are shown in Table 3.4. The coefficients n and m control the period of the error in the radial and azimuthal directions, respectively. In general, inversion errors tend to be higher when m is an even number. If the error function does not vary with azimuth ($m=0$), both methods give more accurate results for even values of n, which agrees with the conclusions of Xu and Tsvankin (2006a). As illustrated by examples in Table 3.4, even for noisy data with $A = 10$ ms VILS produces errors in the interval $\eta^{(1,2,3)}$ not exceeding 0.09, while the Dix-type method distorts the η-parameters by up to 0.22 and the NMO velocities by 10%.

It is interesting that despite the complexity of orthorhombic symmetry, the Dix-type method gives much better results for model 2 than for the VTI model 1 (compare Tables 3.2 and 3.4). This result is particularly surprising given the approximate nature of the layer-stripping procedure used to obtain the interval function $\eta(\alpha)$. Most likely, this improvement is explained by wide azimuthal coverage in 3D inversion, which creates redundancy and makes estimation of the effective moveout parameters more stable.

Any layer-stripping method inevitably becomes less accurate as the layer of interest gets thinner (i.e., the thickness-to-depth ratio decreases). We added the sinusoidal error with $A=3$ ms, $n=3$, and $m=0$ to the traveltimes from the bottom of the target and reduced its thickness until it reached 0.15 km (Table 3.5). VILS gives acceptable results for both the interval $V_{\rm nmo}$ and η when the thickness exceeds 0.25 km (i.e., when the thickness-to-depth ratio exceeds 0.2), while the error in $V_{\rm nmo}$ estimated by the Dix-type method approaches 10%. However, even VILS breaks down for the target layer that is only 0.15 km thick.

The third model includes the target orthorhombic layer beneath an overburden composed of isotropic and orthorhombic layers (Table 3.6). The vertical symmetry planes in the two orthorhombic layers are misaligned, so the azimuthally varying parameter η from the bottom of the target is described by equation 3.42. The vertical variation of the symmetry-plane azimuths does not cause any complications in the

	Layer 1	Layer 2	Layer 3 (target)
Symmetry type	ISO	ORTH	ORTH
Thickness (km)	0.3	0.7	0.5
t_{P0} (s)	0.30	0.58	0.39
$V_{\rm nmo}^{(1)}$ (km/s)	2	2.56	3.68
$V_{\rm nmo}^{(2)}$ (km/s)	2	3.06	2.73
$\eta^{(1)}$	0	0.05	0.24
$\eta^{(2)}$	0	0.07	0.12
$\eta^{(3)}$	0	0.02	−0.10
φ (°)	–	30	70

Table 3.6: Interval parameters of a three-layer model (model 3) that includes two orthorhombic layers with misaligned vertical symmetry planes (Wang and Tsvankin, 2009).

application of VILS, as long as the overburden is laterally homogeneous and has a horizontal symmetry plane.

The accuracy of VILS and the Dix-type method for noise-free data is similar, which was also the case for model 2. The sinusoidal error $t = A\sin(n\pi x/x_{\rm max})$ applied to the traveltimes from the bottom of the target layer produces much more significant distortions in the output of the Dix-type differentiation compared to VILS. For instance, when $A = 6$ ms and $n = 3$, the maximum errors in the interval parameters $V_{\rm nmo}$ and $\eta^{(1,2,3)}$ estimated by VILS are 4% and 0.09 (respectively), while the corresponding errors of the Dix-type method reach 13% and 0.25.

3.4.3 Field-data example

The 3D VILS algorithm was tested by Wang and Tsvankin (2009) on wide-azimuth P-wave data acquired by the Reservoir Characterization Project (RCP) at Rulison field, a basin-centered gas accumulation in the South Piceance Basin, Colorado. The low-porosity reservoir (Williams Fork formation), which produces gas from fractured channel sand lenses, is capped by the UMV (Upper Mesaverde) shale (Figure 3.28).

Xu and Tsvankin (2007) applied a comprehensive anisotropic processing sequence to the data and analyzed the effective and interval NMO ellipses, as well as the azimuthal AVO response (see section 8.3). The same data set, acquired in 2003 and preprocessed for purposes of azimuthal moveout and AVO analysis, is used here. Since the subsurface structure is approximately horizontally layered (Figure 3.29), the moveout equations discussed in this chapter should give an accurate description of reflection traveltimes. As suggested by Xu and Tsvankin (2007), traces with nearby midpoints were combined into 5×5 superbins to improve the offset and azimuthal coverage. The moveout inversion was carried out in the center of the RCP survey

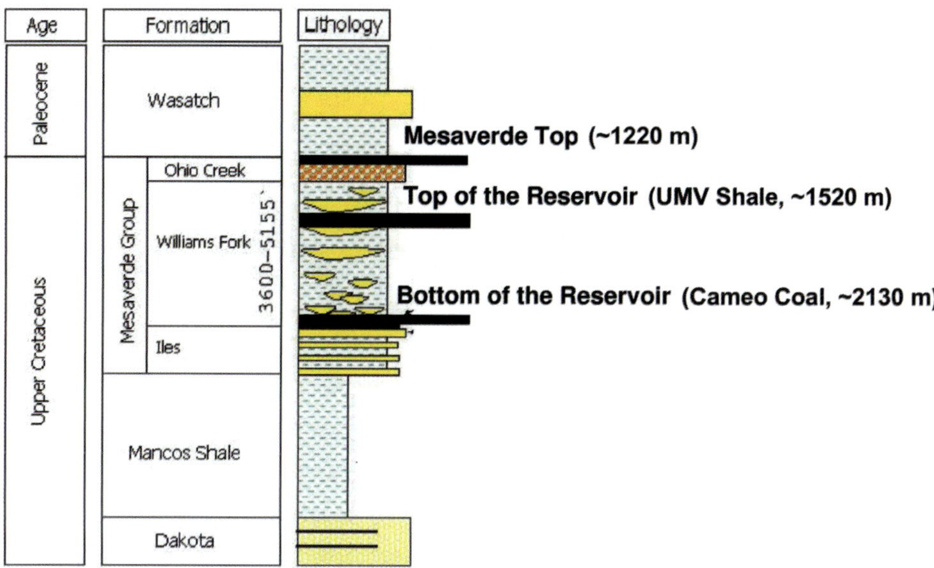

Figure 3.28: Stratigraphic column of Rulison field (Xu and Tsvankin, 2007). The gas-producing reservoir is bounded by the UMV shale (the target layer in this study) and the Cameo coal.

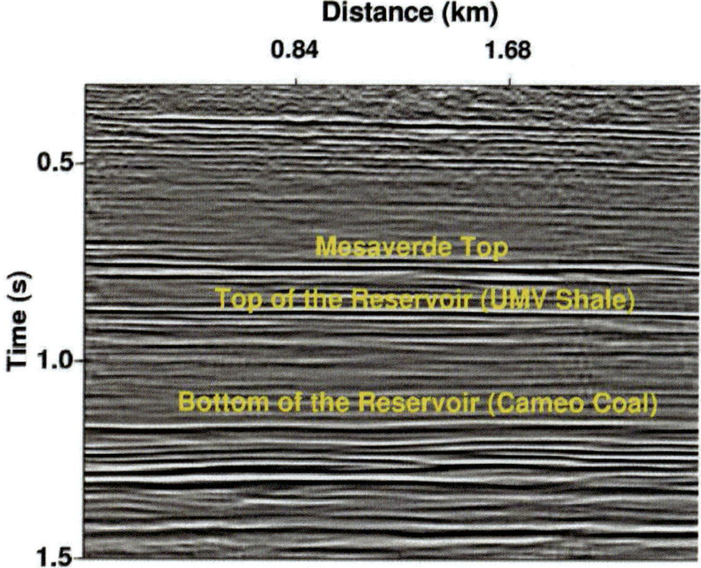

Figure 3.29: Seismic section across the middle of the survey area at Rulison field (Xu and Tsvankin, 2007).

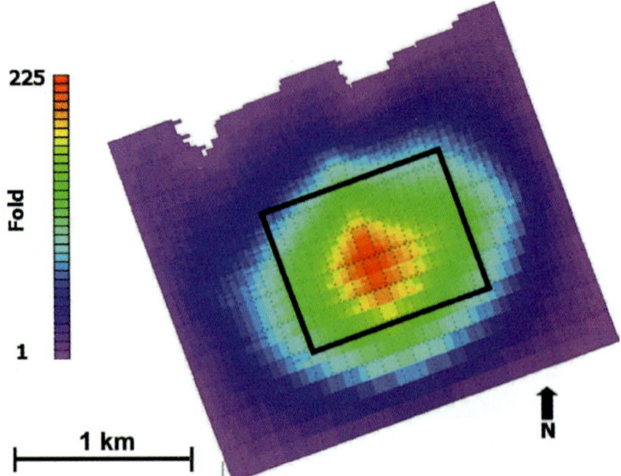

Figure 3.30: P-wave fold for the 17×17 m bin size at Rulison field (Xu and Tsvankin, 2007). The rectangle in the center marks the study area.

area (Figure 3.30), where the coverage is sufficient for minimizing the influence of the acquisition footprint.

The maximum offset-to-depth ratio for the bottom of the reservoir does not exceed 1.6 and is much smaller near the edges of the study area. Therefore, nonhyperbolic moveout inversion was performed for the UMV shale, the layer immediately above the reservoir (Figure 3.28). In the center of the study area, the offset-to-depth ratio at the bottom of the shale is between 1.9 and 2.2. The interval moveout parameters were obtained using the VILS and Dix-type algorithms for layered orthorhombic media.

The NMO ellipticity is small for both the top and bottom of the target shale layer over most of the area. Therefore, the principal directions of the effective and interval NMO ellipses are poorly constrained by the data. Nevertheless, as long as the maximum offset-to-depth ratio reaches two, the parameters $\eta^{(1,2,3)}$ can be estimated in a reliable fashion. The interval values of $\eta^{(1,2,3)}$ obtained for two superbin gathers near the center of the area are listed in Table 3.7.

Despite the absence of independent information about the actual anellipticity parameters in the field, the values produced by VILS are much more plausible than those computed from the Dix-type equations. First, several Dix-derived interval parameters $\eta^{(1,2,3)}$ are too large for shale formations and lie outside the range suggested by laboratory and field studies (Toldi et al., 1999; Wang, 2002). The interval value of $\eta^{(2)}$ for the first superbin even exceeds unity. Also, horizontal shale layers typically exhibit weak (if any) azimuthal anisotropy, unless they are fractured. Available geologic information for Rulison field and the small eccentricity of the NMO ellipses suggest that the symmetry of the UMV shale over most of the study area is close to VTI. This implies that the difference between the parameters $\eta^{(1)}$ and $\eta^{(2)}$, as well as the magnitude of $\eta^{(3)}$, should be relatively small, which is in agreement with the values obtained by VILS. The values of $\eta^{(2)}$ for both superbins produced by the Dix-type method, however, are much larger than those of $\eta^{(1)}$.

Superbin	1			2		
	$\eta^{(1)}$	$\eta^{(2)}$	$\eta^{(3)}$	$\eta^{(1)}$	$\eta^{(2)}$	$\eta^{(3)}$
VILS	0.38	0.47	−0.18	0.24	0.31	−0.15
Dix	0.74	1.24	−0.35	0.31	0.62	−0.19

Table 3.7: Interval parameters $\eta^{(1,2,3)}$ estimated for two superbins in the center of the study area at Rulison field (Wang and Tsvankin, 2009).

Superbin	1			2		
	$V_{\text{nmo}}^{(1)}$ (km/s)	$V_{\text{nmo}}^{(2)}$ (km/s)	φ (°)	$V_{\text{nmo}}^{(1)}$ (km/s)	$V_{\text{nmo}}^{(2)}$ (km/s)	φ (°)
VILS	4.22	3.99	115	4.33	3.84	122
Dix	4.26	3.82	104	4.41	3.91	139

Table 3.8: Interval NMO ellipses for two superbins near the east boundary of the study area (Wang and Tsvankin, 2009).

To test the stability of both methods, a linear time error (from 4 ms at zero offset to −4 ms at the maximum offset for each azimuth) was added to the reconstructed reflection moveout from the bottom of the shale layer in the second superbin. The interval parameters $\eta^{(1,2,3)}$ estimated by VILS change by only −0.06, −0.07, and 0.01, respectively, while the corresponding variations produced by the Dix-type method are much larger (−0.12, −0.21, and 0.13). Hence, VILS is more stable than the Dix-type method in the presence of correlated time errors, as was established above for synthetic data.

The NMO ellipticity in the UMV shale is pronounced only near the east boundary of the study area (Xu and Tsvankin, 2007). Because of the small offset-to-depth ratio (less than 1.3) for the bottom of the shale layer near the area boundary, the inverted parameters $\eta^{(1,2,3)}$ are unstable and likely to contain large errors. The interval velocities $V_{\text{nmo}}^{(1,2)}$ and the azimuth φ estimated by both methods for two adjacent superbin gathers in the area of significant NMO ellipticity are listed in Table 3.8. A linear error (from 2 ms at zero offset to −2 ms at the maximum offset for each azimuth) added to the traveltimes from the bottom of the shale in the second superbin causes a deviation of about 3% in the effective velocities $V_{\text{nmo}}^{(1,2)}$. As a result, the interval $V_{\text{nmo}}^{(1,2)}$ and φ estimated by the Dix-type method change by 13%, 16% and 4°, respectively. The sensitivity of VILS to the time error is much lower, with the interval NMO velocities changing by less than 8% and φ by 1°. Despite the superior performance by VILS, the errors in $V_{\text{nmo}}^{(1,2)}$ are substantial because of the relatively small thickness of the target layer. Still, this result shows that it might be beneficial to apply VILS to interval NMO-velocity estimation from conventional-spread data. Azimuthal hyperbolic moveout analysis for the Rulison reservoir is discussed in sections 8.3 and 9.6.

3.5 Time-processing parameters for orthorhombic media

2D P-wave time processing (i.e., NMO and DMO corrections, prestack and poststack time migration) in each vertical symmetry plane of orthorhombic media is controlled solely by the in-plane zero-dip NMO velocity and the corresponding η-parameter (Tsvankin, 1997a, 2005). As follows from the kinematic equivalence between the symmetry planes of orthorhombic and VTI media, the needed parameter pairs are $V_{\text{nmo}}^{(2)}$ and $\eta^{(2)}$ for the $[x_1, x_3]$-plane, and $V_{\text{nmo}}^{(1)}$ and $\eta^{(1)}$ for the $[x_2, x_3]$-plane. The 2D assumption, however, is valid only when the incidence symmetry plane coincides with the dip plane of the subsurface structure; otherwise, reflected rays and/or the corresponding phase-velocity vectors deviate from the incidence plane.

The results of Chapter 2 indicate that 3D time processing outside the symmetry planes involves at least one additional parameter – $\eta^{(3)}$ or $\delta^{(3)}$. Indeed, while the NMO ellipse for horizontal reflectors (and, therefore, NMO correction) are governed by the symmetry-plane azimuths and velocities $V_{\text{nmo}}^{(1)}$ and $V_{\text{nmo}}^{(2)}$, NMO velocities of dipping events (and DMO correction) also depend on $\eta^{(3)}$.

As shown in this chapter, the symmetry-plane directions and parameters $V_{\text{nmo}}^{(1,2)}$ and $\eta^{(1,2,3)}$ control P-wave nonhyperbolic moveout in a horizontal orthorhombic layer. Because long-spread reflection traveltimes describe the corresponding diffraction curves on zero-offset sections, the parameters responsible for nonhyperbolic moveout should be sufficient to generate an accurate poststack time-migration impulse response, at least up to some maximum dip. The analysis of migration impulse responses below shows that P-wave poststack time migration in orthorhombic media is indeed governed by the same parameters as are NMO ellipses and nonhyperbolic moveout. This conclusion is in agreement with the work by Ikelle (1996), who studied the dispersion relation in weakly anisotropic orthorhombic media.

The impulse responses in Figures 3.31 and 3.32 were computed by the 3D PSPI (phase-shift-plus-interpolation) migration code of Le Rousseau (1997). Models 1 and 2 (used in Figure 2.37) have vastly different elastic properties but identical parameters $V_{\text{nmo}}^{(1,2)}$ and $\eta^{(1,2,3)}$. The 3D impulse responses for these models are practically indistinguishable, both within and outside the symmetry planes; this is illustrated in Figure 3.31 for the vertical section making an angle of 30° with the x_1-axis. The same results were obtained in other tests performed for a representative set of orthorhombic models (Grechka and Tsvankin, 1999b). Any variation in the relevant parameters would lead to changes in the shape of the migration impulse response. For instance, a decrease in $\eta^{(1)}$ and $\eta^{(2)}$ translates into smaller horizontal velocities and a narrower vertical section of the impulse response (Figure 3.32).

Since normal moveout (i.e., NMO and DMO corrections) and poststack time migration are fully controlled by the symmetry-plane orientation, zero-dip NMO velocities $V_{\text{nmo}}^{(1,2)}$, and the anisotropy parameters $\eta^{(1,2,3)}$, this parameter set is responsible for the whole 3D P-wave time-processing sequence. Consequently, the same parameters are sufficient for prestack time migration in homogeneous orthorhombic media.

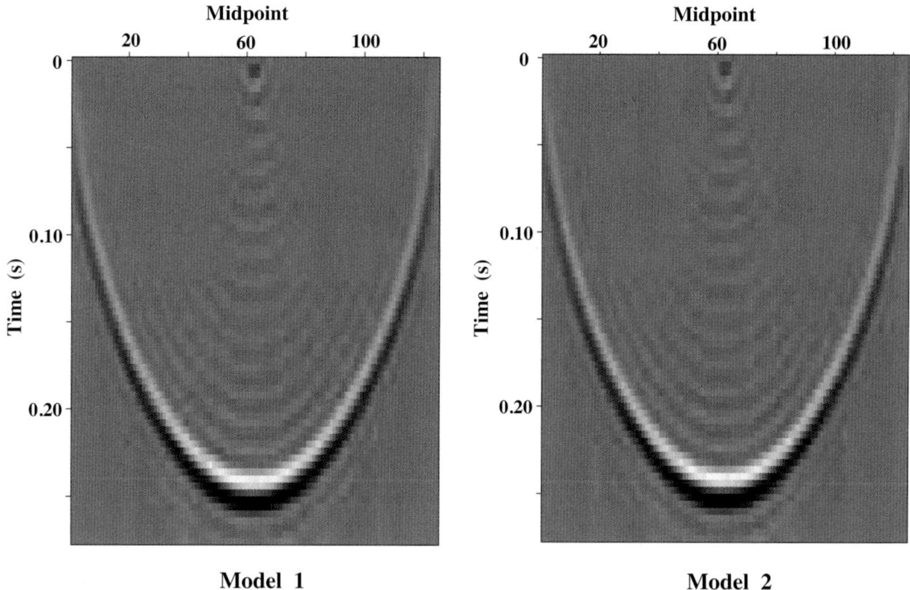

Figure 3.31: Vertical section of the impulse response of 3D poststack time migration for two orthorhombic models (Grechka and Tsvankin, 1999b). The response is computed in the vertical plane that makes an angle of 30° with the symmetry plane $[x_1, x_3]$. The models are taken from Figure 2.37. For model 1, $V_{\rm nmo}^{(1)}$=1.8 km/s, $V_{\rm nmo}^{(2)}$=2.2 km/s, $\eta^{(1)}$=0.2, $\eta^{(2)}$=0.3, $\eta^{(3)}$=0.15, $\epsilon^{(1)}$=0.2, $\epsilon^{(2)}$=0.695, $\delta^{(1)}$=0, $\delta^{(2)}$=0.25, and $\delta^{(3)}=-0.27$. For model 2, all $V_{\rm nmo}^{(i)}$ and $\eta^{(i)}$ are the same as those for model 1, but $\epsilon^{(1)}=-0.031$, $\epsilon^{(2)}$=0.3, $\delta^{(1)}=-0.165$, $\delta^{(2)}$=0, and $\delta^{(3)}=-0.27$.

As follows from the results of this chapter and Chapter 2, time-processing operators for vertically heterogeneous orthorhombic media above a dipping or curved target reflector have to be built using the interval values of these parameters. P-wave depth processing, however, requires knowledge of the vertical velocity V_{P0} and anisotropy parameters $\epsilon^{(1,2)}$ and $\delta^{(1,2,3)}$.

3.6 Summary

Long-spread P-wave reflection traveltime in azimuthally anisotropic media can be accurately described by generalizing the nonhyperbolic moveout equations of Tsvankin and Thomsen (1994) and Alkhalifah and Tsvankin (1995) originally developed for VTI media. In contrast to the conventional Taylor series $t^2(x^2)$, these equations converge at offsets approaching infinity, which substantially increases the accuracy in the intermediate offset range (i.e., for offset-to-depth ratios between two and three) important in reflection seismology. While the NMO velocity used in 2D moveout equations can be simply replaced with the NMO ellipse, analytic description of the

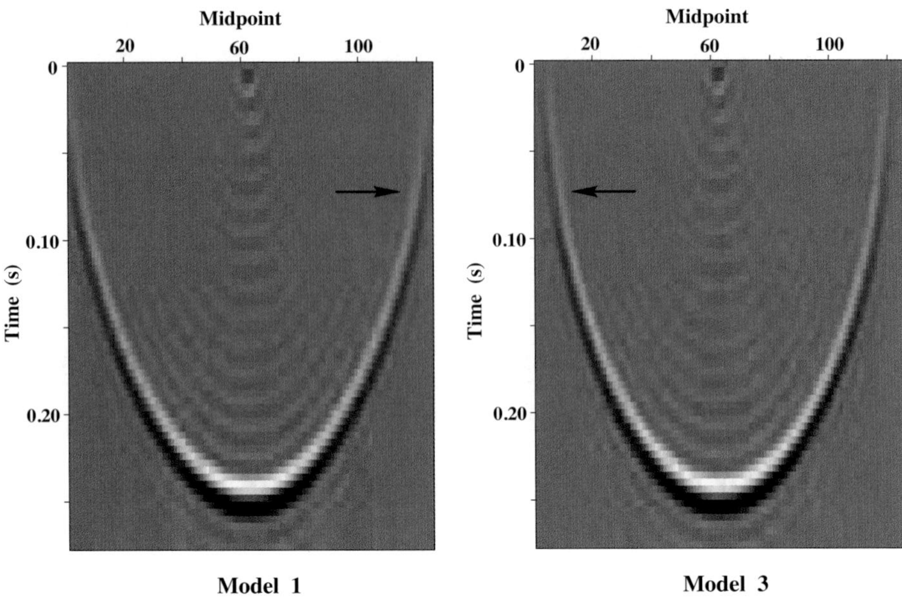

Figure 3.32: Influence of the parameters $\eta^{(1)}$ and $\eta^{(2)}$ on a vertical section of the poststack time-migration impulse response (Grechka and Tsvankin, 1999b). The section makes an angle of 30° with the $[x_1, x_3]$-plane; the plot for model 1 is reproduced from Figure 3.31. For model 3, the parameters $V_{\text{nmo}}^{(1,2)}$ and $\eta^{(3)}$ are the same as those for model 1, but $\eta^{(1)}$=0.1 and $\eta^{(2)}$=0.2.

quartic moveout coefficient A_4 in the presence of azimuthal anisotropy becomes much more involved.

We represented the exact coefficient A_4 in arbitrarily anisotropic, heterogeneous media as a function of the spatial derivatives of the one-way traveltime between the reflector and the surface. This expression for A_4 does not require the model to have a horizontal symmetry plane and accounts for reflection-point dispersal on dipping or irregular interfaces. All quantities needed to calculate the azimuthally varying quartic moveout coefficient are generated by tracing a single (zero-offset) ray. This result can be used to avoid the need for multioffset, multiazimuth ray tracing in modeling long-spread moveout for laterally heterogeneous media.

The general expression for the quartic moveout coefficient was first applied to P-waves in a dipping TI layer with the symmetry axis confined to the dip plane. The weak-anisotropy approximation for A_4 is proportional to the parameter η, so the magnitude of nonhyperbolic moveout increases as the model deviates from elliptical. The azimuthal variation of the quartic coefficient is more complicated compared to the NMO ellipse, and A_4 can go to zero and change sign in several azimuthal directions. In the special case of the symmetry axis orthogonal to the reflector, the dip-line coefficient A_4 decreases rapidly with dip, while the strike-line A_4 is constant (i.e., independent of dip). Therefore, nonhyperbolic moveout for dipping reflectors in this model is significant mostly for azimuths close to the strike direction. For a VTI layer, the linearized coefficient A_4 on the dip line vanishes for a dip of 30°, which explains the well-documented small magnitude of nonhyperbolic moveout for moderately dipping reflectors in VTI media (Tsvankin, 2005).

The magnitude and azimuthal variation of A_4 in an orthorhombic layer with a horizontal symmetry plane is controlled by the anellipticity parameters $\eta^{(1)}$, $\eta^{(2)}$, and $\eta^{(3)}$, which are also responsible for the dip dependence of the P-wave NMO ellipse (Chapter 2). The quartic coefficient in each vertical symmetry plane of a horizontal layer is given by the VTI equation and is proportional to the in-plane η-parameter (i.e., $\eta^{(1)}$ or $\eta^{(2)}$). If the reflector is dipping and the dip plane coincides with the vertical symmetry plane $[x_1, x_3]$, the dip-line coefficient A_4 depends just on $\eta^{(2)}$ and changes sign at a dip of 30°, as in VTI models. The quartic coefficient on the strike line is a function of all three η-parameters, but for mild dips it is mostly governed by $\eta^{(1)}$. Whereas the magnitude of the dip-line A_4 typically becomes small for dips exceeding 45°, the nonhyperbolic moveout on the strike line can remain substantial even for steep reflectors. The influence of $\eta^{(3)}$ generally increases with dip and contributes to the complicated azimuthal pattern of A_4, which is sensitive to the signs and relative magnitudes of the parameters $\eta^{(1,2,3)}$.

The analytic description of the quartic moveout coefficient helped extend nonhyperbolic moveout equations originally developed for VTI media to orthorhombic and HTI models (note that HTI can be treated as a special case of orthorhombic symmetry). In particular, the Alkhalifah-Tsvankin equation with the azimuthally varying parameters V_{nmo} and η gives a concise and accurate description of azimuthally varying, long-spread P-wave traveltimes. The moveout parameters for layered orthorhombic media with a uniform symmetry-plane orientation include the symmetry-plane azimuth φ, the semiaxes $V_{\text{nmo}}^{(1,2)}$ of the NMO ellipse, and the effective anellipticity parameters $\eta^{(1,2,3)}$. If the symmetry-plane azimuths vary with depth, the principal azimuthal direction of the parameter η has to be decoupled from the orientation of the NMO ellipse and described by a separate angle φ_1. The parameters φ, φ_1, $V_{\text{nmo}}^{(1,2)}$, and $\eta^{(1,2,3)}$ can be estimated by a robust algorithm that maximizes semblance for the full range of offsets and azimuths using the generalized Alkhalifah-Tsvankin equation. A modified version of this method (not described here) that properly handles polarity reversals was developed by Yan and Tsvankin (2008) who employed the so-called *AK-semblance* operator.

The moveout-inversion algorithm was successfully tested on wide-azimuth P-wave reflections recorded at Weyburn field in Canada. Taking azimuthal anisotropy into account increased the semblance values for most long-offset reflection events in the overburden, which likely indicates that fracturing is not limited to the reservoir level. The inverted symmetry-plane directions are close to the fracture azimuths determined from borehole data and shear-wave splitting analysis. Nonhyperbolic moveout inversion of wide-azimuth data has several important applications in data processing and reservoir characterization. First, the algorithm accurately flattens long-spread CMP gathers for the entire range of azimuths prior to stacking or azimuthal AVO analysis. Second, inversion of long-spread moveout can be used to build a model for time imaging of P-wave data in orthorhombic media. Third, the effective moveout parameters are also responsible for geometrical spreading in horizontally layered models and, therefore, help remove the directionally-dependent spreading factor from reflection amplitudes (see sections 8.2 and 8.3). For example, the moveout-inversion results in Figure 3.18 allowed Xu and Tsvankin (2006a) to correct the wide-azimuth, long-spread reflection from the top of the reservoir for anisotropic geometrical spreading. Fourth, the anellipticity parameters carry information about the physical properties (e.g., fracturing) of the section above the reflector.

Application of nonhyperbolic moveout analysis in lithology and fracture characterization requires estimation of the interval parameters, which can be accomplished using Dix-type techniques. Inversion for the interval η values, however, suffers from instability caused by the trade-off between the effective moveout parameters and by subsequent error amplification during Dix-type layer stripping. We described an alternative approach based on the velocity-independent layer-stripping (VILS) method, which in theory produces the exact 3D interval traveltime-offset function in the target layer without knowledge of the velocity field. Because effective traveltimes are much better constrained by reflection data compared to effective moveout parameters, VILS gives more stable interval parameter estimates than do Dix-type techniques. The superior accuracy and stability of VILS in the presence of correlated noise was illustrated on synthetic examples and a wide-azimuth data set from Rulison field in Colorado. The interval parameters $\eta^{(1,2,3)}$ estimated by VILS on field data are more plausible and less influenced by noise than those obtained by the Dix-type method.

For orthorhombic media with a fixed symmetry-plane orientation, the parameters $V_{\text{nmo}}^{(1,2)}$ and $\eta^{(1,2,3)}$ control just P-wave nonhyperbolic moveout, but also NMO ellipses of horizontal and dipping events (see Chapter 2). Application of the phase-shift-plus-interpolation (PSPI) migration algorithm showed that 3D poststack time migration is governed by the same parameter set as that for reflection moveout. Therefore, the orientation of the symmetry planes and the parameters $V_{\text{nmo}}^{(1,2)}$ and $\eta^{(1,2,3)}$ are responsible for all 3D P-wave time-processing steps (NMO correction, DMO removal, prestack and poststack time migration) in orthorhombic models. Depth-domain processing, however, requires knowledge of the entire set of P-wave kinematic parameters – the vertical velocity V_{P0} and the anisotropy parameters $\epsilon^{(1,2)}$ and $\delta^{(1,2,3)}$.

Appendices for Chapter 3

3A Linearized P-wave coefficient A_4 in a tilted TI layer

We consider a homogeneous transversely isotropic layer above a plane dipping reflector and assume that the symmetry axis (unit vector $\mathbf{a}$) lies in the dip plane (Figure 3.2). Then the zero-offset ray is confined to the dip plane, which is a vertical plane of symmetry for the whole model. Without losing generality, the x_1-axis can be aligned with the dip (symmetry-plane) direction, so the vector $\mathbf{a}$ can be expressed through the tilt ν as

$$\mathbf{a} = [\sin\nu, 0, \cos\nu] . \tag{3.52}$$

The exact equation 3.5 for the quartic moveout coefficient involves the one-way traveltime τ between the common-midpoint $\mathbf{y}$ and the reflector. For a homogeneous medium,

$$\tau(y_1, y_2, x_1, x_2) = \frac{\sqrt{(x_1 - y_1)^2 + (x_2 - y_2)^2 + z^2(x_1, x_2)}}{\mathcal{G}(y_1, y_2, x_1, x_2)} , \tag{3.53}$$

where $z(x_1, x_2)$ defines the reflecting plane, and $\mathcal{G}(y_1, y_2, x_1, x_2)$ is the group velocity. Using the weak-anisotropy approximations for the P-wave group velocity and group angle in TI media (e.g., Tsvankin, 2005), we find

$$\begin{aligned} \mathcal{G} = \frac{V_{P0}}{4} \Big\{ 4 + m \big[-\delta - \epsilon - \eta \sin^2 a \sin^2 b - \eta \cos 2\nu \, (2 \cos^2 b \sin^2 a \\ + \sin^2 a \sin^2 b - 1) + \eta \cos b \sin 2a \sin 2\nu \big] \Big\} , \end{aligned} \tag{3.54}$$

where

$$\sin a \equiv \frac{\sqrt{(y_1 - x_1)^2 + (y_2 - x_2)^2}}{\sqrt{(y_1 - x_1)^2 + (y_2 - x_2)^2 + z^2}} , \tag{3.55}$$

$$\cos a \equiv \frac{z}{\sqrt{(y_1 - x_1)^2 + (y_2 - x_2)^2 + z^2}} , \tag{3.56}$$

$$\sin b \equiv \frac{(y_2 - x_2)}{\sqrt{(y_1 - x_1)^2 + (y_2 - x_2)^2}} , \tag{3.57}$$

$$\cos b \equiv \frac{(y_1 - x_1)}{\sqrt{(y_1 - x_1)^2 + (y_2 - x_2)^2}} , \tag{3.58}$$

$$\begin{aligned} m \equiv -1 - \sin^2 a \sin^2 b - \cos 2\nu \, (-1 + 2 \sin^2 a \cos^2 b + \sin^2 a \sin^2 b) \\ + \cos b \sin 2a \sin 2\nu . \end{aligned} \tag{3.59}$$

The parameters V_{P0}, ϵ, δ, and η are defined with respect to the symmetry axis, as explained in the main text.

Since the zero-offset traveltime must satisfy Fermat's principle, the minimum value of τ corresponds to the coordinates $x_1^{(0)}$ and $x_2^{(0)}$ of the zero-offset reflection point. This implies that the derivatives of $\tau(y_1, y_2, x_1, x_2)$ with respect to x_1 and x_2 should vanish at the point $[x_1^{(0)}, x_2^{(0)}]$:

$$\left.\frac{\partial \tau(y_1, y_2, x_1, x_2)}{\partial x_1}\right|_{\left[x_1^{(0)}, x_2^{(0)}\right]} = 0\,, \tag{3.60}$$

$$\left.\frac{\partial \tau(y_1, y_2, x_1, x_2)}{\partial x_2}\right|_{\left[x_1^{(0)}, x_2^{(0)}\right]} = 0\,. \tag{3.61}$$

Equations 3.60 and 3.61 can be used to relate the CMP coordinates y_1 and y_2 to $x_1^{(0)}$ and $x_2^{(0)}$. Substituting equations 3.53 and 3.54 into equations 3.60 and 3.61 and dropping quadratic and higher-order terms in the anisotropy parameters yields

$$y_1 = z\left[\epsilon - \eta\cos 2(\phi - \nu)\right]\frac{\sin 2(\phi - \nu)}{\cos^2\phi} + z\tan\phi + x_1^{(0)}\,, \tag{3.62}$$

$$y_2 = x_2^{(0)}\,, \tag{3.63}$$

where $z = z(x_1^{(0)}, x_2^{(0)})$ is the depth of the zero-offset reflection point. Equation 3.63 confirms that the zero-offset ray is confined to the dip plane $x_2 = \text{const}$; if the CMP lies on the x_1-axis (Figure 3.2), $y_2 = x_2^{(0)} = 0$.

Using equations 3.53, 3.62 and 3.63 to evaluate the derivatives in equation 3.5, we obtain the following linearized approximation for the P-wave quartic moveout coefficient:

$$A_4^{\text{TTI}} = -\frac{2\eta}{t_{P0}^2 V_{P0}^4}\,F(\alpha, \phi, \nu)\,; \tag{3.64}$$

$$\begin{aligned} F(\alpha, \phi, \nu) = \frac{1}{128}\Big[&18 - 24\cos 2\alpha + 6\cos 4\alpha + 8\cos 2(3\phi - 2\nu) \\ &+ 4\cos 2(\alpha - 2\nu) - 4\cos 2(2\alpha - \nu) + 24\cos 2(\phi - 2\nu) + 12\cos 2(\alpha + \phi - 2\nu) \\ &+ 8\cos 2(\alpha + 2\phi - 2\nu) + 4\cos 2(\alpha + 3\phi - 2\nu) + \cos 4(\alpha - \nu) + 32\cos 2(\phi - \nu) \\ &+ 32\cos 4(\phi - \nu) - 16\cos 2(\alpha + \phi - \nu) + 8\cos 2\nu + 6\cos 4\nu \\ &+ \cos 4(\alpha + \nu) - 4\cos 2(2\alpha + \nu) - 16\cos 2(\alpha - \phi + \nu) + 4\cos 2(\alpha + 2\nu) \\ &+ 4\cos 2(\alpha - 3\phi + 2\nu) + 8\cos 2(\alpha - 2\phi + 2\nu) + 12\cos 2(\alpha - \phi + 2\nu)\Big]. \end{aligned} \tag{3.65}$$

Chapter 4

Joint processing of PP and PS data

In the presence of anisotropy, supplementing P-wave reflection data with shear waves is often needed for estimating even the parameter set responsible for P-wave kinematics. For example, reflection moveout of P-waves in laterally homogeneous VTI media generally constrains only two parameter combinations – the zero-dip NMO velocity $V_{\mathrm{nmo},P}$ and the anellipticity parameter η (Alkhalifah and Tsvankin, 1995). As discussed in Chapter 5, addition of reflection traveltimes of SV-waves to P-wave moveout helps resolve the vertical P- and S-wave velocities and the anisotropy parameters ϵ and δ, provided the reflector has a mild dip and the data are acquired for a wide range of azimuths. Moreover, joint inversion of P- and S-wave data can be effectively used in lower-symmetry orthorhombic and monoclinic media (see Chapter 6).

Therefore, finding practical ways of combining P- and S-waves is critically important for anisotropic velocity model-building. In theory, such joint processing algorithms might seem to be easy to implement because the techniques discussed in previous chapters are equally valid for both PP and SS pure-mode reflections. (Here we use the double indices "PP" and "SS" to emphasize that the downgoing and upgoing segments of ray trajectories correspond to the same wave type.) The applicability of algorithms originally designed for PP-waves to SS data is ensured by the reciprocity of pure-mode traveltimes with respect to the source and receiver positions. As a result, moveout of any pure-mode reflection in common-midpoint (CMP) gathers can be described by the traveltime series $t^2(x^2)$ that contains only even powers of the offset x. The hyperbolic moveout equation and the NMO ellipse (see Chapters 1 and 2) define a truncated form of this series, which is usually sufficient for moderate offsets (i.e., those not exceeding reflector depth).

Joint processing and inversion of PP and SS data, however, encounters practical difficulties related to the often poor quality of shear data on land and to the absence of reliable marine S-wave sources. Consequently, it has become common to replace pure SS reflections in seismic imaging and parameter estimation with P-to-S converted modes. Reflected PS-waves (the conversion point is typically located at the reflector) are of particular importance in offshore seismic, where they represent the only available type of shear energy. Application of mode-converted data has proved to be highly beneficial in a number of exploration and reservoir monitoring scenarios. Examples include imaging of targets beneath gas clouds and of reflectors with a weak P-wave impedance contrast, characterization of naturally fractured reservoirs, and

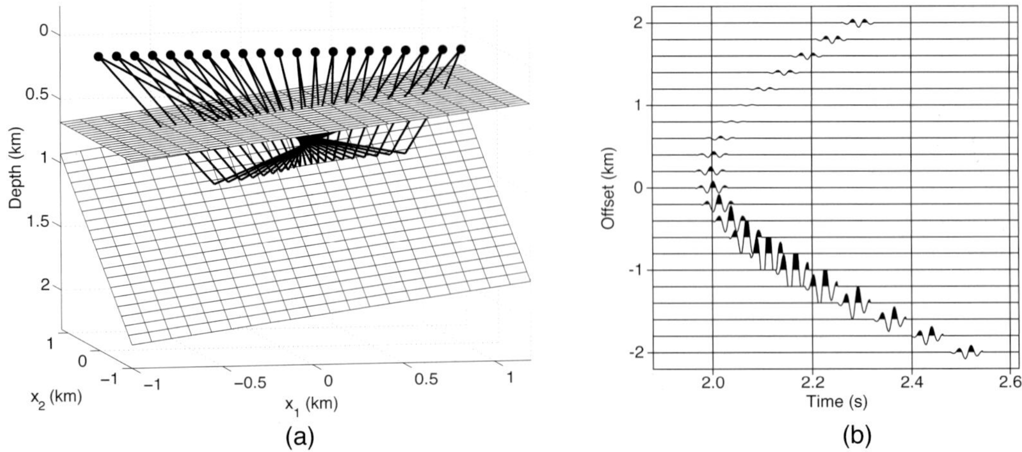

Figure 4.1: (a) Ray trajectories and (b) the horizontal inline displacement component of a reflected PSV-wave recorded in CMP geometry. The P-to-S conversion occurs at the reflector – the plane dipping interface beneath two homogeneous VTI layers. The relevant layer parameters are (the subscript "1" corresponds to the upper layer): $V_{P0,1}$=1.5 km/s, $V_{S0,1}$=0.8 km/s, ϵ_1=0.15, δ_1=0.05, $V_{P0,2}$=2.0 km/s, $V_{S0,2}$=0.9 km/s, ϵ_2=0.20, and δ_2=0.10. The interface dips ϕ and azimuths ψ (with respect to the CMP line) are: ϕ_1=10°, ψ_1=0°, ϕ_2=30°, and ψ_2=70°. The normal distances between the CMP and the interfaces are 0.5 km and 1.2 km.

estimation of rock and fluid parameters (e.g., Granli et al., 1999; Pérez et al., 1999; Thomsen, 1999; Gaiser, 2000).

Velocity-analysis algorithms operating with PS-waves, however, are impeded by several inherent features of mode conversions. Below we analyze the properties of PS reflections that complicate converted-wave processing in both isotropic and anisotropic media. Then we introduce the so-called PP+ PS = SS method designed to construct waves kinematically equivalent to pure SS primary reflections without using velocity information.

4.1 Properties of reflected PS-waves

The main problems associated with mode conversions are illustrated by Figure 4.1, which shows rays and a ray-traced synthetic seismogram of a reflected PSV-wave recorded in a CMP gather above two dipping VTI layers. Three undesirable properties of mode-converted data clearly seen in Figure 4.1 include their moveout asymmetry, polarity reversal (and the related low amplitude), and reflection-point dispersal:

1. CMP moveout of the converted wave in Figure 4.1b is asymmetric with respect to zero offset. (The offset x is conventionally defined as the difference between the receiver and source coordinates on the CMP line.)

2. The behavior of the PS-wave amplitude is significantly different at positive and negative offsets. While for $x < 0$ the amplitude varies smoothly with offset, at positive offsets the event changes polarity at $x \approx 1$ km and has a low amplitude for a wide range of x. If the VTI model in Figure 4.1 were horizontally layered, the PS reflection would vanish at zero offset.

3. The reflection point moves rapidly along the interface with increasing offset (Figure 4.1a). It can be shown that this shift of the reflection point (reflection-point dispersal) is approximately proportional to x. Reflection-point dispersal for pure modes is much smaller because for moderate offsets it is proportional to x^2.

The most significant problem in PS-wave velocity analysis stems from the moveout asymmetry with respect to zero offset, both in CMP geometry (Figure 4.1b) and common-conversion-point (CCP) gathers. Indeed, in contrast to pure modes, converted-wave traveltimes (t_{PS}) generally are not invariant with respect to interchanging the source and receiver positions or, in other words, with respect to the sign of offset [i.e., $t_{PS}(x) \neq t_{PS}(-x)$]. In the presence of moveout asymmetry, the PS-wave traveltime curve in CMP gathers has a nonzero slope at $x = 0$ and its minimum is shifted away from zero offset (Figure 4.1b).

PS-wave moveout is symmetric with respect to $x = 0$ only if the subsurface is composed of laterally homogeneous, horizontal layers with a horizontal symmetry plane. Various aspects of PS-wave kinematics in this special case were discussed for isotropic media by Tessmer and Behle (1988), and for anisotropic media by Seriff and Sriram (1991), Tsvankin and Thomsen (1994), Grechka et al. (1999a), Thomsen (1999), and others; most of these results are summarized in Tsvankin (2005). Conventional-spread PS-wave moveout for such stratified models with a horizontal symmetry plane is controlled primarily by the NMO ellipse. A Dix-type relationship between the NMO ellipses of PP-, PS-, and SS-waves is examined in section 1.6 and Appendix 1D.

If present, the asymmetry of PS-wave traveltimes invalidates the conventional series $t^2(x^2)$ and makes the standard notions of NMO velocity and quartic moveout coefficient meaningless. Therefore, most velocity-analysis methods developed for pure modes cannot be directly applied to converted waves. Forcing the series $t^2(x^2)$ to fit asymmetric PS data in CMP geometry (see Figure 4.1b) results in two different stacking (moveout) velocities corresponding to positive and negative offsets. Thomsen (1999) calls this phenomenon the "diodic velocity" of mode conversions. Although moveout equations specifically tailored to converted waves in anisotropic media have been developed (see chapter 5 in Tsvankin, 2005), their application requires substantial modification of existing processing algorithms.

Another feature of PS reflections, a polarity reversal near the zero-amplitude trace, causes problems for moveout-analysis methods based on coherency measures (e.g., semblance) of the reflected signal. The influence of polarity reversals can be mitigated by fitting the amplitude variation (i.e., the AVO signature) along with the traveltime function (Sarkar et al., 2002; Yan and Tsvankin, 2008). Amplitude analysis, however, increases the number of unknown parameters, which makes velocity estimation more computationally expensive and prone to instability.

Finally, significant conversion-point dispersal in CMP geometry leads to smearing of PS reflections in the presence of lateral heterogeneity and has to be accounted for in parameter estimation and imaging. Generation of common-conversion-point gathers theoretically eliminates the dispersal, but CCP sorting requires knowledge of the velocity model (see Figure 5.23 in Chapter 5).

Because of the severity of the above problems, Grechka and Tsvankin (2002c) suggested abandoning the whole concept of PS-wave moveout analysis and proposed an alternative technique known as the PP+ PS = SS method. It is a velocity-independent algorithm that operates with PP and PS reflections from the same interface and constructs data (called *pseudoshear-waves* or ΨS-waves by Grechka and Dewangan, 2003) that have the kinematics of the corresponding pure SS-wave primaries. Traveltimes of ΨS-waves are described by conventional hyperbolic or nonhyperbolic moveout equations and can be used jointly with P-wave moveout in anisotropic velocity analysis.

We defer discussion of this multicomponent anisotropic parameter-estimation procedure to Chapter 5. Here we concentrate on the kinematic and dynamic features of the PP+ PS = SS method and examine its performance on synthetic and field data.

4.2 Kinematics of the PP+ PS = SS method

4.2.1 2D problem

Suppose that PP- and PS-wave reflection data are acquired along a straight line in the dip direction of the subsurface structure and, if the medium is anisotropic, the vertical incidence plane is a plane of symmetry. Under those conditions, the ray trajectories and the corresponding slowness vectors do not deviate from the incidence plane. To explain the idea of the method of Grechka and Tsvankin (2002c), it is convenient to assume that P-wave sources and two-component (in-plane horizontal and vertical) receivers are distributed continuously along the line. The receivers record both PP reflections and P-to-SV conversions at the reflector. We make the common assumption in joint processing of PP and PS data that PP- and PS-waves are reflected from the *same* interface in the subsurface. The velocity field above this interface is arbitrary (as long as the incidence plane is a plane of symmetry) and, throughout this chapter, does not need to be specified.

Our first goal is to identify the PP and PS rays excited by the same source located at $x^{(1)}$ (Figure 4.2) and reflected from the same point on the interface. Let us examine PP data sorted into the common-receiver gather for a particular receiver located at $x^{(2)}$. The reflection traveltime t_{PP} in the vicinity of the source can be approximated by a linear function of the coordinate $x^{(1)}$. By tracking the PP reflection event in the common-receiver gather, one can find its traveltime $t_{PP}(x^{(1)}, x^{(2)})$, along with the local slope

$$p_{PP}(x^{(1)}, x^{(2)}) = \left. \frac{d\, t_{PP}(x, x^{(2)})}{dx} \right|_{x=x^{(1)}} . \qquad (4.1)$$

It is known from ray theory that the slope $p_{PP}(x^{(1)}, x^{(2)})$ is equal to the horizontal

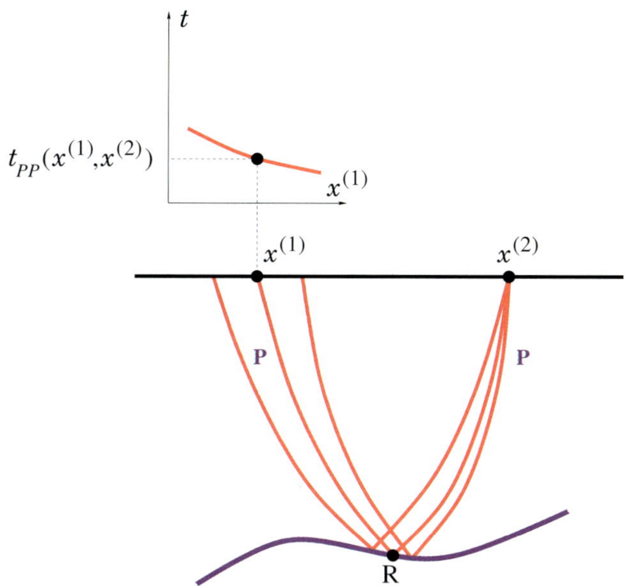

Figure 4.2: Reflected PP rays and traveltimes recorded in a common-receiver gather (Grechka and Tsvankin, 2002c).

inline component of the slowness vector (measured at location $x^{(1)}$) of the ray that travels from the source $x^{(1)}$ to the receiver $x^{(2)}$. This ray satisfies Snell's law at the (generally unknown prior to velocity analysis) reflection point R (Figure 4.2):

$$\left[\mathbf{p}_P(x^{(1)} \to \mathrm{R}) - \mathbf{p}_P(\mathrm{R} \to x^{(2)})\right] \times \mathbf{b}(\mathrm{R}) = 0\,. \tag{4.2}$$

Here $\mathbf{p}_P(x^{(1)} \to \mathrm{R})$ and $\mathbf{p}_P(\mathrm{R} \to x^{(2)})$ are the slowness vectors of the downgoing and upgoing P-wave rays, respectively, and $\mathbf{b}(\mathrm{R})$ is the reflector normal. All quantities in equation 4.2 are evaluated at the reflection point R, where the projections of the slowness vectors $\mathbf{p}_P(x^{(1)} \to \mathrm{R})$ and $\mathbf{p}_P(\mathrm{R} \to x^{(2)})$ onto the reflector should equal one another.

Next, we examine a P-wave excited by the source at $x^{(1)}$ and converted into an S-wave at the same interface (Figure 4.3). The reflected PS data are sorted into common-receiver gathers for different receiver locations $x^{(3)}$ to obtain the traveltimes $t_{PS}(x^{(1)}, x^{(3)})$ and the local reflection slopes at point $x^{(1)}$,

$$p_{PS}(x^{(1)}, x^{(3)}) = \left.\frac{d\,t_{PS}(x, x^{(3)})}{d\,x}\right|_{x=x^{(1)}}\,. \tag{4.3}$$

We are interested in finding a specific receiver location $x^{(3)}$, for which

$$p_{PP}(x^{(1)}, x^{(2)}) = p_{PS}(x^{(1)}, x^{(3)})\,. \tag{4.4}$$

Equation 4.4 can be solved for the unknown coordinate $x^{(3)}(x^{(1)}, x^{(2)})$ by scanning over receiver locations along the line. It is unknown in advance whether this solution

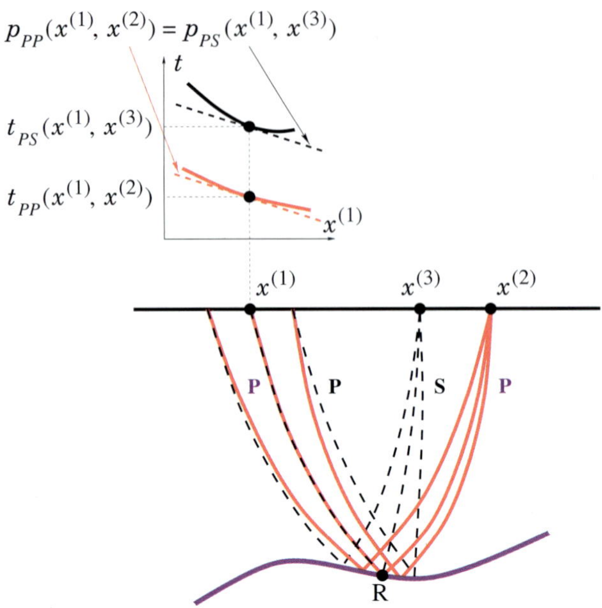

Figure 4.3: Reflection slope $p_{PS}(x^{(1)}, x^{(3)})$ in the common-receiver gather for the PS-wave recorded at $x^{(3)}$ is equal to the slope $p_{PP}(x^{(1)}, x^{(2)})$ in the PP-wave common-receiver gather (Grechka and Tsvankin, 2002c). Both slopes are measured at the source location $x^{(1)}$.

exists for any given coordinates $x^{(1)}$ and $x^{(2)}$ and, if it does, whether it is unique. If equation 4.4 has no solution (see a numerical example in Figure 4.8 below), the whole procedure is simply repeated for another pair of $(x^{(1)}, x^{(2)})$. If equation 4.4 has several solutions, as in the case of shear-wave triplications, it is theoretically possible to find all of them and construct the multivalued traveltime t_{SS}.

Suppose that equation 4.4 has at least one solution $x^{(3)}$. Then the rays $x^{(1)}\mathrm{R}\,x^{(2)}$ and $x^{(1)}\mathrm{R}\,x^{(3)}$ in Figure 4.3 have a common P-wave segment $x^{(1)}\mathrm{R}$ and, therefore, the same reflection point R. Indeed, both rays represent P-waves excited by the same source and initially propagating in the same direction specified by the horizontal slowness $p_{PP}(x^{(1)}, x^{(2)}) = p_{PS}(x^{(1)}, x^{(3)})$ (equation 4.4). Hence, those two rays should have identical trajectories between the source and the reflection point R, where one of the downgoing P-waves gets converted into a shear mode.

The slowness vector of the incident P-wave at the reflection/conversion point R is equal to $\mathbf{p}_P(x^{(1)} \rightarrow \mathrm{R})$, the quantity that appears in Snell's law (equation 4.2) for the pure PP reflection. The reflected PS-wave $x^{(1)}\mathrm{R}\,x^{(3)}$ also obeys Snell's law,

$$\left[\mathbf{p}_P(x^{(1)} \rightarrow \mathrm{R}) - \mathbf{p}_S(\mathrm{R} \rightarrow x^{(3)})\right] \times \mathbf{b}(\mathrm{R}) = 0\,, \tag{4.5}$$

where $\mathbf{p}_S(\mathrm{R} \rightarrow x^{(3)})$ is the slowness vector of the upgoing shear ray at point R.

Next, let us switch the source and receiver positions by placing the source at $x^{(2)}$ and the receiver at $x^{(1)}$. Repeating the same event-tracking procedure in the PP-wave

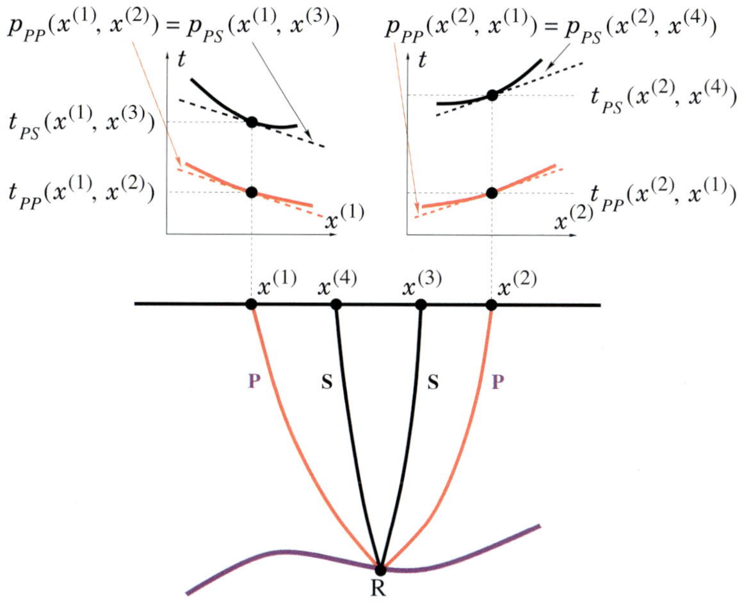

Figure 4.4: Matching reflection slopes in common-receiver PP and PS gathers at locations $x^{(1)}$ and $x^{(2)}$ helps find the source-receiver coordinates $x^{(3)}$ and $x^{(4)}$ of the pure SS reflection $x^{(3)}\mathrm{R}\,x^{(4)}$ (Grechka and Tsvankin, 2002c). This constructed SS-wave has the same reflection point R as the PP-wave $x^{(1)}\mathrm{R}\,x^{(2)}$ and PS-waves $x^{(1)}\mathrm{R}\,x^{(3)}$ and $x^{(2)}\mathrm{R}\,x^{(4)}$.

common-receiver gather yields the reflection slope at location $x^{(2)}$ (Figure 4.4):

$$p_{_{PP}}(x^{(2)}, x^{(1)}) = \left.\frac{d\,t_{PP}(x, x^{(1)})}{dx}\right|_{x=x^{(2)}} . \tag{4.6}$$

Alternatively, $p_{_{PP}}(x^{(2)}, x^{(1)})$ can be found by forming a common-shot gather at $x^{(1)}$ and evaluating the time slope at $x^{(2)}$. Note that although the traveltimes $t_{PP}(x^{(1)}, x^{(2)})$ and $t_{PP}(x^{(2)}, x^{(1)})$ coincide because of reciprocity, the corresponding slopes generally differ:

$$p_{_{PP}}(x^{(1)}, x^{(2)}) \neq p_{_{PP}}(x^{(2)}, x^{(1)}) , \tag{4.7}$$

as illustrated by Figure 4.4.

Repeating the procedure described above, we find the receiver coordinate $x^{(4)}$, for which the slope $p_{_{PS}}(x^{(2)}, x^{(4)})$ of the reflected PS-wave evaluated at $x^{(2)}$ satisfies the condition

$$p_{_{PP}}(x^{(2)}, x^{(1)}) = p_{_{PS}}(x^{(2)}, x^{(4)}) . \tag{4.8}$$

Hence, the converted-wave ray $x^{(2)}\mathrm{R}\,x^{(4)}$ [its traveltime is $t_{PS}(x^{(2)}, x^{(4)})$] in Figure 4.4 has the same reflection point R as that of the two previously discussed PP and PS events. The slowness vectors of the downgoing P-wave and upgoing S-wave for the path $x^{(2)} \mathrm{R}\, x^{(4)}$ obey Snell's law at R:

$$\left[\mathbf{p}_P(x^{(2)} \to \mathrm{R}) \;-\; \mathbf{p}_S(\mathrm{R} \to x^{(4)})\right] \times \mathbf{b}(\mathrm{R}) = 0\,. \tag{4.9}$$

Thus, evaluation of time slopes allowed us to find three ray trajectories – $x^{(1)} \mathrm{R}\, x^{(2)}$, $x^{(1)} \mathrm{R}\, x^{(3)}$, and $x^{(2)} \mathrm{R}\, x^{(4)}$ – with the *same* reflection point R (Figure 4.4). Combining the segments $\mathrm{R}\, x^{(3)}$ and $\mathrm{R}\, x^{(4)}$ yields a pure shear (SS) ray $x^{(3)} \mathrm{R}\, x^{(4)}$.

Still, we need to verify that the trajectory $x^{(3)} \mathrm{R}\, x^{(4)}$ corresponds to a physical SS reflection excited at $x^{(3)}$ and recorded at $x^{(4)}$ or vice versa. Combining equations 4.2, 4.5, and 4.9 shows that the shear-wave slowness vectors at R $[\mathbf{p}_S(x^{(3)} \to \mathrm{R})$ and $\mathbf{p}_S(\mathrm{R} \to x^{(4)})]$ satisfy Snell's law:

$$\left[\mathbf{p}_S(x^{(3)} \to \mathrm{R}) \;-\; \mathbf{p}_S(\mathrm{R} \to x^{(4)})\right] \times \mathbf{b}(\mathrm{R}) = 0\,. \tag{4.10}$$

Therefore, the ray $x^{(3)} \mathrm{R}\, x^{(4)}$ corresponds to a pure SS reflection for source and receiver positions at $x^{(3)}$ and $x^{(4)}$. The reflection traveltime $t_{SS}(x^{(3)}, x^{(4)})$ along the trajectory $x^{(3)} \mathrm{R}\, x^{(4)}$ is given by (Figure 4.4)

$$t_{SS}(x^{(3)}, x^{(4)}) \;=\; t_{PS}(x^{(1)}, x^{(3)}) \;+\; t_{PS}(x^{(2)}, x^{(4)}) \;-\; t_{PP}(x^{(1)}, x^{(2)})\,. \tag{4.11}$$

It should be emphasized that for the considered geometry equation 4.11 is *exact.*

In summary, the traveltime of the pure SS reflection can be found as a simple combination of the measured PP and PS traveltimes. The source and receiver positions ($x^{(3)}$ and $x^{(4)}$) for the SS arrivals were identified by matching time slopes in common-receiver gathers of PP- and PS-waves (equations 4.4 and 4.8) without using any information about the velocity model.

4.2.2 Extension to 3D

The methodology described above can be extended in a straightforward way to 3D wide-azimuth, multicomponent data. Let us assume that the sources and receivers cover a certain area in the acquisition plane $\mathbf{x} = [x_1, x_2]$. Then, for a given receiver location $\mathbf{x}^{(2)} = [x_1^{(2)}, x_2^{(2)}]$ we build a common-receiver gather (data cube) in the vicinity of the source $\mathbf{x}^{(1)} = [x_1^{(1)}, x_2^{(1)}]$. Tracking the PP event inside this common-receiver gather makes it possible to estimate the reflection traveltime $t_{PP}(\mathbf{x}^{(1)}, \mathbf{x}^{(2)})$ and the time slopes

$$p_{PP,i}(\mathbf{x}^{(1)}, \mathbf{x}^{(2)}) = \left.\frac{\partial t_{PP}(\mathbf{x}, \mathbf{x}^{(2)})}{\partial x_i}\right|_{\mathbf{x}=\mathbf{x}^{(1)}}, \qquad (i = 1, 2)\,. \tag{4.12}$$

The slopes $p_{PP,1}$ and $p_{PP,2}$, which represent the horizontal projections of the slowness vector at point $\mathbf{x}^{(1)}$, can be computed by fitting a plane to the traveltime surface $t_{PP}(\mathbf{x}, \mathbf{x}^{(2)})$ or by numerical differentiation of the corresponding traveltime table. Alternatively, time slopes can be found by applying so-called *plane-wave destruction filters* (Fomel, 2002) to reflection data.

In general, an incident P-wave excites two reflected split PS-waves (PS_1 and PS_2), which have to be separated by Alford-type rotation prior to application of the PP+PS=SS method. Then we can obtain the traveltime $t_{PS}(\mathbf{x}^{(1)}, \mathbf{x}^{(3)})$ of the PS-wave (either PS_1 or PS_2) excited at $\mathbf{x}^{(1)}$ and recorded at a certain location $\mathbf{x}^{(3)}$ along with the corresponding time slope $\mathbf{p}_{PS}$ at the source position:

$$p_{PS,i}(\mathbf{x}^{(1)}, \mathbf{x}^{(3)}) = \left.\frac{\partial t_{PS}(\mathbf{x}, \mathbf{x}^{(3)})}{\partial x_i}\right|_{\mathbf{x}=\mathbf{x}^{(1)}} . \tag{4.13}$$

If the receiver coordinates $\mathbf{x}^{(3)} = [x_1^{(3)}, x_2^{(3)}]$ satisfy the condition

$$\mathbf{p}_{PP}(\mathbf{x}^{(1)}, \mathbf{x}^{(2)}) = \mathbf{p}_{PS}(\mathbf{x}^{(1)}, \mathbf{x}^{(3)}) , \tag{4.14}$$

the PP- and PS-waves have the same reflection point R.

Next, by switching the P-wave source and receiver positions exactly as was done in 2D, we estimate the coordinates $\mathbf{x}^{(4)} = [x_1^{(4)}, x_2^{(4)}]$ of the receiver that records the PS-wave excited at $\mathbf{x}^{(2)}$ and reflected at R, which yields the traveltime $t_{PS}(\mathbf{x}^{(2)}, \mathbf{x}^{(4)})$. Finally, the traveltime $t_{SS}(\mathbf{x}^{(3)}, \mathbf{x}^{(4)})$ of the pure SS reflection (either S_1S_1 or S_2S_2) is computed from equation 4.11.

4.2.3 Synthetic test

The performance of the PP+PS=SS method was tested on a model that includes homogeneous VTI layers separated by curved interfaces (Figure 4.5). PP and PSV rays reflected from the bottom of the model were traced for 21 sources and 21 receivers with coordinates $x = i\Delta x$, $\Delta x = 0.1$ km, $i = 0, \ldots, 20$ (Figure 4.5 shows the rays recorded at $x = 0.5$ km). While the computed PP traveltimes (Figure 4.6a) are symmetric with respect to zero offset, the converted-wave times (Figure 4.6b) are not. The asymmetry of the PS-wave moveout, however, does not impede the proposed method designed to produce the corresponding SS traveltimes.

The ray-traced traveltime tables t_{PP} and t_{PS} (each contains $21 \times 21 = 441$ values) represent the input data needed to construct the shear-wave traveltimes t_{SS}. First, reflection slopes in common-receiver gathers were computed by simple two-point linear interpolation. For example, the approximation for the slope $p_{PP}(x^{(1)}, x^{(2)})$ (equation 4.1) is

$$p_{PP}(x^{(1)}, x^{(2)}) \approx \frac{t_{PP}(x^{(1)} + \Delta x, x^{(2)}) - t_{PP}(x^{(1)} - \Delta x, x^{(2)})}{2\Delta x} . \tag{4.15}$$

The coordinates $x^{(3)}$ and $x^{(4)}$ of the SS-wave sources and receivers (Figure 4.7) were found for each pair $(x^{(1)}, x^{(2)})$ by matching the slopes of PP and PS arrivals in common-receiver gathers as described above (equations 4.4 and 4.8). Since $x^{(3)}$ and $x^{(4)}$ do not necessarily coincide with the source/receiver locations used in the modeling, the slopes were linearly interpolated between adjacent common-receiver gathers.

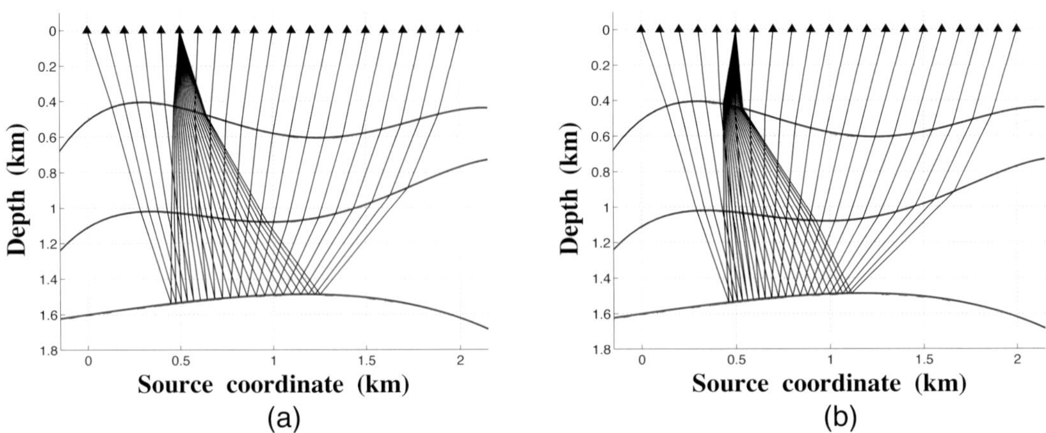

Figure 4.5: Reflected rays of (a) PP-waves, and (b) PSV-waves recorded by the receiver at x=0.5 km in a 2D model composed of three homogeneous VTI layers separated by curved interfaces (Grechka and Tsvankin, 2002c). The interval vertical velocities (from top to bottom) are $V_{P0,1}$=2.0 km/s, $V_{S0,1}$=0.8 km/s, $V_{P0,2}$=2.5 km/s, $V_{S0,2}$=1.25 km/s, $V_{P0,3}$=3.0 km/s, and $V_{S0,3}$=1.8 km/s. The anisotropy parameters are ϵ_1=0.20, δ_1=0.10, ϵ_2=0.25, δ_2=0.05, ϵ_3=0.15, and δ_3=0.10.

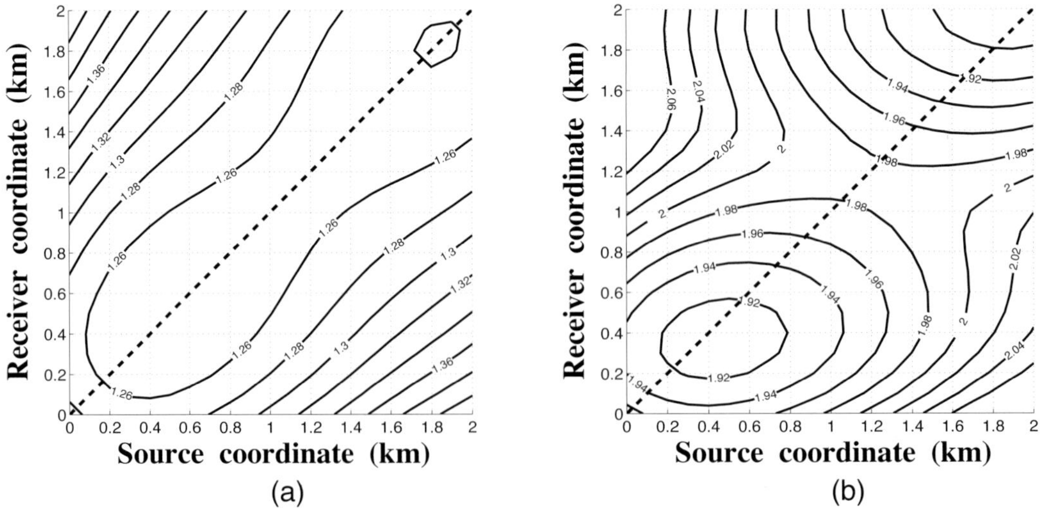

Figure 4.6: Contours of the traveltimes (a) t_{PP}, and (b) t_{PS} for the model in Figure 4.5 (Grechka and Tsvankin, 2002c). The dashed lines mark the zero-offset (a) PP, and (b) PS reflections.

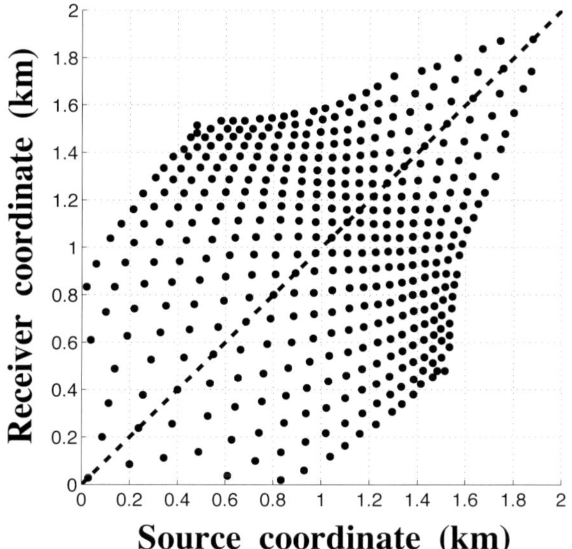

Figure 4.7: Coordinates of the SS-wave sources-receiver pairs computed from the PP and PS traveltimes in Figure 4.6 (Grechka and Tsvankin, 2002c).

In keeping with the reciprocity of pure-mode traveltimes, the source and receiver positions for the computed SS arrivals in Figure 4.7 are symmetric with respect to the zero-offset line (dashed).

Note the absence of the constructed SS-wave sources and receivers for relatively large offsets (the bottom right and top left corners in Figure 4.7). Because the P-to-S velocity ratio is always greater than unity ($V_P > V_S$), the constructed SS rays are closer to the reflector normal than are the corresponding PP rays, as schematically shown in Figure 4.4. Therefore, in the absence of strong lateral heterogeneity in the overburden, the source-receiver offset for a constructed SS reflection is generally smaller than the offsets of the input PP- and PS-waves (Figure 4.7).

The relationship between the offsets of PP- and SS-waves with a common reflection point can be estimated from the approximation of Tessmer and Behle (1988) derived for a single isotropic, homogeneous layer above a horizontal reflector:

$$x_{_{SS}} \approx \frac{2g x_{_{PS}}}{1+g} \approx g x_{_{PP}}, \tag{4.16}$$

where $x_{_{SS}}$, $x_{_{PS}}$, and $x_{_{PP}}$ are the offsets of the SS- PS-, and PP-waves, respectively, and $g \equiv V_S/V_P$. For a typical value $g = 0.5$, approximation 4.16 yields

$$x_{_{SS}} \approx \frac{2}{3} x_{_{PS}} \approx \frac{1}{2} x_{_{PP}}, \tag{4.17}$$

which is reasonably close to the result in Figure 4.7.

Another noteworthy observation is that Figure 4.7 contains only 417 constructed positions of SS-wave sources and receivers, whereas the number of input PP (and PS)

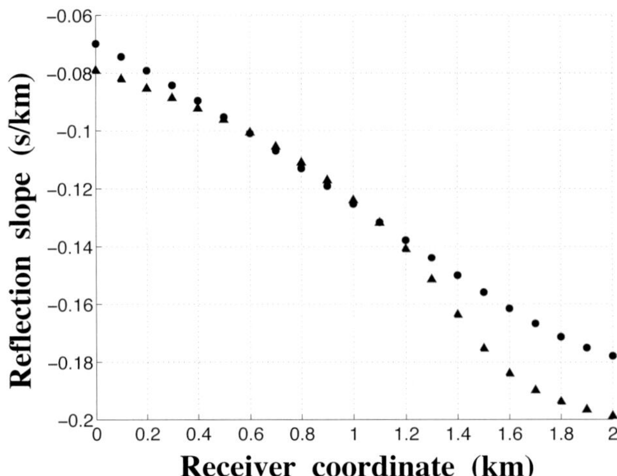

Figure 4.8: Reflection slopes p_{PP} (dots) and p_{PS} (triangles) at the source location $x^{(1)}$=0.1 km computed by ray tracing for the full range of receiver coordinates (Grechka and Tsvankin, 2002c).

data points is 441. This happens because the curved model interfaces (Figure 4.5) produce PP reflection slopes at some source locations that cannot be matched by any available slopes of the PS-wave. Figure 4.8 gives an example of such a situation for the source located at 0.1 km: the slopes p_{PP} for PP reflections recorded at $x^{(2)} < 0.2$ km are greater than the PS-wave slopes for all receiver locations. This means that the locations $x^{(3)}$ sought by the algorithm are outside the actual source-receiver array, and no SS traveltime can be constructed for the PP-wave excited at $x^{(1)} = 0.1$ km and recorded at $x^{(2)} < 0.2$ km. As shown below, however, the absence of SS reflections for some source-receiver pairs does not compromise the quality of the traveltimes t_{SS} constructed elsewhere.

To verify the accuracy of the method, we can compare the SS-wave traveltimes obtained from PP and SS data with those computed by ray tracing. The ray-traced traveltimes of the pure SS reflections for the whole range of the source and receiver coordinates are shown in Figure 4.9. The contours of t_{SS} are symmetric with respect to the zero-offset line (dashed), as should hold for pure-mode reflections. The comparison in Figure 4.10 demonstrates that the constructed traveltimes t_{SS} practically coincide with the ray-traced values within the entire area occupied by the SS-wave sources and receivers in Figure 4.7. The maximum difference between the estimated and ray-traced traveltimes t_{SS} in Figure 4.10 is just 1.6 ms or 0.06%. This error stems primarily from linear approximation 4.15 for the reflection slopes and from interpolation of the PS-wave slopes in the process of searching for receiver positions $x^{(3)}$ and $x^{(4)}$. A small additional error is caused by interpolating the SS traveltimes from the irregular grid in Figure 4.7 to the regular one used in the ray tracing.

Thus, PP and PS traveltimes (Figure 4.6) are sufficient for constructing the corresponding pure shear-wave traveltimes (Figure 4.9). As long as both PP- and PS-waves

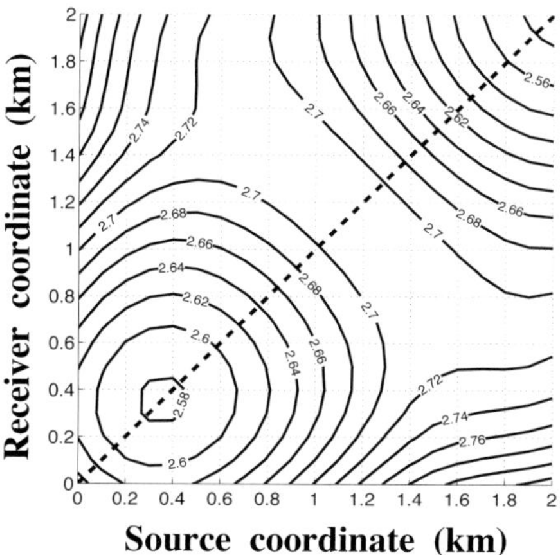

Figure 4.9: Contours of the ray-traced SS-wave traveltimes t_{SS} (Grechka and Tsvankin, 2002c).

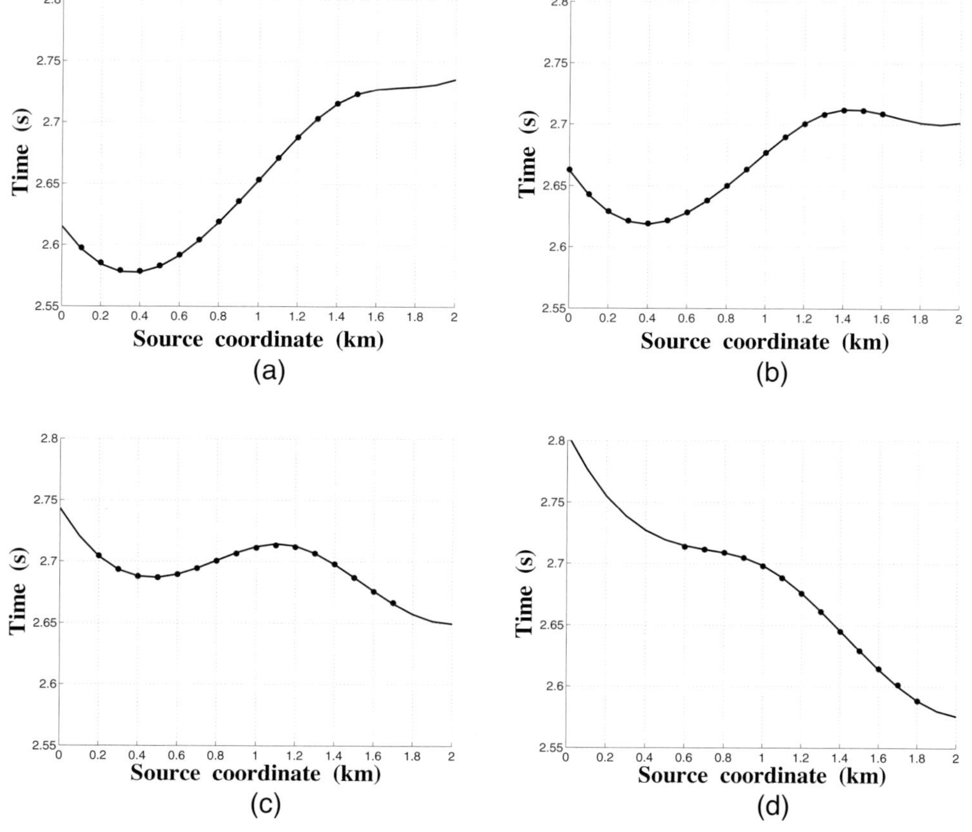

Figure 4.10: Comparison of the ray-traced (solid lines) and constructed (dots) traveltimes t_{SS} for the receivers at (a) 0.4 km, (b) 0.8 km, (c) 1.2 km, and (d) 1.6 km (Grechka and Tsvankin, 2002c).

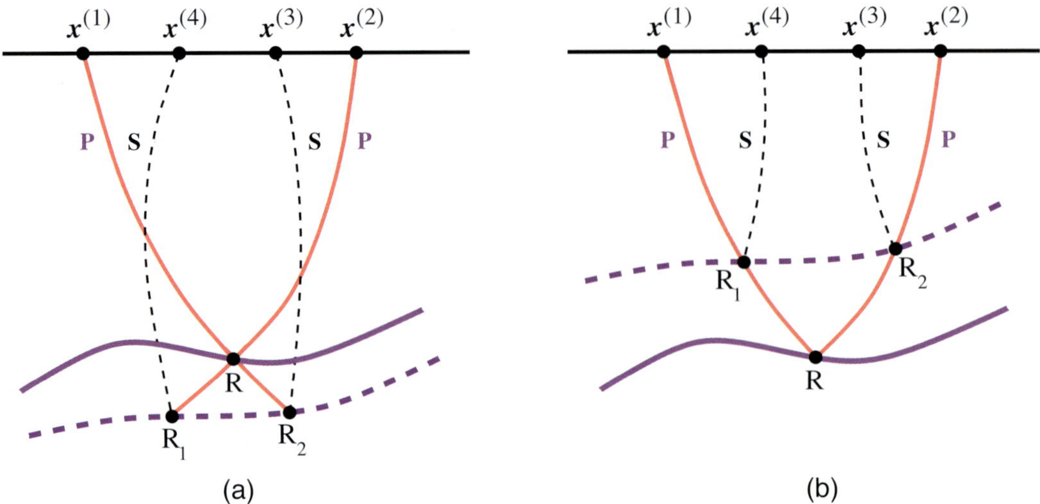

Figure 4.11: Influence of correlation (i.e., event-registration) errors on the output of the PP+PS=SS method (Grechka and Tsvankin, 2002c). If the PP- and PS-waves are reflected from different interfaces, the method will produce the SS traveltimes corresponding to either (a) a peg-leg multiple, or (b) a nonexistent raypath.

are reflected from the same interface, the PS+PS=SS method is exact, with errors caused solely by traveltime picking and numerical implementation.

4.2.4 Influence of errors in PP-PS correlation

An important assumption of our method is that the input PP- and PS-waves are reflected from the *same* interface in the subsurface. Correlation of interpreted PP and PS reflections (event registration), however, is not straightforward and often requires additional information, such as borehole data. Possible consequences of interpretation errors in correlating PP and PS data are demonstrated in Figure 4.11.

If the PS-wave is reflected from a deeper interface than is the PP event (Figure 4.11a), the PP+PS=SS method will construct the trajectory $x^{(4)}\mathrm{R}_1\mathrm{R}\mathrm{R}_2x^{(3)}$ corresponding to the peg-leg multiple SPPS with two mode conversions at points R_1 and R_2. Equation 4.11 in this case yields

$$t_{SS} = t_{x^{(4)}\mathrm{R}_1} + t_{\mathrm{R}_1\mathrm{R}} + t_{\mathrm{R}\mathrm{R}_2} + t_{\mathrm{R}_2x^{(3)}}\,, \tag{4.18}$$

where $t_{x^{(4)}\mathrm{R}_1}$ and $t_{\mathrm{R}_2x^{(3)}}$ are the S-wave traveltimes along the segments $x^{(4)}\mathrm{R}_1$ and $\mathrm{R}_2x^{(3)}$, and $t_{\mathrm{R}_1\mathrm{R}}$ and $t_{\mathrm{R}\mathrm{R}_2}$ are the P-wave traveltimes along $\mathrm{R}_1\mathrm{R}$ and $\mathrm{R}\mathrm{R}_2$ (Figure 4.11a). Clearly, equation 4.18 does not produce the traveltime of the SS primary reflection from either interface.

If the PS-wave is reflected from a more shallow interface (Figure 4.11b), the constructed traveltime cannot be associated with any physical ray. Applying equation 4.11, we obtain

$$t_{SS} = t_{x^{(4)}\mathrm{R}_1} - t_{\mathrm{R}_1\mathrm{R}} - t_{\mathrm{R}\mathrm{R}_2} + t_{\mathrm{R}_2 x^{(3)}} \,, \tag{4.19}$$

where $t_{x^{(4)}\mathrm{R}_1}$ and $t_{\mathrm{R}_2 x^{(3)}}$ correspond to S-waves, and $t_{\mathrm{R}_1\mathrm{R}}$ and $t_{\mathrm{R}\mathrm{R}_2}$ to P-waves. The value of t_{SS} in equation 4.19 can even become negative if the two interfaces in Figure 4.11b are sufficiently separated in depth.

Thus, any errors in identifying and correlating PP and PS reflections inevitably lead to incorrect traveltimes t_{SS}. The difference in the kinematics of the actual and erroneously constructed events (in particular, obtained unrealistic stacking velocities), however, might provide a criterion that can be used to correct such interpretation errors.

4.2.5 Case study from the North Sea

Next, we present application of the 2D version of the PP+PS=SS method to a multicomponent data set (courtesy of Shell) acquired over the Lower Tertiary Siri reservoir in the North Sea. Following Grechka et al. (2002c), we discuss the processing sequence employed to compute the traveltimes of SS-waves from PP and PS data. This case study continues in section 5.3 (Chapter 5), which describes joint anisotropic velocity analysis of the PP and SS traveltimes and improvements achieved by using the obtained VTI model in processing of the PS-waves.

The Siri survey includes three 4C (three-component geophone and hydrophone) seabed seismic lines and an 8.7 km × 17.6 km 3D towed-streamer data set that crosses the Norwegian-Danish North Sea border. A portion of one of the 4C lines was selected for anisotropic velocity analysis. The initial P-wave velocity model was built using the 3D data and then supplemented with shear-wave information obtained from a seabed line (Iversen et al., 2000). A description of acquisition, isotropic processing and interpretation, and the subsurface geology can be found in Signer et al. (2000). The deepest interpreted horizon (top Balder) is immediately above the top of the thin (~30 m) Siri reservoir, and the overburden structure is close to being horizontally layered (Figure 4.12).

Prestack PP and PS (PSV) reflection traveltimes for all interpreted horizons in Figure 4.12 were obtained using Schlumberger's semiautomatic picker. The traveltime picks were made in common-receiver gathers with the receiver increment $\Delta r = 200$ m; the source increment within each gather was $\Delta s = 25$ m. Figure 4.13 shows typical raw PP and PS traveltime picks that exhibit several distinct features important for our study.

1. The PP traveltime contours (Figure 4.13a) are almost symmetric with respect to zero offset. Their mild asymmetry can be attributed to inaccuracies in traveltime picking and to the fact that the sources are located at the ocean surface, whereas the receivers are placed on the sea bottom.

2. The asymmetry of the PS traveltimes is much stronger than that for PP-waves and can be clearly seen in Figure 4.13b. The most pronounced PS-wave asymmetry

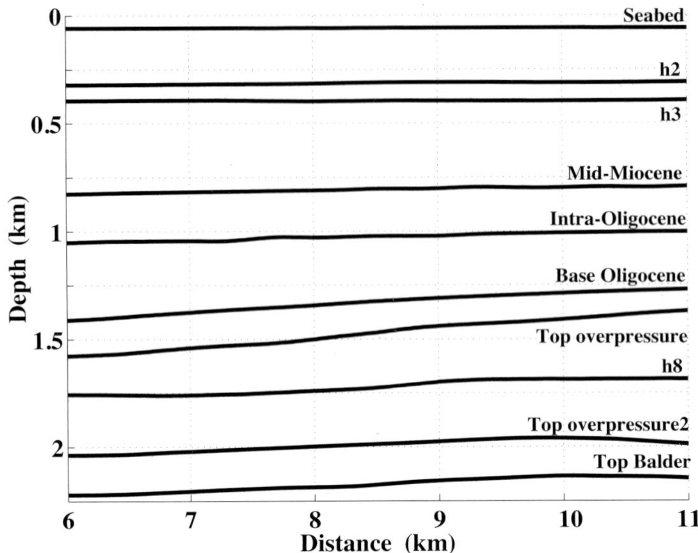

Figure 4.12: Depth section above the Siri reservoir built by Signer et al. (2000).

is observed in the area of increased lateral variation in the traveltimes (indicative of lateral heterogeneity) near CMP location $y_{\text{CMP}} = 8$ km. For example, the difference between the PS times at offsets $x = \pm 3$ km for this CMP reaches 100 ms (Figure 4.13b).

3. The traveltimes of PP- and PS-waves in Figure 4.13 are picked for a wide range of offsets limited by 4 km, which yields a maximum offset-to-depth ratio $x_{\max,PP}/D$ for the top overpressure interface close to 2.5.

If the subsurface structure has only mild lateral heterogeneity, which is the case here, an approximate relationship between the maximum offset $x_{\max,SS}$ of the constructed SS data and that of the acquired PP-waves is given by equation 4.16:

$$x_{\max,SS} \approx g\,x_{\max,PP} \,. \tag{4.20}$$

Although Tessmer and Behle (1988) derived equation 4.20 for isotropic media with the velocity ratio $g = V_S/V_P$, it can be used for qualitative analysis in VTI media, if g is replaced by the ratio of the vertical velocities,

$$g_0 \equiv \frac{V_{S0}}{V_{P0}} = \frac{t_{P0}}{t_{S0}} \,, \tag{4.21}$$

where t_{P0} and t_{S0} are the zero-offset traveltimes of P- and S-waves, respectively. For the top overpressure interface, $g_0 \approx 0.33$ and $x_{\max,SS}/D \approx 0.82$. Thus, to obtain reliable estimates of SS-wave stacking (moveout) velocities, it is necessary to acquire long-spread PP and PS data. This conclusion also follows from the synthetic example discussed above.

The PP+PS=SS method (with a minor modification needed to account for the difference in depth between the surface sources and seabed receivers) was used to obtain the locations of the pure shear-wave sources and receivers along with the

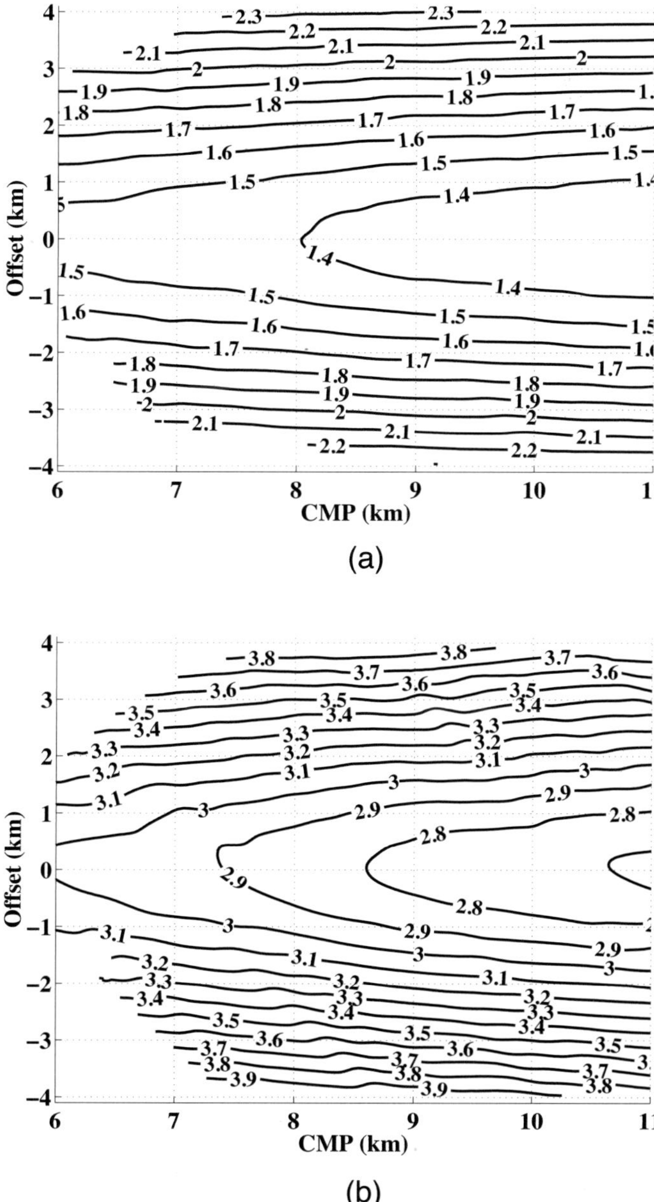

Figure 4.13: Raw picked traveltimes (in s) of the (a) PP and (b) PS reflections from the top overpressure interface (see Figure 4.12) above the Siri reservoir (Grechka et al., 2002c). The offset $x = r - s$ is either positive or negative depending on the sign of the difference between the receiver (r) and source (s) coordinates.

corresponding SS traveltimes for all horizons marked in Figure 4.12. More details about the processing flow can be found in Grechka et al. (2002c).

The results for the top overpressure interface in Figure 4.14 illustrate the main properties of the computed SS (SVSV) traveltime field. Despite the mild asymmetry of the PP-traveltimes in Figure 4.13a, both the SS-wave source and receiver positions and traveltimes in Figure 4.14 are symmetric with respect to zero offset. To remove the influence of the PP-wave moveout asymmetry (caused primarily by noise), equation 4.11 was applied with the average of the reciprocal PP traveltimes:

$$t_{SS}(x^{(3)}, x^{(4)}) = t_{PS}(x^{(1)}, x^{(3)}) + t_{PS}(x^{(2)}, x^{(4)}) - \frac{t_{PP}(x^{(1)}, x^{(2)}) + t_{PP}(x^{(2)}, x^{(1)})}{2}. \tag{4.22}$$

The constructed shear-wave offsets cluster in the range 0.4 km – 1.2 km, with almost no coverage for larger offsets (Figure 4.14a). The maximum offset-to-depth ratio reaches 0.8 (close to the approximate value from equation 4.20), which is marginally acceptable for estimation of the SS-wave moveout velocity. The constructed SS data for other horizons (not shown) have similar offset-to-depth ratios.

The traveltimes of PP- and SS-waves (such as those in Figures 4.13a and 4.14b) can be used to compute their zero-offset times (t_{P0} and t_{S0}) and stacking (moveout) velocities ($V_{\text{nmo},P}$ and $V_{\text{nmo},S}$). Hyperbolic moveout analysis of the PP and SS data followed by interval anisotropic parameter estimation is discussed in section 5.3.

4.3 Full-waveform version of the PP+PS=SS method

The implementation of the PP+PS=SS method desribed above involves two key preprocessing elements: identifying PP and PS events reflected from the same interfaces (the only step that requires indirect velocity information) and picking traveltimes from PP and PS prestack data. While the interpretive step of establishing the correspondence of PP and PS primaries cannot be avoided, the (usually laborious and time-consuming) traveltime picking can. In this section, based on the results of Grechka and Dewangan (2003), we describe a specially designed convolution of PP and PS traces that produces so-called *pseudoshear* (ΨS) data. If the recorded PP- and PS-waves are reflected from the same horizon in the subsurface, ΨS data have the kinematics of the pure shear-wave primary reflections from that horizon. Therefore, ΨS events can be processed by conventional velocity-analysis procedures, and their stacking velocities can be used in parameter estimation (see Chapter 5).

For 2D data, ΨS-wave seismograms can be computed from the following integral:

$$w_{\Psi S}(t, x^{(3)}, x^{(4)}) = \iint \Big[w_{PS}(t, x^{(1)}, x^{(3)}) * w_{PP}(-t, x^{(1)}, x^{(2)}) * w_{PS}(t, x^{(2)}, x^{(4)}) \Big] \, dx^{(1)} dx^{(2)} , \tag{4.23}$$

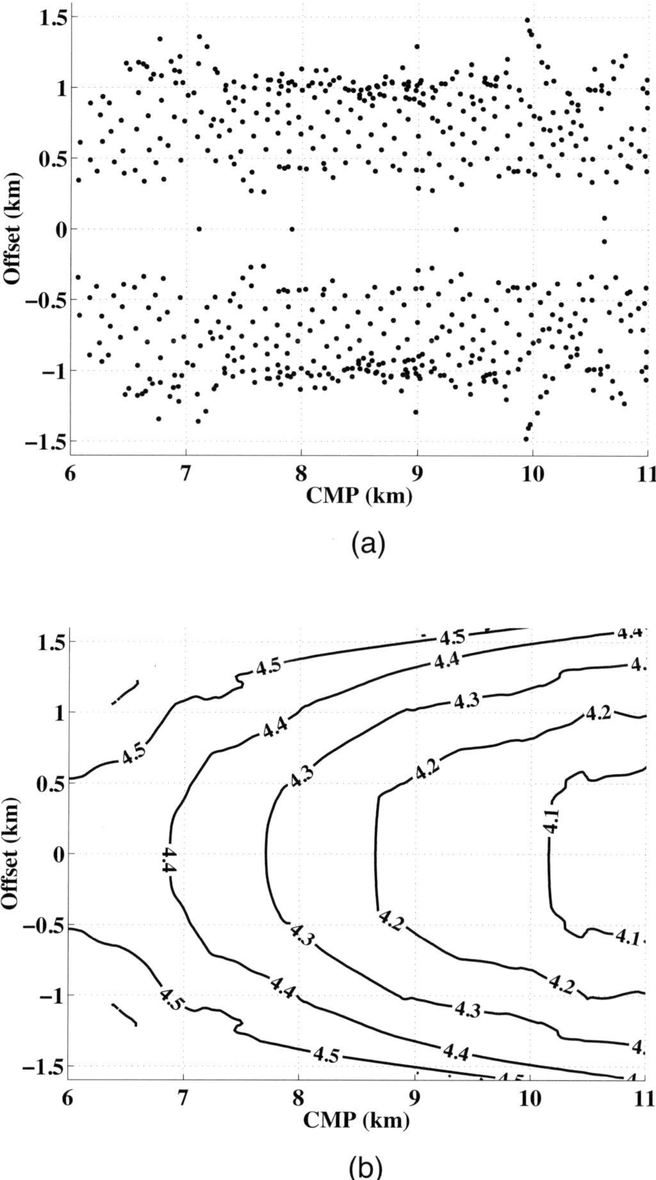

Figure 4.14: (a) Constructed positions of the shear-wave sources and receivers and (b) SS traveltimes (in s) for the top overpressure interface (Grechka et al., 2002c).

where $x^{(3)}$ and $x^{(4)}$ are the source and receiver coordinates, w_{PP} and w_{PS} are the input traces of PP- and PS-waves, and asterisk denotes time convolution. The trace w_{PP} is taken in reverse time because the PP-wave traveltime is subtracted in equation 4.11. Integration is performed over the P-wave source and receiver coordinates $x^{(1)}$ and $x^{(2)}$. Equation 4.23 can be viewed as one of many techniques that comprise the general methodology of seismic interferometry. For more details on existing interferometric approaches, the reader is referred to the papers collected in Wapenaar et al. (2008).

Evaluation of integral 4.23 using the stationary-phase method (Appendix 4A) shows that if traces w_{PP} and w_{PS} consist of the PP and PS primaries for the same reflector, the trace $w_{\Psi S}$ represents the pure SS-wave primary from that reflector. The fact that the proof in Appendix 4A is possible only for an isolated interface indicates the need to window the input w_{PP} and w_{PS} traces around the events of interest. This, in turn, requires establishing the correspondence of PP and PS reflections prior to generating ΨS-waves.

Similarly to the kinematic version of the PP+PS=SS method, integral 4.23 can be extended to 3D wide-azimuth reflection data. When sources and receivers are distributed over a given area, their coordinates $\mathbf{x} \equiv [x_1, x_2]$ become two-dimensional vectors. ΨS-wave traces for 3D data are formally obtained from an integral that has the same form as equation 4.23:

$$w_{\Psi S}(t, \mathbf{x}^{(3)}, \mathbf{x}^{(4)}) = \iiiint \Big[w_{PS}(t, \mathbf{x}^{(1)}, \mathbf{x}^{(3)}) * w_{PP}(-t, \mathbf{x}^{(1)}, \mathbf{x}^{(2)}) * w_{PS}(t, \mathbf{x}^{(2)}, \mathbf{x}^{(4)}) \Big] d\mathbf{x}^{(1)} d\mathbf{x}^{(2)} . \tag{4.24}$$

Since each vector $\mathbf{x}^{(j)}$ $(j = 1, 2, 3, 4)$ has two components, equation 4.24 involves four-fold integration.

4.3.1 Amplitudes of ΨS-waves

Even though integrals 4.23 and 4.24 are supposed to produce ΨS data that have the kinematics of SS reflections, ΨS-wave amplitudes differ from those of the actual shear-wave primaries. Indeed, the full-waveform version of the PP+PS=SS method is *not* a true-amplitude procedure because, as follows from the reciprocity theorem, PP and PS data generally do not contain enough information for computing the true SS-wave amplitudes. For instance, consider normal-incidence PP and PS reflections from a horizontal interface separating two isotropic halfspaces. The amplitude of the zero-offset reflected P-wave in this model is proportional to the P-wave impedance contrast across the interface, while the PS-wave amplitude is zero (assuming that the source and receiver are sufficiently far from the reflector). The vertical traveltime of SS-waves can be easily found from those of PP- and PS-waves, provided the PS traveltime was estimated from moveout analysis. Zero-offset PP and PS traces, however, contain no information about the contrast in the shear-wave impedance, which governs the amplitude of the normal-incidence SS reflection.

On the other hand, since ΨS-wave amplitudes are obtained in a deterministic way from those of PP and PS reflections, they potentially might be attractive for purposes of amplitude-variation-with-offset (AVO) analysis. The benefits of using ΨS data instead of mode conversions include the simplicity of their moveouts and a higher (in most cases) signal-to-noise ratio than that of the original PP- and PS-waves (see Figure 4.19). An analytic description of ΨS amplitudes, however, still has to be developed.

Below we mostly ignore amplitude-related issues and concentrate primarily on velocity analysis of ΨS-waves. We shall see that the unphysical amplitudes of ΨS data do not cause any problems for shear-wave velocity estimation.

4.3.2 Synthetic examples

Here, the theory outlined above is illustrated by 2D numerical tests. We begin by examining the time function τ (equation 4.33 in Appendix 4A),

$$\tau = t_{PS}(x^{(1)}, x^{(3)}) + t_{PS}(x^{(2)}, x^{(4)}) - t_{PP}(x^{(1)}, x^{(2)}), \tag{4.25}$$

whose extrema determine the kinematics of ΨS-waves. Then the main features of ΨS data are studied by computing integral 4.23. Finally, we analyze the performance of the procedure in the presence of random noise.

Time function τ

As follows from the results of Appendix 4A, the ability of integral 4.23 to reproduce the kinematics of SS-wave primaries depends upon the existence of the solutions $x^{(1)}$ and $x^{(2)}$ of equations 4.34 and 4.35,

$$\frac{\partial \tau}{\partial x^{(1)}} = \frac{\partial \tau}{\partial x^{(2)}} = 0\,. \tag{4.26}$$

Clearly, the PP and PS data have to be physically recorded at the source and receiver locations $x^{(1)}$ and $x^{(2)}$ to contribute to the integral. Because the number (and the very existence) of source-receiver pairs satisfying equation 4.26 is unknown in advance, we present two examples that demonstrate expected outcomes.

It is convenient to express the function τ through the P-wave half-offset h_{PP} and midpoint y_{PP}:

$$h_{PP} = \frac{x^{(2)} - x^{(1)}}{2} \quad \text{and} \quad y_{PP} = \frac{x^{(2)} + x^{(1)}}{2}\,.$$

For a homogeneous, isotropic medium above a horizontal reflector (Figure 4.15), y_{PP} coincides with the midpoint for ΨS-waves,

$$y_{\Psi S} = \frac{x^{(3)} + x^{(4)}}{2}\,.$$

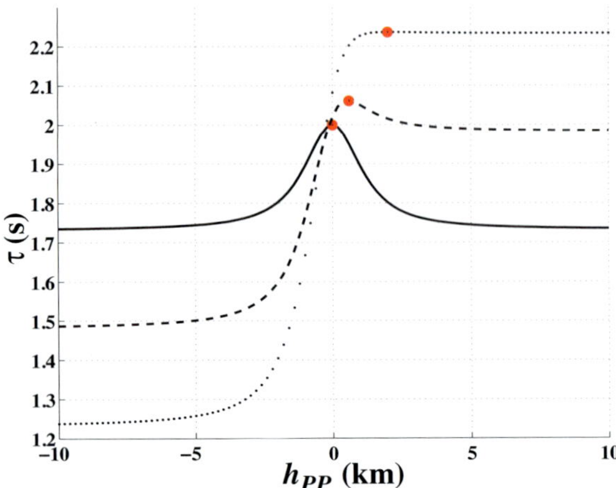

Figure 4.15: Time functions τ in a horizontal, homogeneous, isotropic layer 1-km thick with velocities V_P=2 km/s and V_S=1 km/s (Grechka and Dewangan, 2003). The half-offsets of ΨS-waves are: $h_{\Psi S}$=0 (solid line), $h_{\Psi S}$=0.25 km (dashed), and $h_{\Psi S}$=0.5 km (dotted). The large dots mark the stationary values of τ.

As a consequence, the curves $\tau(h_{PP})$ in Figure 4.15 are computed for the correct (i.e., stationary) values $y_{PP} = y_{\Psi S}$. Figure 4.15 displays two clear maxima[1] of τ for half-offsets $h_{\Psi S} = (x^{(3)} - x^{(4)})/2 = 0$ and $h_{\Psi S} = 0.25$ km as well as a poorly defined maximum for $h_{\Psi S} = 0.5$ km. Those maxima (marked with large dots) produce stationary points that yield pure shear-wave reflection traveltimes, which can be easily verified by computing the SS-wave moveout.

The curves in Figure 4.15 flatten out at large half-offsets h_{PP}, suggesting that the time function τ has extrema at $h_{PP} \to \pm\infty$. Although these extrema seem irrelevant for the problem at hand because the source-receiver offset cannot be infinite, both the flatness of the function τ and the finite-frequency bandwidth of seismic data might lead to noticeable contributions associated with those distant extrema even for moderate h_{PP}.

The extrema at $h_{PP} \to \pm\infty$ are related to the critical offsets of ΨS-waves (Figure 4.16). In our example, the P-wave incidence and S-wave reflection angles (θ_P and θ_S, respectively) satisfy isotropic Snell's law,

$$\frac{\sin\theta_P}{V_P} = \frac{\sin\theta_S}{V_S} \,. \tag{4.27}$$

Therefore, the maximum reflection angle of the shear wave generated by mode conversion is $\theta_S^{\text{crit}} = \sin^{-1}(V_S/V_P)$, and the half-offset $h_{\Psi S}$ cannot exceed the critical value,

$$h_{\Psi S}^{\text{crit}} = D \tan\left[\sin^{-1}\left(\frac{V_S}{V_P}\right)\right] ; \tag{4.28}$$

[1]They actually are saddle points. To show this, one needs to plot $\tau(h_{PP}, y_{PP})$ in the $[h_{PP}, y_{PP}]$ coordinates as done in Figure 4.17 below.

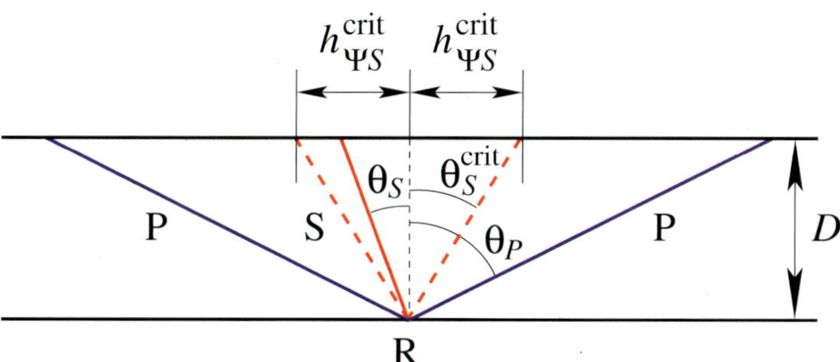

Figure 4.16: Critical offset of ΨS-waves in a horizontal layer (Grechka and Dewangan, 2003).

D is the layer's thickness. The P-wave half-offset h_{PP} corresponding to $h_{\Psi S}^{\rm crit}$ is infinite, so equation 4.28 establishes the physical upper limit for the ΨS-wave offset that can be obtained from PP and PS reflection data. Clearly, a low V_S/V_P velocity ratio results in a small spreadlength of the ΨS data and uncertainty in shear-wave velocity analysis. Hence, the PP+PS=SS method reveals the dependence of the accuracy of the estimated shear-wave moveout velocity on the V_S/V_P ratio, a point that often remains hidden in direct velocity analysis of mode-converted waves.

Equations 4.25 and 4.26 for a horizontal isotropic layer indicate that the waves corresponding to the contributions of the flat segments of the time function $\tau(h_{PP})$ at large offsets (see the dotted line in Figure 4.15) propagate with the P-wave rather than S-wave velocity. For that reason, such events can be called ΨP-waves. The extrema of $\tau(h_{PP})$ corresponding to ΨS- and ΨP-waves approach one another in a continuous fashion with increasing $h_{\Psi S}$ and become indistinguishable as $h_{\Psi S} \rightarrow h_{\Psi S}^{\rm crit}$. Since ΨS and ΨP arrivals have different kinematics, mixing them together (under the improper name ΨS) leads to errors in the estimated S-wave velocities. To avoid this complication, integration in equation 4.23 should be carried out for half-offsets $h_{\Psi S} < h_{\Psi S}^{\rm crit}$.

An example of the time function $\tau(h_{PP}, y_{PP})$ for a dipping layer is given in Figure 4.17. The stationary (saddle) point corresponding to the arrival of the ΨS-wave is located at $y_{PP} = -0.1$ km and $h_{PP} = 0.5$ km (the cross in Figure 4.17); the common midpoints $y_{\Psi S}$ and y_{PP} differ due to the influence of reflector dip. The flattening of the time function in the bottom corners of Figure 4.17 is indicative of approaching the extrema associated with ΨP-waves.

CMP gathers of ΨS-waves

According to the above analysis, the integration limits in equation 4.23 need to be chosen properly to avoid mutual contamination of ΨS and ΨP data. Also, the convolutions have to be performed within time gates that ensure PP-PS event correspondence. Figure 4.18 displays input data and computed ΨS-wave CMP gathers for an isotropic medium with dipping reflectors. The input PP and PS traces (Figures 4.18a,b) have

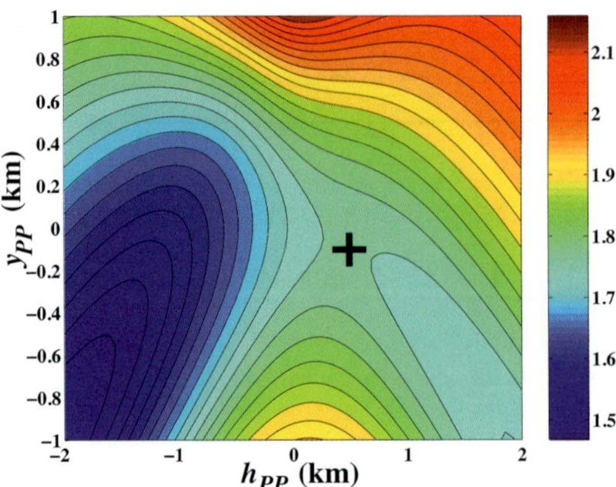

Figure 4.17: Time function $\tau(h_{PP}, y_{PP})$ (in s) for a dipping isotropic layer with vertical thickness D=1 km (measured under the coordinate origin h_{PP}=y_{PP}=0), dip of 30°, V_P=2 km/s, and V_S=1 km/s (Grechka and Dewangan, 2003). The midpoint and half-offset of the ΨS-wave are $y_{\Psi S}$=0 and $h_{\Psi S}$=0.2 km, respectively. The cross marks the stationary (saddle) point.

the following main features:

1. The moveouts of PS reflections from the two dipping interfaces are asymmetric with respect to interchanging the source and receiver positions (Figure 4.18b). This asymmetry, as discussed above, precludes application of conventional velocity analysis to PS data.

2. The polarity of the reflected PS-wave flips at zero offset.

3. By design, and as can be expected in practice, the PP and PS wavelets differ in both shape and frequency content; the ratio of the dominant frequencies of the PP- and PS-waves is 1.5.

Although such features are usually troublesome for conventional converted-wave processing, they do not prevent the PP+PS=SS method from constructing meaningful ΨS data (Figure 4.18c). The observation of primary importance is that the ΨS events have the correct SS-wave moveouts. Other noteworthy properties of the ΨS data include:

1. As expected from the convolution in equation 4.23, the ΨS wavelets are longer and have more lobes compared to the input PP and PS data. The dominant frequency of the ΨS-waves lies between those of the PP- and PS-waves. In principle, it is possible to deconvolve the PP or PS wavelets from the ΨS traces and produce spike-like ΨS arrivals.

2. There are no polarity reversals in the ΨS-wave CMP gather, which is favorable for moveout analysis. Indeed, PS data are used twice in equation 4.23, and the sign of the PS-wave polarity cancels out.

3. The near-offset ΨS arrivals have small amplitudes because of the weakness of the PS reflections near the polarity reversal. In particular, the PS-wave amplitude

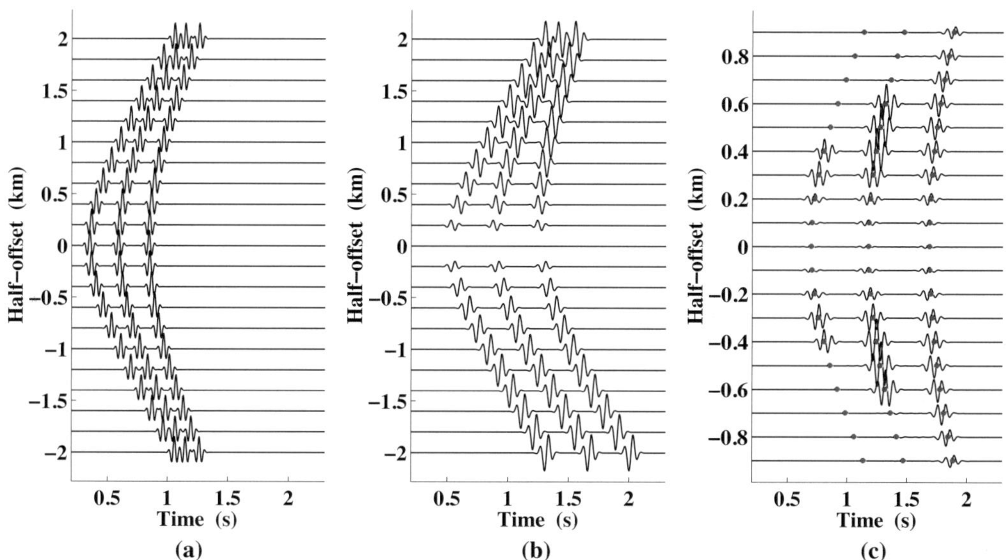

Figure 4.18: Computation of ΨS-waves from PP and PS data for three plane reflectors embedded in a homogeneous isotropic medium with V_P=4 km/s and V_S=2 km/s (Grechka and Dewangan, 2003). CMP gathers of (a) PP-waves, (b) PS-waves, and (c) ΨS-waves. The reflector depths beneath the CMP location are 0.7, 1.2, and 1.8 km, and the dips are 0°, 10°, and 20°. The dots on plot (c) mark the ray-traced SS reflection traveltimes.

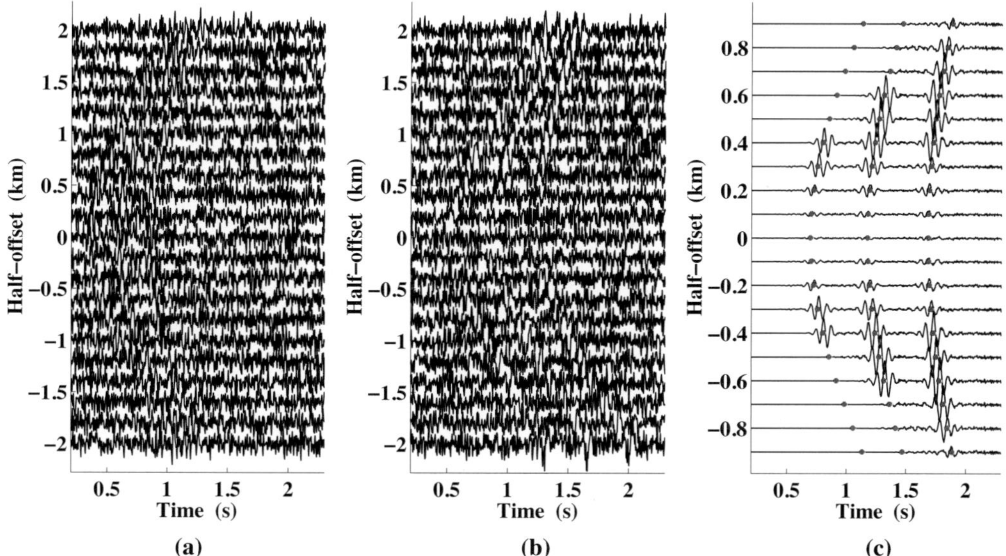

Figure 4.19: Same as Figure 4.18 but the input PP and PS data are contaminated with Gaussian noise; the signal-to-noise ratio is equal to two (Grechka and Dewangan, 2003).

goes to zero at normal incidence. Since $x^{(1)} = x^{(2)} = 0$ is the stationary point for $x^{(3)} = x^{(4)} = 0$, the leading (zero-order) approximation of the stationary-phase method predicts a vanishing ΨS-wave amplitude at zero offset. The low-amplitude ΨS arrivals observed at $h_{\Psi S} = 0$ in Figure 4.18c correspond to the contributions of higher-order terms in the stationary-phase expansion.

3. ΨS-waves are absent at large offsets. The algorithm evaluates integral 4.23 only if $h_{\Psi S} < h_{\Psi S}^{\text{crit}}$; otherwise, it fills the output trace with zeros. The termination points for the two shallow ΨS events in Figure 4.18c correspond to offset-to-depth ratios slightly smaller than unity, in agreement with the theoretical result in equation 4.28.

Figures 4.19a,b show PP and PS traces from Figures 4.18a,b contaminated with Gaussian noise that has the standard deviation equal to one-half of the maximum signal amplitude. Even though the PP and PS reflections in Figures 4.19a,b are almost invisible, Figure 4.19c displays remarkably clean ΨS traces. This is a consequence of employing convolutions that efficiently attenuate random noise. Application of the full-waveform version of the PP+PS=SS method to field data from the Gulf of Mexico is discussed in section 5.4.

4.4 Summary

In this chapter, we described a data-driven method (PP+PS=SS) that combines PP and PS reflection data to construct ΨS-waves, which have the kinematics of pure SS primary reflections and can be processed by conventional techniques to estimate shear-wave NMO velocities. Generation of ΨS-waves circumvents problems associated with the asymmetry of converted-wave moveout and conversion-point dispersal. Both kinematic and full-waveform versions of the algorithm can be implemented in either 2D or 3D by tracking PP and PS events inside prestack data volumes. The main features of the method are summarized below:

1. The sole assumption used to generate ΨS-waves is that the input data (PP and PS primary reflections) correspond to the *same* reflector. Correlation of PP and PS events requires some knowledge of the V_S/V_P ratio in the subsurface and might encounter practical difficulties, especially when reflectors are close to horizontal. Other than that, the method needs no quantitative information about the velocity field and anisotropy parameters.

2. The kinematic version of the PP+PS=SS method matches *local* reflection slopes (horizontal slownesses) in common-receiver gathers. The same local slopes are implicitly present in the full-waveform (interferometric) formulation of the method because the first-order contribution to the ΨS-wavefield is made by the source and receiver locations that satisfy the kinematic condition of matching slopes. This locality helps avoid a number of complications inherent in processing of converted-wave data:

- *Asymmetry of PS traveltimes and "diodic" nature of PS-wave NMO velocity.* Since the algorithm operates only with local slopes in common-receiver gathers, the asymmetry of PS moveout is irrelevant.

- *Reflection-point dispersal for converted waves.* Common-midpoint gathers of PS-waves, which suffer from reflection-point dispersal, are not used in the computation of SS traveltimes and ΨS data. The reflection points of the input PS-waves and constructed ΨS-waves are guaranteed to be exactly the same as those of the corresponding PP-waves. (The actual reflection-point locations, however, remain unknown until a depth-domain velocity model is built.) Reflection-point dispersal in the constructed CMP gathers of ΨS-waves is much smaller than that in the input PS-wave gathers.
- *Polarity reversal.* Small amplitudes of PS-waves in the vicinity of the polarity reversal hamper traveltime picking needed in the kinematic implementation of the PP+ PS = SS method. The locality of the method, however, makes it possible to mute out low-amplitude PS traces without compromising the quality of results obtained from the remaining data. The full-waveform version of the method requires no muting. While ΨS data inherit the low amplitudes of the input PS traces, they do not have polarity reversals.

3. The SS-wave traveltimes constructed by the PP+ PS = SS method are *exact.* If the input PP- and PS-waves are reflected from the same interface, the results might be distorted only by noise in the data and, in the kinematic implementation of the method, by interpolation or extrapolation of reflection slopes and traveltimes.

4. The generated ΨS data can be processed by means of any velocity-analysis technique developed for pure modes. The ΨS-wave (i.e., SS-wave) NMO velocities and ellipses are especially attractive for anisotropic stacking-velocity tomography because they provide information complementary to that contained in P-waves (see Chapter 5).

5. To ensure that ΨS data are constructed for a sufficiently wide range of offsets, it is necessary to use long-offset PP and PS data. Still, the maximum offset-to-depth ratio of ΨS-waves is governed by the effective V_S/V_P ratio above the reflector and, therefore, is not guaranteed to be large enough for robust shear-wave velocity analysis.

6. The PP+ PS = SS method is not a "true-amplitude" algorithm and cannot produce correct SS-wave amplitudes. Without better understanding of dynamic aspects of the full-waveform PP+ PS = SS algorithm, AVO analysis for ΨS data is not feasible. Note, however, that Ursin et al. (2009) employed surface-to-surface propagator matrices in the PP+ PS = SS method to evaluate the second-order traveltime derivatives and geometrical spreading of ΨS-waves.

Adapting the idea of the PP+ PS = SS method, Dewangan and Tsvankin (2006c) devised a velocity-independent layer-stripping method (VILS) to compute the interval moveout of PP- or PS-waves in a target layer overlain by a laterally homogeneous overburden with a horizontal symmetry plane. As discussed in section 3.4, application of VILS to nonhyperbolic moveout inversion of P-wave data significantly increases the accuracy of interval parameter estimation in VTI and orthorhombic media.

Appendices for Chapter 4

4A Kinematics of ΨS-waves

In this appendix we prove that integral 4.23 produces waves that have the kinematics of the pure SS-wave primaries if the input traces contain only PP and PS primary reflections. Applying the Fourier transform to equation 4.23 yields

$$W_{\Psi S}(\omega, x^{(3)}, x^{(4)}) = \iint \Big[W_{PS}(\omega, x^{(1)}, x^{(3)})\, W^{\star}_{PP}(\omega, x^{(1)}, x^{(2)}) \times W_{PS}(\omega, x^{(2)}, x^{(4)}) \Big]\, dx^{(1)}\, dx^{(2)}\,, \tag{4.29}$$

where ω is the frequency, W_{PP}, W_{PS}, and $W_{\Psi S}$ are the spectra of the PP, PS, and ΨS traces, respectively, and the star denotes complex conjugate.

If the input PP and PS traces include only the primaries reflected from a given interface, their spectra have the form

$$W^{\star}_{PP}(\omega, x^{(1)}, x^{(2)}) = F^{\star}_{PP}(\omega)\, A^{\star}_{PP}(x^{(1)}, x^{(2)})\, e^{-i\,\omega\, t_{PP}(x^{(1)}, x^{(2)})}\,, \tag{4.30}$$

$$W_{PS}(\omega, x^{(s)}, x^{(r)}) = F_{PS}(\omega)\, A_{PS}(x^{(s)}, x^{(r)})\, e^{i\,\omega\, t_{PS}(x^{(s)}, x^{(r)})}\,. \tag{4.31}$$

Here A_{PP} and A_{PS} are the amplitudes of the reflected PP- and PS-waves, F_{PP} and F_{PS} are the spectra of their wavelets, t_{PP} and t_{PS} are the PP and PS reflection traveltimes, and the indices r and s correspond to the two PS-wave source-receiver pairs in equation 4.29: $\{s=1,\, r=3\}$ and $\{s=2,\, r=4\}$. Substituting equations 4.30 and 4.31 into integral 4.29, we obtain

$$W_{\Psi S}(\omega, x^{(3)}, x^{(4)}) = F^{2}_{PS}(\omega)\, F^{\star}_{PP}(\omega) \times \iint A_{PS}(x^{(1)}, x^{(3)})\, A^{\star}_{PP}(x^{(1)}, x^{(2)})\, A_{PS}(x^{(2)}, x^{(4)})\, e^{i\,\omega\,\tau}\, dx^{(1)}\, dx^{(2)}, \tag{4.32}$$

where the time τ is given by

$$\tau \equiv \tau(x^{(1)}, x^{(2)}, x^{(3)}, x^{(4)}) = t_{PS}(x^{(1)}, x^{(3)}) + t_{PS}(x^{(2)}, x^{(4)}) - t_{PP}(x^{(1)}, x^{(2)})\,. \tag{4.33}$$

To show that the ΨS event has the kinematics of the pure SS-wave primary, we evaluate integral 4.32 in the limit $\omega \to \infty$. According to the stationary-phase method, the main contributions to the integral at high frequencies are made by the coordinate pairs $\{x^{(1)}, x^{(2)}\}$ that satisfy the conditions of stationarity,

$$\frac{\partial \tau}{\partial x^{(1)}} = 0\,, \tag{4.34}$$

$$\frac{\partial \tau}{\partial x^{(2)}} = 0\,. \tag{4.35}$$

Note that equations 4.34 and 4.35 coincide with equations 4.4 and 4.8, which represent the requirement of matching time slopes of the reflected PP- and PS-waves (Figure 4.4). Therefore, if equations 4.34 and 4.35 have a solution,

$$\left\{x^{(1)}, x^{(2)}\right\} \equiv \left\{x^{(1)}(x^{(3)}, x^{(4)}),\, x^{(2)}(x^{(3)}, x^{(4)})\right\}, \tag{4.36}$$

the time function τ defined by equation 4.33 becomes identical to the traveltime t_{SS} in equation 4.11. We conclude that integral 4.32 in the high-frequency limit indeed reproduces the kinematics of the SS-wave primary.

Chapter 5

Moveout inversion of multicomponent data for TI media

Continued progress in acquiring and processing high-quality multicomponent data has provided clear evidence of the influence of anisotropy on reflection traveltimes and moveout inversion. In particular, conventional isotropic imaging methods routinely produce depth misties between PP and PS (converted-wave) sections (e.g., Nolte et al., 2000), which can be removed by joint anisotropic velocity analysis of PP and PS data volumes. In this chapter, PP-wave reflection moveout is combined with traveltimes of mode-converted (PS) and shear (SS) waves in parameter estimation for transversely isotropic media.

We begin by examining joint inversion of PP and PS (PSV) data for the simple model of a horizontal VTI layer. Although the addition of PS traveltimes makes it possible to obtain the ratio of the vertical velocities of P- and S-waves and the shear-wave NMO velocity, inversion for the Thomsen parameters and reflector depth remains nonunique, even for uncommonly large spreadlength-to-depth ratios. Reconstruction of the depth scale of horizontally layered VTI models from surface data requires generation of shear waves and recording of wide-angle SS reflections, as demonstrated by Tsvankin and Thomsen (1995).

In the second section we extend P-wave stacking-velocity inversion (*tomography*) described in Chapter 2 to the combination of NMO ellipses, zero-offset traveltimes, and reflection time slopes of PP- and SS-waves (SS traveltimes can be computed from PP and PS data). Application of the inversion algorithm to a homogeneous TI layer above a dipping reflector shows that for a range of dips and tilt angles of the symmetry axis conventional-spread PP and PS data constrain the symmetry-direction velocities and the parameters ϵ and δ. Then multicomponent stacking-velocity tomography is tested on synthetic noise-contaminated data from multilayered TI media (each layer can be VTI, HTI, or TTI) with dipping and curved interfaces.

To demonstrate the feasibility of joint velocity analysis of PP and PS data and the advantages of anisotropic imaging of mode-converted waves, we use two multicomponent 2D field data sets. In the first case study, reflection traveltimes of PP and PS waves acquired above the Siri reservoir in the North Sea are processed by the PP+ PS = SS method (as described in Chapter 4) to generate SS traveltimes and compute the shear-wave NMO velocities. To build a layered VTI depth model, stacking-velocity tomography of the PP and SS data is combined with the time-depth curve obtained from check shots. Taking anisotropy into account dramatically improves the image of the reservoir on the PS-wave stacked section and helps focus and properly position faults in the overburden. The second case study, from the Gulf of Mexico, corroborates the effectiveness of the full-waveform version of the PP+ PS = SS method. Much of the section is shown to possess significant anisotropy, although the interval Thomsen parameters cannot be resolved solely from PP and PS reflection traveltimes.

5.1 Joint inversion of PP and PS data for a horizontal VTI layer

5.1.1 Mode conversions versus SS-waves in parameter estimation

A major complication caused by anisotropy in velocity analysis is the uncertainty in estimating the vertical velocity and depth scale of the model from surface P-wave data. For a horizontal VTI layer, P-wave reflection moveout is fully determined by the vertical (zero-offset) time, the NMO velocity $V_{\text{nmo},P}$ and the anellipticity parameter η (Alkhalifah and Tsvankin, 1995; Tsvankin, 2005). Therefore, even though P-wave phase and group velocities depend individually on the vertical velocity V_{P0} and Thomsen parameters ϵ and δ, none of these three parameters can be obtained from P-wave reflection moveout. This conclusion remains valid for horizontally layered VTI media above a horizontal or a dipping reflector, in which P-wave traveltime is controlled by the interval parameters $V_{\text{nmo},P}$ and η averaged over the stack of layers. As a result, velocity analysis of P-wave reflection data does not provide enough information for reconstructing such models in depth. Although the presence of lateral heterogeneity (e.g., dipping interfaces) makes P-wave moveout dependent on V_{P0}, ϵ, and δ, inversion for these parameters is feasible only in special cases (see sections 2.2 and 2.3).

Because the velocity of SV-waves in TI media also depends on the parameters V_{P0}, ϵ, and δ (in addition to V_{S0}), it is natural to combine P- and SV-wave traveltimes in velocity-analysis algorithms. As shown by Tsvankin and Thomsen (1995), joint inversion of nonhyperbolic moveout of PP- and SS-waves[1] for a horizontal VTI layer yields all four relevant parameters (V_{P0}, V_{S0}, ϵ, and δ). In stratified VTI media, long-spread PP- and SS-wave traveltimes can be inverted for the interval Thomsen parameters and the layer thicknesses.

Shear waves, however, are not excited in offshore surveys and are only occasionally used on land because of cost considerations and problems with data quality (e.g.,

[1]Hereafter in this section, "S" denotes SV-waves; P- and SH-waves in horizontally layered VTI media are decoupled.

statics). The question addressed in this section is whether or not pure SS reflections can be replaced in moveout inversion for horizontally layered VTI media with converted PS-waves. We start by combining nonhyperbolic PP-wave moveout with the zero-offset traveltime and NMO velocity of PS-waves and then analyze joint inversion of long-spread PP and PS traveltimes.

5.1.2 Feasibility of the inversion procedure

Nonhyperbolic moveout of PP-waves in a horizontal VTI layer can be inverted for the NMO velocity $V_{\text{nmo},P}$ and the anellipticity parameter η:

$$V_{\text{nmo},P} = V_{P0}\sqrt{1+2\delta}\,, \tag{5.1}$$

$$\eta \equiv \frac{\epsilon - \delta}{1+2\delta}\,. \tag{5.2}$$

As discussed in Chapter 3, however, the trade-off between the parameters $V_{\text{nmo},P}$ and η may lead to sizable errors in η reaching or exceeding ± 0.1, even for this simple model (also, see Grechka and Tsvankin, 1998b; Tsvankin, 2005). This trade-off has serious implications for the inverse problem analyzed here.

Because the layer is horizontal and has a horizontal symmetry plane, PS moveout is symmetric with respect to zero offset and can be described by the $t^2(x^2)$-function. Hence, moveout analysis on moderate spreads can provide the vertical traveltime (t_{PS0}) and the NMO velocity ($V_{\text{nmo},PS}$) of PS-waves. The NMO velocities of pure and converted modes are related by the following Dix-type equation (Seriff and Sriram, 1991; Tsvankin, 2005):

$$2t_{PS0}V^2_{\text{nmo},PS} = t_{P0}V^2_{\text{nmo},P} + t_{S0}V^2_{\text{nmo},S}\,, \tag{5.3}$$

where t_{P0} and t_{S0} are the two-way vertical traveltimes of PP and SS reflections ($2t_{PS0} = t_{P0} + t_{S0}$). Equation 5.3 represents a special (2D) case of the general relationship between the NMO ellipses of pure and converted waves in laterally homogeneous media with a horizontal symmetry plane (equation 1.66). Therefore, the combination of PP and PS data can be used to compute the SS-wave NMO velocity $V_{\text{nmo},S}$,

$$V_{\text{nmo},S} = V_{S0}\sqrt{1+2\sigma}\,, \tag{5.4}$$

$$\sigma \equiv \left(\frac{V_{P0}}{V_{S0}}\right)^2 (\epsilon - \delta)\,.$$

Alternatively, the velocity $V_{\text{nmo},S}$ can be found from SS traveltimes generated by the PP+PS=SS method. As shown in Chapter 4, robust estimation of $V_{\text{nmo},S}$ requires acquisition of long-spread PP and PS data. Also, the vertical-velocity ratio can be found from the traveltimes t_{P0} and t_{S0}:

$$\frac{V_{S0}}{V_{P0}} = \frac{t_{P0}}{t_{S0}}\,. \tag{5.5}$$

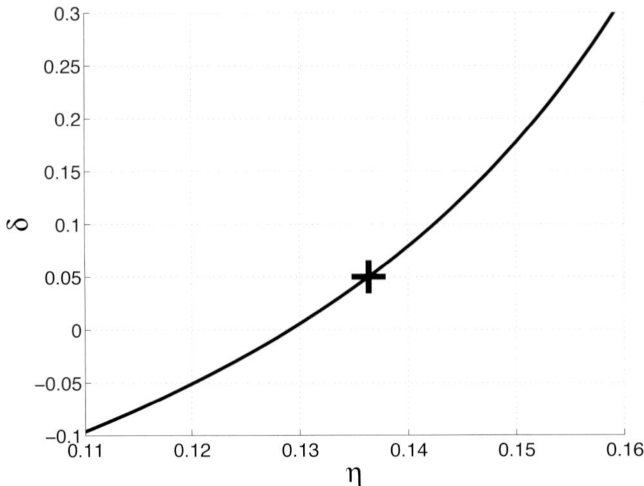

Figure 5.1: Parameter δ of a horizontal VTI layer computed from equation 5.6 for a range of η-values and the exact parameters t_{P0}, t_{S0}, $V_{\text{nmo},P}$, and $V_{\text{nmo},S}$ (Grechka and Tsvankin, 2002d). The VTI parameters are: V_{P0}=2.5 km/s, V_{S0}=1.0 km/s, ϵ=0.2, and δ=0.05; the layer thickness D=1.0 km. The cross marks the correct values of η and δ. Note that small errors in η get amplified in the computation of δ.

In principle, the parameters $V_{\text{nmo},P}$, η, $V_{\text{nmo},S}$, and the ratio V_{S0}/V_{P0} are sufficient for recovering all four unknown Thomsen parameters (V_{P0}, V_{S0}, ϵ, and δ). For example, δ can be found from (see equations 5.1, 5.2, and 5.4):

$$1 + 2\delta = \left(\frac{V_{S0}}{V_{P0}}\right)^2 \frac{1}{\left(V_{\text{nmo},S}/V_{\text{nmo},P}\right)^2 - 2\eta}\,. \tag{5.6}$$

Then equations 5.1, 5.2, and 5.5 yield V_{P0}, ϵ, and V_{S0}. This procedure can be extended to interval parameter estimation in multilayered media by employing velocity-independent layer stripping (VILS, see section 3.4) or Dix-type differentiation.

As discussed by Grechka and Tsvankin (2002d), however, the inversion for the Thomsen parameters is too unstable to be employed in practice. Equation 5.6 provides insight into the reasons for the failure of this approach. In the presence of realistic errors in the NMO velocities and the parameter η, the term $[(V_{\text{nmo},S}/V_{\text{nmo},P})^2 - 2\eta]$ may become small or even vanish, which leads to dramatic amplification of measurement errors in the estimation of δ. If η deviates from the correct value by just ± 0.02, the error in δ becomes unacceptably large (Figure 5.1). Because η can seldom be estimated with accuracy higher than $\pm(0.05 - 0.1)$, the corresponding errors in δ can far exceed those shown in Figure 5.1. This conclusion is confirmed by the numerical examples below.

It has been suggested in the literature (e.g., Li and Yuan, 1999) that the instability in the estimation of the VTI parameters from PP and PS data can be overcome by including in the inversion not just t_{PS0} and $V_{\text{nmo},PS}$, but also PS traveltimes at large

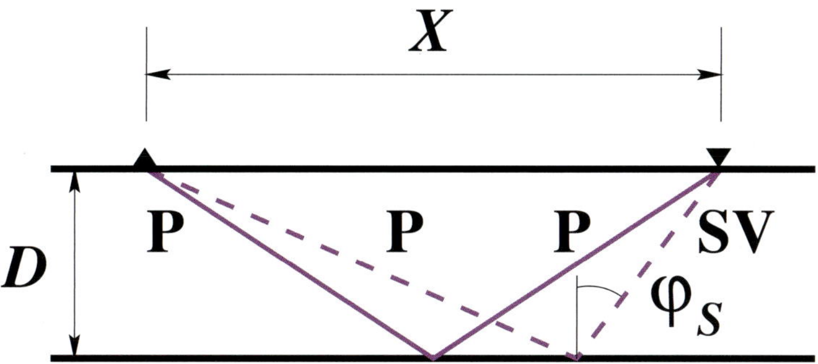

Figure 5.2: PP-wave (solid) and PS-wave (dashed) reflections from the bottom of a horizontal, homogeneous layer (Grechka and Tsvankin, 2002d). The shear-wave reflection group angle is denoted by φ_S.

(compared with the reflector depth) offsets. Grechka and Tsvankin (2002d), however, demonstrated that the combination of long-spread moveouts of PP- and PS-waves is still insufficient for obtaining V_{P0}, V_{S0}, ϵ, and δ without additional constraints.

The basic difference between the long-spread moveout of pure and converted waves is illustrated for a single horizontal layer in Figure 5.2. For PS-waves in isotropic media, the maximum shear-wave reflection angle φ_S that corresponds to infinitely large offsets is defined by Snell's law (equation 4.27):

$$\max \varphi_S = \sin^{-1}\left(\frac{V_S}{V_P}\right), \tag{5.7}$$

where V_P and V_S are the P- and S-wave velocities, respectively. If the velocity ratio V_S/V_P takes a typical value of 0.5, the angle φ_S reaches only 30°, and it can be much smaller for loosely consolidated (e.g., underwater) sediments with low shear-wave velocities.

This estimate of the maximum angle φ_S remains qualitatively valid in the presence of moderate anisotropy. For example, in a typical VTI model with the parameters $V_{P0}/V_{S0} = 2.5$, $\epsilon = 0.20$, and $\delta = 0.05$, the group angle φ_S for the offset-to-depth ratio $X/D = 3$ is equal to 38°, while the corresponding phase angle is just $\theta_S = 16°$. Such angles φ_S and θ_S are insufficient to create pronounced deviations of the S-wave portion of converted-wave moveout from being hyperbolic (Tsvankin and Thomsen, 1994).

Because the success of the Tsvankin-Thomsen (1995) joint inversion of PP and SS traveltimes was ensured by including wide-angle (group and phase angles up to and beyond 45°) shear-wave reflections with strong nonhyperbolic moveout, we can expect the replacement of SS data with PS-waves to be inadequate. In fact, the above estimates of the reflection angles suggest that the long-offset portion of PS moveout is governed largely by the wide-angle PP traveltimes (i.e., by $V_{\text{nmo},P}$ and η) and might not provide additional information for the inversion. Below, we substantiate this conclusion by numerical testing and quantify the uncertainty in the inverted VTI parameters.

To generate a family of kinematically equivalent models, a typical VTI medium (see the top row of Tables 5.1 and 5.2) was perturbed by introducing small errors into the parameter η. To keep the long-spread PP-wave moveout almost unchanged, the errors in η were compensated by variations of up to 1% in $V_{\mathrm{nmo},P}$, while the vertical traveltimes and the S-wave NMO velocity were fixed at the actual values. For each perturbed (erroneous) model, the VTI parameters and reflector depth were computed from equations 5.1, 5.2, 5.4, and 5.5 (Table 5.2).

Because all four erroneous models have the correct vertical traveltimes and close values of the PS-wave NMO velocity, they produce practically identical PS-wave moveout for offsets limited by the reflector depth. As expected, however, the small variations in $V_{\mathrm{nmo},P}$ and η lead to significant errors in the vertical velocities (greater than 10%), anisotropy parameters (e.g., ϵ changes from 0.06 to 0.42), and reflector depth D.

The next question is whether or not it is possible to distinguish between the models listed in Table 5.2 using *long-spread* moveout of PS-waves. Reflection traveltimes computed by anisotropic ray tracing for the actual and erroneous models are compared in Figure 5.3. The maximum offset in this test was fixed at 3 km, so the maximum offset-to-depth ratio varies from 2.7 to 3.4 in accordance with the reflector depth (Table 5.2). All four vastly different models produce virtually the same moveout of both PP- and PS-waves as the actual medium, with traveltime errors barely exceeding 1 ms for models 1 and 4, and much smaller for models 2 and 3.

Thus, even for large offsets-to-depth ratios that are seldom attained in practice, the combination of PP and PS reflection traveltimes does not constrain the velocities V_{P0} and V_{S0} and the anisotropy parameters ϵ and δ. Because the range of shear-wave reflection angles for converted waves is limited, PS traveltimes at large offsets do not provide independent information for inversion. Long-offset PS arrivals, however, help increase the accuracy of η estimation. Indeed, for a fixed source-receiver offset the P-wave incidence angle for the PS conversion is larger than that for the PP reflection (Figure 5.2), which makes PS traveltime more sensitive to η.

5.2 Multicomponent stacking-velocity tomography for TI media

Although the addition of PS traveltimes to PP data proved to be insufficient for depth-domain velocity analysis in horizontally layered VTI media, the presence of dip might increase the stability of parameter estimation (see Chapter 2). Here, using the results of Grechka et al. (2002b), we describe the methodology of anisotropic multicomponent stacking-velocity inversion (*tomography*) for models composed of homogeneous TI layers separated by dipping or curved interfaces. This algorithm represents an extension of P-wave stacking-velocity tomography for layered VTI media

Data	t_{P0} (s)	t_{S0} (s)	$V_{\mathrm{nmo},P}$ (km/s)	$V_{\mathrm{nmo},S}$ (km/s)	η
Actual	0.800	2.000	2.622	1.696	0.136
Erroneous					
1	0.800	2.000	2.648	1.696	0.116
2	0.800	2.000	2.635	1.696	0.126
3	0.800	2.000	2.609	1.696	0.146
4	0.800	2.000	2.596	1.696	0.156

Table 5.1: Actual and erroneous sets of the moveout parameters of a horizontal VTI layer (Grechka and Tsvankin, 2002d). The four erroneous sets have slightly distorted values of the NMO velocity $V_{\mathrm{nmo},P}$ and parameter η.

Model parameters	V_{P0} (km/s)	V_{S0} (km/s)	ϵ	δ	σ	D (km)
Actual	2.500	1.000	0.200	0.050	0.938	1.000
Erroneous						
1	2.787	1.115	0.057	−0.049	0.657	1.115
2	2.646	1.058	0.121	−0.004	0.783	1.058
3	2.349	0.939	0.298	0.117	1.128	0.939
4	2.190	0.876	0.421	0.202	1.372	0.876

Table 5.2: Actual VTI parameters and those corresponding to the four erroneous models from Table 5.1 (Grechka and Tsvankin, 2002d).

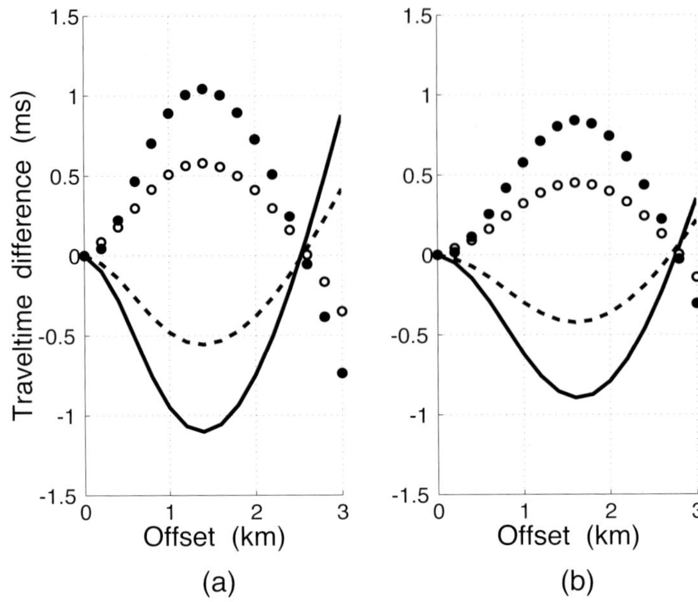

Figure 5.3: Differences between the ray-traced reflection traveltimes (traveltime residuals) computed for the four erroneous models and for the actual model from Table 5.2 (Grechka and Tsvankin, 2002d). Plot (a) shows the residuals for PP-waves, and (b) for PS-waves. The solid line corresponds to model 1, dashed line to model 2, empty circles to model 3, and filled circles to model 4.

introduced in section 2.3.

Rather than working with PS-waves directly, we combine them with PP-waves to obtain the traveltimes of the pure SS reflections from the same interface using the PP+ PS = SS method discussed in Chapter 4. Because this method does not require knowledge of the velocity model, it can be applied prior to anisotropic parameter estimation. In contrast to the more complicated moveout of mode conversions, traveltime of the constructed SS-waves in CMP gathers is symmetric with respect to zero offset and, for conventional spreadlength-to-depth ratios, can be described by the NMO ellipse.

5.2.1 Inversion methodology

The implementation described below is designed for homogeneous TI layers separated by plane or smooth curved interfaces. An extension to multicomponent data from layered orthorhombic media is presented in Chapter 6. The algorithm includes the following main steps:

1. Tracking PP and PS arrivals in prestack 3D data volumes and identifying the PP and PS events reflected from the same interfaces. In general, both split converted waves (PS_1 and PS_2) can be included.

2. Computation of the traveltimes of the pure SS (S_1S_1 and S_2S_2) reflections using the PP+ PS = SS method introduced in Chapter 4. Alternatively, it is possible to avoid time picking by generating ΨS data with the full-waveform version of the PP+ PS = SS method.

3. Application of azimuthal velocity analysis to obtain the NMO ellipses of the acquired PP-waves and constructed SS-waves (see Chapter 2).

4. Inversion of the NMO ellipses, zero-offset traveltimes, and reflection time slopes for the interval anisotropy parameters and the shapes of layer boundaries.

The data vector for an N-layered TI medium is given by

$$\mathbf{d}(Q, \mathbf{Y}, n) \equiv \{\tau_Q(\mathbf{Y}, n), \mathbf{p}_Q(\mathbf{Y}, n), \mathbf{W}_Q(\mathbf{Y}, n)\}, \qquad (5.8)$$

where Q is the mode type (i.e., PP or SS; the algorithm discussed here operates only with SV-waves), $\mathbf{Y} = [Y_1, Y_2]$ is the CMP location, $n = 1, 2, ..., N$ is the reflector number, τ is the zero-offset traveltime, $\mathbf{p} = [p_1, p_2]$ is the two-component vector of the reflection slopes measured in the x_1- and x_2-directions on zero-offset time sections (i.e., p_1 and p_2 are the horizontal slownesses of the zero-offset ray), and $\mathbf{W}$ is the 2×2 matrix describing the NMO ellipse.

In general, the parameter-estimation algorithm is organized in the same way as that introduced for P-waves in section 2.3. The model vector $\mathbf{m}$ contains the interval TI parameters and the polynomial coefficients (see equation 2.41) defining the model interfaces. For TI media with an unknown tilt of the symmetry axis, traveltimes of P- and SV-waves depend on six interval parameters – the P- and S-wave symmetry-direction velocities V_{P0} and V_{S0}, anisotropy parameters ϵ and δ defined with respect to the symmetry axis, and two angles (tilt ν and azimuth β) responsible for the symmetry-axis orientation.

For a given set of trial interval parameters, the zero-offset traveltimes τ_Q and reflection slopes $\mathbf{p}_Q$ are used to trace the one-way zero-offset rays for all reflection events. Then the interfaces for the trial model are reconstructed by fitting 2D polynomials to the termination points of the zero-offset rays. Finally, the interval parameters are obtained by computing the NMO ellipses for the trial model and minimizing the following objective function:

$$\mathcal{F}(\mathbf{m}) \equiv \sum_{Q,\mathbf{Y},n} ||\mathbf{W}_Q^{\text{calc}}(\mathbf{Y}, n, \mathbf{m}) - \mathbf{W}_Q^{\text{meas}}(\mathbf{Y}, n)||, \qquad (5.9)$$

where $\mathbf{W}_Q^{\text{calc}}$ and $\mathbf{W}_Q^{\text{meas}}$ correspond to the calculated and measured NMO ellipses for all available modes and reflectors at each CMP location. As shown in Chapter 1, the effective NMO ellipse of any pure-mode reflection in heterogeneous anisotropic media can be computed using the results of tracing just the zero-offset ray.

First, we examine homogeneous TTI media with the full range of symmetry-axis directions and establish the conditions needed for stable parameter estimation. The method is then applied to multicomponent data from models composed of VTI, HTI, and TTI layers.

P	SV	SH
Kinematic parameters		
V_{P0}	V_{S0}	V_{S0}
ϵ	0	γ
δ	σ	γ
Time-processing parameters (VTI)		
$V_{\text{nmo},P}$	$V_{\text{nmo},SV}$	$V_{\text{nmo},SH}$
η	$-\sigma$	0

Table 5.3: Correspondence between the parameters responsible for the kinematic signatures of P-, SV-, and SH-waves in weakly anisotropic TI media (Grechka et al., 2002b).

5.2.2 Parameter estimation for a homogeneous layer

Consider a homogeneous TI medium with an arbitrary orientation of the symmetry axis overlying a plane reflector (horizontal or dipping). To gain insight into the information provided by shear-wave NMO ellipses, it is convenient to apply the "substitution rule" based on the similarity in form of the weak-anisotropy approximation for the P- and S-wave phase-velocity functions. Any kinematic signature of SV-waves for weak transverse isotropy can be obtained from the corresponding P-wave signature by making the following substitutions, listed in Table 5.3: $V_{P0} \rightarrow V_{S0}$, $\delta \rightarrow \sigma$, and $\epsilon \rightarrow 0$ (Tsvankin, 2005). A similar rule for elliptically anisotropic SH-waves is $V_{P0} \rightarrow V_{S0}$, $\delta \rightarrow \gamma$, and $\epsilon \rightarrow \gamma$. SH-wave kinematic signatures, however, often can be obtained in closed form without resorting to the weak-anisotropy approximation.

VTI layer

Because VTI models are azimuthally isotropic, the axes of the NMO ellipse are parallel to the dip and strike directions of the reflector. The exact P-wave NMO ellipse is controlled by the NMO velocity for a horizontal reflector ($V_{\text{nmo},P}$) and the parameter η. For moderate dips, the strike-line semiaxis is dependent largely on $V_{\text{nmo},P}$, whereas the dip-line semiaxis is sensitive to η (Grechka and Tsvankin, 1998a; Tsvankin, 2005).

According to the substitution rule, the SV-wave NMO ellipse can be used to constrain (at least when anisotropy is weak) the zero-dip NMO velocity $V_{\text{nmo},S}$ (equation 5.4) and the parameter σ because $[-\sigma/(1+2\sigma)]$ plays the role of η. Knowledge of $V_{\text{nmo},S}$ and σ is sufficient for computing the shear-wave vertical velocity V_{S0}. Then the zero-offset traveltime and reflection slope of the SS reflection can be used to reconstruct the depth and dip of the reflector. Adding this information to that obtained from PP-waves should help estimate V_{P0}, ϵ, and δ (Grechka et al., 2002b).

These analytic predictions are confirmed by the numerical example for a mildly dipping VTI layer (Figure 5.4). Nonlinear minimization (the Gauss-Newton method)

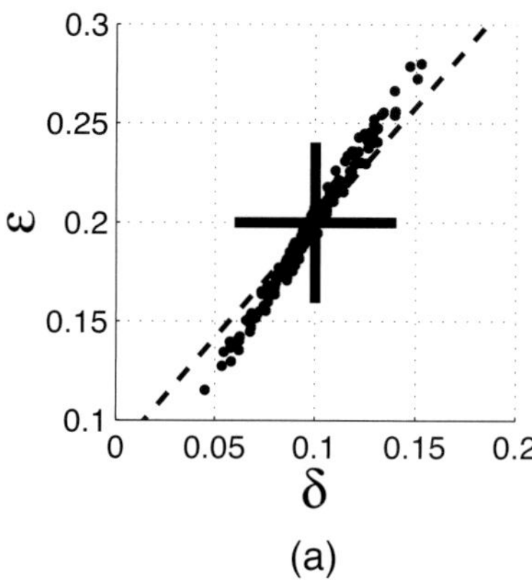

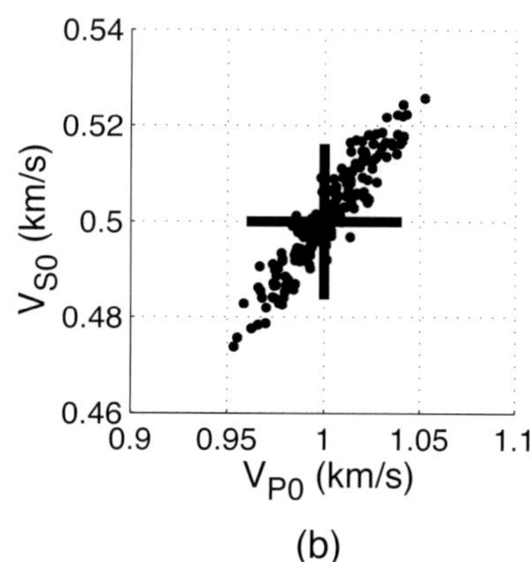

Figure 5.4: Results (dots) of the joint inversion of PP and SS (SVSV) data for a VTI layer above a plane reflector dipping at 15° (Grechka et al., 2002b). The actual parameters are marked by the crosses. The dashed line on plot (a) shows the correct value of η. The data were contaminated by Gaussian noise with the standard deviations equal to 2% for the NMO velocities and 1% for the zero-offset traveltimes and reflection slopes.

was applied to objective function 5.9 computed using exact equations for the NMO ellipses, zero-offset traveltimes, and reflection slopes. The dots in Figure 5.4 mark the estimated VTI parameters for different realizations of the noise added to the input data. The standard deviations of the inverted parameters (2% for V_{P0} and V_{S0}, 0.03 for ϵ, and 0.02 for δ) indicate that noise does not get amplified by the parameter-estimation procedure, so the inversion is reasonably stable. Note that the obtained ϵ and δ cluster near the line of the well-constrained parameter $\eta \approx \epsilon - \delta$.

Although stable inversion of the P-wave NMO ellipse for η requires dips of at least $25° - 30°$ (Grechka and Tsvankin, 1998a), the dip in Figure 5.4 was just 15°. As discussed by Grechka et al. (2002b), the NMO ellipse and zero-offset traveltime of SS-waves become sufficiently sensitive to the parameter σ at relatively small dips, which ensures the convergence of the inversion algorithm.

The only parameter of VTI media that cannot be obtained from P and SV data is γ because it is responsible just for the elliptical anisotropy of SH-waves. Reflected PSH-waves are generated by P-to-SH conversion for source-receiver azimuths outside the dip plane. The methodology introduced here can be applied to γ-estimation from the NMO ellipses of pure SH reflections computed from PP and PSH data.

HTI layer

For a horizontal HTI layer, the symmetry-axis direction (specified by the azimuth β) coincides with one of the axes of the NMO ellipse. Moveout analysis in the isotropy plane yields the true vertical velocity of the mode at hand and, therefore, the depth scale of the medium. Then the NMO velocity in the orthogonal, symmetry-axis plane can be used to estimate the pertinent anisotropy parameters, as discussed in detail by Tsvankin (1997b).

In particular, the PP-wave NMO ellipse in a horizontal HTI layer with symmetry axis x_1 constrains the vertical velocity and the parameter $\delta^{(2)} = \delta^{(V)} \approx 2\epsilon - \delta$ defined in Appendix 1B (see equation 3.71 in Tsvankin, 2005). Combination of the NMO ellipses of all three pure modes is sufficient for estimating the P- and S-wave symmetry-direction velocities V_{P0} and V_{S0} and the anisotropy parameters ϵ, δ, and γ. A more general version of this algorithm for a horizontal orthorhombic layer is described in section 6.1 (HTI can be treated as a special case of orthorhombic symmetry).

The analysis here is restricted to the inversion of the NMO ellipses of PP- and SVSV-waves for the parameters V_{P0}, V_{S0}, ϵ, and δ. By the "SV-wave" we always mean the mode polarized in the plane formed by the slowness vector and the symmetry axis. If the symmetry axis is horizontal or tilted, this plane is not necessarily vertical, but for the sake of consistency we retain the classification of shear waves commonly used for VTI or isotropic media. Typical results of inverting noise-contaminated PP and SS traveltimes for the parameters of a horizontal HTI layer are shown in Figure 5.5. The standard deviations of all estimated parameters are quite small; for the symmetry-axis azimuth β (not shown on the plot) the standard deviation is just 0.6°.

The moveout inversion for an HTI layer produces accurate results for any reflector dip from 0° to 90° (see an example in Figure 5.6). The dots of estimated parameter values in Figure 5.6 form smaller clouds than those in Figure 5.4, which indicates that the inversion is more stable compared to that for a dipping VTI layer. (For a horizontal VTI layer, V_{P0}, V_{P0}, ϵ, and δ are not constrained by the input data.) Also, the inversion algorithm for HTI media converges much more rapidly towards the actual model than it does when the symmetry axis is vertical. If two distinctly different dips (e.g., a horizontal and a dipping reflector) of the base of an HTI layer are available, the parameters V_{P0}, ϵ, δ, and β can be found from PP-wave NMO ellipses alone (Contreras et al., 1999).

TTI layer

The parameter-estimation problem for transverse isotropy with a tilted symmetry axis (TTI) includes only one additional unknown compared to the HTI model – the tilt ν. This, however, makes the inversion substantially more ill-posed than that for HTI media, in part because NMO velocity is a nonlinear function of ν, even for weak anisotropy. Grechka and Tsvankin (2000) found a nonlinear dependence on the tilt in the weak-anisotropy approximation for the PP-wave NMO ellipse in TTI media, and adaptation of their equations for shear modes leads to the same conclusion for both SV- and SH-waves. As a result, the objective function typically is multimodal (i.e., it has local minima), which complicates the minimization procedure and may require performing several inversions starting from different points in the model space.

The schematic summary of numerical results in Figure 5.7 illustrates the influence of the tilt of the symmetry axis and reflector dip on the stability of parameter esti-

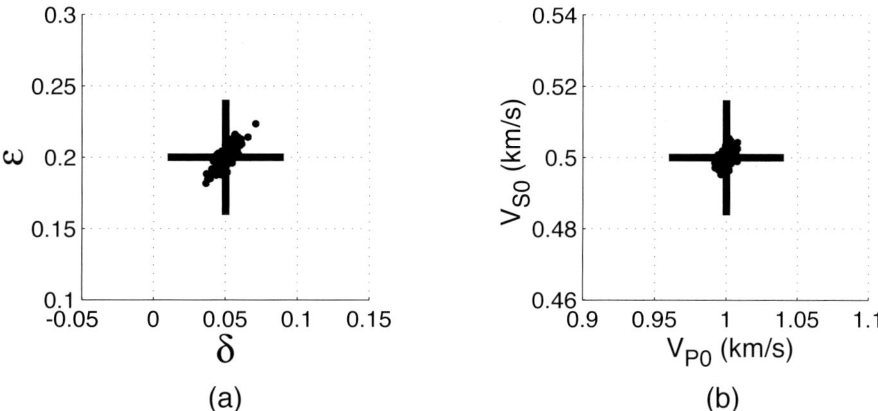

Figure 5.5: Results of the inversion (dots) of PP and SS data for a horizontal HTI layer (Grechka et al., 2002b). The actual parameters are marked by the crosses. The data were contaminated by Gaussian noise with the same standard deviations as those in Figure 5.4.

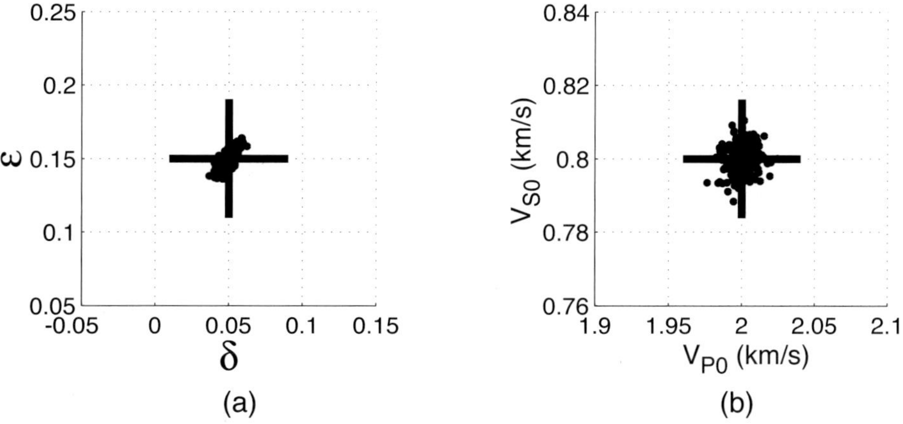

Figure 5.6: Same as Figure 5.5 but for an HTI layer with the lower boundary dipping at 25° (Grechka et al., 2002b). The azimuth of the symmetry axis with respect to the dip plane is β=40°.

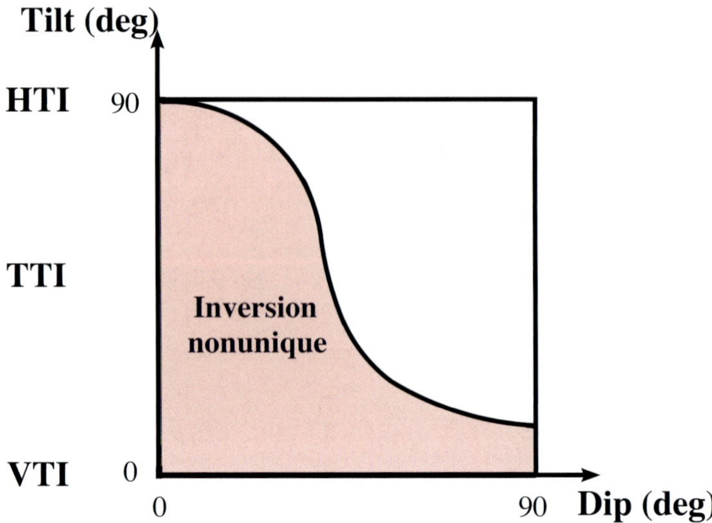

Figure 5.7: Illustration of the uniqueness of depth-domain parameter estimation in a homogeneous TI layer for the full range of reflector dips and symmetry-axis orientations (Grechka et al., 2002b). The tilt and azimuth of the symmetry axis were considered unknown prior to the inversion for all models including VTI and HTI.

mation in homogeneous TTI media. This plot should not be interpreted in a strict quantitative sense because the criteria used to identify the areas of "unique" and "nonunique" inversion were somewhat loose. In general, the line dividing those areas corresponds to the standard deviations of ϵ and δ close to 0.03 (for the errors in the input data listed in the caption of Figure 5.4). However, the quality of the inversion results also depends on such parameters not shown in Figure 5.7 as the azimuth of the symmetry axis with respect to the dip plane and the magnitude of ϵ and δ. In particular, the inversion becomes more stable with increasing absolute values of the anisotropy parameters.

It should be emphasized that in generating Figure 5.7 the orientation of the symmetry axis was assumed to be unknown, even if the model is HTI or VTI. Therefore, the stability of the inversion results for both vertical and horizontal transverse isotropy is lower than that in the two previous sections where we fixed the tilt of the symmetry axis at the correct value (see Figures 5.4 – 5.6). Still, even if the HTI model is not assumed in advance, it can be accurately reconstructed from reflection data despite some problems caused by the more complicated topology (e.g., multiple local minima) of the objective function.

In contrast, when the tilt of the symmetry axis is small (i.e., the model is close to VTI) and unknown, the inversion remains ambiguous for all reflector dips. As the symmetry axis deviates further from the vertical, the inversion becomes more stable and, for a fixed tilt, can be performed for an increasingly wide range of dips (Figure 5.7). Parameter estimation for HTI media is feasible even for a horizontal reflector and without a priori knowledge of tilt.

Figure 5.8 displays inversion results typical for models located within the area of nonuniqueness in Figure 5.7 near the separation line. As in the previous examples, the input data included noise-contaminated NMO ellipses, zero-offset traveltimes and reflection slopes of PP- and SS-waves. Although the obtained parameters are not biased, the standard deviations (4% for V_{P0} and V_{S0}, 0.05 for ϵ, and 0.04 for δ) indicate a substantial error amplification in the inversion. Still, it is interesting that the tilt ν of the symmetry axis is relatively well-constrained. It should be mentioned that the stability of parameter estimation might be enhanced by adding SH-waves to the input data.

5.2.3 Interval parameter estimation for layered media

The multicomponent inversion procedure was first tested on the two-layer VTI medium with curved interfaces in Figure 5.9. The input wide-azimuth reflection traveltimes of PP- and PSV-waves were processed by the PP+PS=SS method to compute the moveout of the pure SS-waves for both reflectors. Then the PP and SS data were collected into CMP gathers for azimuthal velocity analysis and estimation of the NMO ellipses. The data vector (equation 5.8) was obtained at the four CMP locations marked in Figure 5.9. The objective function (equation 5.9) was minimized under the assumption that both layers were known to be VTI. The inversion results for 100 realizations of the input data in Figure 5.10 indicate that the noise does not get enhanced by the parameter-estimation procedure. The standard deviations are about only 0.01 for ϵ and δ, and less than 1% for V_{P0} and V_{S0} (not shown).

Note that neither interface in the model from Figure 5.9 has steeply dipping segments. In agreement with the results for a single VTI layer, the high sensitivity of the SS-wave traveltimes to the parameter σ ensures the stability of parameter estimation for dips as mild as $15° - 20°$.

Next, the inversion was applied to multicomponent data from a more complicated model that includes layers with different tilts of the symmetry axis (Figures 5.11 and 5.12). The symmetry axis was known to be vertical in the VTI layer and horizontal in the HTI layer. Despite the larger error bars for the bottom layer, the overall stability of the algorithm is satisfactory. A general increase in errors with depth, caused mostly by the relatively small contributions of the deeper layers to the effective reflection traveltimes, is typical for all kinematic inversion algorithms. Also, although the method is not based on Dix-type differentiation, errors in the parameters of the upper layers influence the inversion results for the deeper part of the section.

Stacking-velocity inversion (tomography) of PP and SS data operates only with hyperbolic moveout and excludes information contained in long-offset traveltimes. Despite this limitation, it has significant advantages over conventional reflection tomography (e.g., Le Stunff and Grenié, 1998). The first advantage, which is particularly important in anisotropic media, is related to computational efficiency. Because the NMO ellipse (and, therefore, the multiazimuth, multioffset hyperbolic moveout

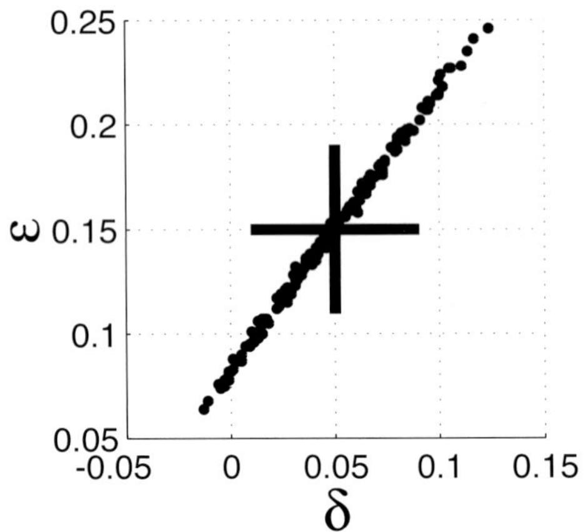

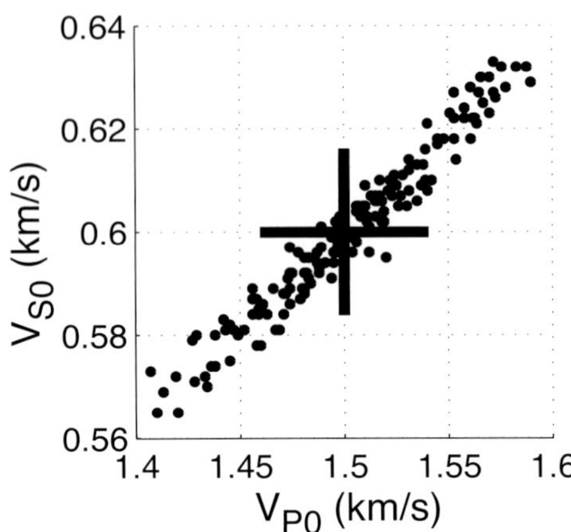

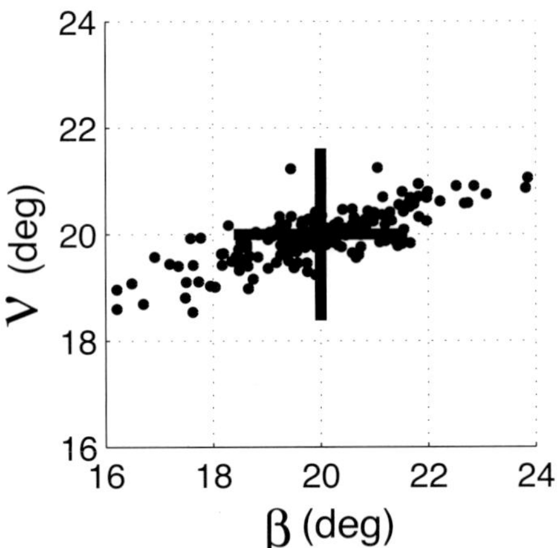

Figure 5.8: Results of the inversion (dots) of PP and SS data for a TTI layer with the bottom dipping at 30° (Grechka et al., 2002b). The actual parameters are marked by the crosses. The azimuth β of the symmetry axis is measured with respect to the dip plane. The standard deviations of Gaussian noise are the same as those in Figure 5.4.

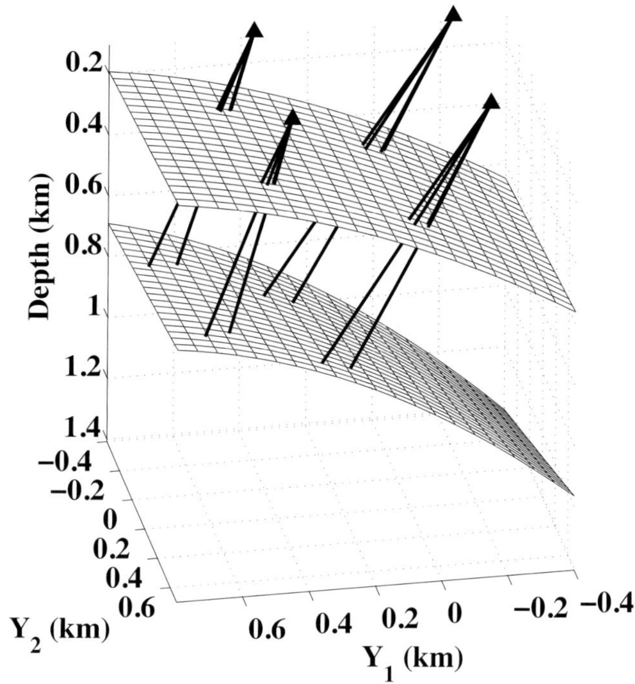

Figure 5.9: Zero-offset rays of PP- and SS-waves recorded at four CMP locations over a model composed of two homogeneous VTI layers (Grechka et al., 2002b). The parameters of the top layer are $V_{P0,1}$=2.0 km/s, $V_{S0,1}$=0.8 km/s, ϵ_1=0.15, and δ_1=0.05. For the bottom layer, $V_{P0,2}$=2.5 km/s, $V_{S0,2}$=0.9 km/s, ϵ_2=0.20, and δ_2=0.10.

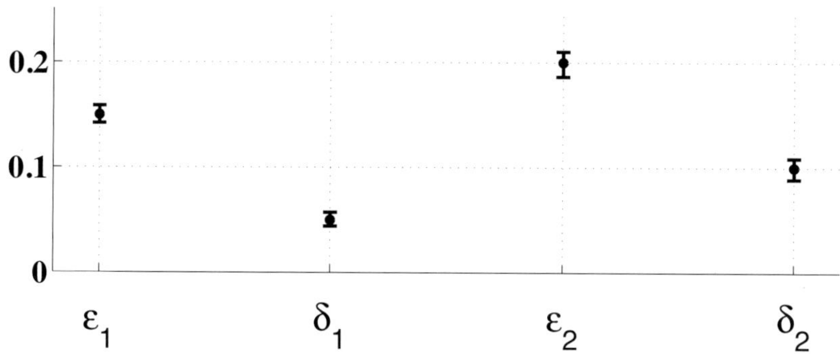

Figure 5.10: Inversion results for the model from Figure 5.9 (Grechka et al., 2002b). The input data were distorted by Gaussian noise with the standard deviations equal to 2% for the NMO velocities and 1% for the zero-offset traveltimes and reflection slopes. The starting model was isotropic. The dots mark the actual anisotropy parameters, the bars correspond to the $\pm$ standard deviation in each parameter.

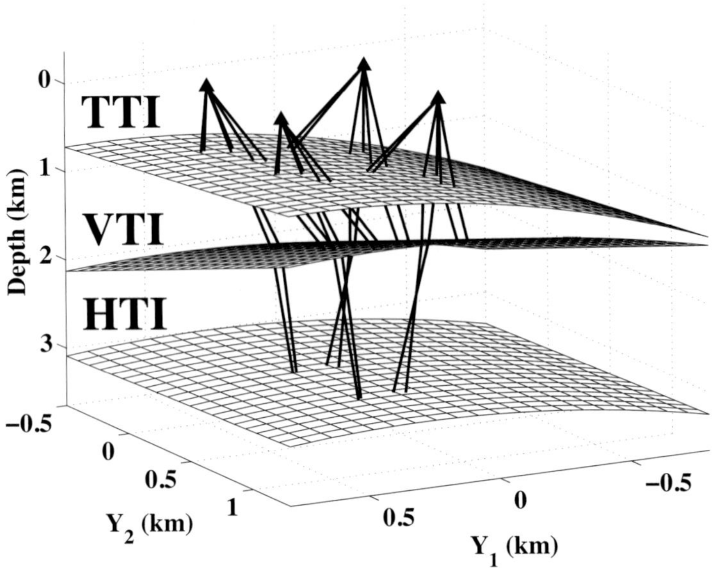

Figure 5.11: Zero-offset rays of PP- and SS-waves for a model composed of TTI, VTI, and HTI layers (Grechka et al., 2002b).

as a whole) can be computed by tracing only one (zero-offset) ray for each reflection event at a given CMP location, the number of rays to be generated in forward modeling is reduced by orders of magnitude. Second, it is possible to derive approximate expressions for the NMO ellipse even in arbitrarily anisotropic media, if the model is structurally simple (see Chapters 1 and 6). Such analytic solutions help identify the parameter combinations constrained by NMO-velocity inversion. Third, restricting the range of source-receiver offsets reduces the influence of lateral heterogeneity on reflection moveout. When the velocity field is estimated separately for blocks of relatively small lateral extent, the layers typically can be treated as homogeneous, and the interfaces can be approximated by simple smooth surfaces, such as low-order polynomials.

The laterally varying anisotropic velocity field, along with reflecting interfaces, can be built by global smoothing of the solutions obtained for each block. The output of stacking-velocity inversion can serve as an accurate initial model for migration velocity analysis and anisotropic post-migration reflection tomography.

Influence of errors in the symmetry type

In the above tests of the parameter-estimation algorithm it was assumed that the tilt of the symmetry axis in the VTI and HTI layers was known in advance. Because this does not necessarily hold in practice, it is instructive to examine the inversion for an intentionally incorrect symmetry type in one of the layers. The model in Figure 5.13 is composed of two VTI layers on top of an HTI layer (Table 5.4). Parameter estimation

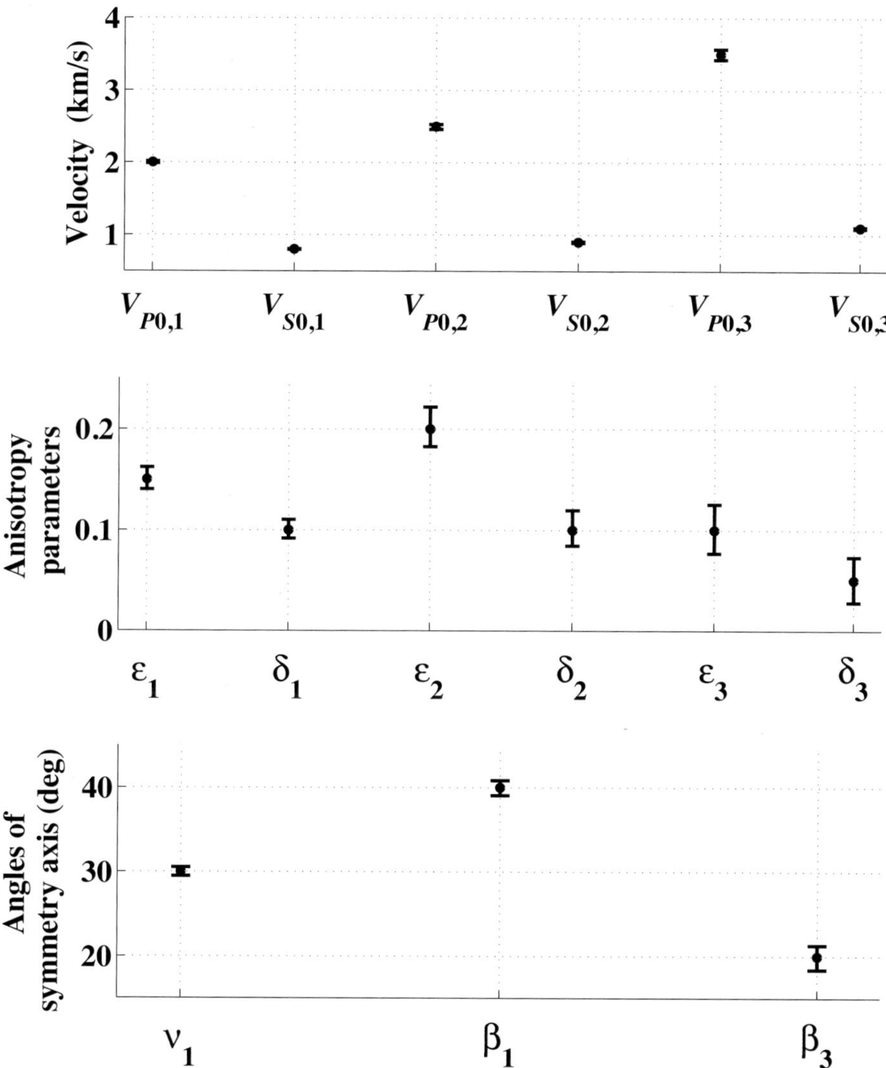

Figure 5.12: Inversion results for the model from Figure 5.11 (Grechka et al., 2002b). The dots mark the actual anisotropy parameters, the bars correspond to the $\pm$ standard deviation in each parameter. The anisotropy parameters of the starting model were set to zero. The parameters ν_1 and β_1 are the tilt and azimuth of the symmetry axis in the top (TTI) layer; β_3 is the symmetry-axis azimuth in the bottom (HTI) layer.

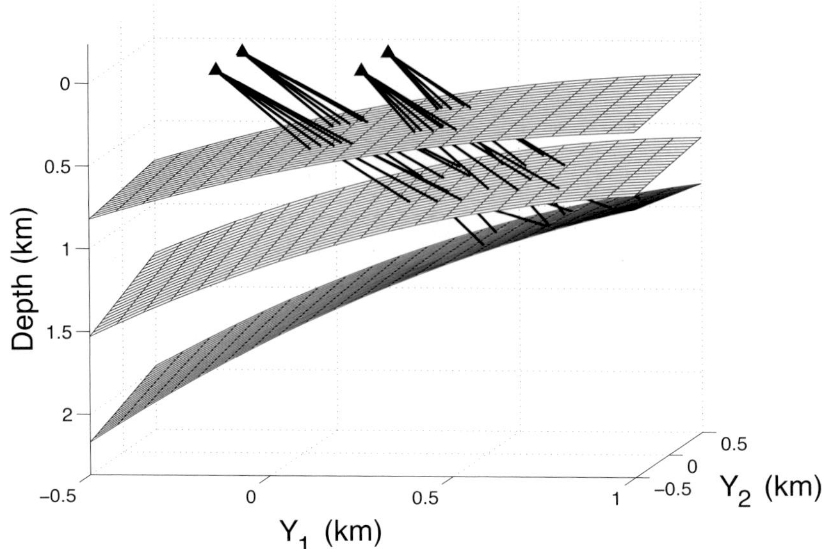

Figure 5.13: Zero-offset rays of PP- and SS-waves for a model composed of two VTI layers above an HTI layer (Grechka et al., 2002b). The interval parameters are listed in the top row of Table 5.4.

was performed under the erroneous assumption that the bottom (HTI) layer has VTI symmetry (the second row in Table 5.4). As expected, the inversion produced strongly distorted values of the symmetry-direction velocities $V_{P0,3}$ and $V_{S0,3}$ and the parameters ϵ_3 and δ_3 in that layer. It is interesting that the parameters of the two top layers are slightly inaccurate because the error in the bottom layer gets distributed throughout the model to minimize the objective function. Hence, errors can propagate not only downward (accumulate with depth) but also upward, albeit with a much smaller amplification.

Another implication of this observation is that it might be preferable to perform the inversion in the layer-stripping mode, starting with estimation of the parameters of the uppermost layer using the reflections from its bottom. Then, fixing the obtained values, one can determine the interval parameters of the second layer and continue the procedure downward. In addition to eliminating upward error propagation, this approach is computationally efficient because only a few unknowns need to be found at each stage (i.e., for each layer). The main shortcoming of layer stripping is its implicit reliance on the assumption that the parameters of each layer can be obtained using the reflections from its bottom. However, as illustrated by the examples for PP-waves in section 2.2, reflections from deeper interfaces can provide critically important information for inversion (e.g., when the interval beneath the layer of interest is known to be isotropic). Therefore, for some models the layer-stripping method might increase ambiguity in the joint inversion of PP and PS data.

$V_{P0,1}$	$V_{S0,1}$	ϵ_1	δ_1	β_1	$V_{P0,2}$	$V_{S0,2}$	ϵ_2	δ_2	$V_{P0,3}$	$V_{S0,3}$	ϵ_3	δ_3	β_3
Model parameters													
2.00	0.80	0.15	0.05	–	2.50	0.90	0.20	0.10	3.50	1.10	0.20	0.05	30
Inversion assuming the bottom layer to be VTI													
1.98	0.79	0.16	0.06	–	2.53	0.90	0.18	0.09	3.00	1.28	0.35	0.34	–
Inversion assuming the top layer to be HTI													
2.23	0.94	0.01	0.03	101	2.61	0.92	0.15	0.02	3.55	0.92	0.24	−0.09	34

Table 5.4: Parameters of the model from Figure 5.13 composed of two VTI layers above an HTI layer (top row) and the inversion results based on erroneous assumptions about the symmetry in one of the layers (Grechka et al., 2002b). The velocities are in km/s, angles are in degrees.

In the second test, the top (VTI) layer was assumed to have HTI symmetry. Then, in addition to the expected significant errors for that layer, the inversion produced distorted parameters in both deeper layers (see the third row in Table 5.4). This test underscores the importance of choosing the right type of anisotropy in the overburden, because any errors in the shallow part of the section propagate through the entire model.

An obvious alternative to *assuming* a specific orientation of the symmetry axis is to adopt the more general tilted TI (or even lower-symmetry) model from the outset of the inversion. The correct type of anisotropy could then be established from the determined symmetry-axis direction (for TTI media) or from the estimated anisotropy parameters or stiffness coefficients. For example, as discussed in section 7.1, under favorable circumstances the inversion of VSP (vertical seismic profiling) data can be carried out without any assumptions about the symmetry (Dewangan and Grechka, 2003; Kochetov and Slawinski, 2009a, 2009b, 2009c). According to the above numerical results, however, the need to estimate the tilt of the symmetry axis typically reduces the stability of the algorithm introduced in this section, especially for models close to VTI.

Influence of heterogeneity

Accurate reconstruction of heterogeneous velocity fields is a major challenge in reflection data processing. For example, even for isotropic media certain types of vertical velocity variations can never be resolved from reflection traveltimes, no matter how the inversion is performed (e.g., Goldin, 1986). Given the complexity of this problem for anisotropic media, the discussion here is limited to a numerical example illustrating the errors in the estimated anisotropy parameters caused by heterogeneity unaccounted for in the inversion.

Suppose one attempts to estimate the parameters of the VTI overburden for the model from Figure 5.13 and Table 5.4 using only the reflections from the bottom of the second layer. The effective parameters of the overburden then change both vertically (because it actually consists of two layers) and laterally (because the uppermost interface is not horizontal). While the vertical variations of the VTI parameters cannot be resolved without including reflections from the bottom of the first layer, one can try to evaluate lateral parameter variations by performing the inversion for a range of CMP locations.

For purposes of this test, PP and SS reflection data (equation 5.8) from the bottom of the second and third layer were generated for the following CMP coordinates: $Y_1 = [-0.6, -0.4, \ldots, 0.8, 1.0]$ km; and $Y_2 = [-0.2, 0.2]$ km (Figure 5.13). Each inversion result was obtained using four adjacent common midpoints that form the corners of a rectangle, and the estimated anisotropy parameters were assigned to the center of the rectangle. Treating the heterogeneous, two-layer overburden as a single homogeneous VTI layer yields the laterally varying parameters displayed in Figure 5.14. It is interesting that while the effective vertical velocities V_{P0} and V_{S0} can be regarded as certain averages of the corresponding interval velocities, the best-fit parameters ϵ and δ often lie *outside* the range bounded by the interval values.

Although this result seems counterintuitive, the "average" or effective parameters produced by such algorithms as stacking-velocity inversion are not necessarily bounded by the minimum and maximum interval values. As an example, Appendix 5A shows that the effective parameter δ derived from the PP-wave NMO velocity for a stack of horizontal, homogeneous VTI layers can exceed the maximum interval value δ_n. In the special case of isotropy, when all $\delta_n = 0$, the effective δ above a vertically heterogeneous medium is always positive. The model in Figure 5.13 contains curved interfaces, and the averaging procedure is more complicated than that in Appendix 5A. In both cases, however, the effective values can lie outside the range of the corresponding interval quantities.

Thus, replacing the actual heterogeneous medium with the "equivalent" homogeneous model that provides the best fit to the data can result in effective properties not obtainable by straightforward averaging of the corresponding interval (local) properties. In general, the anisotropy parameters estimated by stacking-velocity inversion inevitably bear an imprint of the adopted subsurface model and available reflection events.

5.3 Case study from the North Sea

In section 4.2 we discussed application of the PP+PS=SS method to 2D multicomponent data acquired above the Siri reservoir in the North Sea. Here, following the work of Grechka et al. (2002c), the SS traveltimes generated by the PP+PS=SS method for that data set are combined with the PP moveout to estimate the interval medium parameters using both stacking-velocity tomography and the Dix equation. The layered VTI model obtained by multicomponent velocity analysis greatly improves the quality of common-conversion-point stacks and migrated sections of the acquired PS-waves.

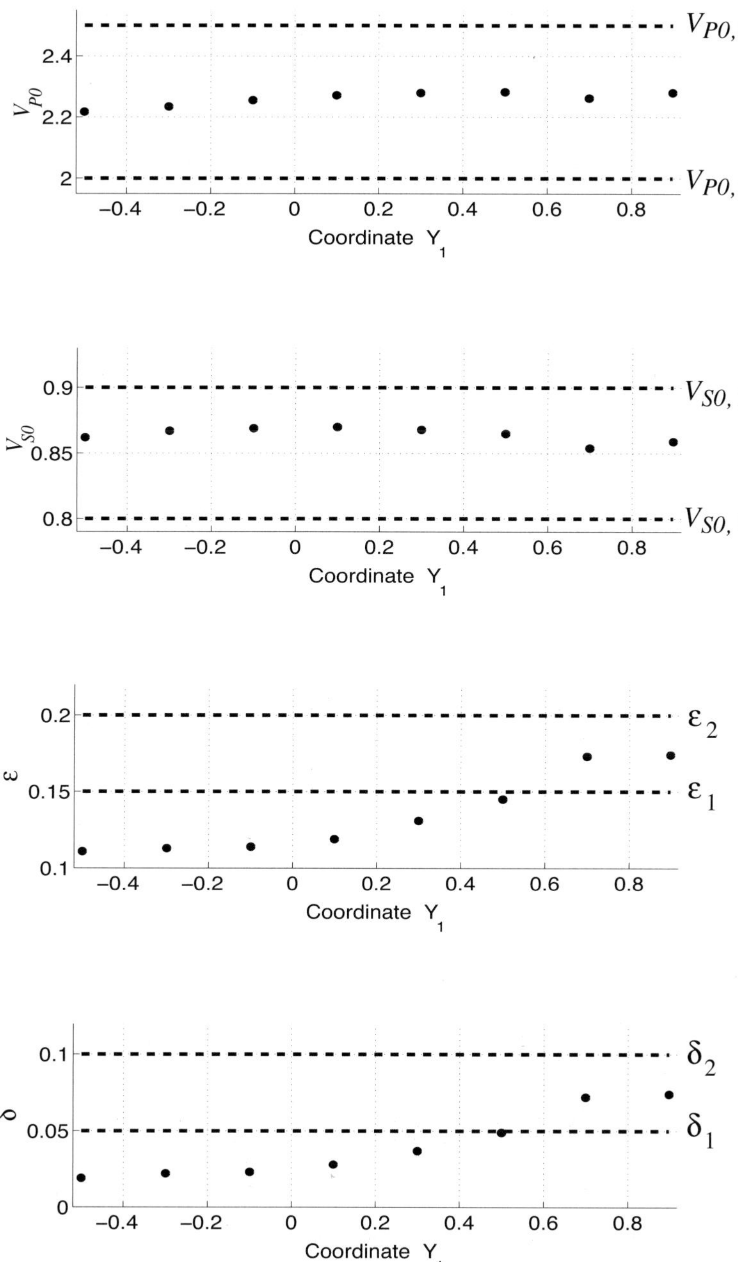

Figure 5.14: Effective parameters (dots) of the two-layer VTI overburden for the model from Figure 5.13 estimated using the reflections from the bottom of the second layer (Grechka et al., 2002b). The interval parameters of the VTI layers are marked by the dashed lines. The velocities are in km/s.

5.3.1 Velocity analysis of PP and SS data

Although the section above the Siri reservoir has mild lateral heterogeneity (Figure 4.12), reflection traveltimes of PS-waves exhibit nonnegligible asymmetry with respect to zero offset. In contrast, the SS traveltimes computed by the PP+ PS = SS method are symmetric (Figure 4.14b) and, because of the limited offset range, are adequately described by the hyperbolic moveout equation. Therefore, hyperbolic moveout analysis of the PP- and SS-waves can be used to estimate their zero-offset traveltimes (t_{P0} and t_{S0}) and NMO (stacking) velocities ($V_{\mathrm{nmo},P}$ and $V_{\mathrm{nmo},S}$).

Because the sources and receivers for the SS-waves are obtained on an irregular grid (Figure 4.14a), the SS data were collected into composite CMP gathers that include source-receiver pairs with the midpoints within a specified interval Δx_{CMP}. Clearly, increasing Δx_{CMP} improves the stability of velocity estimates but reduces the number of gathers and lateral velocity resolution. Based on a series of numerical tests, $\Delta x_{\mathrm{CMP}} = 0.5$ km was judged to provide both sufficient stability and acceptable lateral resolution.

Figure 5.15 shows the zero-offset traveltimes and NMO velocities for the top of the Balder formation (i.e., the top of the Siri reservoir). Velocity analysis of SS traveltimes was carried out for only those gathers containing ten or more source-receiver pairs. Note the absence of the SS-wave results for $x_{\mathrm{CMP}} = 6$ km, where the fold was too low. The SS traveltimes are also missing near the beginning of the line where the traveltime picks required by the PP+ PS = SS method were not available.

The error bars in Figure 5.15 were calculated assuming random (Gaussian) distribution of the traveltime picks around the best-fit hyperbolas. Although the data might be also distorted by systematic picking errors (for example, those associated with incorrect identification of PP and PS events), such errors are difficult to evaluate or remove without additional information. CMP gathers with errors significantly higher than those at adjacent locations (e.g., the shear-wave velocity pick at $x_{\mathrm{CMP}} = 7.5$ km) were excluded, and only the most reliable portion of the data was used for parameter estimation.

Before describing the inversion results, it is instructive to identify the properties of the conventional-spread moveout of PP- and SS-waves indicative of the presence of anisotropy. The zero-offset traveltimes for the top Balder (Figure 5.15) and other interfaces can be used to find the ratio of the vertical velocities V_{S0} and V_{P0} (assuming negligible dips):

$$g_0 \equiv \frac{V_{S0}}{V_{P0}} = \frac{t_{P0}}{t_{S0}} \,. \tag{5.10}$$

Moveout analysis yields the ratio of the corresponding NMO velocities:

$$g_{\mathrm{nmo}} \equiv \frac{V_{\mathrm{nmo},S}}{V_{\mathrm{nmo},P}} \,. \tag{5.11}$$

The velocities in equations 5.10 and 5.11 represent effective quantities for a particular reflection event.

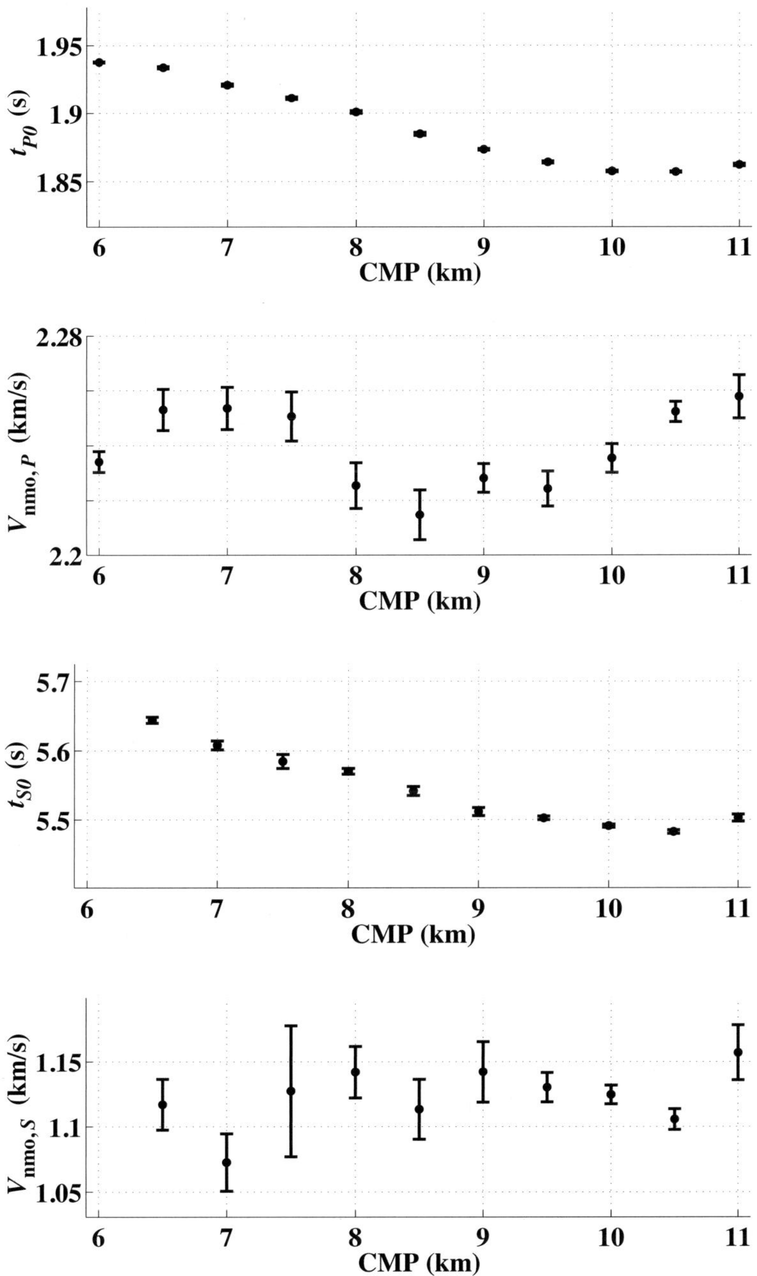

Figure 5.15: Results of velocity analysis for the top of the Balder formation (Grechka et al., 2002c). The parameters t_{P0}, $V_{\text{nmo},P}$, t_{S0}, and $V_{\text{nmo},S}$ are shown for a range of CMP locations. The error bars correspond to the 95% confidence intervals.

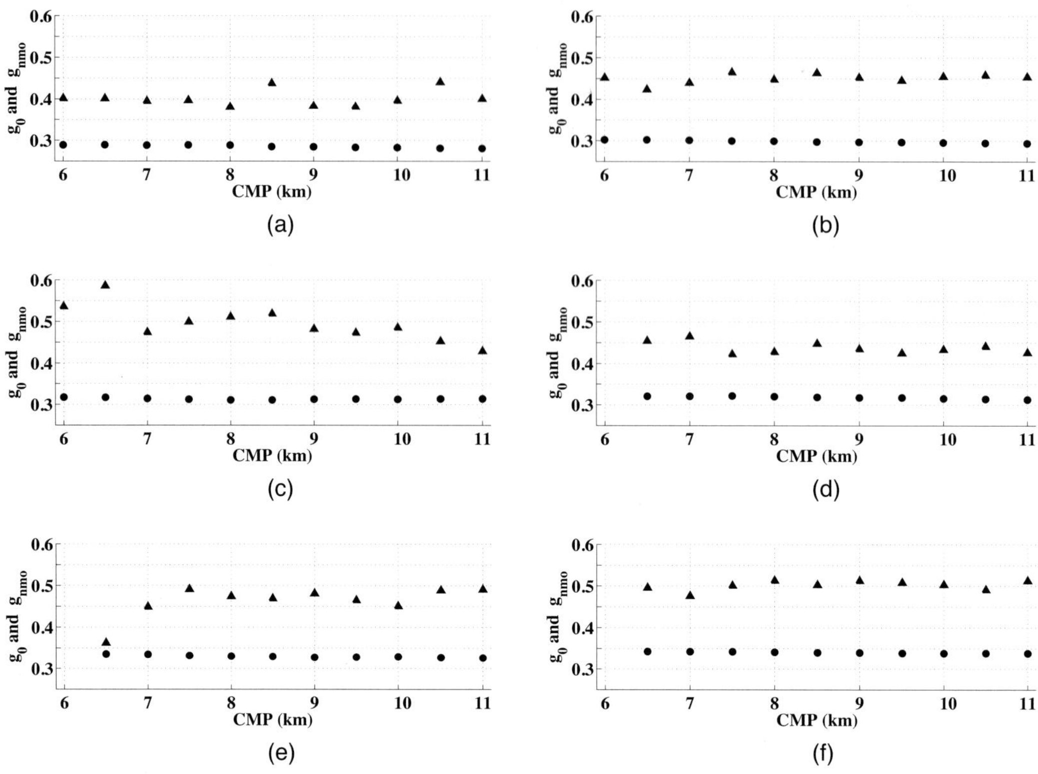

Figure 5.16: The ratios g_0 (circles) and g_{nmo} (triangles) computed from equations 5.10 and 5.11 for the following horizons: (a) mid-Miocene, (b) intra-Oligocene, (c) base Oligocene, (d) top overpressure, (e) h8, and (f) top Balder (Grechka et al., 2002c).

Figure 5.16 shows the ratios g_0 and g_{nmo} for the reflections from several interfaces above the reservoir. Because $g_0 = g_{\text{nmo}}$ in a homogeneous isotropic medium, it is remarkable that for all examined events $g_0 \approx 0.3$ is consistently and significantly smaller than $g_{\text{nmo}} \approx 0.45$. In principle, this difference can be caused by anisotropy, heterogeneity, or a combination of these two factors. Appendix 5B shows, however, that the obtained condition,

$$\frac{g_0}{g_{\text{nmo}}} \approx \frac{2}{3}, \tag{5.12}$$

cannot be satisfied in realistic vertically heterogeneous, isotropic models. The influence of anisotropy is the only plausible explanation for the consistent deviation of g_0 from g_{nmo} in Figure 5.16.

Under the assumption that the subsurface has VTI symmetry, the ratios g_0 and g_{nmo} can be used to find an approximate relationship between the effective anisotropy parameters. Treating the medium above each reflector as a horizontal, homogeneous

VTI layer, equation 5.11 can be rewritten as

$$g_{\rm nmo} = \frac{V_{S0}\sqrt{1+2\sigma}}{V_{P0}\sqrt{1+2\delta}} = g_0\,\frac{\sqrt{1+2\sigma}}{\sqrt{1+2\delta}}\,. \tag{5.13}$$

Substituting the estimates $g_0 = 0.3$ and $g_{\rm nmo} = 0.45$ into equation 5.13, taking into account that $\sigma \equiv (\epsilon - \delta)/g_0^2$, and dropping quadratic and higher-order terms in ϵ and δ yields

$$\epsilon \approx 0.06 + 1.25\,\delta\,. \tag{5.14}$$

Clearly, only one of the anisotropy parameters in equation 5.14 can vanish, so the combination of conventional-spread PP and SS data cannot be interpreted without making the effective model anisotropic. Below, we refine the crude estimate in equation 5.14 by performing a more elaborate inversion for the interval VTI parameters.

5.3.2 Anisotropic parameter estimation

An important issue to be addressed prior to inversion is identification of the medium parameters constrained by the available data. The low energy observed on the transverse displacement component (not shown here) suggests that either the medium is azimuthally isotropic (i.e., VTI or purely isotropic) or the line coincides with a vertical symmetry plane of the subsurface. In the absence of wide-azimuth acquisition, it is natural to assume VTI symmetry in the inversion. Another indication that the VTI model is likely to be adequate is the predominantly horizontal layering above the reservoir, with no evidence of fracturing.

For a structure composed of subhorizontal VTI layers, PP and PS reflection data can be inverted for the NMO velocities of PP- and SS-waves, the vertical-velocity ratio g_0, and the parameter η, but not for V_{P0}, V_{S0}, ϵ, and δ. As discussed in section 5.1, reflection traveltimes of PP- and PS-waves in horizontally layered VTI media do not constrain the vertical velocities, even if the model parameters are selected in such a way that PP and PS sections are tied in depth. Therefore, in the moveout-inversion procedure described below, the parameter δ is set to a certain predetermined value. Then check-shot traveltimes from a nearby borehole are employed to resolve the interval Thomsen parameters.

Inversion for the interval VTI parameters

The input data included the PP traveltime picks and computed SS traveltimes for the mid-Miocene, intra-Oligocene, base Oligocene, h8, and top Balder horizons. The model was assumed to include homogeneous VTI layers separated by plane interfaces of arbitrary dip.

First, the data were processed using 2D multicomponent stacking-velocity tomography (see section 5.2). For each trial set of the interval VTI parameters, the dips and depths of all reflectors were reconstructed using the zero-offset traveltimes and reflection slopes. Then the best-fit solution was found by minimizing the least-squares objective function 5.9, in which the matrices describing NMO ellipses were replaced with NMO velocities:

$$\mathcal{F}(\mathbf{m}) \equiv \sum_{Q,\, x_{\text{CMP}},\, n} \left[V_{Q,\,\text{nmo}}^{\text{calc}}(x_{\text{CMP}},\, n,\, \mathbf{m}) \; - \; V_{Q,\,\text{nmo}}^{\text{meas}}(x_{\text{CMP}},\, n) \right]^2 , \qquad (5.15)$$

where $\mathbf{m}$ is the vector of the unknown interval parameters V_{P0}, V_{S0}, ϵ, and δ; Q denotes the mode (PP or SS); $n = 1, 2, \ldots, 5$ is the reflector number; $V_{Q,\,\text{nmo}}^{\text{calc}}(x_{\text{CMP}},\, n,\, \mathbf{m})$ are the PP- and SS-wave NMO velocities calculated for a given vector $\mathbf{m}$; and $V_{Q,\,\text{nmo}}^{\text{meas}}(x_{\text{CMP}},\, n)$ are the NMO velocities measured from the data.

The interval parameters of five layers above the top Balder were computed for two different fixed values of δ (Figures 5.17 and 5.18). The error bars, corresponding to 95% confidence intervals, were obtained numerically from the errors in the NMO velocities and zero-offset traveltimes (see Figure 5.15). Despite the substantial differences in all inverted parameters except for the reflector dips (found to be small), the models from Figures 5.17 and 5.18 fit the measured NMO velocities equally well, with the minimum values of the objective function differing by only 2%.

The inversion results can be verified further by comparing PP and PS traveltimes computed for the obtained VTI models with the input time picks (Figure 5.19). Figures 5.19a,c confirm that PP traveltimes for two distinctly different models from Figures 5.17 and 5.18 practically coincide with the picks in the entire offset range. The average standard deviation of the computed traveltimes from the picked values for CMP locations between 7 and 10 km is only 6 ms, or about 0.3% of the average zero-offset traveltime. Although the inversion was based solely on the NMO velocities and zero-offset traveltimes, the quality of fit does not deteriorate at offsets reaching $1.7D$ (D is the reflector depth).

The inverted VTI models closely match the picked PS traveltimes as well (Figures 5.19b,d). The standard deviation of the modeled traveltimes from the picks is 12 ms, which is also 0.3% of the average PS-wave zero-offset traveltime. The PS data are noisier and have a lower frequency content, which explains the larger absolute standard deviation compared with that for the PP-waves. Hence, the family of inverted VTI models provides a good fit (given the level of noise) to the PP and PS time picks in the full range of available offsets. It is clearly impossible to distinguish between these models using the combination of PP and PS traveltimes, which confirms the conclusions of section 5.1.

With the reflector dips being so small, the interval NMO velocities were also computed from the Dix equation and combined with the ratios of the interval vertical velocities V_{P0} and V_{S0} to estimate the VTI parameters in each layer. As before, the parameter δ was fixed at a known constant value to ensure a unique result. After lateral smoothing of the effective NMO velocities, this inversion was performed for all common midpoints between 7 and 10 km. Figure 5.20 shows the parameter ϵ of several equivalent anisotropic models obtained at four CMP locations. These models are close to those obtained by stacking-velocity tomography (Figures 5.17 and 5.18) for the same values of δ. Although the interval parameter ϵ in each layer

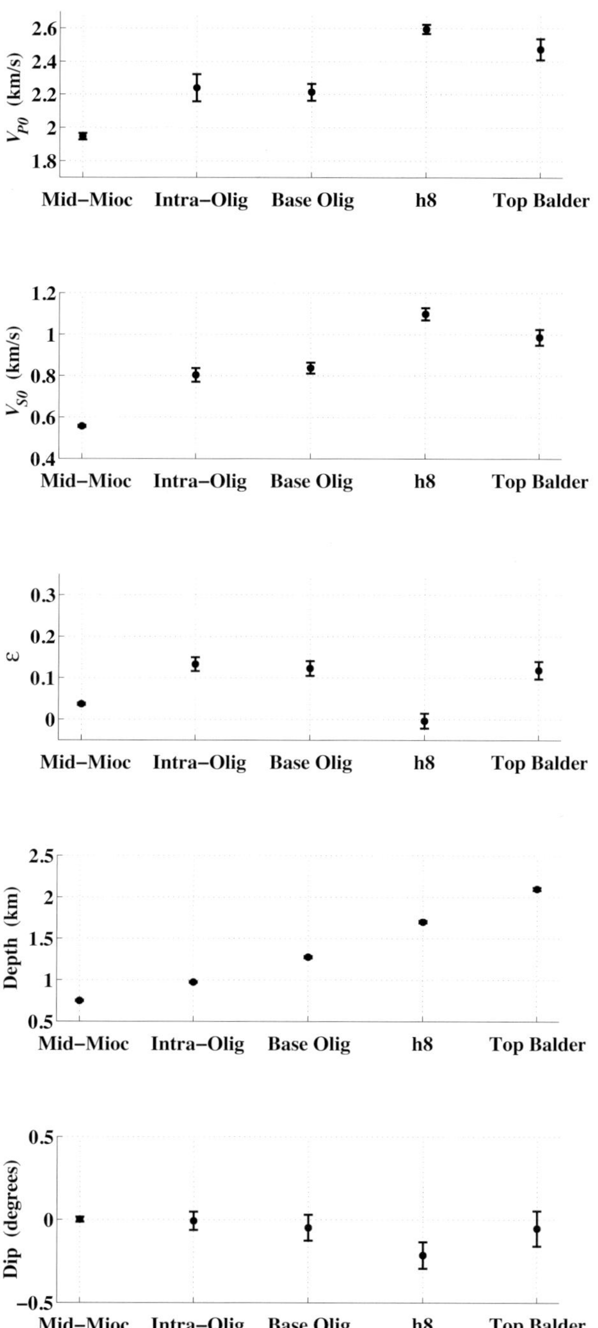

Figure 5.17: Inverted interval VTI parameters, along with the depths and dips of the interfaces, for CMP locations $7\,\mathrm{km} \leq x_{\mathrm{CMP}} \leq 10\,\mathrm{km}$ (Grechka et al., 2002c). The parameter δ=0 is fixed throughout the section.

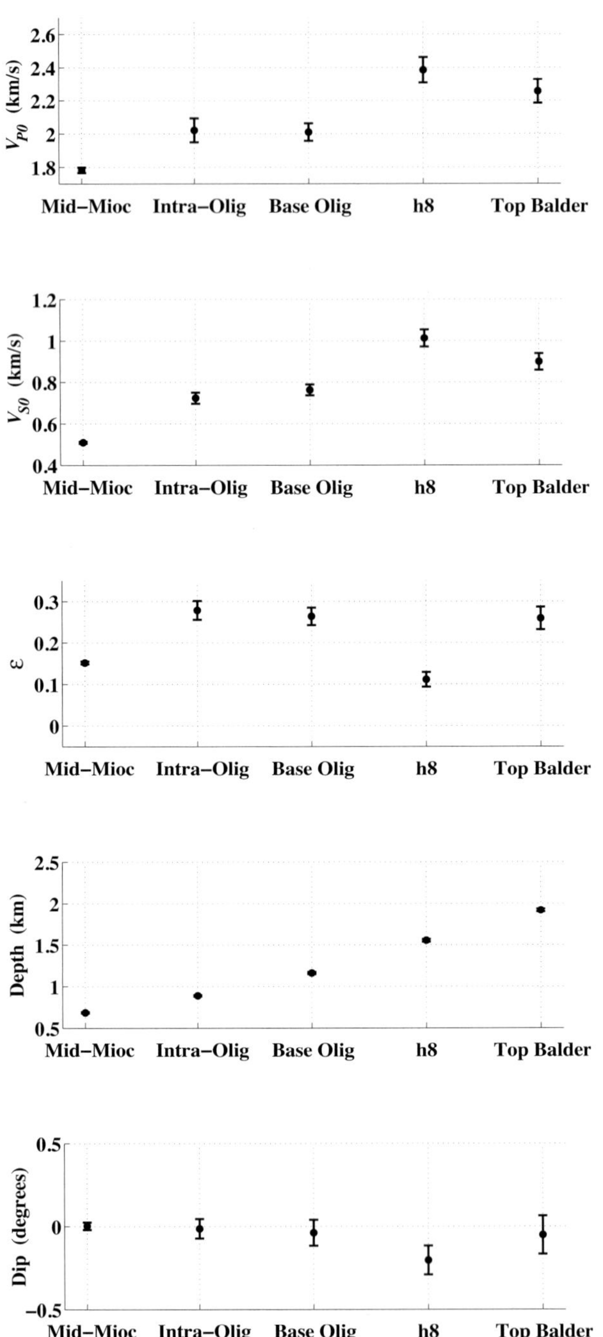

Figure 5.18: Same as Figure 5.17 but for a different fixed parameter δ=0.1 (Grechka et al., 2002c).

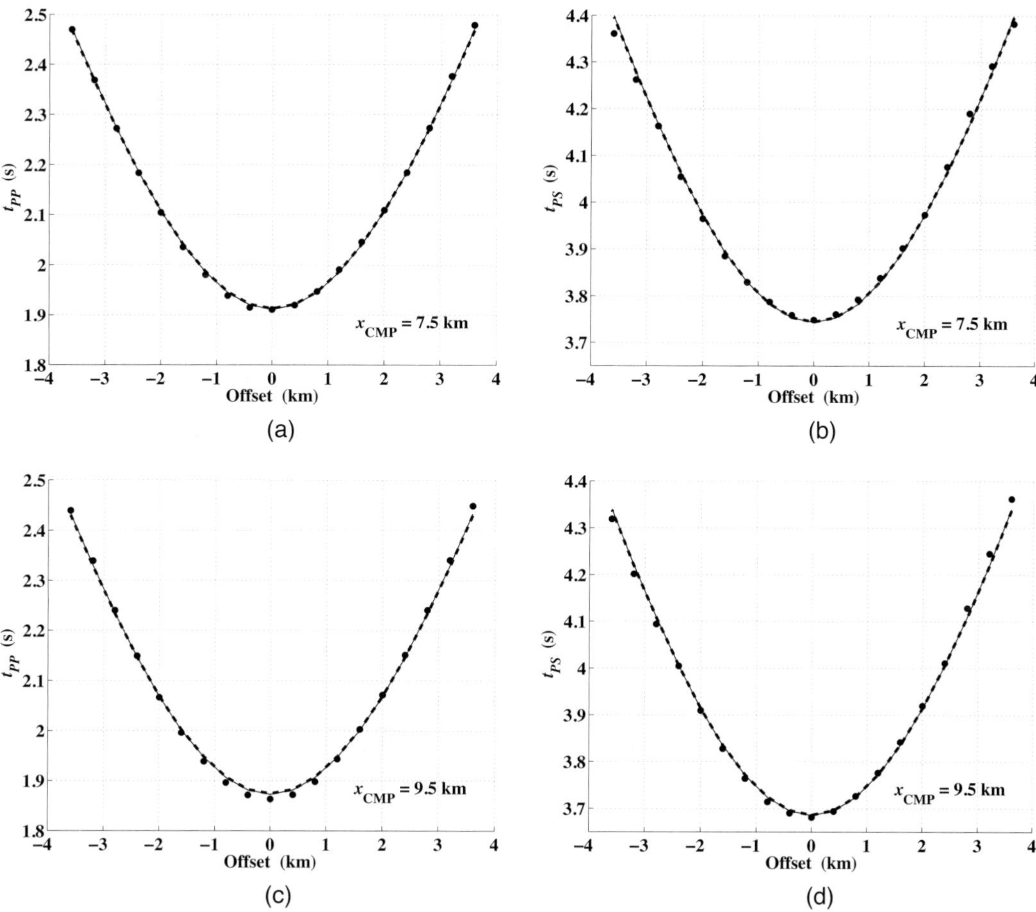

Figure 5.19: Comparison of PP-wave (a and c) and PS-wave (b and d) traveltime picks (dots) for the top Balder reflection with modeled traveltimes at two CMP locations, 7.5 km and 9.5 km (Grechka et al., 2002c). The solid and dashed lines are computed by ray tracing for the inverted models from Figures 5.17 (δ=0) and 5.18 (δ=0.1), respectively.

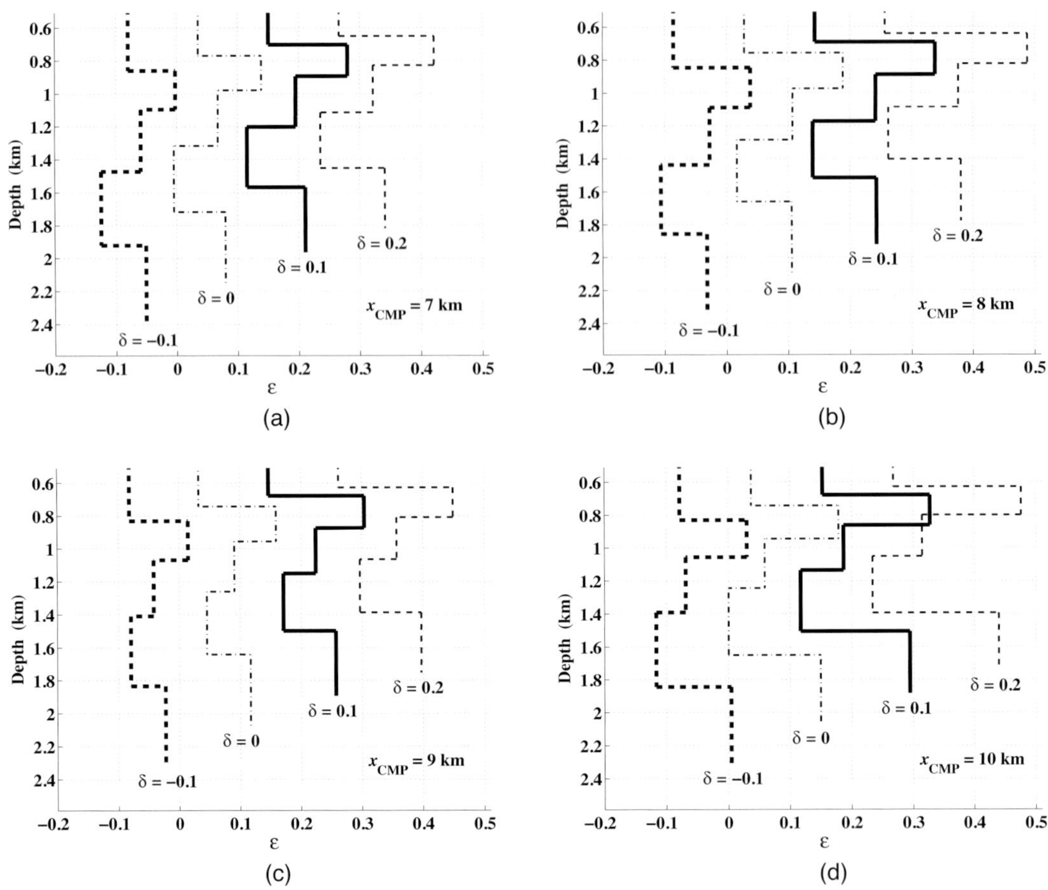

Figure 5.20: Interval parameter ϵ estimated using Dix-based inversion for the following CMP locations: (a) 7 km, (b) 8 km, (c) 9 km, and (d) 10 km (Grechka et al., 2002c). The values of δ used to generate each ϵ-curve are marked on the plots.

changes somewhat along the line, the statistical significance of this lateral variation is questionable (the standard deviation of ϵ for a fixed δ is about 0.03). On the whole, although stacking-velocity inversion provides some useful information about the subsurface, building a depth-domain VTI model requires additional data.

Constraining the inversion using check-shot data

To overcome the inherent nonuniqueness of the inversion of PP and PS reflection traveltimes for laterally homogeneous VTI media, it is sufficient to obtain an independent estimate of at least one model parameter (V_{P0}, V_{S0}, ϵ, δ, or reflector depth). In practice, velocity analysis of reflection data is often constrained by using check shots or well logs. In the Siri case study, P-wave check-shot traveltimes were measured in a

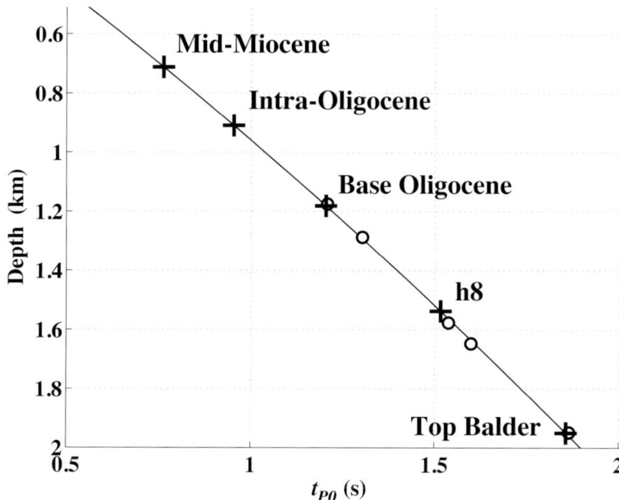

Figure 5.21: P-wave depth-time relationship (circles) obtained from check shots in well Siri-1 (Grechka et al., 2002c). The traveltimes were doubled for comparison with reflection data. The interpolated and extrapolated depth-time function is marked by the solid line. The crosses correspond to the zero-offset traveltimes t_{P0} estimated from reflection data for several horizons at common midpoint x_{CMP}=10 km.

borehole located close to $x_{\text{CMP}} = 10$ km.

The time-to-depth conversion curve used for parameter estimation is marked by the solid line in Figure 5.21. Because no check-shot receivers were placed above the base of Oligocene, the traveltimes were extrapolated to the shallow part of the section and also interpolated between the measurements. The standard deviation of the check-shot data points from this curve is 5 ms, which is close to the standard deviation of the PP-wave reflection traveltime picks from the best-fit hyperbolas.

Once the reflector depths (corresponding to the crosses in Figure 5.21) were determined, the NMO velocities and zero-offset traveltimes of PP- and SS-waves could be inverted for the interval parameters V_{P0}, V_{S0}, ϵ, and δ. The results in Figure 5.22 were obtained by combining the interval NMO velocities (estimated using the Dix equation) and zero-offset traveltimes for $x_{\text{CMP}} = 10$ km with the reflector depths from Figure 5.21. Clearly, the parameter δ varies with depth, which violates the artificial assumption of constant δ used in the inversion above.

The largest values of ϵ and δ are observed in the middle part of the section, with ϵ reaching almost 0.25 in the Oligocene layer. Although δ in the layer immediately above the reservoir (i.e., above the top Balder) is relatively small, the value of $\epsilon = 0.17$ indicates that anisotropy in that layer is quite significant for both P-waves and, particularly, for shear and converted waves (the corresponding interval $\sigma = 0.5$). Also, throughout the section $\epsilon > \delta$, which means that the parameters η and σ are positive. This result is consistent with the predominantly positive values of η reported in the majority of case studies for VTI media (e.g., Alkhalifah et al., 1996; Tsvankin, 2005).

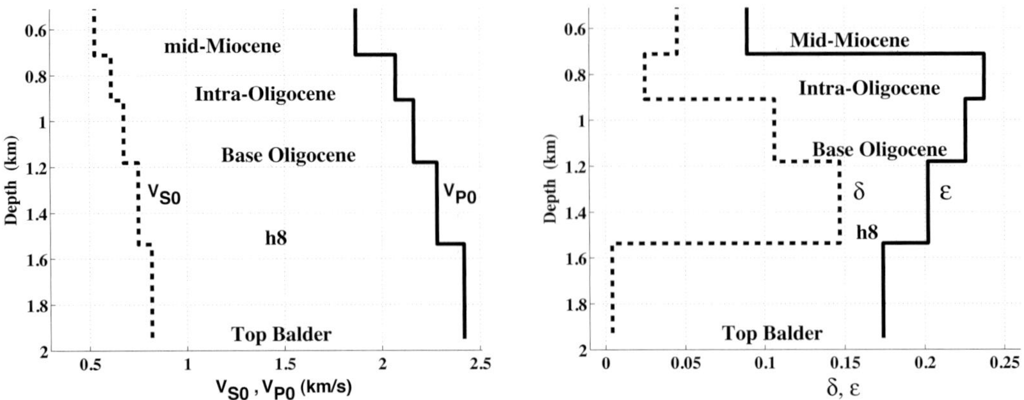

Figure 5.22: Interval VTI parameters estimated from check shots in well Siri-1 (see Figure 5.21) and reflection data at x_{CMP}=10 km (Grechka et al., 2002c). The left plot shows the vertical velocities V_{S0} (dashed) and V_{P0} (solid); on the right are the parameters δ (dashed) and ϵ (solid). The standard deviations of the interval values of ϵ and δ do not exceed 0.03.

In section 7.2, however, we present an example of VSP inversion from the Gulf of Mexico that produced a negative parameter η in subsalt sediments.

While the interval parameters V_{P0}, V_{S0}, ϵ, and δ in Figure 5.22 adequately describe the VTI model at the well location, they cannot be extrapolated along the seismic line without using additional borehole data or making assumptions about the lateral variation of the velocity field. Because the dips in this section are small, a plausible assumption is to fix the reflector depths at the values in Figure 5.21 (i.e., make the interfaces horizontal) and estimate the vertical velocities and anisotropy parameters along the line. Alternatively, the weak lateral variation of the parameter ϵ (Figure 5.20) might suggest that its interval values in Figure 5.22 can be used to estimate the parameters V_{P0}, V_{S0}, and δ for the full range of CMP locations. Neither of these assumptions, however, can be fully verified based on the available data.

5.3.3 Anisotropic imaging of PS data

Because of its velocity-independent nature, the PP+PS=SS method does not produce correct reflection amplitudes of SS-waves. Instead of operating with the "unphysical" SS data, Grechka et al. (2002d) processed the acquired converted waves and compared the results obtained for isotropic and VTI models of the overburden.

Common-conversion-point (CCP) stacks of PS-waves in Figure 5.23 were generated using the inverted VTI model for $\delta = 0$ from Figure 5.17 (the check shots became available after the processing had been completed). The CCP ray trajectories and traveltime curves were computed for each reflector and used to generate CCP stacks. The

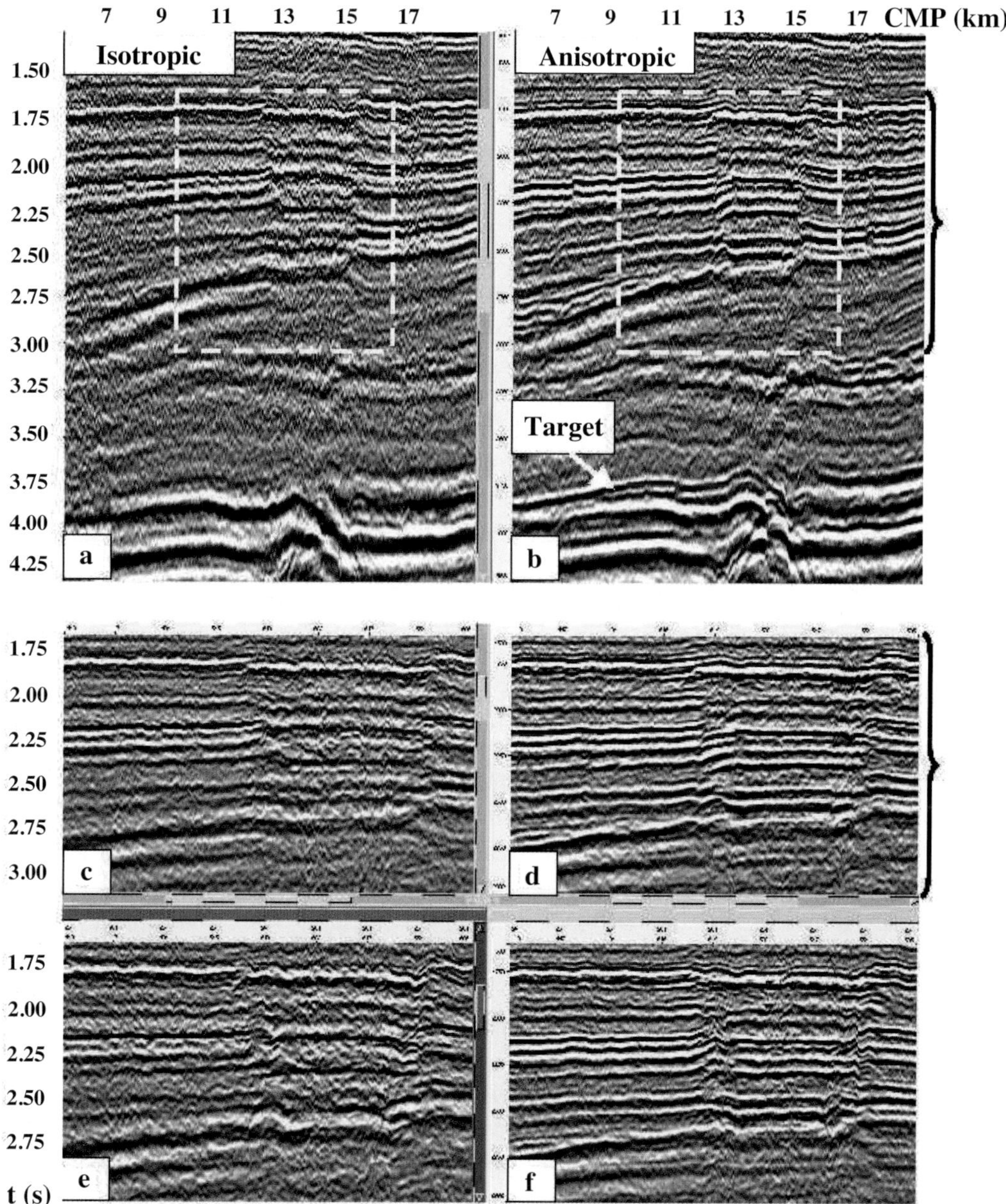

Figure 5.23: Model-based common-conversion-point (CCP) time stacks of PS-waves computed using the inversion results in Figure 5.17: (a) isotropic (i.e., for $\epsilon = \delta = 0$) and (b) VTI (Grechka et al., 2002d). Plots (c–f) show the area inside the dashed line on plots (a) and (b) obtained with separate moveout computation and CCP sorting: (c) isotropic moveout and isotropic CCP sorting; (d) anisotropic moveout and anisotropic CCP sorting; (e) isotropic moveout and anisotropic CCP sorting; and (f) anisotropic moveout and isotropic CCP sorting.

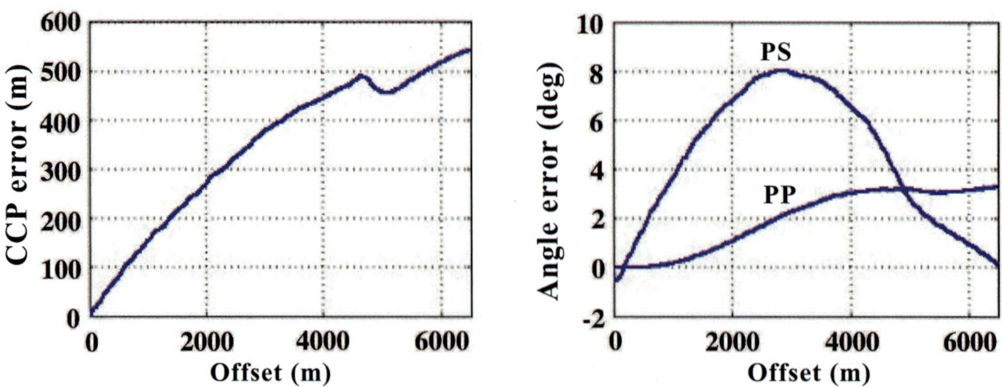

Figure 5.24: Errors in the lateral position of the PS-wave conversion point (left plot) and in the incidence angle of the P-leg for the PP- and PS-waves (right) at the target horizon caused by neglecting anisotropy (Grechka et al., 2002d).

reference isotropic model was obtained by setting $\epsilon = \delta = 0$, while keeping the other parameters in Figure 5.17 unchanged. Comparison of the isotropic (Figure 5.23a) and VTI (Figure 5.23b) stacked sections reveals major improvements achieved by accounting for anisotropy. First, anisotropic processing provided a much better image of the reservoir top and a crisp picture of faulting in the shallow part of the section. It is likely that employing the depth-constrained VTI model from Figure 5.22 would further enhance the image quality. Second, application of accurate NMO velocities in the VTI model substantially boosted higher frequencies in the stacked reflections and, therefore, increased temporal resolution. Although the anisotropic parameter estimation was performed for only the left part of the line, the improvements are observed for the whole range of CMP locations in Figure 5.23.

Figures 5.23c–f further clarify the main reasons for the superior quality of the anisotropic result. As illustrated by Figures 5.23d,e, the higher temporal resolution of the VTI section is indeed ensured by the better moveout correction. To enhance fault imaging, it is necessary to account for anisotropy in computing CCP ray trajectories (compare Figures 5.23d and 5.23f). Poor focusing and positioning of fault-plane reflections on the isotropic section are explained by the smearing of the conversion point caused by depth-varying anisotropy. This smearing at the target level (top Balder) exceeds 500 m for the largest offset in the data (Figure 5.24) and is about 340 m for the maximum offset (2600 m) used for CCP stacking.

Such a significant shift of the conversion point not only reduces lateral resolution but also biases the AVO response for PS-waves because each CCP gather generated for the isotropic model includes reflections from a relatively wide range of subsurface points. In addition, neglecting anisotropy in AVO analysis of PS-waves introduces an error in the offset-to-angle transformation that reaches 8° for an offset of 3 km; the corresponding error for PP-waves is only about 2.5° (Figure 5.24).

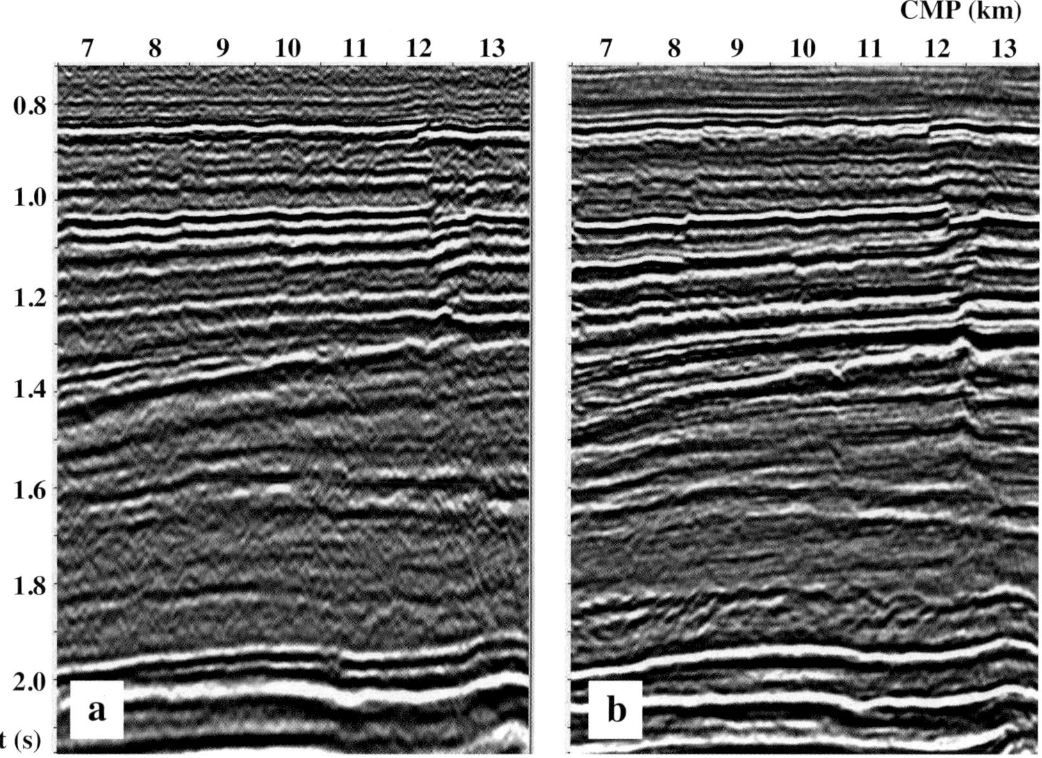

Figure 5.25: Time-migrated sections of (a) PS-waves and (b) PP-waves computed for the VTI model from Figure 5.17 (Grechka et al., 2002d). The PS image is obtained from the CCP stack in Figure 5.23b, and the PP image from the CRP (common-reflection-point) stack. The time scale of the PS section is compressed by a factor of two.

Figure 5.25 shows time-migrated images of PP- and PS-waves obtained using the anisotropic inversion results. Although the general appearance of the two sections is similar, there are some noticeable differences in the images of the shallow faults and subhorizontal events at the reservoir level. Because reflection amplitudes of PP- and PS-waves depend on different combinations of the medium parameters, PP and PS sections can provide complementary information about the subsurface.

It should be mentioned that distortions in isotropic processing of PS-waves are particularly severe for model-based methods (e.g., Figure 5.23a), including prestack depth migration. Some alternative approaches attempt to optimize sorting and stacking of PS data. Flattening of PS events is often achieved by using a moveout equation with best-fit parameters that are not derived from the estimated subsurface model. To choose ray trajectories for generating CCP gathers, positive and negative offsets can be processed and stacked separately. The main criterion is the lateral alignment of the structure between the stacks for positive and negative offsets, which requires the model to have some degree of structural complexity. Another way of searching for the optimal sorting is based on maximizing semblance in CCP gathers for several key reflectors.

Although such methods can improve the quality of the isotropic image to some extent, they cannot fully correct for the influence of anisotropy. The algorithm described here does not rely on the presence of structure, which makes it especially attractive for reservoirs in stratigraphic traps. Also, with the advent of prestack depth migration it becomes necessary to build a velocity model suitable for accurate depth imaging of both PP and PS data.

5.4 Case study from the Gulf of Mexico

The full-waveform (interferometric) implementation of the PP+PS = SS method introduced in section 4.3 was tested by Grechka and Dewangan (2003) on multicomponent 2D data (hydrophone, vertical geophone, and inline horizontal geophone) acquired in the Gulf of Mexico. The PP-to-PS event correlation (registration) was established on the near-offset stacks displayed in Figure 5.26. This procedure is more reliable in the deeper part of the section where similar structural features can be identified on both PP and PS data (red lines). Lateral heterogeneity is not as pronounced for the rest of the section, which makes event correlation (yellow lines) more difficult. Nevertheless, the two-way zero-offset SS-wave traveltime was obtained for a number of reflectors as a function of the CMP coordinate y_{CMP}:

$$t_{S0}(y_{\text{CMP}}) = 2t_{PS0}(y_{\text{CMP}}) - t_{P0}(y_{\text{CMP}}) . \tag{5.16}$$

Then the effective vertical-velocity ratio g_0 for all correlated reflections was computed from equation 5.10.

The smoothed and interpolated sections of t_{S0} and g_0 are shown in Figure 5.27. An important observation from Figure 5.27b is that the ratio g_0 at shallow depths is quite small (only about 0.15). Ignoring the possible influence of anisotropy, reflector dip, and lateral heterogeneity and using $g_0 = 0.15$ to estimate the critical (maximum) offset-to-depth ratio of ΨS-waves from equation 4.28 yields $x_{\Psi S}^{\text{crit}}/D = 2h_{\Psi S}^{\text{crit}}/D \approx 0.3$. Clearly, shear-wave NMO velocities cannot be picked from such short-spread moveouts with sufficient accuracy. The ratio g_0, however, rapidly increases with depth and reaches 0.35 at t_{P0} close to 5.5 s. The corresponding ratio $x_{\Psi S}^{\text{crit}}/D$ approaches 0.75, which is marginally acceptable for hyperbolic velocity analysis of the constructed ΨS-waves.

The computed ΨS-wave gathers in Figure 5.28 corroborate the above analysis because the maximum offset rapidly increases with the time t_{S0}. Moveout analysis of the ΨS data produces poorly resolved semblance maxima when $t_{S0} < 6$ s, which corresponds to $t_{P0} < 2.5$ s. This problem, however, is a consequence of the low V_{S0}/V_{P0} ratio in the shallow layers rather than a shortcoming of the PP+PS = SS method. The semblance maxima of the ΨS-waves are better focused for deeper events (i.e., when t_{P0} exceeds 2.5 s).

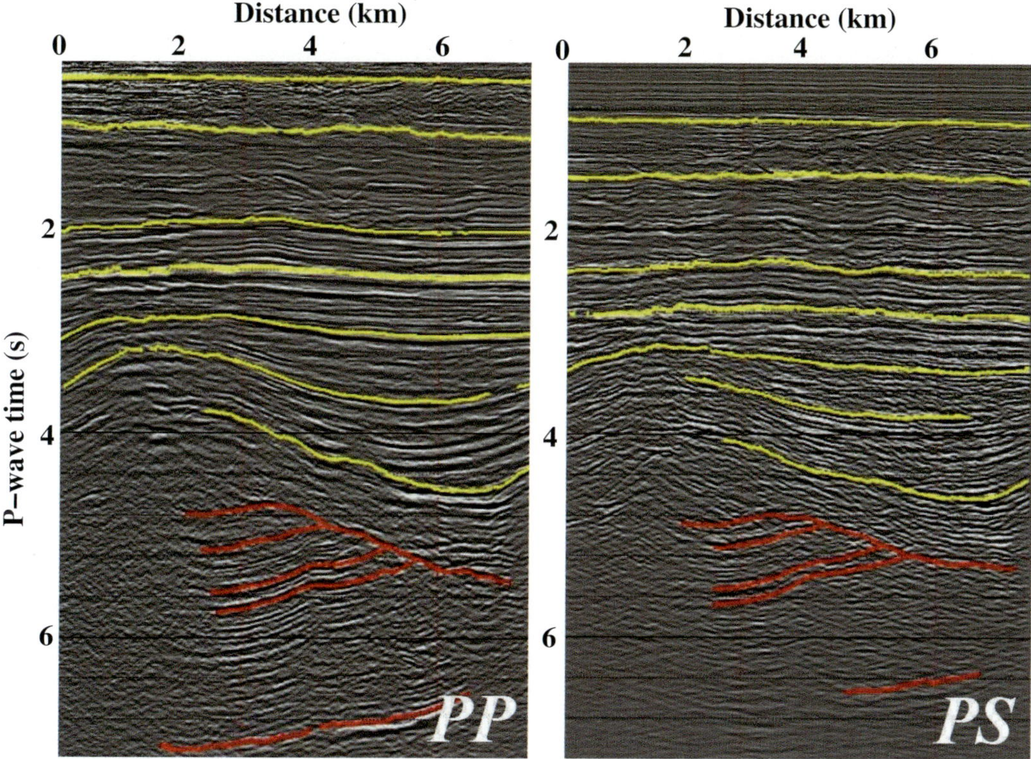

Figure 5.26: Near-offset stacks of PP and PS data used for event correlation (yellow and red lines) in the Gulf of Mexico case study. The time axis on the PS section is compressed (approximately) to the PP time so that the horizons have similar shapes on both sections.

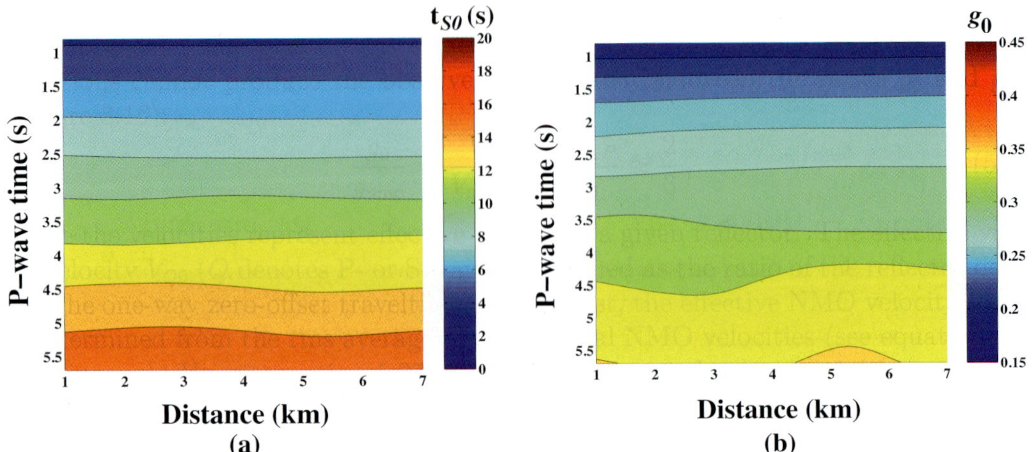

Figure 5.27: Sections of (a) the two-way zero-offset SS traveltime t_{S0} and (b) the vertical-velocity ratio g_0 (Grechka and Dewangan, 2003). The vertical axis is the two-way zero-offset PP traveltime t_{P0}.

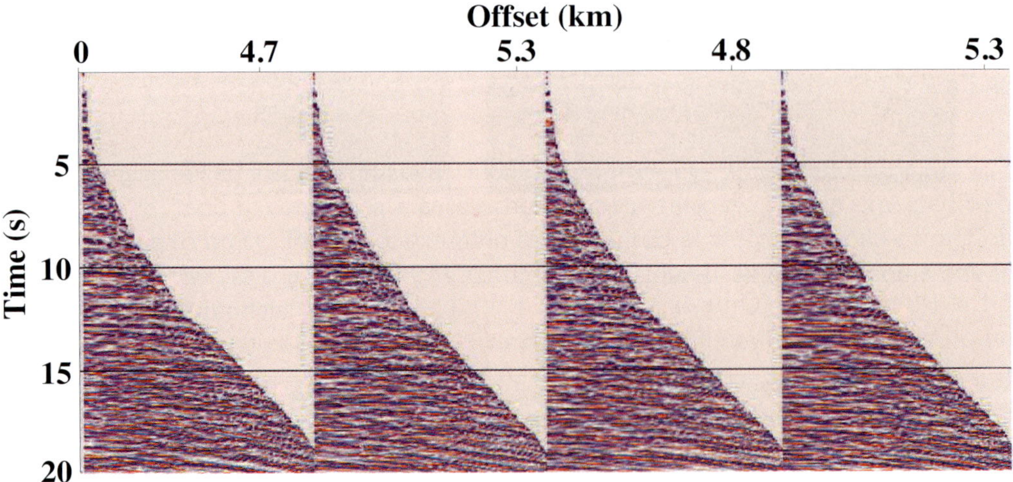

Figure 5.28: Typical CMP gathers of the constructed ΨS-waves (Grechka and Dewangan, 2003). The ΨS traveltimes reach 20 s.

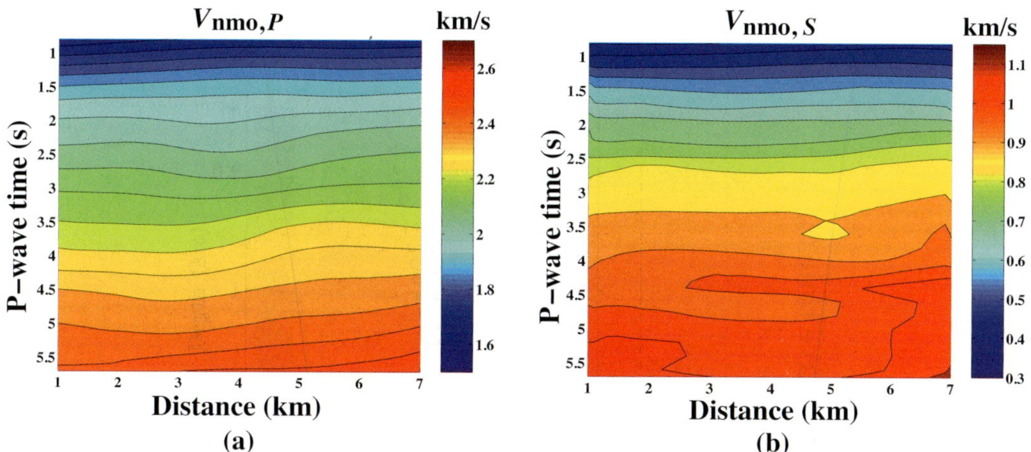

Figure 5.29: Estimated NMO (stacking) velocities of (a) PP-waves and (b) ΨS-waves (Grechka and Dewangan, 2003).

Joint hyperbolic velocity analysis of PP and ΨS data (Figure 5.29) helps identify the presence of effective anisotropy. As was the case in the Siri case study, the ratio $g_{\text{nmo}} = V_{\text{nmo},S}/V_{\text{nmo},P}$ in Figure 5.30 is consistently larger than the corresponding vertical-velocity ratio g_0 (Figure 5.27b). For the deeper part of the section where $t_{P0} > 2.5$ s and the ΨS-wave NMO velocities should be sufficiently accurate, $g_0 \approx 0.3$, whereas $g_{\text{nmo}} \approx 0.4$. This difference is too pronounced to be explained by vertical heterogeneity in isotropic media.

To find an approximate relationship between the effective anisotropy parameters, it is convenient to make the simplifying assumption of a homogeneous VTI overburden at each zero-offset time. Substituting the values g_0 and g_{nmo} into equation 5.13 and linearizing the result in ϵ and δ leads to

$$\epsilon \approx 0.04 + 1.16\delta \,. \tag{5.17}$$

Equation 5.17 is similar to the relationship between ϵ and δ for the Siri data set (equation 5.14). In both case studies, the subsurface has to be anisotropic to fit both PP and SS (ΨS) moveouts.

Unfortunately, the vertical velocities and anisotropy parameters could not be constrained without borehole or other additional information, which was not available. Instead, the effective anisotropy for each reflection event was quantified by computing the following parameter:

$$\chi \equiv \frac{1}{2}\left(\frac{g_{\text{nmo}}^2}{g_0^2} - 1\right) = \frac{\sigma - \delta}{1 + 2\delta} \,. \tag{5.18}$$

The parameter χ vanishes for a homogeneous isotropic medium. In this case study, however, $g_{\text{nmo}} > g_0$, so χ is positive and reaches large values close to 0.5 in the middle of the section (Figure 5.31). Thus, the combination of PP and PS data provided unambiguous evidence of substantial anisotropy, although the parameters ϵ and δ could not be estimated individually.

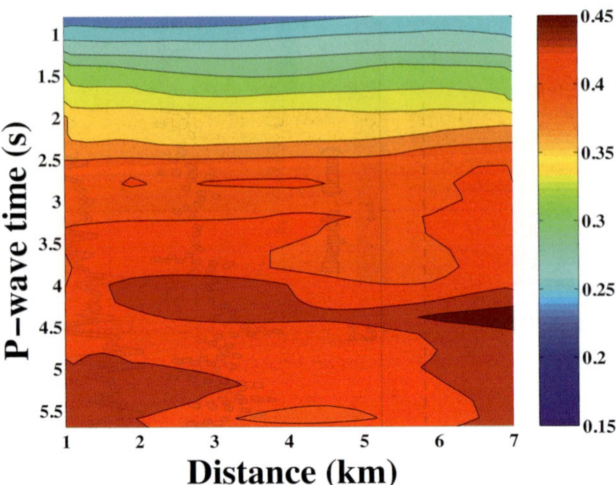

Figure 5.30: Ratio g_{nmo} of the NMO velocities of the ΨS- and PP-waves from Figure 5.29 (Grechka and Dewangan, 2003). To simplify comparison with the ratio g_0, the color scale is the same as that in Figure 5.27b.

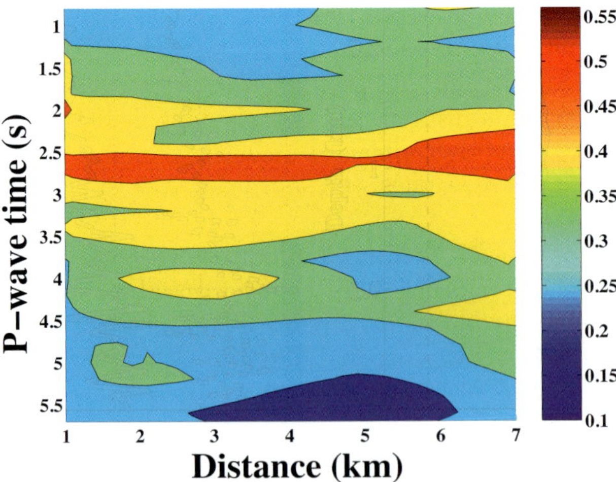

Figure 5.31: Effective anisotropy parameter χ computed from equation 5.18 (Grechka and Dewangan, 2003).

5.5 Summary

Mode-converted PS-waves recorded in multicomponent surveys are sensitive to the anisotropy parameters and can provide valuable information for anisotropic velocity analysis. For a horizontal VTI layer, supplementing PP-wave moveout with traveltimes of PSV-waves helps estimate the vertical-velocity ratio V_{P0}/V_{S0} and the NMO velocity of pure SS (SVSV) reflections (which do not have to be physically excited), as well as increase the accuracy in the inverted parameter η. The velocities V_{P0} and V_{S0} and anisotropy parameters ϵ and δ, however, remain poorly constrained even by long-offset PP and PS reflection traveltimes recorded on spreadlengths as large as three times the reflector depth. Therefore, although horizontally layered VTI models obtained from moveout inversion of PP and PS data remove misties between PP and PS sections and are suitable for time processing, they can have distorted vertical velocities and reflector depths. This ambiguity in depth-domain velocity analysis of reflection data can be overcome by combining nonhyperbolic moveout of PP-waves and pure SS-waves (Tsvankin and Thomsen, 1995).

Next, we discussed joint traveltime inversion of PP and PS (PSV) data for more complicated, layered TTI models with dipping or curved interfaces. The PP+ PS = SS method introduced in Chapter 4 can be used to replace PS-wave traveltimes, which are generally asymmetric with respect to zero offset, with the symmetric (hyperbolic on conventional spreads) moveout of the corresponding SS reflections. Then the NMO ellipses, zero-offset traveltimes, and reflection time slopes of the recorded PP-waves and computed SS-waves are inverted for the medium parameters. For a homogeneous TI layer, the stability of parameter estimation usually increases with the tilt ν of the symmetry axis and reflector dip. Whereas the inversion is quite stable when the symmetry axis is horizontal (i.e., the medium is HTI), for small and moderate values of ν it is essential to have a priori constraints on the tilt. If the layer is known to be VTI, a mild dip of $15° - 20°$ is sufficient to obtain the parameters V_{P0}, V_{S0}, ϵ, and δ and, therefore, the layer thickness. To carry out the inversion for multiple TI layers, we employed multicomponent stacking-velocity tomography, which represents a generalization of the P-wave tomographic algorithm introduced in Chapter 2. The feasibility of interval parameter estimation depends on the tilt ν, the range of available reflector dips, and information about the symmetry-axis orientation.

The PP+ PS = SS method and multicomponent stacking-velocity tomography were applied to 2D PP and PS data acquired over the Siri reservoir in the North Sea. Comparison of the ratios of the PP- and SS-wave NMO velocities with the corresponding vertical-velocity ratios provided clear evidence of anisotropy in the section above the reservoir. Because the subsurface was well-approximated by a stack of VTI layers with gentle dips, reflection data could not constrain the vertical velocities and the parameters ϵ and δ, and the inversion produced a family of equivalent anisotropic models that fit PP and PS reflection traveltimes equally well. Independent depth information from check shots helped remove the nonuniqueness and allowed estimation of the interval Thomsen parameters near the borehole location. Throughout the section, $\epsilon > \delta$ (i.e., $\eta > 0$), with the largest values of both parameters observed in the Oligocene layer (at depths 0.7 – 1.5 km) where ϵ reaches almost 0.25. PS-wave time sections computed using the obtained VTI model have a much higher quality than those produced by conventional isotropic processing.

In the second case study of this chapter, PP and PS reflections recorded on a 2D line in the Gulf of Mexico were processed using the full-waveform version of the PP+PS = SS method. The quality of CMP gathers of the generated ΨS-waves was sufficient for hyperbolic semblance analysis, although the narrow offset range of shallow ΨS events precluded accurate estimation of their moveout velocities. For the deeper part of the section, the NMO velocities of the recorded PP-waves and computed ΨS-waves could not be reconciled with the P-to-S vertical-velocity ratios without making the model anisotropic, a conclusion also drawn for the Siri data set. Due to the absence of borehole information needed to resolve the Thomsen parameters, anisotropy was quantified by computing the effective parameter $\chi \approx \sigma - \delta$ from the NMO velocities and vertical traveltimes of the PP- and ΨS-waves. The values of χ are positive throughout the section and reach 0.5, which indicates substantial anisotropy in the sedimentary layers.

While replacing acquired PS-waves with constructed SS-waves is advantageous from the processing viewpoint and provides input data for NMO-velocity inversion, the PP+PS = SS method does not preserve information about the moveout asymmetry of mode conversions. Dewangan and Tsvankin (2006a,b; their results are not reproduced here) demonstrated that the addition of moveout-asymmetry attributes of PS-waves to conventional-spread PP and SS moveout data increases the stability of traveltime inversion for tilted TI media. For a horizontal TI layer with a tilted symmetry axis (which describes a system of dipping fractures in isotropic host rock), the method of Dewangan and Tsvankin (2006a) accurately estimates the parameters ν, V_{P0}, V_{S0}, ϵ, and δ using 2D data in the symmetry-axis plane. These theoretical results were corroborated by Dewangan et al. (2006), who conducted a physical-modeling experiment on a phenolic sample manufactured to simulate the effective TTI medium formed by a dipping fracture set. The contribution of the PS-wave asymmetry attributes, however, is less beneficial for a dipping TI layer with symmetry axis orthogonal to its boundaries (Dewangan and Tsvankin, 2006b). The combination of long-spread reflection traveltimes of PP- and PS-waves for that practically important model constrains the symmetry-direction velocities and parameter δ only when the dip (equal to tilt) exceeds $30° – 40°$, whereas the accuracy in ϵ remains marginal even for large dips.

Appendices for Chapter 5

5A Effective parameter δ for layered VTI media

As an example of the relationship between the effective and interval anisotropy parameters, consider the effective anisotropy parameter $\langle\delta\rangle$ for a stack of horizontal, homogeneous VTI layers. (In this appendix, $\langle a\rangle$ denotes the effective value of the parameter a.) The effective P-wave NMO velocity $\langle V_{\text{nmo},P}\rangle$ can be found by Dix (1955) averaging of the interval values $V_{\text{nmo},P,n}$:

$$\langle V_{\text{nmo},P}\rangle^2 = \frac{1}{\langle\tau\rangle}\sum_n \tau_n\, V^2_{\text{nmo},P,n}, \tag{5.19}$$

where τ_n is the interval one-way zero-offset traveltime, and $\langle\tau\rangle$ is the total (effective) traveltime:

$$\langle\tau\rangle = \sum_n \tau_n\,. \tag{5.20}$$

$V_{\text{nmo},P,n}$ depends on the interval vertical velocity $V_{P0,n}$ and anisotropy parameter δ_n:

$$V^2_{\text{nmo},P,n} = V^2_{P0,n}\,(1+2\delta_n)\,. \tag{5.21}$$

The products of the velocities $V_{P0,n}$ and zero-offset traveltimes τ_n yield the layer thicknesses,

$$d_n = V_{P0,n}\,\tau_n\,. \tag{5.22}$$

The effective vertical velocity can be written as

$$\langle V_{P0}\rangle = \frac{\langle d\rangle}{\langle\tau\rangle} = \frac{\sum\limits_n d_n}{\sum\limits_n \tau_n}\,. \tag{5.23}$$

Using equations 5.22 and 5.23, we find

$$\left[\frac{1}{\langle V_{P0}\rangle}\right]^{-1} = \langle V_{P0}\rangle = \frac{\langle d\rangle}{\langle\tau\rangle} = \frac{\langle d\rangle}{\sum\limits_n \dfrac{d_n}{V_{P0,n}}} = \left[\frac{1}{\langle d\rangle}\sum_n \frac{d_n}{V_{P0,n}}\right]^{-1}. \tag{5.24}$$

Equation 5.24 shows that the effective vertical velocity $\langle V_{P0}\rangle$ expressed through the layer thicknesses is the *harmonic* average of the interval vertical velocities $V_{P0,n}$.

The effective parameter $\langle\delta\rangle$ is defined using equation 5.21,

$$\langle V_{\text{nmo},P}\rangle^2 = \langle V_{P0}\rangle^2\,(1+2\,\langle\delta\rangle)\,. \tag{5.25}$$

To find the relationship between $\langle\delta\rangle$ and the interval coefficients δ_n, we substitute equations 5.21, 5.22, 5.23, and 5.25 into the Dix formula 5.19:

$$\langle V_{P0}\rangle\,(1+2\,\langle\delta\rangle) = \frac{1}{\langle d\rangle}\sum_n d_n\,V_{P0,n}\,(1+2\delta_n)\,, \tag{5.26}$$

or

$$1+2\,\langle\delta\rangle \;=\; \frac{\dfrac{1}{\langle d\rangle}\displaystyle\sum_n d_n\,V_{P0,n}}{\langle V_{P0}\rangle} + 2\,\frac{\dfrac{1}{\langle d\rangle}\displaystyle\sum_n d_n\,V_{P0,n}\,\delta_n}{\langle V_{P0}\rangle}\,. \tag{5.27}$$

Note that equation 5.27 is exact. The first term on the right-hand side is the ratio of the arithmetic and harmonic averages of the vertical velocities (scaled by the layer thicknesses). From the Cauchy-Schwarz inequality, this ratio is always greater than or equal to unity (Grechka and Tsvankin, 2002b). Therefore, $\langle\delta\rangle$ satisfies the following inequality:

$$\langle\delta\rangle \ge \frac{\dfrac{1}{\langle d\rangle}\displaystyle\sum_n d_n\,V_{P0,n}\,\delta_n}{\langle V_{P0}\rangle}\,. \tag{5.28}$$

In particular, if δ_n is constant throughout the section,

$$\delta_1 = \delta_2 = \ldots = \delta_n = \tilde{\delta}\,, \tag{5.29}$$

inequality 5.28 yields

$$\langle\delta\rangle \ge \tilde{\delta}\,. \tag{5.30}$$

The equality $\langle\delta\rangle = \tilde{\delta}$ is reached only if the interval velocity $V_{P0,n}$ is constant, which means that the stack of the layers degenerates into a homogeneous medium. Hence, the effective $\langle\delta\rangle$ *overestimates* the interval value of this parameter.

In the special case of isotropy ($\tilde{\delta}=0$),

$$\langle\delta\rangle \ge 0\,. \tag{5.31}$$

According to inequality 5.31, isotropic layering creates an effective VTI medium with a positive parameter δ. For a more detailed discussion of the model with $\tilde{\delta}=0$, see the paper on "processing-induced anisotropy" by Grechka and Tsvankin (2002b).

The results of this appendix may partially explain the known discrepancy between laboratory measurements on shale cores, which give both positive and negative δ values (Vernik and Liu, 1997; Wang, 2002), and larger, predominantly positive estimates of δ obtained from reflection seismic data (Tsvankin, 2005).

5A.1 Two VTI layers

Next, we discuss the special case of two horizontal VTI layers with equal thickness ($d_1 = d_2$). Then equation 5.27 reduces to

$$1+2\,\langle\delta\rangle = \frac{V_{P0,1}+V_{P0,2}}{4V_{P0,1}V_{P0,2}}\left[V_{P0,1}+V_{P0,2}+2\,(V_{P0,1}\,\delta_1+V_{P0,2}\,\delta_2)\right]. \tag{5.32}$$

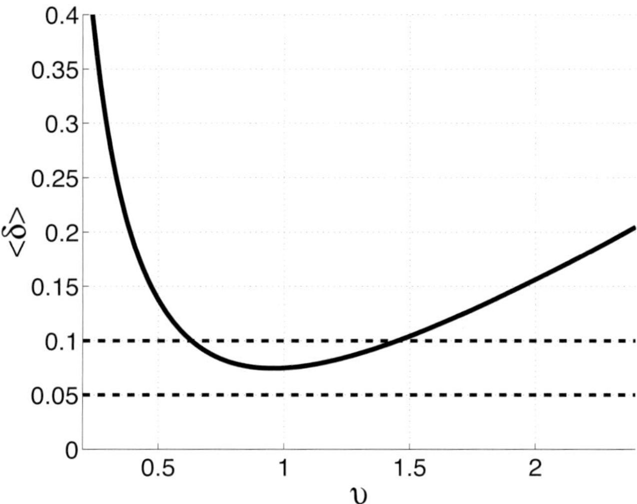

Figure 5.32: Function $\langle \delta(\upsilon) \rangle$ (solid line) computed from equation 5.35 for two horizontal VTI layers with the interval parameters δ_1=0.05 and δ_2=0.1 marked by the dashed lines (Grechka et al., 2002b).

Introducing the ratio of the vertical velocities

$$\upsilon = \frac{V_{P0,2}}{V_{P0,1}}, \tag{5.33}$$

we rewrite equation 5.32 as

$$1 + 2\langle \delta \rangle = \frac{1+\upsilon}{4\upsilon} \left[1 + \upsilon + 2\,(\delta_1 + \upsilon\,\delta_2) \right], \tag{5.34}$$

and

$$\langle \delta \rangle = \frac{(1-\upsilon)^2}{8\upsilon} + \frac{1+\upsilon}{4\upsilon}\,(\delta_1 + \upsilon\,\delta_2)\,. \tag{5.35}$$

For a positive υ, the function $\langle \delta \rangle$ from equation 5.35 has only one minimum, at

$$\upsilon = \sqrt{\frac{1+2\delta_1}{1+2\delta_2}}\,. \tag{5.36}$$

The asymptotic values of $\langle \delta \rangle$ are:

$$\lim_{\upsilon \to 0} \langle \delta \rangle = \infty \quad \text{and} \quad \lim_{\upsilon \to \infty} \langle \delta \rangle = \infty\,. \tag{5.37}$$

Therefore, for a large velocity contrast between the two layers, the effective parameter $\langle \delta \rangle$ significantly exceeds both interval values $[\langle \delta \rangle > \max(\delta_1, \delta_2)]$.

This analysis is confirmed by the numerical example in Figure 5.32, which shows that $\langle \delta \rangle$ is bounded by the interval parameters δ_1 and δ_2 only for moderate velocity variations when υ does not deviate from unity by more than ± 0.4. For the velocity ratio $\upsilon = 0.5$, $\langle \delta \rangle \approx 0.14$, which is much larger than the maximum interval parameter $\delta_2 = 0.1$.

5B Influence of layering on the velocity ratios

Suppose the overburden above the Siri reservoir (section 5.3) is isotropic and vertically heterogeneous. The goal of this appendix is to show that realistic vertical velocity variations cannot produce the observed discrepancy between the ratios g_0 and $g_{\rm nmo}$ (Figure 5.16):

$$\frac{g_0}{g_{\rm nmo}} = \frac{V_{S0}}{V_{{\rm nmo},S}} \frac{V_{{\rm nmo},P}}{V_{P0}} \approx \frac{2}{3}\,, \tag{5.38}$$

where the velocities represent effective values for a given reflector. The effective vertical velocity V_{Q0} (Q denotes P- or S-waves) is defined as the ratio of the reflector depth and the one-way zero-offset traveltime. In contrast, the effective NMO velocity $V_{{\rm nmo},Q}$ is determined from the rms average of the interval NMO velocities (see equation 5.19 for P-waves). Using equation 5.27 with the parameter δ set to zero (the medium is assumed to be isotropic), we find

$$\sqrt{\frac{V_{{\rm nmo},P}}{V_{P0}}} = \frac{\dfrac{1}{\langle d \rangle} \displaystyle\sum_n d_n V_{P0,\,n}}{\left[\dfrac{1}{\langle d \rangle} \displaystyle\sum_n \dfrac{d_n}{V_{P0,\,n}}\right]^{-1}}\,. \tag{5.39}$$

Equation 5.39 remains valid for S-waves if V_{P0} is replaced with V_{S0}. As discussed in Appendix 5A, the right-hand side of equation 5.39 cannot be smaller than unity. Therefore, the effective NMO velocity in isotropic media is always greater than or equal to the effective vertical velocity:

$$\frac{V_{S0}}{V_{{\rm nmo},S}} \le 1 \quad \text{and} \quad \frac{V_{{\rm nmo},P}}{V_{P0}} \ge 1\,. \tag{5.40}$$

In the presence of vertical heterogeneity alone, the terms $V_{S0}/V_{{\rm nmo},S}$ and $V_{{\rm nmo},P}/V_{P0}$ in equation (5.38) tend to compensate each other, and the ratio $g_0/g_{\rm nmo}$ is expected to be relatively close to unity. Below we examine several interval vertical-velocity functions and show that the ratio $g_0/g_{\rm nmo} \approx 2/3$ should be considered as highly anomalous for realistic isotropic models.

5B.1 Numerical examples

We begin with the simplest type of vertically heterogeneous, isotropic media in which both interval velocities V_{P0} and V_{S0} are linear functions of depth (Figure 5.33a). The P-wave velocity at the ocean bottom is $V_{P0}(0) = 1.5$ km/s, while the corresponding value for S-waves, $V_{S0}(0) = 0.2$ km/s, was obtained by Muyzert (2000) for underwater sediments. The P- and S-wave velocity gradients are selected in such a way that V_{P0} and V_{S0} at the depth $z = 2$ km are close to the interval velocities for the horizon immediately above the top of the Balder formation.

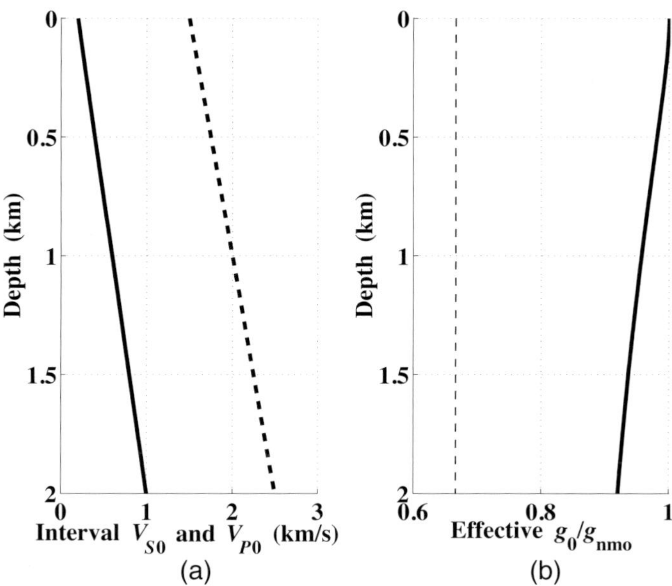

Figure 5.33: (a) Linear interval velocity functions $V_{P0}(z) = 1.5 + 0.5z$ (dashed line) and $V_{S0}(z) = 0.2 + 0.4z$ (solid). (b) The ratio $g_0/g_{\rm nmo}$ (solid) computed for the velocities on plot (a) and the reference value $g_0/g_{\rm nmo} = 2/3$ (dashed) (Grechka et al., 2002c).

For the depth range in Figure 5.33a, the S-wave interval velocity increases by a factor of five compared to just 1.7 for the P-wave velocity. Although the vertical heterogeneity for S-waves is much greater than that for P-waves, it still cannot make the ratio $g_0/g_{\rm nmo}$ smaller than 0.92 (Figure 5.33b). Therefore, vertical-velocity gradients are clearly insufficient to bring the ratio $g_0/g_{\rm nmo}$ down to values close to 2/3.

To make the interval velocity variations more realistic, we added a random component to the linear velocity functions (Figure 5.34a). The random variations, however, have a relatively minor influence on the ratio $g_0/g_{\rm nmo}$ (compare Figures 5.33b and 5.34b).

The ratio $g_0/g_{\rm nmo}$ can be reduced by further increasing the magnitude of the shear-wave velocity heterogeneity. For example, it is possible for the velocity V_{S0} near the ocean bottom to be as small as 25 m/s (Figure 5.35a). Such uncommonly low velocities were reported by Ayres and Theilen (1999) in unconsolidated ocean-bottom sediments. Although the ratio $g_0/g_{\rm nmo}$ does drop below 2/3 immediately under the low-velocity layer, it quickly increases with depth to values exceeding 0.8, as the influence of the rapid velocity variation in the shallow part of the section becomes less pronounced (Figure 5.35b). Therefore, we conclude that it is hardly possible to find a realistic vertically heterogeneous, isotropic model for which the ratio $g_0/g_{\rm nmo}$ stays as small as 2/3 over a wide depth range.

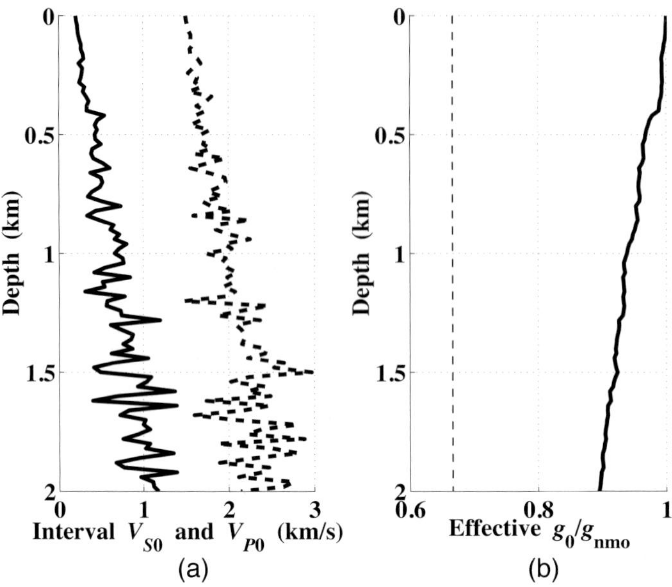

Figure 5.34: Same as Figure 5.33, but with depth-dependent random variations added to the interval velocities (Grechka et al., 2002c).

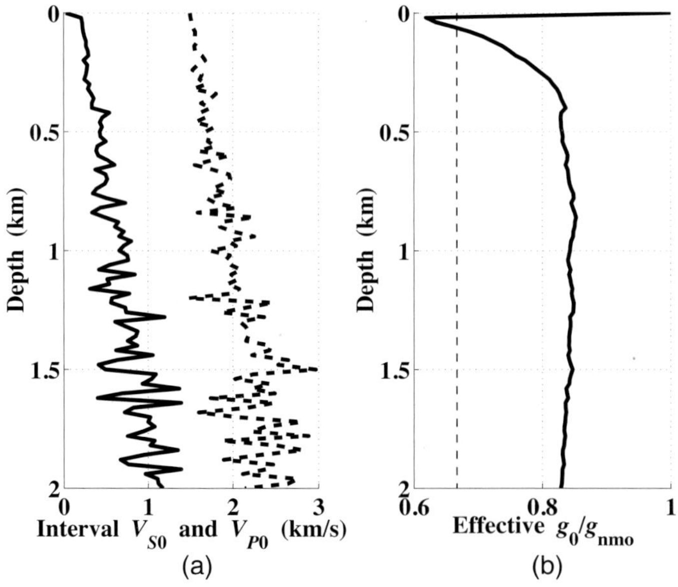

Figure 5.35: Same as Figure 5.34, but the shear-wave velocity model contains a thin underwater layer ($0 \leq z \leq 20$ m) with an extremely low interval velocity, V_{S0}=25 m/s (Grechka et al., 2002c).

Chapter 6

Moveout inversion of multicomponent data for lower symmetries

The results of Chapter 5 demonstrate the benefits of combining P-wave data with reflection moveout of shear or mode-converted waves in parameter estimation for TI media. The advantages of multicomponent velocity analysis are even more crucial for lower-symmetry models described by a larger number of independent anisotropy parameters. This chapter is focused on moveout inversion of wide-azimuth, multicomponent data from layered orthorhombic (sections 6.1 – 6.3) and monoclinic (section 6.4) media.

Velocity model building using PP and PS traveltimes is especially attractive because it does not require shear-wave excitation. We begin the first section by reviewing the properties of reflection moveout of converted waves for models composed of horizontal layers with a horizontal symmetry plane. As demonstrated in Chapter 1, PS traveltimes for such media are *reciprocal* with respect to the source and receiver positions and can be described by the conventional $t^2(x^2)$ series. Furthermore, there exists a simple relationship between the NMO ellipses of pure and converted waves which, combined with the generalized Dix differentiation, makes it possible to compute the interval NMO ellipses of the split shear waves S_1S_1 and S_2S_2 from PP and PS (PS_1 and PS_2) data. Then the interval PP- and SS-wave NMO ellipses are inverted for the parameters of orthorhombic or monoclinic layers.

This algorithm is tested on multicomponent (PP and PS) physical-modeling data acquired in several azimuthal directions over a block of orthorhombic composite material. The computed symmetry-plane NMO velocities of PP- and SS-waves are combined with the known reflector depth to estimate eight (out of nine) medium parameters. The obtained orthorhombic model accurately predicts the moveout curves of the pure SS-waves and is in excellent agreement with direct measurements of the horizontal velocities.

Next, we explore the possibility of using reflections from dipping interfaces in multicomponent velocity analysis for orthorhombic media. If the medium above the dipping reflector is homogeneous, under certain conditions it is possible to invert wide-azimuth PP and PS (or PP and SS) data for Tsvankin's parameters and reflector depth. To perform parameter estimation for a stack of homogeneous layers separated by plane or curved interfaces, we generalize multicomponent stacking-velocity tomography (see Chapter 5) for orthorhombic models and study its performance on noise-contaminated input data.

A certain class of anisotropic formations that do not have vertical planes of symmetry can be described by an effective monoclinic medium with a horizontal symmetry plane. In the last section we introduce Thomsen-style notation for monoclinic symmetry and develop a moveout-inversion algorithm for horizontally layered monoclinic models. If the vertical scale of the section is known, the NMO ellipses of PP-waves and two split SS-waves constrain all but one interval anisotropy parameters.

6.1 Inversion of PP and PS data for layered orthorhombic media

Orthorhombic models have three mutually orthogonal symmetry planes (Figure 2.36) in which the phase velocities and plane-wave polarizations can be obtained from the corresponding VTI equations. This equivalence allowed Tsvankin (1997a, 2005) to introduce the Thomsen-style notation for orthorhombic media described in section 2.5 and Appendix 1B. Tsvankin's notation is particularly advantageous for moveout inversion because it reduces the number of parameters responsible for P-wave traveltimes from nine to six and simplifies NMO-velocity equations both within and outside the symmetry planes.

In this section, we consider a stack of horizontal orthorhombic layers with a horizontal symmetry plane. Then the semiaxes of the interval P-wave NMO ellipse in each layer are aligned with the vertical symmetry planes (this is the case for other modes as well) and depend on just the vertical velocity V_{P0} and anisotropy parameters $\delta^{(1)}$ and $\delta^{(2)}$ (equation 2.62):

$$V^{(1)}_{\text{nmo},P} = V_{P0}\sqrt{1+2\delta^{(1)}}\,, \tag{6.1}$$

$$V^{(2)}_{\text{nmo},P} = V_{P0}\sqrt{1+2\delta^{(2)}}\,. \tag{6.2}$$

The superscript "1" corresponds to the symmetry plane $[x_2, x_3]$ and "2" to the $[x_1, x_3]$-plane (the superscripts denote the axis normal to each plane). Without knowledge of the vertical velocity, the NMO ellipse can be used to estimate only the orientation of the symmetry planes and the NMO velocities ($V^{(1,2)}_{\text{nmo}}$) within them.

As discussed in Chapter 3, additional information about anisotropy is provided by long-spread P-wave traveltimes, which depend on the anellipticity parameters $\eta^{(1,2,3)}$. The interval values of $V^{(1,2)}_{\text{nmo}}$ and $\eta^{(1,2,3)}$ can be obtained from semblance-based nonhyperbolic moveout inversion of wide-azimuth P-wave data. In combination with the azimuths of the vertical symmetry planes, the parameters $V^{(1,2)}_{\text{nmo}}$ and $\eta^{(1,2,3)}$ are sufficient to carry out time-domain processing of P-waves in orthorhombic media (see section 3.5). The vertical velocity V_{P0} and anisotropy parameters $\epsilon^{(1,2)}$ and $\delta^{(1,2,3)}$, however, cannot be found from P-wave reflection traveltimes alone. Also, P-wave kinematic signatures are independent of the parameters V_{S0}, $\gamma^{(1)}$, and $\gamma^{(2)}$ responsible for the shear-wave velocities in the coordinate directions. Therefore, it is important to investigate the possibility of supplementing P-wave moveout with traveltimes of mode-converted waves.

6.1.1 Properties of PS-wave moveout

In principle, information about shear-wave velocity can be extracted from PS data by applying the PP+PS=SS method and generating ΨS-waves, which have the kinematics of pure SS-wave primaries; this approach was explored in Chapters 4 and 5. Here, however, we consider horizontally layered models with a horizontal symmetry plane, in which reflection traveltime of PS- and SP-waves is an even function of offset and can be described by the conventional traveltime series $t^2(x^2)$ (see section 1.6 and Appendix 1D). The absence of moveout asymmetry allows us to work directly with recorded PS (PS_1 and PS_2) data without computing ΨS-waves. Although mode conversion at the reflector generally enhances deviations from hyperbolic moveout, PS-wave NMO velocity can be estimated in a reliable fashion by applying nonhyperbolic moveout analysis.

NMO velocity of converted waves has the same azimuthal dependence as for pure modes (equation 1.65):

$$V_{\text{nmo}}^{-2}(\alpha) = W_{11}^{(PS)} \cos^2\alpha \,+\, 2\,W_{12}^{(PS)} \sin\alpha\,\cos\alpha \,+\, W_{22}^{(PS)} \sin^2\alpha\,, \tag{6.3}$$

where α is the source-to-receiver azimuth. In contrast to pure modes, the 2×2 symmetric matrix $\mathbf{W}^{(PS)}$ is defined through the *two-way* reflection traveltime of the converted wave. If the eigenvalues of $\mathbf{W}^{(PS)}$ are positive (a typical case), equation 6.3 describes an ellipse in the horizontal plane.

The matrix $\mathbf{W}^{(PS)}$ can be expressed through the corresponding matrices of the pure PP ($\mathbf{W}^{(P)}$) and SS ($\mathbf{W}^{(S)}$) reflections as follows (equation 1.66):

$$2t_{PS0}\left[\mathbf{W}^{(PS)}\right]^{-1} = t_{P0}\left[\mathbf{W}^{(P)}\right]^{-1} + t_{S0}\left[\mathbf{W}^{(S)}\right]^{-1}, \tag{6.4}$$

where $t_{PS0} = (t_{P0} + t_{S0})/2$ is the zero-offset traveltime of the PS-wave, and t_{P0} and t_{S0} are the two-way zero-offset times of the pure modes. Equation 6.4 helps compute the effective NMO ellipse of either split SS-wave ($\mathbf{W}^{(S)}$) from PP and PS (PS_1 and PS_2) data in layered models with a horizontal symmetry plane. Then the interval NMO ellipses of PP- and SS-waves are found by applying the generalized Dix differentiation (equation 1.32) to the effective ellipses for the top and bottom of the layer of interest.

Alternatively, the azimuthally varying interval PP and PS traveltimes can be obtained from the velocity-independent layer-stripping method (VILS) of Dewangan and Tsvankin (2006c), described for pure modes in section 3.4. After estimating the interval matrices $\mathbf{W}^{(P)}$ and $\mathbf{W}^{(PS)}$ from azimuthal moveout analysis, they are substituted into equation 6.4 to compute the interval NMO ellipse of SS-waves.

6.1.2 Interval parameter estimation

The techniques discussed above produce the interval PP- and SS-wave NMO ellipses, which can be jointly inverted for the interval medium parameters. This methodology remains valid for orthorhombic models with *depth-varying* symmetry-plane azimuths, as well as for monoclinic layers with a horizontal symmetry plane.

Pure and converted waves in a homogeneous medium

Here, we describe P-to-S mode conversions and the relationship between the NMO velocities of PP-, PS-, and SS-waves for a single orthorhombic layer. In a medium with the vertical symmetry planes $[x_1, x_3]$ and $[x_2, x_3]$, the NMO ellipse of any mode is fully defined by the symmetry-plane NMO velocities:

$$W_{11}^{(Q)} = \frac{1}{\left[V_{\text{nmo},Q}^{(2)}\right]^2}, \tag{6.5}$$

$$W_{22}^{(Q)} = \frac{1}{\left[V_{\text{nmo},Q}^{(1)}\right]^2}, \tag{6.6}$$

where Q corresponds to PP-, SS-, or PS-waves; the off-diagonal element $W_{12}^{(Q)} = 0$. When all matrices $\mathbf{W}^{(Q)}$ are diagonal, equation 6.4 splits into two separate equations for the symmetry-plane NMO velocities:

$$2t_{PS0}\left[V_{\text{nmo},PS}^{(i)}\right]^2 = t_{P0}\left[V_{\text{nmo},P}^{(i)}\right]^2 + t_{S0}\left[V_{\text{nmo},S}^{(i)}\right]^2, \qquad (i = 1, 2). \tag{6.7}$$

As could be expected from the kinematic equivalence between the symmetry planes of orthorhombic and VTI media, equation 6.7 has the same form as the relationship between the corresponding NMO velocities for vertical transverse isotropy (equation 5.3). Equation 6.7 helps obtain the symmetry-plane shear-wave NMO velocities from PP and PS data:

$$\left[V_{\text{nmo},S}^{(i)}\right]^2 = \frac{2t_{PS0}\left[V_{\text{nmo},PS}^{(i)}\right]^2 - t_{P0}\left[V_{\text{nmo},P}^{(i)}\right]^2}{2t_{PS0} - t_{P0}}. \tag{6.8}$$

When applying equation 6.8, one has to take into account that only one converted wave (PSV) can be generated in each vertical symmetry plane. To better explain the nature of mode conversions in orthorhombic media, we need to review some relevant polarization properties of the split shear waves. A cartoon of typical phase-velocity sheets in orthorhombic media is shown in Figure 2.36. The outer (P-wave) phase-velocity surface is usually separated from the two sheets corresponding to the S-waves. The two shear-wave phase velocities coincide in certain directions corresponding to so-called "point" (also called "conical") singularities (e.g., Crampin, 1991), such as point A in Figure 2.36. The shear wavefronts near point singularities have complicated shapes and cannot be accurately described by ray theory. We assume that singularities are far enough from the vertical so that the fast shear wave S$_1$ (we denote its vertical velocity by V_{S1}) can be separated from the slow wave S$_2$ (vertical velocity V_{S2}) for a range of source-receiver offsets sufficient to estimate the NMO velocities.

In contrast to VTI media, for orthorhombic and HTI symmetry S$_1$- and S$_2$-waves cannot be classified as "SV" or "SH" modes. Suppose that the vertically traveling fast wave S$_1$ is polarized in the x_2-direction and, therefore, represents a pure transverse (SH) mode for any phase direction in the $[x_1, x_3]$-plane. As we move along the phase-velocity surface of the S$_1$-wave around the x_3-axis to the $[x_2, x_3]$-plane (Figure 2.36), its polarization changes from transverse (cross-plane) to in-plane (in other words, from SH in the $[x_1, x_3]$-plane to SV in the $[x_2, x_3]$-plane). Likewise, the polarization of the S$_2$-wave changes from SV in the $[x_1, x_3]$-plane to SH in the $[x_2, x_3]$-plane.

Since the phase-velocity sheets (and the corresponding wavefronts) of the waves P, S$_1$, and S$_2$ are continuous, from the kinematic viewpoint the converted waves PS$_1$ and PS$_2$ can be recorded in all azimuthal directions. However, P-waves propagating in either vertical symmetry plane of a horizontal orthorhombic layer cannot generate the converted PSH-wave because the particle motion should be confined to the incidence plane. Thus, the PS$_1$-wave does not exist in the $[x_1, x_3]$-plane, while the PS$_2$-wave cannot be excited in the $[x_2, x_3]$-plane. Moreover, PSH-type reflections will also be weak near those symmetry planes because shear-wave polarizations change in a continuous fashion. Hence, the combination of PP and PS-waves is insufficient to directly determine the symmetry-plane NMO velocities for all three pure reflection modes. Nonetheless, the NMO velocities of these nonexistent converted waves in the symmetry planes can be reconstructed from moveout measurements in other azimuthal directions by taking advantage of the known (elliptical) azimuthal dependence of the NMO-velocity function.

Parameter estimation using NMO ellipses

Exact expressions for the interval NMO ellipses of the split SS-waves can be found from the kinematic analogy with vertical transverse isotropy. The S$_1$-wave propagating in the $[x_1, x_3]$-plane and polarized in the x_2-direction is equivalent to the SH-wave in VTI media, while in the $[x_2, x_3]$-plane it is equivalent to the SV-wave. In contrast, the slow mode S$_2$ is equivalent to the SV-wave in the $[x_1, x_3]$-plane and to the SH-wave in the $[x_2, x_3]$-plane. These properties have a direct bearing on the form of the shear-wave NMO velocities listed below (as before, the superscript "1" corresponds to the $[x_2, x_3]$-plane and "2" to the $[x_1, x_3]$-plane).

$$V^{(1)}_{\text{nmo},S1} = V_{S1}\sqrt{1 + 2\sigma^{(1)}}\,, \tag{6.9}$$

$$V^{(2)}_{\text{nmo},S1} = V_{S1}\sqrt{1 + 2\gamma^{(2)}}\,, \tag{6.10}$$

$$V^{(1)}_{\text{nmo},S2} = V_{S2}\sqrt{1 + 2\gamma^{(1)}}\,, \tag{6.11}$$

$$V^{(2)}_{\text{nmo},S2} = V_{S2}\sqrt{1 + 2\sigma^{(2)}}\,, \tag{6.12}$$

where

$$\sigma^{(1)} = \left(\frac{V_{P0}}{V_{S1}}\right)^2 \left(\epsilon^{(1)} - \delta^{(1)}\right), \qquad \sigma^{(2)} = \left(\frac{V_{P0}}{V_{S2}}\right)^2 \left(\epsilon^{(2)} - \delta^{(2)}\right); \tag{6.13}$$

the notation is described in Appendix 1B. Since we assume the slow S-wave at vertical incidence to be polarized in the x_1-direction, its vertical velocity V_{S2} is equal to V_{S0} in Tsvankin's notation. According to the parameter definitions, the fast shear-wave vertical velocity can be expressed as

$$V_{S1} = V_{S0} \sqrt{\frac{1 + 2\gamma^{(1)}}{1 + 2\gamma^{(2)}}} \,. \tag{6.14}$$

Combined with equations 6.1 and 6.2 for the NMO ellipse of PP-waves, equations 6.9 – 6.12 provide an analytic basis for interval moveout inversion. Note that the NMO ellipses in a horizontal orthorhombic layer depend on eight (out of nine) medium parameters. The parameter $\delta^{(3)}$ (or the stiffness c_{12}), which is defined in the horizontal symmetry plane, cannot be found from near-vertical measurements of normal moveout. Another important observation, which follows from equation 6.14, is that the SH-wave NMO velocities in the symmetry planes (equations 6.10 and 6.11) are identical,

$$V^{(2)}_{\mathrm{nmo},S1} = V^{(1)}_{\mathrm{nmo},S2} = \sqrt{c_{66}/\rho} \,. \tag{6.15}$$

Equation 6.15 can be used to verify the validity of the assumed orthorhombic model.

Suppose the interval symmetry-plane NMO velocities and zero-offset traveltimes of all three pure modes have been estimated from moveout analysis (in our case, using PP- and PS-waves). Since the ratios of the vertical velocities of different waves are equal to the corresponding ratios of the zero-offset traveltimes, the velocities V_{S0} and V_{S1} can be found from V_{P0}. Still, for a total of *six* remaining unknowns (V_{P0}, $\epsilon^{(1,2)}$, $\delta^{(1,2)}$, and either $\gamma^{(1)}$ or $\gamma^{(2)}$), there are only *five* independent NMO-velocity equations (equations 6.10 and 6.11 are equivalent). The four moveout equations (6.1, 6.2, 6.9, and 6.12) for in-plane polarized modes include five unknown parameters (V_{P0}, $\epsilon^{(1,2)}$, and $\delta^{(1,2)}$). Evidently, the inversion for general orthorhombic media cannot be carried out without additional information. However, parameter estimation becomes feasible for some special types of fracture-induced orthorhombic symmetry with a smaller number of independent stiffnesses (see sections 9.5 and 9.6).

The same underdetermined inverse problem arises for conventional-spread multicomponent data in VTI media or, equivalently, in each vertical symmetry plane of orthorhombic media (see our Chapter 5 or section 7.3 in Tsvankin, 2005). To estimate the anisotropy parameters in the physical-modeling experiment below, the known layer thickness is used to find the vertical velocities from the zero-offset traveltimes. Possible alternatives for field-data applications are to employ either NMO ellipses for dipping interfaces (section 6.3) or long-spread (nonhyperbolic) moveout of PP and SS reflections.

6.2 Physical-modeling example for an orthorhombic layer

Scaled physical modeling over anisotropic materials has proved to be useful in testing theoretical predictions of various wave-propagation phenomena (e.g., see section 2.2). In a series of experiments with orthorhombic phenolic laminate, Brown et al. (1991) and Cheadle et al. (1991) recorded transmitted waves for a range of propagation directions and estimated the stiffness coefficients. Sayers and Ebrom (1997) generated wide-azimuth P-wave data for an orthorhombic model in VSP geometry and used a nonhyperbolic moveout equation to reconstruct the P-wave group-velocity surface. Here, we describe the results of Grechka et al. (1999a), who simulated a multicomponent reflection survey over a horizontal orthorhombic layer and carried out moveout inversion for the medium parameters.

6.2.1 Laboratory setup and data processing

The model for the ultrasonic experiment was constructed from six blocks of Phenolite XX-324, a composite material with orthorhombic symmetry. Other experiments with the same material were described by Gibson and Theophanis (1996). The "slow" (in terms of P-wave velocity) axes of the blocks were aligned in the direction that will be called the x_1-axis of the model. The blocks were bonded with epoxy to make a sample with the dimensions $30.0 \times 30.0 \times 14.81$ cm ($z = 14.81$ cm is the vertical thickness). The reflection coefficient for the epoxy joints between layers was tested with both P- and S-waves of appropriate frequencies and found to be vanishingly small.

The acoustic wavelength excited by ultrasonic transducers was close to one-sixth of the model thickness. This wavelength corresponds to frequencies of approximately 200 kHz for P-waves and 100 kHz for S-waves. Vertical and horizontal contact transducers, employed as sources and receivers, had a diameter of one inch. Traces were recorded directly from the receiving transducer with an oscilloscope, which had real-time signal-averaging capability for noise reduction and a disk drive for data storage.

A wide-azimuth survey was simulated by recording arrivals reflected from the bottom of the block (free surface) along differently oriented CMP lines. Data were collected sequentially with P-P transducer pairs at azimuths 0°, 45°, 90°, and 135° with respect to the x_1-axis and with P-S transducer pairs (the receiver polarization was horizontal inline) at azimuths 0°, 30°, 60°, and 90°. Source-receiver offsets for each line ranged from 5 to 25 cm, with an increment of 2 cm.

Processing of PP-wave data

Reflection data recorded by vertical transducers are displayed in Figure 6.1. The most prominent, clearly visible feature of the seismograms is a significant azimuthal variation of the first-arrival traveltime. The moveout velocity of this arrival (easily identified as the PP reflection) rapidly increases as the azimuth changes from 0°

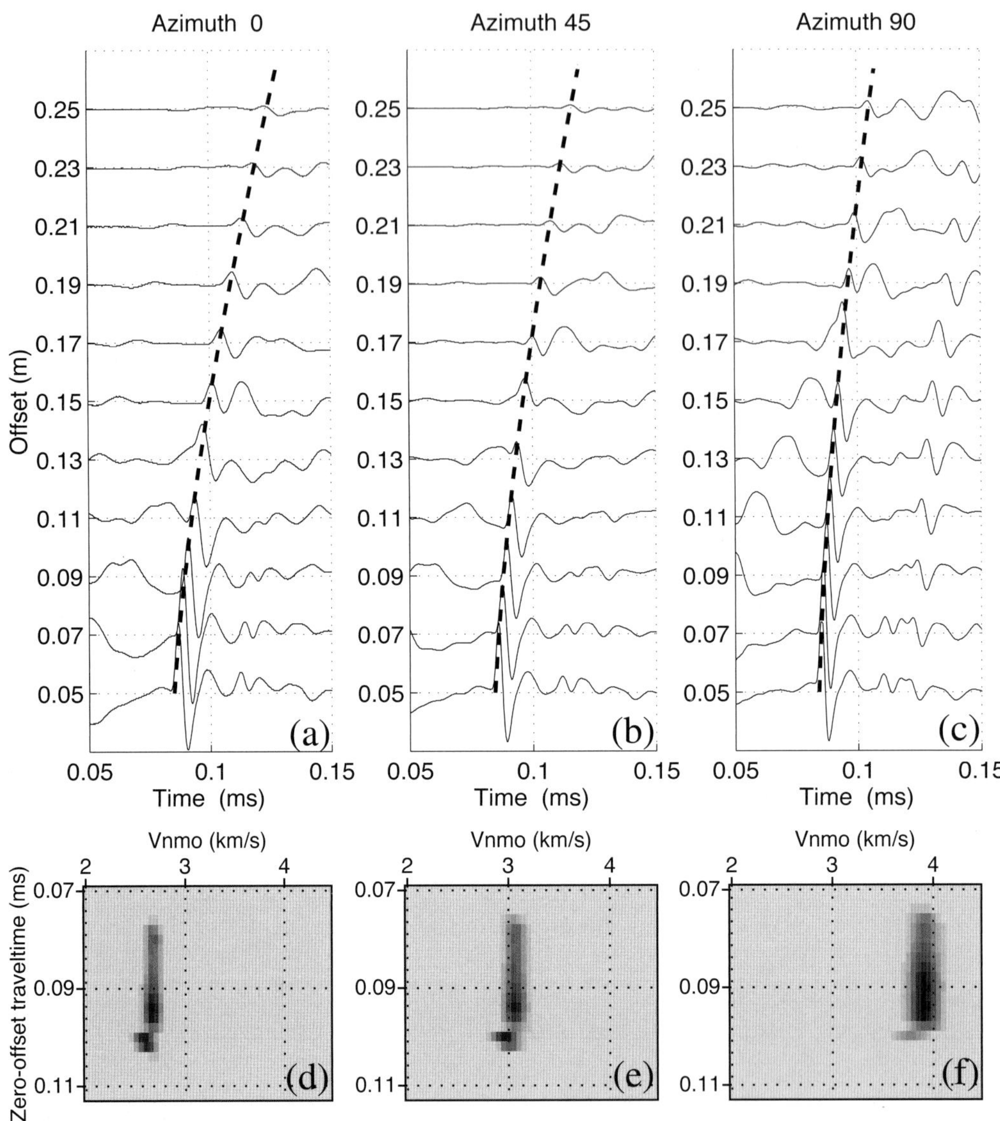

Figure 6.1: Seismograms of the vertical displacement component recorded at azimuths (a) 0°, (b) 45°, and (c) 90° (Grechka et al., 1999a). Plots (d–f) display the corresponding semblance panels. The seismogram for azimuth 135° is not shown because it is similar to that for azimuth 45°. The dashed lines mark the PP-wave hyperbolic moveout curves corresponding to the semblance maxima. All semblance values below a certain level were muted out, and the semblance maxima were enhanced using a nonlinear amplitude scale.

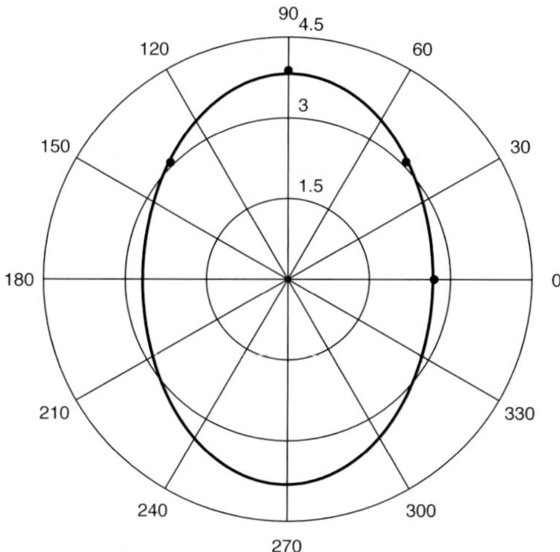

Figure 6.2: PP-wave moveout velocities (dots) corresponding to the semblance maxima in Figure 6.1 and the best-fit NMO ellipse marked by the solid line (Grechka et al., 1999a).

to 90°, which leads to a corresponding decrease in traveltime for any fixed offset. The best-fit PP-wave moveout velocity was estimated from conventional hyperbolic semblance analysis for each azimuth (Figure 6.1d-f). The position of the semblance maximum on the velocity axis varies with azimuth by more than 1.0 km/s (42%), indicating a high magnitude of P-wave azimuthal anisotropy.

The hyperbolic moveout curves computed with the estimated moveout velocities for each azimuth (dashed lines) are superimposed on the seismograms. Since the traveltimes of the semblance maxima were somewhat larger than the times of the first breaks, the picked values of t_{P0} were corrected to match the actual arrival times. Except for relatively small deviations at the very far offsets, the hyperbolic moveout approximation is close to the recorded traveltimes in all azimuthal directions. Therefore, despite the wide offset range (the maximum offset-to-depth ratio is 1.7), the magnitude of nonhyperbolic moveout is relatively small. This is explained by the weak anellipticity of the model, as discussed below.

The PP-wave moveout velocities measured in the four azimuthal directions were used to reconstruct the NMO ellipse (Figure 6.2). The data points, which correspond to *finite-spread* moveout velocities, lie close to the best-fit elliptical curve (the maximum deviation is just 1.6% at azimuth 90°). The azimuths of the axes of the ellipse in the coordinate frame of the experiment are 0.6° and 90.6°, close to the expected symmetry-plane directions for the Phenolite model. The semiaxes of the NMO ellipse (i.e., the symmetry-direction NMO velocities; see Table 6.1) were combined with the known vertical velocity $V_{P0} = 2z/t_{P0} = 3.57$ km/s to compute the anisotropy parameters $\delta^{(1,2)}$ from equations 6.1 and 6.2 ($\delta^{(1)} = 0.073$, $\delta^{(2)} = -0.22$). No other information can be obtained from PP-wave conventional-spread moveout for a horizontal reflector.

Reflection	Azimuth	
	0°	90°
PP	2.68	3.82
PS_1	1.92	2.84
PS_2	2.10	2.33
S_1S_1	1.36	2.14
S_2S_2	1.83	1.35

Table 6.1: Moveout velocities (in km/s) in the vertical symmetry planes of the model (Grechka et al., 1999a). The velocities of the PP- and PS-waves were obtained from the estimated NMO ellipses; the velocities of the pure shear reflections were computed using equation 6.8.

Processing of converted-wave data

The moveout velocities of converted waves were estimated on the records of inline horizontal transducers at azimuths 0°, 30°, 60°, and 90° (Figures 6.3 and 6.4). PP-waves are visible even on the horizontal displacement component but the largest semblance maximum corresponds to the second arrival, which can be identified as a converted wave. This interpretation is based on the zero-offset traveltimes of these modes, their relatively low moveout velocities, and predominantly horizontal polarization. Since the wavefield was excited by a vertical force, pure shear-wave reflections are relatively weak at small and moderate offsets. Although S_1S_1 and S_2S_2 arrivals are observed on the horizontal displacement component, their semblance maxima are not sufficiently focused for accurate velocity picking. The reflection moveouts of the pure SS-waves were computed for the inverted model and plotted against the actual arrivals to verify the accuracy of the estimated parameters (see below).

Semblance analysis for each section provided the moveout velocity of a single (PS) mode corresponding to the most prominent semblance maximum. It is clear from both the seismograms and semblance panels that we observe two *different* converted waves, each one dominating the radial component of the wavefield in a certain range of azimuthal angles. While the PS arrival at azimuths 0° and 30° (Figure 6.3) has a zero-offset traveltime of 0.148 ms, the time t_{PS0} for the PS-wave at 60° and 90° (Figure 6.4) is noticeably smaller (0.119 ms). This difference between the zero-offset traveltimes allows us to identify the PS reflection in Figure 6.3 as the slow converted wave (PS_2) and the corresponding event in Figure 6.4 as the fast wave (PS_1). The magnitude of shear-wave splitting in the vertical direction is sufficient for the two converted waves to be well separated in time. The shear-wave vertical velocities were calculated from the zero-offset traveltimes of the PP- and PS-waves and the known layer thickness: $V_{S1} = 1.91$ km/s and $V_{S2} = 1.39$ km/s.

Figures 6.3 and 6.4 confirm our expectation that only one converted wave (close to the PSV type) can be observed in the vicinity of each symmetry plane. The fast

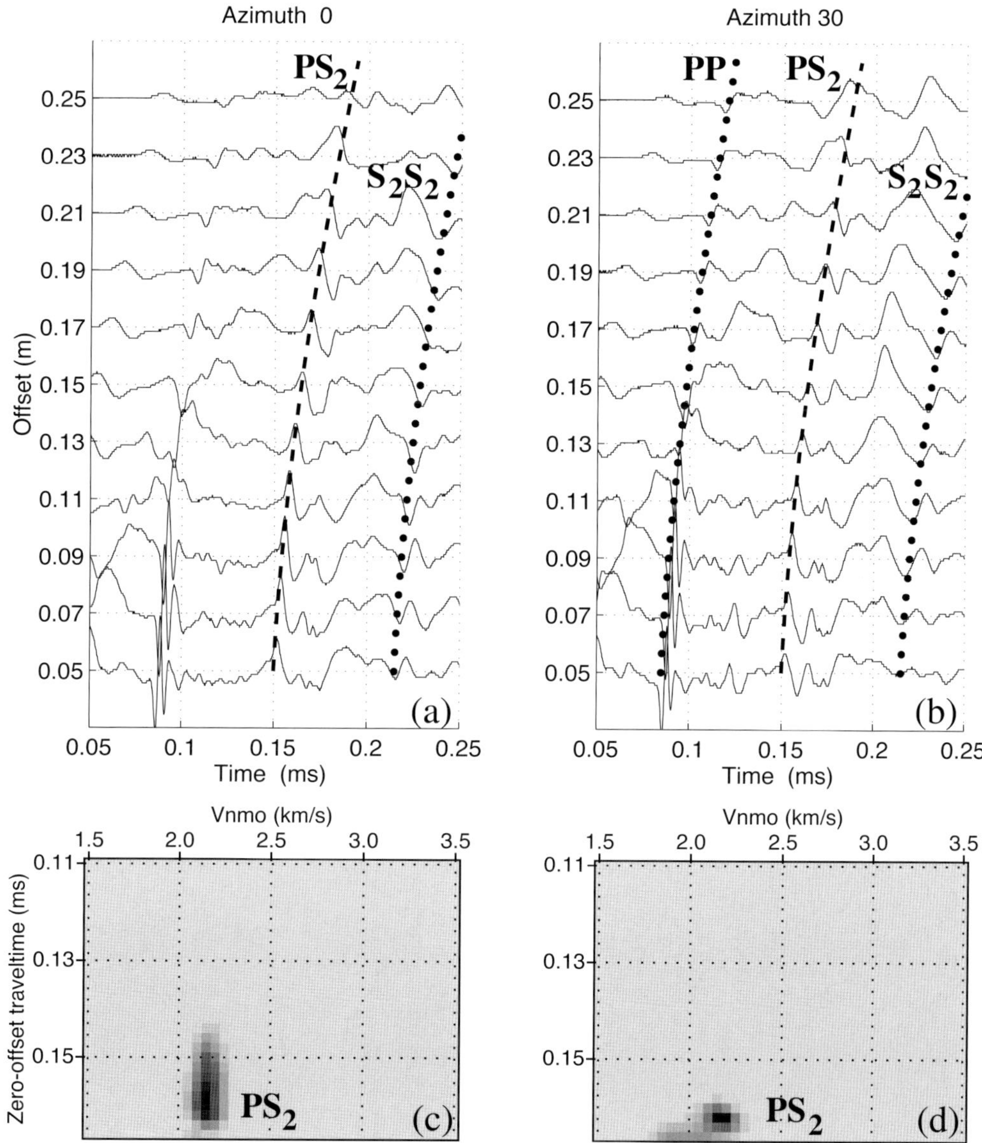

Figure 6.3: Seismograms of the horizontal inline displacement component at azimuths (a) 0° and (b) 30°; (c,d) are the corresponding semblance panels (Grechka et al., 1999a). The dashed lines mark the hyperbolic moveout curves of the PS_2-wave picked from the semblance panels. The moveouts of the pure PP and S_2S_2 reflections, computed using the inversion results (see below), are marked by the dots.

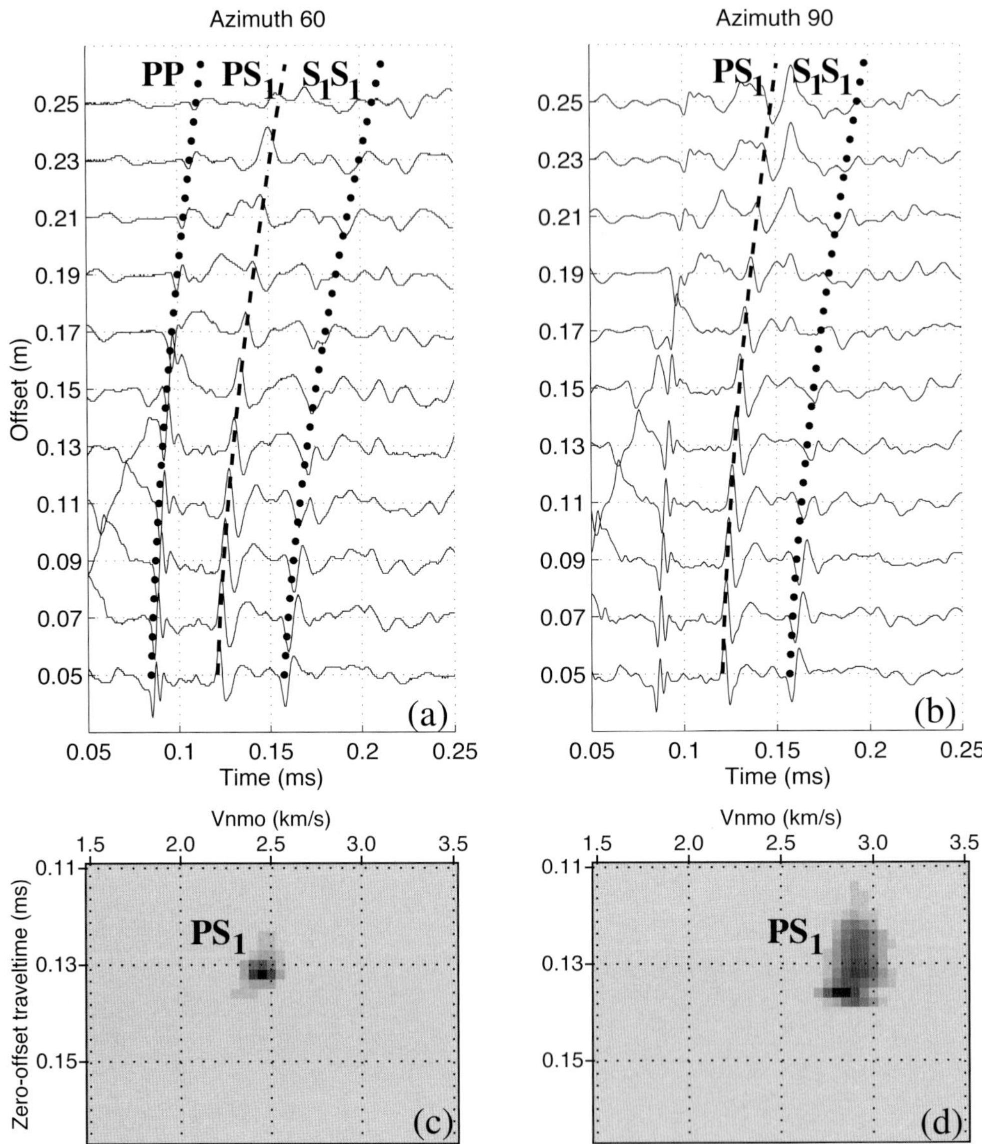

Figure 6.4: Seismograms of the horizontal inline displacement component at azimuths (a) 60° and (b) 90°; (c,d) are the corresponding semblance panels (Grechka et al., 1999a). The dashed lines mark the hyperbolic moveout curves of the PS_1-wave picked from the semblance panels. The moveouts of the pure PP and S_1S_1 reflections, computed using the inversion results (see below), are marked by the dots.

wave PS_1 has the SV polarization in the $[x_2, x_3]$-plane (azimuth 90°) and, therefore, is recorded at azimuths 60° and 90°. In contrast, the sections at 0° and 30° contain the intensive PS_2 arrival, which represents the SV mode in the $[x_1, x_3]$-plane. It is likely that the event with the zero-offset traveltime close to 0.12 ms that can be seen at small offsets in Figure 6.3b is the PS_1-wave, but its amplitude is not sufficient to generate a focused semblance maximum.

Note that even if SH-type waves were excited by mode conversion at the reflector, they would not be recorded by our *inline* horizontal receivers. In general, PS-wave processing in azimuthally anisotropic media requires employing two orthogonal horizontal geophones (inline and crossline). Then the inline and crossline displacement components should be simultaneously rotated to separate the converted arrivals. Because of the large values of the shear-wave splitting coefficient in our model, the rotation was not necessary. A crossline receiver, however, could have helped to enhance the converted waves with the "quasi-SH" polarization near the vertical symmetry planes.

The hyperbolic moveout curves of both PS-waves computed with the moveout velocities picked from the semblance panels are marked by the dashed lines in Figures 6.3a,b and 6.4a,b. As was the case for PP-waves, the best-fit hyperbolic moveout function is close to the reflection traveltimes, with deviations becoming noticeable only at the farthest offsets. The velocities corresponding to the semblance maxima for each converted wave exhibit a pronounced azimuthal dependence. For instance, there is a significant increase in the moveout velocity of the PS_1 reflection between azimuths of 60° and 90°.

In general, a minimum of three azimuthal moveout measurements is necessary to recover the NMO ellipse for each mode. However, since the orientation of the NMO ellipses (i.e., the azimuths of the symmetry planes) has already been established from P-wave data (Figure 6.2), two sufficiently separated CMP lines provide enough information to find the elliptical semiaxes.[1]

The reconstructed NMO ellipses of both converted waves are shown in Figure 6.5. The most nontrivial part of the converted-wave processing was estimation of the moveout velocities of the physically *nonexistent* PSH modes in the symmetry planes (PS_1 in the $[x_1, x_3]$-plane and PS_2 in the $[x_2, x_3]$-plane). These velocities were obtained essentially by *extrapolating* moveout measurements in other directions using the known elliptical form of the azimuthal dependence of NMO velocity.

6.2.2 Estimation of the anisotropy parameters

Substitution of the symmetry-plane NMO velocities of the PP- and PS-waves into equation 6.8 yielded the corresponding velocities of the pure shear modes (Table 6.1), which were used to compute their NMO ellipses (Figure 6.5). As an aside, although the time delay between the *vertically* traveling shear and converted waves is rather

[1]If the medium does not have vertical symmetry planes, the NMO ellipses of pure and converted modes can have different orientations.

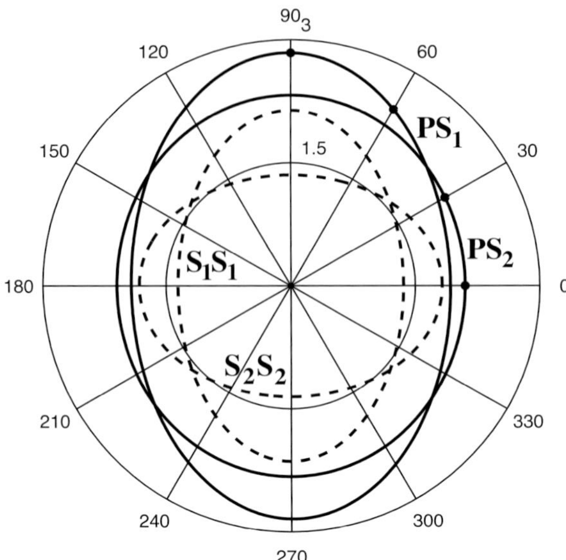

Figure 6.5: NMO ellipses of the converted waves (solid lines) reconstructed from the picked moveout velocities (dots) and the NMO ellipses of the pure SS reflections (dashed) computed from the PP and PS data (Grechka et al., 1999a).

significant, the influence of the anisotropy parameters in equations 6.9 – 6.12 makes the shear-wave *NMO ellipses* (as well as the ellipses of the converted waves) intersect each other.

Knowledge of the symmetry-plane NMO velocities of the S_1S_1- and S_2S_2-waves, along with estimates of the vertical velocities V_{S1} and V_{S2}, was sufficient to find the parameters $\gamma^{(1,2)}$ and $\sigma^{(1,2)}$ from equations 6.9 – 6.12. Then the parameters $\epsilon^{(1,2)}$ were computed from equations 6.13 using the known values of $\delta^{(1,2)}$. In summary, the moveout inversion produced the following parameter set of the Phenolite model: $V_{P0} = 3.57$ km/s, $V_{S0} = 1.91$ km/s, $\delta^{(2)} = -0.22$, $\delta^{(1)} = 0.073$, $\epsilon^{(2)} = -0.16$, $\epsilon^{(1)} = 0.11$, $\gamma^{(2)} = -0.25$, and $\gamma^{(1)} = -0.03$.

The difference between $\delta^{(1)}$ and $\delta^{(2)}$ is responsible for the azimuthal variation of the PP-wave NMO velocity, while $\epsilon^{(1)} - \epsilon^{(2)}$ determines the magnitude of the variation of the P-wave horizontal velocity between the x_2- and x_1-axes. The parameters $\gamma^{(1)}$ and $\gamma^{(2)}$ govern the SH-wave velocity anisotropy in the symmetry planes (i.e., the anisotropy of the S_1-wave in the $[x_1, x_3]$-plane and S_2-wave in the $[x_2, x_3]$-plane).

The only medium parameter that could not be found is $\delta^{(3)}$ because it has no influence on either the vertical or NMO velocity of any mode. The parameter $\delta^{(3)}$, however, is constrained by the traveltime of the direct P-wave, which travels along the top of the block in the horizontal symmetry plane. This direct P-wave arrival, visible (but not marked) at large offsets in Figures 6.3a,b and 6.4a,b, is shown for the full offset range in Figure 6.6. Since $\delta^{(3)}$ contributes to the P-wave horizontal velocity only outside the symmetry planes, its value ($\delta^{(3)} = -0.21$) was estimated from the

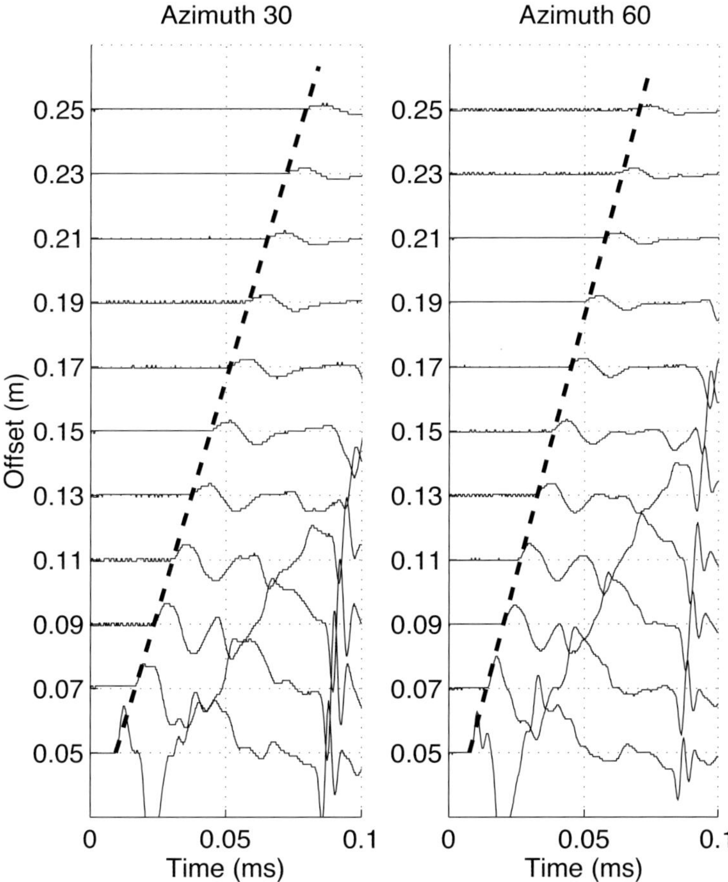

Figure 6.6: Seismograms of the horizontal displacement component at azimuths 30° and 60° (Grechka et al., 1999a). The straight dashed lines mark the direct P-wave arrival. The group velocities determined from the slope of the direct-wave moveout are 2.86 km/s (30°) and 3.20 km/s (60°).

group-velocity measurements at azimuths 30° and 60° (Figure 6.6).

The obtained set of Tsvankin's parameters is sufficient to fully characterize the model. Still, for completeness we computed the corresponding density-normalized stiffness matrix (in km^2/s^2):

$$\frac{c_{ij}}{\rho} = \begin{pmatrix} 8.58 & 2.74 & 2.01 & 0 & 0 & 0 \\ 2.74 & 15.51 & 9.76 & 0 & 0 & 0 \\ 2.01 & 9.76 & 12.74 & 0 & 0 & 0 \\ 0 & 0 & 0 & 3.65 & 0 & 0 \\ 0 & 0 & 0 & 0 & 1.93 & 0 \\ 0 & 0 & 0 & 0 & 0 & 1.84 \end{pmatrix}. \qquad (6.16)$$

Although independent measurements of the stiffness tensor of Phenolite were not available, the accuracy of the inversion results could be checked in several other ways. First, there exists a redundancy in the moveout measurements since the NMO velocity of the fast mode S_1S_1 in the $[x_1, x_3]$-plane (azimuth 0°) should be equal to the S_2S_2-wave NMO velocity in the $[x_2, x_3]$-plane (azimuth 90°; see equation 6.15). These velocities correspond to the transversely polarized (SH) modes in the symmetry planes and, therefore, could not be measured from the data directly because of the absence of PSH reflections. Nevertheless, the values computed using the extrapolated NMO ellipses of both converted waves are remarkably close to each other (Table 6.1).

Another way of verifying the inversion results is to compute the traveltimes of reflection arrivals not used for parameter estimation and check whether the predicted moveouts match the data. PP-wave moveout curves at azimuths 30° and 60° calculated for the obtained model parameters are shown in Figures 6.3b and 6.4a. Although the PP-wave NMO ellipse was built from measurements at azimuths 0°, 45°, 90°, and 135°, it accurately predicts PP traveltimes in these intermediate directions. It is also possible to model the traveltimes of the pure SS-wave reflections, which essentially are a byproduct of the experiment. The predicted moveout curves of the S_1S_1- and S_2S_2-waves in Figures 6.3a,b and 6.4a,b lie close to the relatively weak shear-wave arrivals.

The most direct check of the inversion procedure is a comparison between the predicted and measured group velocities in certain directions. The reflection data used in our experiment illuminate a limited range of group (ray) angles with the vertical, with the maximum incidence angle for the pure-mode reflections reaching about 40°. Still, including the shear-wave NMO velocities helped estimate the parameters $\epsilon^{(1)}$ and $\epsilon^{(2)}$ responsible for the P-wave horizontal velocity in the x_1- and x_2-directions ($\sqrt{c_{11}/\rho}$ and $\sqrt{c_{22}/\rho}$, respectively). These results are in excellent agreement with the direct measurements of the P-wave horizontal velocities (Table 6.2). Also, the SH-wave NMO velocities in the vertical symmetry planes should be equal to the horizontal velocity $\sqrt{c_{66}/\rho}$ (equation 6.15). Although the moveout velocities of both SH-waves were estimated by extrapolating the NMO ellipses of the corresponding converted modes, the inverted and directly measured SH-wave horizontal velocities are close. There was no need to verify the horizontal velocities of the SV-waves in

	Inverted velocity (km/s)	Measured velocity (km/s)	Difference (%)
$\sqrt{c_{11}/\rho}$	2.93	2.92	0.2
$\sqrt{c_{22}/\rho}$	3.94	4.02	−1.9
$\sqrt{c_{66}/\rho}$	1.36	1.39	−2.6

Table 6.2: Comparison of the horizontal velocities computed for the inverted model with the directly measured values (Grechka et al., 1999a).

the vertical symmetry planes because they are equal to the corresponding vertical velocities.

We conclude that the underlying assumption about the orthorhombic symmetry of the model is valid, and multicomponent moveout inversion supplemented with depth information provides sufficient accuracy in estimating the medium parameters.

6.2.3 Implications for field data processing

Despite the excellent results of the inversion procedure, the model used in the experiment was relatively simple (a single layer) and strongly anisotropic, which helped separate the split converted waves in a straightforward fashion. Below, we discuss possible problems in the application of this algorithm to processing of multicomponent data acquired over vertically heterogeneous, azimuthally anisotropic formations.

Nonhyperbolic moveout

The inversion scheme operated with the moveout velocities obtained by conventional hyperbolic semblance analysis. The magnitude of nonhyperbolic moveout for PP reflections is mostly governed by the anellipticity parameters $\eta^{(1)}$ and $\eta^{(2)}$ defined in the vertical symmetry planes (see Chapter 3). Because neither parameter exceeded 10% ($\eta^{(1)}=0.042$, $\eta^{(2)}=0.098$), PP-wave moveout on the spreadlengths used in the experiment (the maximum offset-to-depth ratio was 1.7) did not deviate much from a hyperbola. The weak anellipticity of the model also explains the small magnitude of nonhyperbolic moveout of the converted waves PS_1 and PS_2. Another factor that mitigates the influence of nonhyperbolic moveout for both PP- and PS-waves is a rapid decrease of the reflection amplitudes with offset (Figures 6.3a,b and 6.4a,b), most likely related to the source radiation pattern.

In more realistic layered media or for orthorhombic models with larger anellipticity, deviations from hyperbolic moveout for PP- and PS-waves could be much more substantial. Stable recovery of NMO ellipses then requires application of nonhyperbolic moveout analysis (see Chapter 3).

Required azimuthal coverage

The physical modeling confirmed that each of the converted modes is almost undetectable near the vertical symmetry plane, in which it has the SH polarization. This fact imposes additional requirements on the number of azimuthal measurements needed to reconstruct the PS-wave NMO ellipses in layered orthorhombic media with uniform orientation of the vertical symmetry planes. While in principle pure-wave azimuthal velocity analysis can be applied to three well-separated CMP lines, there is no guarantee that *both* PS reflections will be sufficiently intensive along at least two of those lines. (Two measurements of the NMO velocity of each PS-wave are needed to build the NMO ellipse, assuming that the symmetry-plane orientation has been determined from PP data.) Acquisition of two horizontal displacement components of converted waves along four well-separated CMP lines should be considered as a minimum requirement. Useful redundancy in moveout measurements for both converted and pure modes is provided by 3D surveys with a wide range of source-receiver azimuths.

For layered media with depth-varying symmetry-plane azimuths, mode-converted waves do not belong to PSV or PSH types, and both split PS-waves could be excited for any azimuthal direction. Still, the complexity of PS-wave processing in such models (see below) requires wide azimuthal coverage and deployment of two-component receivers.

Strength of azimuthal anisotropy and shear-wave splitting

The inversion procedure is based on combining PP data with the zero-offset traveltimes and azimuthally-dependent NMO velocities of the converted waves PS_1 and PS_2. Measurements of PS moveout require prior separation of the split converted waves on multicomponent data. The magnitude of shear-wave splitting near the vertical is described by the splitting parameter $\gamma^{(S)}$, which is close to the fractional difference between the vertical velocities V_{S1} and V_{S2} (equation 1.87). For the Phenolite sample, the splitting coefficient turned out to be uncommonly large ($\gamma^{(S)} = 0.44$), so the PS_1 and PS_2 arrivals were well separated at all offsets used in the experiment.

For smaller values of $\gamma^{(S)}$, the converted waves interfere near the vertical, making their moveout velocities much more difficult to estimate. Then it is necessary to deploy two horizontal geophones and use an Alford-type rotation algorithm to separate the waves PS_1 and PS_2 (Thomsen, 1988; Gaiser, 2000). This operation becomes more robust if mode conversions are acquired for a wide range of azimuths. Still, the need to apply receiver rotation can limit the accuracy of moveout inversion.

PS-wave processing becomes much more complicated in stratified media, especially if the target layer is overlain by an azimuthally anisotropic overburden. Multiple splitting in the overburden impedes reliable identification of converted waves and requires application of polarization layer stripping. Another potentially difficult processing step is PP-PS event correlation (see the discussion in Chapter 4), which is particularly prone to errors in the presence of azimuthal anisotropy.

Shear-wave point singularities

The problem of shear-wave singularities is directly related to the issue of the magnitude of shear-wave splitting discussed above. We assumed that the phase-velocity (or slowness) sheets of the waves S_1 and S_2 do not intersect in the vicinity of the vertical direction. In other words, shear-wave point singularities, which always exist in orthorhombic media (Figure 2.36), are presumed to be sufficiently far from the vertical. The position of singularities with respect to the vertical axis is determined by the value of the parameter $\gamma^{(S)}$. The separation of the shear-wave velocity sheets in the vertical direction increases with $\gamma^{(S)}$, which moves singularities toward the horizontal plane.

Because of the large splitting parameter in the Phenolite model, the singularity closest to the vertical corresponds to a phase angle of 69° with the vertical axis. Clearly, the assumption that shear-wave singularities do not hamper moveout analysis is satisfied for that model. If, however, $\gamma^{(S)}$ were smaller and reflected rays for some range of offsets and azimuthal angles crossed a singularity region, the wavefronts of converted modes would be much more complicated and include multiple arrivals and areas of nonlinear polarization (Crampin and Yedlin, 1981; Grechka and Obolentseva, 1993). Separation of wavefields near singularities into the distinct PS_1 and PS_2 reflections is expected to be extremely difficult.

Sorting into CCP gathers

One of the essential steps in conventional processing of mode-converted reflections is sorting seismic traces into common-conversion-point (CCP) gathers. If the medium is laterally heterogeneous, the offset dependence ("dispersal") of the conversion point can cause distortions in velocity analysis and degrade stacking quality (see Chapters 4 and 5). The models used in sections 6.1 and 6.2 were composed of horizontal, homogeneous layers, so CCP sorting was unnecessary.

The subsurface, however, always possesses some degree of lateral heterogeneity, and in field-data applications it is preferable to carry out CCP sorting before velocity estimation. Unfortunately, the position of the conversion point in orthorhombic media depends on the anisotropy parameters, which are seldom known in advance. Hence, azimuthal velocity analysis and moveout inversion should be first carried out in CMP gathers. The results of parameter estimation could then be used to sort the data into CCP gathers and refine the model by repeating the inversion procedure. If necessary, this parameter updating can be continued in iterative fashion.

Alternatively, the problems caused by conversion-point dispersal can be avoided by applying the PP+ PS = SS method and constructing ΨS-waves from PP and PS data (see Chapter 4). Then the interval NMO ellipses of the acquired PP-waves and generated ΨS-waves (which have the kinematics of the SS primary reflections) can be obtained from the generalized Dix differentiation or velocity-independent layer stripping (VILS).

6.3 Stacking-velocity tomography for orthorhombic media

The previous two sections were devoted to moveout inversion for horizontally layered orthorhombic models. Here, we follow Grechka et al. (2005) in extending multicomponent stacking-velocity inversion (*tomography*) introduced in Chapter 5 to a stack of orthorhombic layers separated by dipping or curved interfaces. As in sections 6.1 and 6.2, one of the symmetry planes in each layer is assumed to be horizontal.

6.3.1 Interval parameter-estimation algorithm

Reflection moveout of PS-waves in laterally heterogeneous media is asymmetric with respect to zero offset and cannot be described by the NMO ellipse even at small offsets. Hence, the methodology based on the relationship between the NMO ellipses of pure and converted waves (equation 6.4) employed in sections 6.1 and 6.2 is not applicable here. Instead, PP and PS data can be processed by the PP+PS=SS method to generate ΨS-waves, which have the moveout of the pure SS primaries.

Although the PP+PS=SS method requires no explicit velocity information, it operates with reflections from the same interface and, therefore, should be preceded by establishing PP-PS event correspondence. Identification of PS arrivals can be difficult for traces with low PS amplitudes (and, correspondingly, a low signal-to-noise ratio) caused by small conversion coefficients in areas near polarity reversals. To evaluate the seriousness of this issue for a typical orthorhombic model with moderate anisotropy, we use a synthetic test.

The multicomponent seismograms in Figure 6.7 are computed by anisotropic dynamic ray tracing for a single orthorhombic layer with a dipping lower boundary. As expected, the PP reflection has the highest amplitude on the vertical component, whereas the two split converted waves PS_1 and PS_2 dominate the horizontal inline and crossline components. Although we observe some dimming of the PS_1- and PS_2-waves at certain azimuths, the offset range of those low-amplitude arrivals is relatively narrow. Therefore, Figure 6.7 suggests that, despite the presence of polarity reversals, converted-wave reflections over orthorhombic media can be efficiently used for generating SS data. This conclusion remains valid for models with a smaller shear-wave splitting coefficient, where PS_1 and PS_2 arrivals interfere and have to be separated by means of Alford-style rotation.

Once PP, PS_1, and PS_2 reflections from a certain interface have been correlated, they are processed by the PP+PS=SS method to compute seismograms of the split ΨS-waves, which have the kinematics of the reflections S_1S_1 and S_2S_2. Then the NMO ellipses of the recorded PP-waves and generated SS-waves are found from 3D semblance analysis (see section 2.1). An accurate reconstruction of the NMO ellipse might require a correction for the influence of nonhyperbolic moveout; a nonhyperbolic moveout-inversion algorithm for orthorhombic media is described in section 3.3.

The NMO ellipses of all three pure modes, along with the corresponding zero-offset

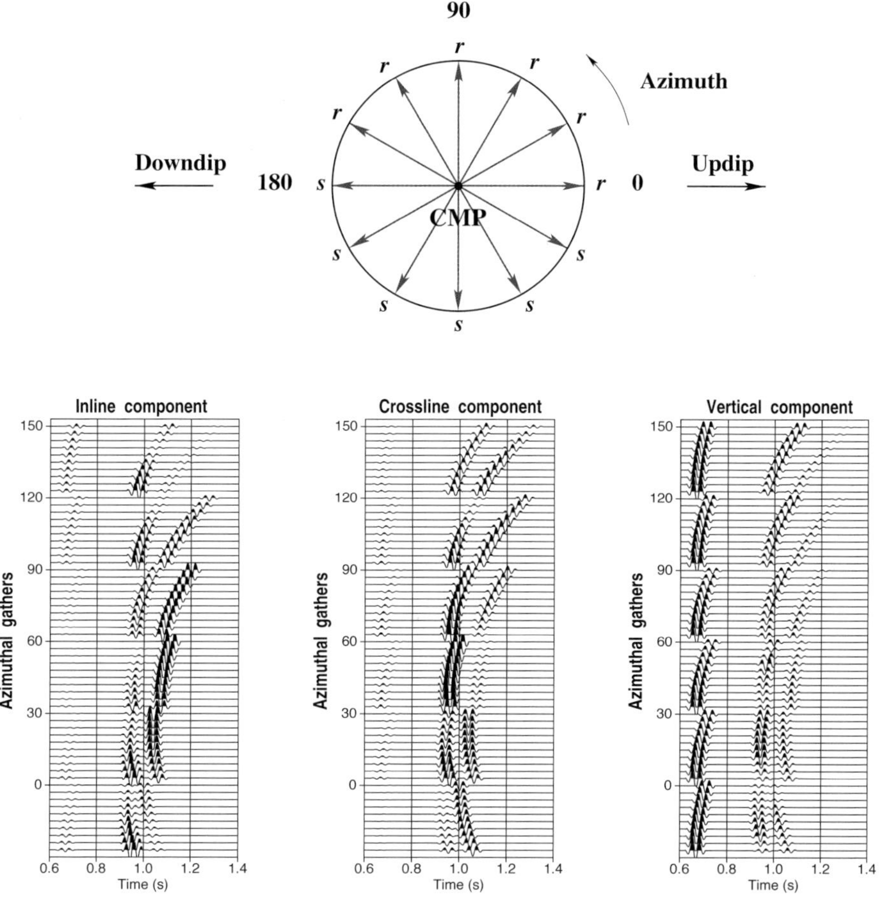

Figure 6.7: Ray-traced three-component synthetic seismograms (bottom row of plots) of PP-, PS_1-, and PS_2-waves reflected from a dipping interface beneath an orthorhombic layer (Grechka et al., 2005). The wavefield is recorded in a multiazimuth CMP gather (see the plan view on top), with sources (a vertical force at the free surface) located on the side marked by s, and receivers on the side marked by r. The numbers 0, 30, 60, 90, 120, and 150 on the seismograms' vertical axis indicate the azimuth (in degrees) for the block of traces below this number (e.g., all traces between numbers 150 and 120 are recorded on the CMP line with the azimuth 150°). The source-receiver offsets for each block of traces (i.e., CMP line) vary from 0.1 km to 1 km in 0.1 km increments. The azimuth of the reflector dip plane is ψ=0° (see the plan view), the dip ϕ=30°, and the depth under the common midpoint z=1 km. The azimuth of the vertical symmetry plane $[x_1, x_3]$ with respect to the dip plane is β=60°. The medium parameters are V_{P0}=2.9 km/s, V_{S0}=1.4 km/s, $\epsilon^{(1)}$=0.25, $\delta^{(1)}$=0.15, $\gamma^{(1)}$=−0.20, $\epsilon^{(2)}$=0.15, $\delta^{(2)}$=0.05, $\gamma^{(2)}$=−0.25, $\delta^{(3)}$=−0.05, and the density ρ=1.9 g/cm^3. The reflecting halfspace is isotropic with P-wave velocity V_P=3.0 km/s, S-wave velocity V_S=1.5 km/s, and density ρ=2.1 g/cm^3.

traveltimes and reflection time slopes (i.e., the horizontal slownesses of the zero-offset ray), are used as the input data for stacking-velocity inversion. The data vector for an N-layered orthorhombic medium has the form

$$\mathbf{d}(Q, \mathbf{Y}, n) \equiv \{\tau_Q(\mathbf{Y}, n),\, \mathbf{p}_Q(\mathbf{Y}, n),\, \mathbf{W}_Q(\mathbf{Y}, n)\}\,, \tag{6.17}$$

where Q denotes the mode (PP, S_1S_1, or S_2S_2), $\mathbf{Y} = [Y_1, Y_2]$ is the CMP location, $n = 1, 2, ..., N$ is the reflector number, τ is the zero-offset traveltime, $\mathbf{p}$ is the horizontal projection of the slowness vector of the zero-offset ray, and $\mathbf{W}$ is the 2×2 matrix that describes the NMO ellipse. The only difference between the input data in equation 6.17 and the data vector used in multicomponent stacking-velocity tomography for TI media (equation 5.8) is that the algorithm for orthorhombic symmetry operates with both split shear waves.

As discussed in section 5.2, one of the main advantages of restricting the inversion to hyperbolic moveout is computational efficiency because the NMO ellipse for a given reflection event can be computed by tracing a single (zero-offset) ray. Also, NMO ellipses provide valuable analytic insight into the constrained parameter combinations for different anisotropic symmetries.

The model vector $\mathbf{m}$ includes the interval Tsvankin's parameters of orthorhombic layers and the coefficients of the polynomials specifying the model interfaces. The inversion is based on the two-step procedure described in sections 2.3 and 5.2. First, for a given set of trial interval parameters, the zero-offset traveltimes and reflection slopes are used to compute the one-way zero-offset rays and reconstruct the medium interfaces (i.e., build the trial model in depth). Second, the NMO ellipses $\mathbf{W}$ are calculated in the trial model for all recorded PP, S_1S_1, and S_2S_2 events at each CMP location. Then the interval parameters are updated to minimize the misfit between the calculated and measured NMO ellipses (equation 5.9), and the inversion continues in iterative fashion. Below we analyze the uniqueness and stability of this multicomponent moveout-inversion algorithm.

6.3.2 Single layer above a dipping reflector

We start by establishing the conditions needed to invert the data vector $\mathbf{d}$ for the parameters of a homogeneous orthorhombic medium above a plane dipping interface. When the reflector is horizontal, without additional information the NMO ellipses and zero-offset traveltimes of PP-, S_1S_1-, and S_2S_2-waves are insufficient for estimating the vertical velocities and anisotropy parameters (see sections 6.1 and 6.2).

Note that moveout inversion of conventional-spread multicomponent data is also nonunique for the simpler model of a horizontal VTI layer. The results of section 5.2, however, show that the parameters V_{P0}, V_{S0}, ϵ, and δ of a homogeneous VTI medium can be estimated from the hyperbolic moveout of PP- and SS-waves (only SV data are needed), if the reflector has a dip of at least $15^\circ - 20^\circ$.

Plane dipping reflector co-oriented with a vertical symmetry plane

First, we consider a plane dipping reflector oriented in such a way that the dip plane coincides with the $[x_1, x_3]$ symmetry plane of the orthorhombic medium. Since the slowness vector of the zero-offset ray for any pure mode is orthogonal to the reflector, the horizontal slowness projection **p** is confined to the dip plane. Therefore, reflection time slopes of any mode constrain both the reflector orientation and the azimuth of the $[x_1, x_3]$ symmetry plane.

Then the model vector **m** to be estimated from moveout data contains 11 unknowns: the medium parameters V_{P0}, V_{S0}, $\epsilon^{(1,2)}$, $\delta^{(1,2,3)}$, $\gamma^{(1,2)}$, the reflector dip ϕ, and the distance D between the origin of the Cartesian coordinate system and the reflector. The data vector 6.17 has 18 components, but six of them (two for each mode Q) vanish:

$$p_{2,Q} = 0 \quad \text{and} \quad W_{12,Q} = 0\,, \tag{6.18}$$

because $[x_1, x_3]$ is a symmetry plane for the model as a whole. The remaining 12 data components are the zero-offset traveltimes τ_Q, horizontal slownesses $p_{1,Q}$, and diagonal elements of the matrices $\mathbf{W}_Q$ for the three modes. Below, however, we demonstrate that the expressions for the traveltimes τ_Q are not independent.

Any point **x** on the plane reflecting interface satisfies the equation

$$\mathbf{n} \cdot \mathbf{x} = D\,, \tag{6.19}$$

where $\mathbf{n} \equiv [\sin\phi,\ 0,\ \cos\phi]$ is a unit vector perpendicular to the reflector. Assuming that the origin of the Cartesian coordinate system is at the CMP location, the reflection point $\mathbf{R}_Q$ of the zero-offset ray can be found from

$$\mathbf{R}_Q = \boldsymbol{\mathcal{G}}_Q\, \tau_Q\,, \tag{6.20}$$

where $\boldsymbol{\mathcal{G}}_Q$ is the group-velocity vector of the zero-offset ray and τ_Q is the one-way zero-offset traveltime. Since the point $\mathbf{R}_Q$ lies on the reflector, setting $\mathbf{x} = \mathbf{R}_Q$ in equation 6.19 gives

$$\tau_Q = \frac{D}{\mathbf{n} \cdot \boldsymbol{\mathcal{G}}_Q}\,. \tag{6.21}$$

Snell's law requires that the slowness vector $\mathbf{p}_Q$ of pure-mode zero-offset reflections be orthogonal to the reflector:

$$\mathbf{n} = \frac{\mathbf{p}_Q}{|\mathbf{p}_Q|}\,. \tag{6.22}$$

Combining equations 6.21 and 6.22 and taking into account that $\mathbf{p}_Q \cdot \boldsymbol{\mathcal{G}}_Q = 1$ (this equality plays the role of the eikonal equation in anisotropic media) yields

$$\tau_Q = \frac{D|\mathbf{p}_Q|}{\mathbf{p}_Q \cdot \boldsymbol{\mathcal{G}}_Q} = D|\mathbf{p}_Q| = \frac{D\,p_{1,Q}}{\sin\phi}\,. \tag{6.23}$$

Equation 6.23 shows that the zero-offset traveltimes of the three pure-mode reflections constrain only *one* combination of the model parameters $(D/\sin\phi)$. Therefore, the number of independent equations to be solved for the 11 unknown parameters reduces from 12 to 10. Clearly, this inverse problem does not have a unique solution.

Arbitrarily oriented plane dipping reflector

If the dip plane of the reflector does not coincide with either vertical symmetry plane, the NMO ellipses of all three modes generally have different orientations. This increases the number of independent components of the data vector 6.17 and, as demonstrated below, can make the inversion unique. The model vector for a single layer generally includes 13 elements:

$$\mathbf{m} = \left\{V_{P0},\, V_{S0},\, \epsilon^{(1)},\, \delta^{(1)},\, \gamma^{(1)},\, \epsilon^{(2)},\, \delta^{(2)},\, \gamma^{(2)},\, \delta^{(3)},\, D,\, \phi,\, \psi,\, \beta\right\}, \tag{6.24}$$

where ψ is the azimuth of the dip plane of the reflector and β is the azimuth of the symmetry plane $[x_1, x_3]$; both ψ and β are evaluated with respect to the axis Y_1 of the observational coordinate frame.

Although the data vector (equation 6.17) includes 18 components, the three traveltimes τ_Q still depend on a single parameter combination because equation 6.23 remains valid for any reflector orientation (with $\sqrt{p_{1,Q}^2 + p_{2,Q}^2}$ replacing $p_{1,Q}$). In addition, the ratio of the horizontal slowness components for each mode is fixed by the reflector azimuth ψ ($p_{2,Q}/p_{1,Q} = \tan\psi$), which eliminates two more independent equations. Despite this reduction in the number of independent data elements from 18 to 14, potentially estimation of the 13 model parameters is feasible.

The inverse problem, however, is nonlinear, and the properties of its solution must be studied by numerical testing on noise-contaminated data. Such testing performed by Grechka et al. (2005) shows that for a range of typical orthorhombic models parameter estimation is not only unique but also quite stable. The results in Figure 6.8 are obtained from noise-contaminated data generated for all three modes (PP, S_1S_1, and S_2S_2) at a single CMP location. The inversion was repeated 100 times for different realizations of the noise using a nonlinear least-squares algorithm. Because of the multimodal nature of the misfit function, for each realization of the input data the search was carried out several times starting from different, well-separated points in the model space. The resulting mean values of the estimated anisotropy parameters are unbiased, and the maximum standard deviation does not exceed 0.025 (the value for $\delta^{(1)}$).

For the model from Figure 6.8, the dip plane of the reflector deviates by 30° from the nearest vertical symmetry plane, which ensures the success of the inversion procedure. Figure 6.9 shows the results for a similar model, but the angle between the dip plane and a vertical symmetry plane is just 15°, and the dip ϕ is reduced from 30° to 20°. Both changes have a negative influence on the performance of the inversion algorithm, and the error bars for most parameters become noticeably larger. Still, the moderate magnitude of the standard deviations indicates that the inversion remains sufficiently stable for practical applications.

Further reduction in the difference $|\psi - \beta|$ or in the reflector dip leads to a rapid increase in the errors (i.e., in the standard deviations). Indeed, as shown above, the inversion becomes nonunique when $|\psi - \beta|$ approaches $k\pi/2$ (k is an integer). Numerical testing indicates that stable parameter estimation requires the dip direction

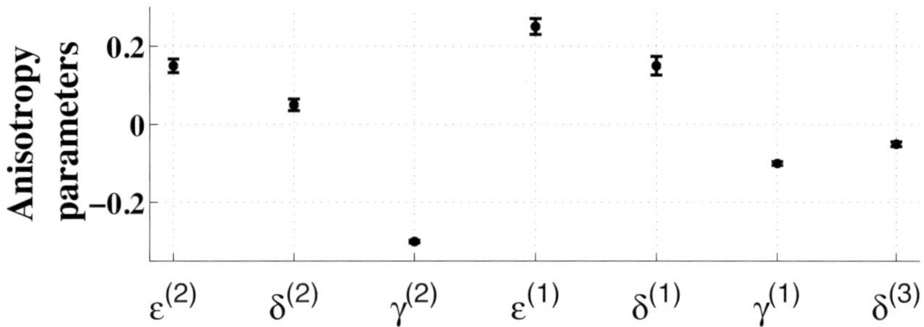

Figure 6.8: Inversion results for a single orthorhombic layer above a dipping reflector (Grechka et al., 2005). The reflector dip ϕ is 30°, the azimuth of the dip plane of the reflector ψ=0°, and the azimuth of the $[x_1, x_3]$ symmetry plane of the layer β=60°. The data were contaminated by Gaussian noise with standard deviations of 1% for τ_Q and $\mathbf{p}_Q$ and 2% for $\mathbf{W}_Q$. The dots mark the actual parameters, and the bars correspond to two (±one) standard deviations in each parameter. The standard deviations for the parameters, not shown on the plot, are 1.2% and 0.6% for V_{P0} and V_{S0}, respectively, and less than 1° for the angles ϕ, ψ, and β.

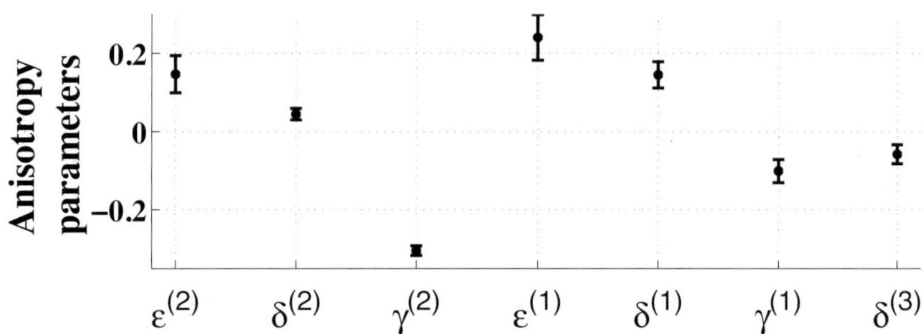

Figure 6.9: Same as Figure 6.8, but the reflector dip ϕ is 20°, and the azimuth of the $[x_1, x_3]$ symmetry plane β=15°; the azimuth of the dip plane remains ψ=0° (Grechka et al., 2005). The standard deviations for the parameters, not shown on the plot, are 1.3% and 0.7% for V_{P0} and V_{S0}, respectively, and less than 1.5° for the angles ϕ, ψ, and β.

to deviate from the nearest symmetry plane by more than 10°. Likewise, the inversion becomes unstable if the reflector dip is less than 15° because for subhorizontal reflectors the model vector **m** cannot be resolved without a priori knowledge of the vertical velocities or reflector depth. A similar dependence of parameter-estimation results on reflector dip was found in section 5.2 for multicomponent data from VTI media.

Curved reflector

If the reflector beneath an orthorhombic layer has variable curvature, the uniqueness of the inversion procedure is much more difficult to assess analytically. The feasibility of estimating the medium parameters in this case strongly depends on the reflector shape and the range of dips as well as on the spatial distribution of the common midpoints (see the discussion for TI media in sections 2.3 and 5.2).

Extensive testing on synthetic traveltimes shows that reflector curvature generally helps in parameter estimation, provided the reflecting interface contains dips exceeding 15° and some of the dip directions deviate from the vertical symmetry planes. Typical inversion results for noise-contaminated data from a curved reflector are displayed in Figure 6.10. As in the previous test, the inversion was repeated 100 times to study the distribution of the estimated parameters. The standard deviations in Figure 6.10 are even smaller than those in Figure 6.8. This increase in stability is explained by the relatively wide range of reflector dips and azimuths sampled by the zero-offset rays. In the presence of reflector curvature, each CMP provides independent information for parameter estimation. The advantages of curved interfaces, however, can be fully exploited only if the medium above the reflector can be treated as homogeneous.

6.3.3 Layered media with dipping or curved interfaces

The above analysis for a homogeneous medium helps evaluate the potential and limitations of the multicomponent inversion algorithm for multiple orthorhombic layers. As demonstrated by Grechka et al. (2005), if *noise-free* data $\{\tau_Q, \mathbf{p}_Q, \mathbf{W}_Q\}$ for all three pure-mode reflections are available, and the reflectors satisfy the conditions established for a single layer (i.e., the dips exceed 15° – 20° and dip directions deviate by more than 10° from the vertical symmetry planes), the algorithm can recover the interval parameters along with the shapes of the interfaces. However, because of the lower sensitivity of surface data to the parameters of deeper layers and downward error propagation through the model, even moderate noise may cause substantial inversion errors.

A practical way to increase the stability of parameter estimation for field-data applications is to put constraints on the interval vertical velocities or layer thicknesses (e.g., if check shots are available). Application of the inversion algorithm with a priori constraints to a model composed of three horizontal orthorhombic layers is illustrated in Figure 6.11. The data $\{\tau_Q, \mathbf{p}_Q, \mathbf{W}_Q\}$ were generated for nine common midpoints

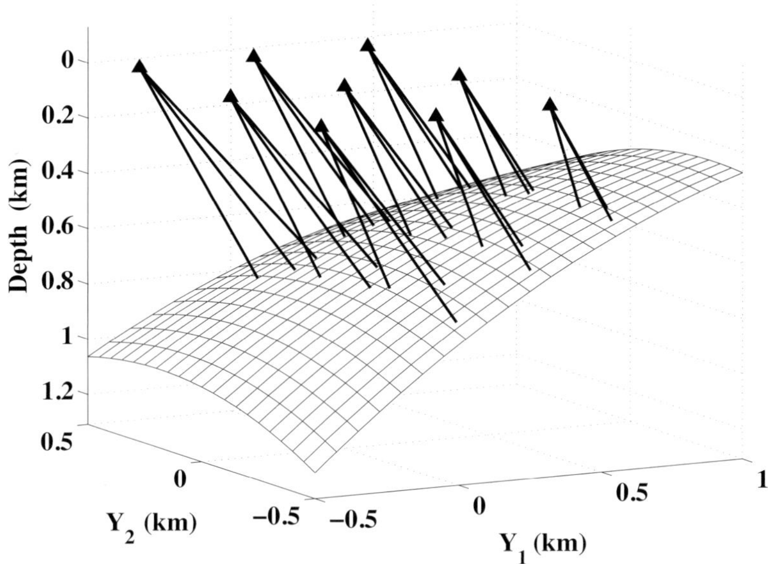

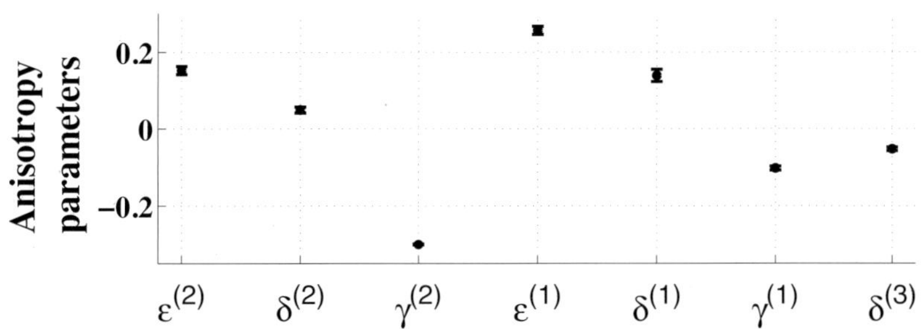

Figure 6.10: Zero-offset rays of PP-, S_1S_1-, and S_2S_2–waves reflected from a curved interface beneath an orthorhombic layer and recorded at nine CMP locations (top plot); the estimated anisotropy parameters are shown on the bottom (Grechka et al., 2005). The $[x_1, x_3]$ symmetry plane makes an angle of 60° with the Y_1-axis. The medium parameters and the standard deviations of the Gaussian noise are the same as in Figure 6.8.

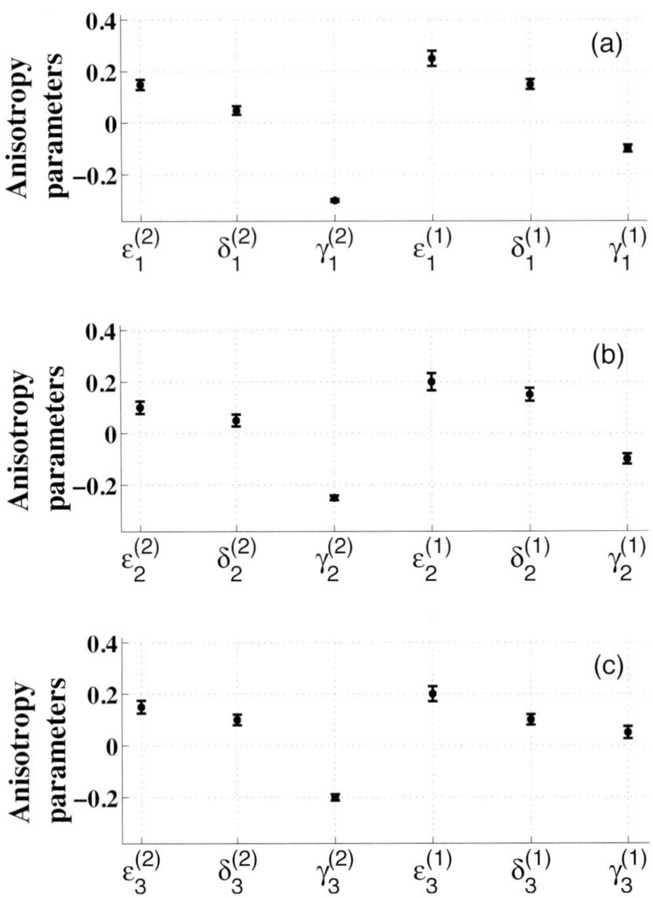

Figure 6.11: Inversion results for a model composed of three horizontal layers [(a) top layer, (b) middle layer, and (c) bottom layer] with different azimuths of the vertical symmetry planes (Grechka et al., 2005). The standard deviations of the Gaussian noise are the same as in Figure 6.8. As before, each error bar corresponds to two standard deviations. The vertical velocities were constrained to be within $\pm 3\%$ of the correct values; from top to bottom, $V_{P0,1}$=1.0 km/s, $V_{S0,1}$=0.5 km/s, $V_{P0,2}$=1.5 km/s, $V_{S0,2}$=0.6 km/s, $V_{P0,3}$=1.8 km/s, and $V_{S0,3}$=1.0 km/s. The layer thicknesses are z_1=0.4 km, z_2=0.6 km, and z_3=0.8 km, and the azimuths of the symmetry plane $[x_1, x_3]$ are $\beta_1=40°$, $\beta_2=70°$, and $\beta_3=10°$.

and contaminated with Gaussian noise. Because the inversion for horizontally layered orthorhombic models requires information about the vertical velocities or reflector depth (see sections 6.1 and 6.2), the interval velocities V_{P0} and V_{S0} were allowed to deviate by no more than ±3% from the correct values. Under these constraints, all interval anisotropy parameters except for $\delta^{(3)}$ were found with high accuracy (the standard deviations did not exceed 0.03), and the standard deviations in the azimuths of the vertical symmetry planes were less than 1.5°. As discussed above, the parameter $\delta^{(3)}$ has no influence on the NMO ellipses in a horizontal orthorhombic layer. Note that $\delta^{(3)}$ contributes to nonhyperbolic reflection moveout and can be constrained by long-offset PP-wave traveltimes (Chapter 3).

Figure 6.12 shows the inversion results for a more complicated layered model with curved interfaces. Using the input data recorded at nine common midpoints (Figure 6.12a), the algorithm estimated the interval anisotropy parameters, the azimuths β_n of the vertical symmetry planes, and the shapes of all three interfaces (the vertical velocities were fixed at the correct values). As expected for any technique based on reflection traveltimes, parameter estimation (Figures 6.12b,c,d) becomes less stable with depth. Nevertheless, the maximum standard deviation in the anisotropy parameters was smaller than 0.1 (the value for $\delta_3^{(1)}$ in Figure 6.12d). Taking into account that the error amplification for a given layer is proportional to its depth-normalized thickness, we conclude that the parameters were sufficiently well constrained. The symmetry-plane directions and the shapes of the layer boundaries were also estimated accurately, with the standard deviations in the azimuths β_1, β_2, and β_3 equal to 1.0°, 1.6°, and 4.6°, respectively.

6.4 Inversion of multicomponent data for monoclinic media

Monoclinic media have a single plane of mirror symmetry, whose orientation we assume to be horizontal. The most common physical reason for monoclinic symmetry discussed in the literature is the presence of multiple fracture sets with the fracture normals confined to the same plane (e.g., Bakulin et al., 2000c). If the background matrix is isotropic, however, the deviation of fracture-related monoclinic models from orthotropy is expected to be small (see Chapter 9). Still, a combination of aligned fractures and nonhydrostatic stresses can lower the symmetry to monoclinic or even triclinic. As an example, consider the effective medium formed by a vertical fracture set embedded in an isotropic or VTI matrix subjected to unequal horizontal stresses, which are oriented obliquely to the fractures. Such a model becomes monoclinic because it has a horizontal symmetry plane but possesses no symmetry with respect to any vertical plane.

Note that if the background matrix is TTI (or tilted orthorhombic), a vertical fracture set might produce a monoclinic medium with a *vertical* symmetry plane. A model of this kind was inferred by Winterstein and Meadows (1991) from multiaz-

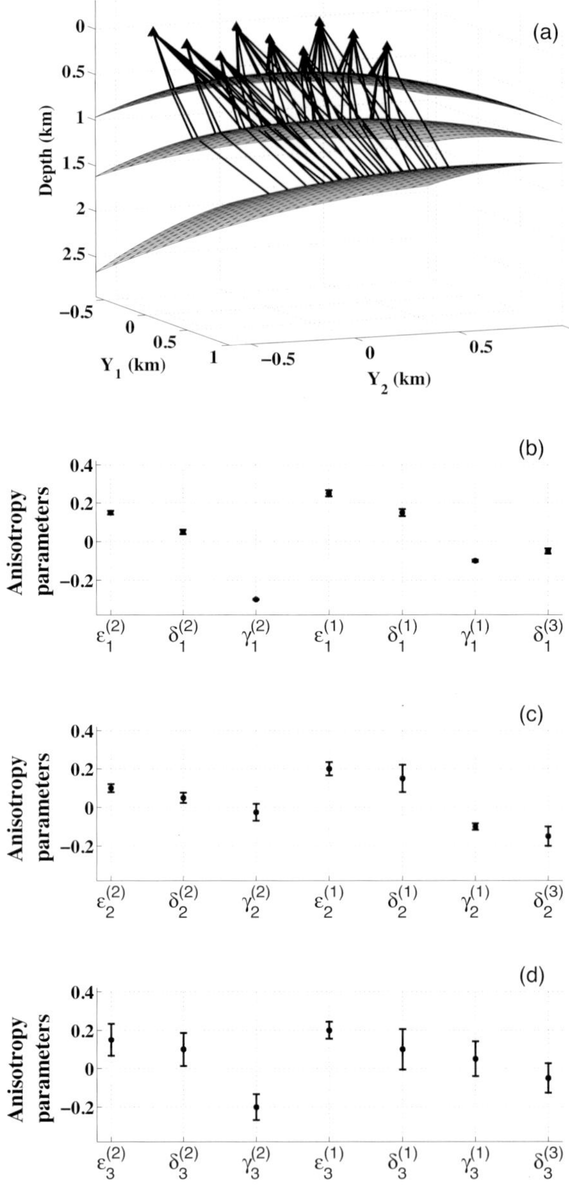

Figure 6.12: (a) Zero-offset rays of reflected PP-, S_1S_1-, and S_2S_2-waves in a layered orthorhombic medium and the inversion results for the (b) top layer, (c) middle layer, and (d) bottom layer (Grechka et al., 2005). The vertical velocities ($V_{P0,1}$=1.0 km/s, $V_{S0,1}$=0.6 km/s, $V_{P0,2}$=1.5 km/s, $V_{S0,2}$=0.8 km/s, $V_{P0,3}$=1.8 km/s, and $V_{S0,3}$=1.0 km/s) were assumed to be known. The azimuths of the symmetry plane $[x_1, x_3]$ are β_1=40°, β_2=50°, and β_3=10°. The standard deviations of the Gaussian noise are the same as in Figure 6.8.

imuth, walkaway VSP measurements of shear-wave polarization.

Because of the large number of the independent stiffness coefficients for a monoclinic medium, "blind" inversion of seismic signatures is expected to be nonunique and suffer from trade-offs among stiffnesses. To focus the inversion procedure on the parameters constrained by seismic data, Grechka et al. (2000; this section is based on their results) introduced Thomsen-style notation for monoclinic symmetry and employed it in a parameter-estimation algorithm operating with wide-azimuth multicomponent data.

6.4.1 Analytic background

Monoclinic stiffness tensor

The stiffness tensor $c_{ijk\ell}$ for a monoclinic medium with the horizontal symmetry plane $[x_1, x_2]$ is represented in a two-index (Voigt) notation as follows (e.g., Musgrave, 1970):

$$\mathbf{c}^{(\mathrm{mnc})} = \begin{pmatrix} c_{11} & c_{12} & c_{13} & 0 & 0 & c_{16} \\ c_{12} & c_{22} & c_{23} & 0 & 0 & c_{26} \\ c_{13} & c_{23} & c_{33} & 0 & 0 & c_{36} \\ 0 & 0 & 0 & c_{44} & c_{45} & 0 \\ 0 & 0 & 0 & c_{45} & c_{55} & 0 \\ c_{16} & c_{26} & c_{36} & 0 & 0 & c_{66} \end{pmatrix}. \tag{6.25}$$

According to equation 6.25, monoclinic symmetry is characterized by four additional nonzero stiffness elements (c_{45}, c_{16}, c_{26}, and c_{36}) compared to the stiffness matrices for VTI, HTI, and orthorhombic media. While the vertical axis x_3 is fixed in the direction orthogonal to the symmetry plane, the axes x_1 and x_2 can be rotated in arbitrary fashion, which changes the nonzero stiffness elements.

For purposes of parameter estimation, it is important to choose a coordinate frame in which the mathematical description of wave propagation has the simplest possible form. As discussed in Appendix 6A, for a certain rotation angle of the horizontal axes the element c_{45} vanishes and the Christoffel matrix becomes diagonal for vertical propagation (i.e., for waves with the slowness vector $\mathbf{p} = \{0, 0, q\}$). Therefore, the axes x_1 and x_2 of this coordinate system coincide with the polarization directions of the vertically traveling shear waves. In the following, we assume that the x_1-axis is parallel to the polarization vector of the fast shear wave S_1, which implies that[2]

$$c_{55} > c_{44}\,. \tag{6.26}$$

Then the polarization vectors $\mathbf{U}$ and vertical slownesses q for waves propagating in the x_3-direction are (Appendix 6A):

$$\mathbf{U}^P = [0, 0, 1]\,, \tag{6.27}$$

[2]Note that in sections 6.1 and 6.2 the polarization vector of the fast S-wave was taken to be parallel to the x_2-axis.

$$\mathbf{U}^{S1} = [1, 0, 0] \,, \tag{6.28}$$

$$\mathbf{U}^{S2} = [0, 1, 0] \,, \tag{6.29}$$

$$q^P = \sqrt{\frac{\rho}{c_{33}}} \,, \tag{6.30}$$

$$q^{S1} = \sqrt{\frac{\rho}{c_{55}}} \,, \tag{6.31}$$

$$q^{S2} = \sqrt{\frac{\rho}{c_{44}}} \,. \tag{6.32}$$

Equations 6.27 – 6.29 define the coordinate system used throughout this section. The corresponding stiffness coefficients are specified by equation 6.25 with $c_{45} = 0$ and $c_{55} > c_{44}$ (equation 6.26).

NMO ellipses for horizontal reflectors

For a horizontal anisotropic layer with arbitrary symmetry, the 2×2 matrix $\mathbf{W}$ responsible for the NMO ellipse was expressed through the slowness vector in equation 1.18:

$$\mathbf{W} = \frac{-q}{q_{,11}\, q_{,22} - q_{,12}^2} \begin{pmatrix} q_{,22} & -q_{,12} \\ -q_{,12} & q_{,11} \end{pmatrix}, \tag{6.33}$$

where $q \equiv q(p_1, p_2) \equiv p_3$ is the vertical component of the slowness vector, p_1 and p_2 are the horizontal slownesses, and $q_{,ij} \equiv \partial^2 q / (\partial p_i \partial p_j)$. Equation 6.33 is evaluated for the zero-offset ray, which has a vertical slowness vector (i.e., $p_1 = p_2 = 0$).

The matrix $\mathbf{W}$ for a given mode (PP, S_1S_1, or S_2S_2) in a horizontal monoclinic layer is obtained from equation 6.33 by substituting the corresponding value of q (equations 6.30 – 6.32) and the derivatives $q_{,ij}$ determined from the Christoffel equation. The exact expressions for the matrices $\mathbf{W}^P$, $\mathbf{W}^{S1}$, and $\mathbf{W}^{S2}$, given in Appendix 6B, help identify the stiffness coefficients and their combinations that influence the NMO ellipses. First, note that the element c_{12} is not present in equations 6.66 – 6.77 and, therefore, cannot be found from conventional-spread moveout of pure reflection modes in a horizontal monoclinic layer. The same conclusion regarding c_{12} was drawn in sections 6.1 and 6.2 for an orthorhombic layer with a horizontal symmetry plane.

Second, the coefficients c_{16}, c_{26}, and c_{36}, which vanish in orthorhombic media, contribute to the diagonal elements W_{11} and W_{22} only through their products. Furthermore, the semiaxes of the NMO ellipses do not contain terms linear in c_{i6} ($i = 1, 2, 3$) either. Therefore, the semiaxes linearized in the (typically small) stiffnesses c_{i6} should have the same form as the corresponding exact expressions in orthorhombic media. This is a strong indication that Tsvankin's notation for orthorhombic symmetry can also be useful for monoclinic models.

In contrast, the off-diagonal matrix elements W_{12}, which govern the rotation of the NMO ellipses with respect to the horizontal coordinate axes (see section 1.1), are *approximately linear* in c_{16}, c_{26}, and c_{36} (equations 6.67, 6.71, and 6.75). Therefore, the coefficients c_{i6} are primarily responsible for the deviation of the axes of the NMO ellipses from the polarization directions of the vertically propagating shear waves.

Notation for monoclinic media

Analysis of the NMO ellipses helped Grechka et al. (2000) extend Tsvankin's (1997a) notation (see Appendix 1B) to monoclinic media. Expressions for their parameters in terms of the stiffness coefficients c_{ij} and density ρ are given below. Note that the horizontal coordinate axes are aligned with the polarization directions of the vertically traveling shear waves.

- V_{P0} is the P-wave vertical velocity:

$$V_{P0} \equiv \sqrt{\frac{c_{33}}{\rho}}. \tag{6.34}$$

- V_{S0} is the velocity of the vertically propagating shear wave polarized in the x_1-direction:

$$V_{S0} \equiv \sqrt{\frac{c_{55}}{\rho}}. \tag{6.35}$$

- $\epsilon^{(1)}$ is defined in the $[x_2, x_3]$-plane by analogy with the VTI parameter ϵ (the superscript denotes the axis x_1 orthogonal to the $[x_2, x_3]$-plane):

$$\epsilon^{(1)} \equiv \frac{c_{22} - c_{33}}{2c_{33}}. \tag{6.36}$$

- $\delta^{(1)}$ is defined in the $[x_2, x_3]$-plane by analogy with the VTI parameter δ:

$$\delta^{(1)} \equiv \frac{(c_{23} + c_{44})^2 - (c_{33} - c_{44})^2}{2c_{33}(c_{33} - c_{44})}. \tag{6.37}$$

- $\gamma^{(1)}$ is defined in the $[x_2, x_3]$-plane by analogy with the VTI parameter γ:

$$\gamma^{(1)} \equiv \frac{c_{66} - c_{55}}{2c_{55}}. \tag{6.38}$$

- $\epsilon^{(2)}$ is defined in the $[x_1, x_3]$-plane by analogy with the VTI parameter ϵ:

$$\epsilon^{(2)} \equiv \frac{c_{11} - c_{33}}{2c_{33}}. \tag{6.39}$$

- $\delta^{(2)}$ is defined in the $[x_1, x_3]$-plane by analogy with the VTI parameter δ:

$$\delta^{(2)} \equiv \frac{(c_{13} + c_{55})^2 - (c_{33} - c_{55})^2}{2c_{33}(c_{33} - c_{55})}. \tag{6.40}$$

- $\gamma^{(2)}$ is defined in the $[x_1, x_3]$-plane by analogy with the VTI parameter γ:

$$\gamma^{(2)} \equiv \frac{c_{66} - c_{44}}{2c_{44}}. \tag{6.41}$$

- $\delta^{(3)}$ is defined in the $[x_1, x_2]$-plane (x_1 plays the role of the symmetry axis) by analogy with the VTI parameter δ:

$$\delta^{(3)} \equiv \frac{(c_{12}+c_{66})^2-(c_{11}-c_{66})^2}{2c_{11}(c_{11}-c_{66})}\,. \tag{6.42}$$

- $\zeta^{(1)}$ is responsible for the rotation of the NMO ellipse of S_1S_1-waves:

$$\zeta^{(1)} \equiv \frac{c_{16}-c_{36}}{2c_{33}}\,. \tag{6.43}$$

- $\zeta^{(2)}$ is responsible for the rotation of the NMO ellipse of S_2S_2-waves:

$$\zeta^{(2)} \equiv \frac{c_{26}-c_{36}}{2c_{33}}\,. \tag{6.44}$$

- $\zeta^{(3)}$ is responsible for the rotation of the NMO ellipse of PP-waves:

$$\zeta^{(3)} \equiv \frac{c_{36}}{c_{33}}\,. \tag{6.45}$$

The parameters V_{P0}, V_{S0}, $\epsilon^{(1,2)}$, $\delta^{(1,2,3)}$, and $\gamma^{(1,2)}$ (equations 6.34 – 6.42) are defined exactly in the same way as the corresponding Tsvankin's parameters for orthorhombic media. One of the advantages of Tsvankin's notation is in providing a simple way to adapt VTI velocity and polarization equations (expressed through Thomsen parameters) for the vertical symmetry planes of orthorhombic media. Unfortunately, the equivalence with VTI media is no longer valid for monoclinic symmetry. Still, the "orthorhombic" parameters proved useful in simplifying the NMO equations in a monoclinic layer.

The remaining parameters ($\zeta^{(1,2,3)}$; equations 6.43 – 6.45) depend on the stiffnesses that vanish in orthorhombic media. One of them, $\zeta^{(3)}$, becomes identical to the parameter χ_z introduced by Mensch and Rasolofosaon (1997) if V_{P0} is chosen as the reference velocity in their model and the coordinate axes x_1 and x_2 are rotated to eliminate the coefficient c_{45}. The parameters $\zeta^{(1)}$ and $\zeta^{(2)}$ differ from their counterparts in the Mensch-Rasolofosaon (1997) notation because the latter is based on the weak-anisotropy approximation for phase velocity.

Approximate NMO ellipses

To simplify the exact equations for the NMO ellipses given in Appendix 6B, it is convenient to linearize them in the parameters $\zeta^{(1,2,3)}$. Additional linearizations in the parameters $\epsilon^{(1,2)}$ $\delta^{(1,2)}$ and $\gamma^{(1,2)}$ are also performed for the off-diagonal matrix elements W_{12} (but not for W_{11} and W_{22}). This yields the following approximations for the NMO ellipses of PP-, S_1S_1-, and S_2S_2-waves in a horizontal monoclinic layer:

$$W_{11}^P = \frac{1}{V_{P0}^2\,(1+2\delta^{(2)})}\,, \tag{6.46}$$

$$W_{12}^P = -2\,\frac{\zeta^{(3)}}{V_{P0}^2}\,, \tag{6.47}$$

$$W_{22}^P = \frac{1}{V_{P0}^2\,(1+2\delta^{(1)})}\,, \tag{6.48}$$

$$W_{11}^{S1} = \frac{1}{V_{S1}^2\,(1+2\sigma^{(2)})}\,, \tag{6.49}$$

$$W_{12}^{S1} = -2\,\frac{\zeta^{(1)}}{V_{S1}^2}\left(\frac{V_{P0}}{V_{S1}}\right)^2, \tag{6.50}$$

$$W_{22}^{S1} = \frac{1}{V_{S1}^2\,(1+2\gamma^{(1)})}\,, \tag{6.51}$$

$$W_{11}^{S2} = \frac{1}{V_{S2}^2\,(1+2\gamma^{(2)})}\,, \tag{6.52}$$

$$W_{12}^{S2} = -2\,\frac{\zeta^{(2)}}{V_{S2}^2}\left(\frac{V_{P0}}{V_{S2}}\right)^2, \tag{6.53}$$

$$W_{22}^{S2} = \frac{1}{V_{S2}^2\,(1+2\sigma^{(1)})}\,. \tag{6.54}$$

Here

$$\sigma^{(2)} \equiv \left(\frac{V_{P0}}{V_{S1}}\right)^2 \left(\epsilon^{(2)}-\delta^{(2)}\right), \qquad \sigma^{(1)} \equiv \left(\frac{V_{P0}}{V_{S2}}\right)^2 \left(\epsilon^{(1)}-\delta^{(1)}\right). \tag{6.55}$$

V_{S1} and V_{S2} are the vertical velocities of the fast and slow shear waves:

$$V_{S1} = V_{S0}\,, \qquad V_{S2} = V_{S0}\sqrt{\frac{1+2\gamma^{(1)}}{1+2\gamma^{(2)}}}\,. \tag{6.56}$$

It is interesting that the approximate diagonal elements W_{11} and W_{22} for all three modes become identical to the corresponding *exact* equations 6.1, 6.2, and 6.9–6.12 for the semiaxes of the NMO ellipses in orthorhombic media, if we take into account that in section 6.1 the vertically traveling S_1-wave was assumed to be polarized in the x_2-direction. As mentioned above, the "monoclinic" coefficients $\zeta^{(1,2,3)}$ contribute to W_{11} and W_{22} only through their products, which were dropped during the linearization procedure.

In contrast, the off-diagonal matrix elements W_{12} are linear in $\zeta^{(1,2,3)}$ and quadratic in other anisotropy parameters. If the medium is made orthorhombic by setting $\zeta^{(1,2,3)}=0$, the elements W_{12} vanish, which means that the axes of the NMO ellipses of all three modes are aligned with the coordinate directions x_1 and x_2 (i.e., with the shear-wave polarizations for vertical propagation). Then the diagonal elements W_{11} and W_{22} determine the NMO velocities in the vertical symmetry planes of orthorhombic media (see section 6.1).

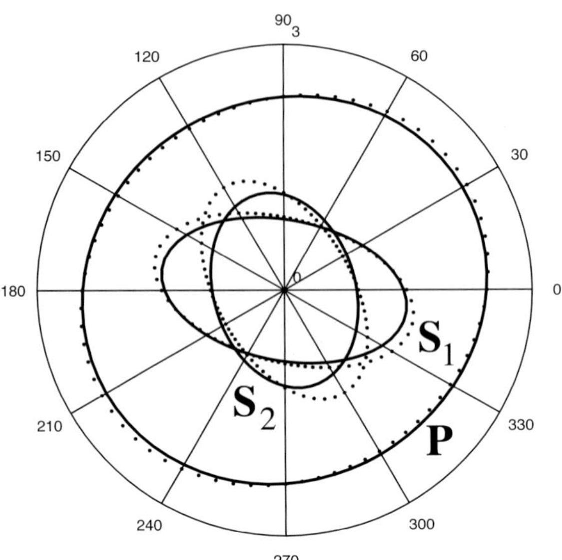

Figure 6.13: Exact (solid curves) and approximate (dotted) NMO ellipses of pure reflection modes in a horizontal monoclinic layer (Grechka et al., 2000). The exact ellipses are computed from equations 6.66 – 6.77, and the approximate ellipses from equations 6.46 – 6.54. The medium parameters are V_{P0}=2.0 km/s, V_{S0}=1.0 km/s, $\epsilon^{(1)}$=0.3, $\epsilon^{(2)}$=0.4, $\delta^{(1)}$=0.2, $\delta^{(2)}$=0.25, $\gamma^{(1)}$=−0.1, $\gamma^{(2)}$=0.15, $\zeta^{(1)}$=−0.03, $\zeta^{(2)}$=−0.02, and $\zeta^{(3)}$=0.04; zero azimuth (axis x_1) corresponds to the polarization vector of the S_1-wave at vertical incidence.

For general monoclinic media, all three off-diagonal elements $W_{12} \neq 0$, and the axes of the NMO ellipses of PP-, S_1S_1- and S_2S_2-waves deviate from the coordinate directions (i.e., from the S-wave polarization vectors) and have different orientations. Sayers (1998) drew similar conclusions from his study of the principal azimuthal directions of the P-wave phase-velocity function for monoclinic symmetry. Experimental evidence of the misalignment of the PP-wave NMO ellipse and shear-wave polarization directions was presented by Pérez et al. (1999) who analyzed multicomponent data acquired over a fractured carbonate reservoir.

Numerical tests show that the accuracy of equations 6.46 – 6.54 depends mostly on the magnitudes of $\zeta^{(1,2,3)}$ and is less sensitive to the other anisotropy parameters. As follows from the results of Bakulin et al. (2000c), the absolute values of the parameters $\zeta^{(1,2,3)}$ for the simplest monoclinic models resulting from two nonorthogonal vertical fracture sets seldom exceed 0.05 – 0.06. Then equations 6.46 – 6.54 give an adequate approximation for the exact NMO ellipses (Figure 6.13). The error of the approximate equations, however, rapidly increases when the magnitudes of $\zeta^{(1,2,3)}$ reach 0.1.

As predicted by the analytic results, the axes of the NMO ellipses of PP-, S_1S_1-, and S_2S_2-waves in Figure 6.13 are parallel neither to one another nor to the coordinate directions. The azimuths of their major axes with respect to the polarization vector of the fast shear wave (azimuth 0°) are $\beta^P = 32°$, $\beta^{S1} = 169°$, and $\beta^{S2} = 106°$. The misalignment of the NMO ellipses is a distinctive feature of monoclinic media that can be used to identify the symmetry and constrain the anisotropy parameters.

6.4.2 Inversion of NMO ellipses for a single layer

Analysis of the weak-anisotropy approximation

Although equations 6.46 – 6.54 lose accuracy with increasing $|\zeta^{(1,2,3)}|$, they provide useful insight into the influence of anisotropy and help in designing the inversion procedure. Just one of the anisotropy parameters, $\delta^{(3)}$, does not appear in any of these equations. Since $\delta^{(3)}$ is the only parameter dependent on the stiffness element c_{12}, this result is consistent with our earlier observation that c_{12} does not contribute to the exact NMO ellipses for horizontal reflectors.

The diagonal matrix elements W_{11} and W_{22}, which yield the NMO velocities along the x_1- and x_2-axes, do not include terms linear in the "monoclinic" parameters $\zeta^{(1,2,3)}$; the same is true for the semiaxes of the NMO ellipses. Neglecting the contributions of the ζ-parameters makes the inversion of the elements W_{11} and W_{22} (supplemented by the zero-offset traveltimes) completely analogous to the parameter-estimation problem for orthorhombic media discussed in sections 6.1 and 6.2. To resolve the anisotropy parameters $\epsilon^{(1,2)}$, $\delta^{(1,2)}$, and $\gamma^{(1,2)}$, either the reflector depth or one of the vertical velocities has to be known. In section 6.2 we successfully tested this depth-constrained inversion procedure on physical-modeling reflection data from an orthorhombic material.

The additional anisotropy parameters of monoclinic media ($\zeta^{(1,2,3)}$) can be estimated from the off-diagonal matrix elements W_{12}, which provide three linear (in the weak-anisotropy approximation) equations for the three unknowns. Typically, the parameters $\zeta^{(1,2,3)}$ determine the orientation of the NMO ellipses because the rotation angles β of the elliptical axes with respect to the shear-wave polarization directions are governed mostly by W_{12} (unless W_{11} is close to W_{22}). Hence, the vertical velocities and NMO ellipses of PP-, S_1S_1-, and S_2S_2-waves contain sufficient information for estimating 11 (out of 12) parameters of a horizontal monoclinic layer. Below, this conclusion is confirmed by performing numerical inversion based on exact NMO equations.

Synthetic tests

Input data for the parameter-estimation procedure include the vertical velocities and NMO ellipses of the three pure modes (PP, S_1S_1, and S_2S_2). If shear-wave sources are not available, reflection moveouts of S_1S_1- and S_2S_2-waves can be computed by applying the PP+ PS = SS method (see Chapters 4 and 5) to PP, PS_1, and PS_2 data. Since here we consider only horizontally layered monoclinic models with a horizontal symmetry plane, it is also possible to obtain the SS-wave NMO ellipses directly from those of PP and PS reflections using equation 6.4. Knowledge of the vertical velocity of a single mode (e.g., estimated from check shots) is sufficient because the other two vertical velocities can be determined from the zero-offset traveltimes.

The coordinate frame needed for moveout inversion can be established by performing Alford (1986) rotation of small-offset shear data to identify the S-wave polarization directions at vertical incidence. Such a rotation is necessary anyway since the split mode-converted or pure shear waves have to be separated prior to moveout analysis.

For inversion purposes, the exact NMO equations from Appendix 6B were rewritten through the anisotropy parameters $\epsilon^{(1,2)}$, $\delta^{(1,2)}$, $\gamma^{(1,2)}$, and $\zeta^{(1,2,3)}$ defined in equations 6.34 – 6.45. An additional constraint on the parameters $\gamma^{(1,2)}$ is provided by the shear-wave vertical velocities (equation 6.56). The inversion is performed by minimizing the following objective function:

$$\begin{aligned}\mathcal{F} \equiv & \left[\frac{V_{P0}^{\text{meas}}}{V_{P0}} - 1\right]^2 + \left[\frac{V_{S1}^{\text{meas}}}{V_{S1}} - 1\right]^2 + \left[\frac{V_{S2}^{\text{meas}}}{V_{S2}} - 1\right]^2 \\ & + \int_0^{2\pi} \left[\frac{V_{\text{nmo},P}^{\text{meas}}(\alpha)}{V_{\text{nmo},P}(\alpha)} - 1\right]^2 d\alpha + \int_0^{2\pi} \left[\frac{V_{\text{nmo},S1}^{\text{meas}}(\alpha)}{V_{\text{nmo},S1}(\alpha)} - 1\right]^2 d\alpha \\ & + \int_0^{2\pi} \left[\frac{V_{\text{nmo},S2}^{\text{meas}}(\alpha)}{V_{\text{nmo},S2}(\alpha)} - 1\right]^2 d\alpha\,, \end{aligned} \tag{6.57}$$

where each term contains the ratio of a measured quantity (denoted by "meas") and the same quantity computed for a trial set of the anisotropy parameters. The NMO velocities are obtained by substituting the exact expressions for the matrices $\mathbf{W}$ (Appendix 6B) into the equation of the NMO ellipse. The measured vertical velocities are included in the objective function to allow for adjustments in their values that might be needed to better approximate the NMO ellipses in the presence of noise in the input data. The function $\mathcal{F}$ is minimized using the simplex method (Press et al., 1987), with the initial parameter values determined from the weak-anisotropy approximations 6.46 – 6.54.

The inversion results for the monoclinic layer from Figure 6.13 are shown in Figure 6.14. The exact NMO velocities of PP-, S_1S_1-, and S_2S_2-waves were computed for four azimuths (0°, 45°, 90°, and 135°) and supplemented by the vertical velocities V_{P0}, V_{S1}, and V_{S2}. Measurement errors were simulated by adding Gaussian noise with a variance of 2% to the vertical and NMO velocities. Then the NMO ellipses were reconstructed for 200 realizations of the noise-contaminated data and, in combination with the vertical velocities, inverted for the anisotropy parameters. Overall, the stability of the algorithm is satisfactory, but there is a substantial variation in the results from one anisotropy parameter to another. In particular, the error bars for $\zeta^{(1)}$ and $\zeta^{(2)}$ are much smaller than those for $\zeta^{(3)}$ and for the other parameters, which is explained by the structure of equations 6.47, 6.50, and 6.53 for the off-diagonal matrix elements W_{12}. The expressions for W_{12}^{S1} and W_{12}^{S2} have large scaling factors, $(V_{P0}/V_{S1})^2$ and $(V_{P0}/V_{S2})^2$, equal to 4.0 and 6.5 for the model used in the numerical test. Therefore, the shear-wave elements W_{12}^{Sk} make a more significant contribution to the respective NMO ellipses than does W_{12}^{P}, which helps estimate $\zeta^{(1,2)}$ with higher accuracy.

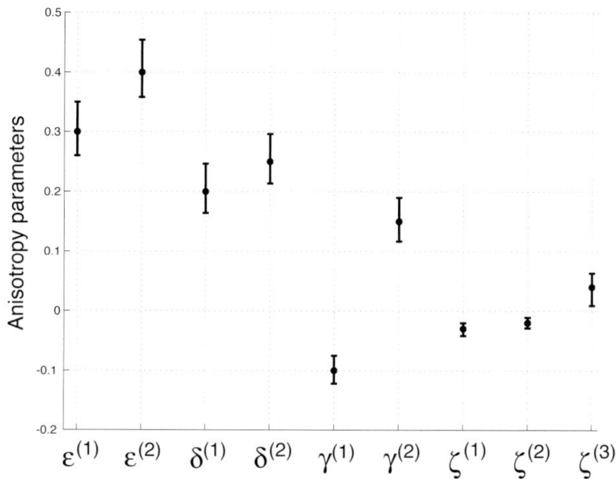

Figure 6.14: Results of the moveout inversion for the monoclinic layer from Figure 6.13 (Grechka et al., 2000). The dots are the actual parameters, and the bars mark two standard deviations in each parameter. The standard deviations for the velocities V_{P0} and V_{S0} (not shown) are 2.1% and 2.0%, respectively.

It might seem that the parameters $\zeta^{(1,2)}$ in Figure 6.14 are so well resolved because the shear-wave NMO ellipses are noticeably elongated (Figure 6.13), and their orientations are well-constrained. The next test, however, indicates that the ζ-parameters can be estimated even if the ellipticity is small. For the model in Figure 6.15 the eccentricity of the NMO ellipses of PP-, S_1S_1-, and S_2S_2-waves reaches only 1.9%, 1.7%, and 2.1%, respectively. Therefore, for all three modes the azimuth β is poorly constrained by the data. One might expect that adding a 2% velocity error can produce significant rotation of the ellipses and, consequently, sizeable errors in $\zeta^{(1,2,3)}$. The error bars for all anisotropy parameters in Figure 6.16 (again, the inversion was repeated for 200 realizations of Gaussian noise), however, are similar to those in Figure 6.14. To explain this result, recall that the inversion algorithm operates with the matrices W_{ij} rather than the angles β. The elements W_{12}, which depend primarily on the parameters $\zeta^{(1,2,3)}$, can be resolved from wide-azimuth data even when NMO ellipticity is relatively weak.

6.4.3 Parameter estimation for layered media

Extension of the multicomponent moveout inversion to horizontally layered monoclinic models is based on the approach discussed in section 6.1 for orthorhombic media. The effective NMO ellipses of PP-, S_1S_1-, and S_2S_2-waves for the top and bottom of the layer of interest can be obtained from PP and PS data using either the PP+PS=SS method or equation 6.4. Then the interval pure-mode NMO ellipses are computed by generalized Dix differentiation (equation 1.32) and inverted for the anisotropy parameters using the single-layer algorithm. Alternatively, inter-

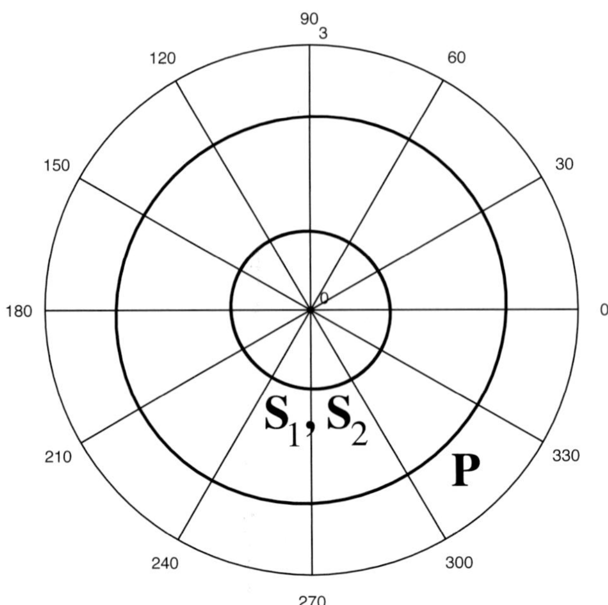

Figure 6.15: Exact NMO ellipses in a monoclinic layer with the parameters V_{P0}=2.0 km/s, V_{S0}=1.0 km/s, $\epsilon^{(1)}$=0.123, $\epsilon^{(2)}$=0.075, $\delta^{(1)}$=$\delta^{(2)}$=0.1, $\gamma^{(1)}$=−0.1, $\gamma^{(2)}$=0.15, $\zeta^{(1)}$=−0.002, $\zeta^{(2)}$=−0.003, and $\zeta^{(3)}$=0.02 (Grechka et al., 2000).

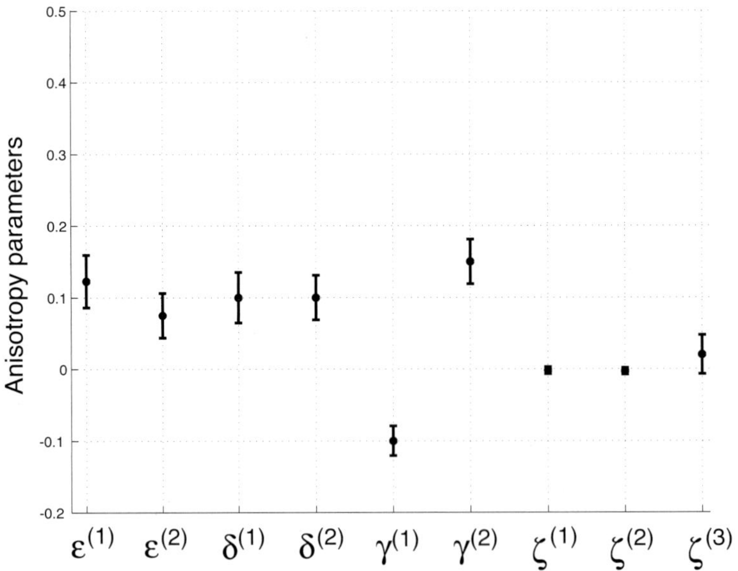

Figure 6.16: Results of the moveout inversion for the layer from Figure 6.15 (Grechka et al., 2000). The dots are the actual parameters, and the bars mark two standard deviations in each parameter. The standard deviations for the velocities V_{P0} and V_{S0} (not shown) are 2.0%.

val reflection traveltimes can be found by the velocity-independent layer-stripping method (VILS, see section 3.4). The coordinate frame in each layer is defined by the shear-wave polarization directions obtained from Alford rotation of reflected SS or PS arrivals followed by polarization layer stripping (Thomsen et al., 1999). For laterally homogeneous models with a horizontal symmetry plane, the NMO ellipses of pure modes can be also measured in a straightforward way from walkaway VSP surveys (see Chapter 7 and the example below).

The parameter-estimation procedure for stratified monoclinic media was tested on the three-layer model from Table 6.3. The polarization directions of the vertically propagating S_1- and S_2-waves, which define the coordinate axes x_1 and x_2, were the same in all layers. To reproduce a walkaway VSP experiment in horizontally layered media, three "downhole" receivers were placed at the layer boundaries. The traveltimes of the direct P-, S_1-, and S_2-waves (Figure 6.17) were computed between the receivers and surface sources placed on six lines making angles of 0°, 30°, 60°, 90°, 120°, and 150° with the x_1-axis. The maximum source-receiver offset reaches 1 km, which corresponds to a ray angle of 45° with the vertical for the most shallow receiver. The traveltimes for that receiver noticeably vary with azimuth and exhibit pronounced deviations from hyperbolic moveout (Figure 6.17). To mitigate the influence of nonhyperbolic moveout on estimation of the NMO ellipses, modeled traveltimes for all offsets and azimuths were approximated by the quartic moveout equation. Using general equation 3.5, it can be shown that the quartic coefficient A_4 in a monoclinic layer depends on five independent parameters (also see Sayers and Ebrom, 1997).

The effective "finite-spread" NMO ellipses were substituted into the generalized Dix equation 1.32 to compute the interval ellipses in all three layers (Figure 6.18). Along with the vertical velocities obtained from the vertical times and exact receiver depths, the interval NMO ellipses of the pure modes (Figures 6.18d,e,f) represent the input data for the inversion procedure (see equation 6.57). The estimated parameters listed in Table 6.4 are close to the actual values in Table 6.3, with the maximum error reaching only 0.03 for the parameters $\gamma^{(1,2)}$ in the second layer. The inversion errors are caused primarily by the influence of nonhyperbolic moveout, which was not completely removed by the quartic moveout equation. Muting out far offsets or applying the more accurate Tsvankin-Thomsen equation 3.32 reduces the errors to negligible values of less than 0.01.

6.5 Summary

The first two sections of this chapter were devoted to joint inversion of wide-azimuth reflection traveltimes of PP-waves and split converted waves PS_1 and PS_2 for horizontally layered orthorhombic media. If each layer has a horizontal symmetry plane, converted-wave moveout is symmetric with respect to zero offset and, for moderate-length spreads, is governed primarily by the NMO ellipse. The inversion algorithm is based on the relationship between the normal moveout of pure and converted waves

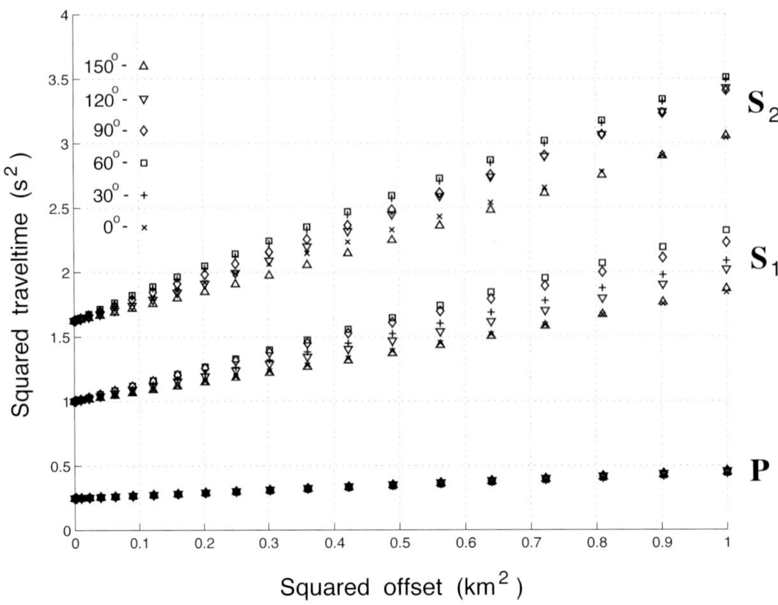

Figure 6.17: Squared ray-traced traveltimes of the direct P-, S_1-, and S_2-waves excited at the surface and recorded by a receiver placed at a depth of 1.0 km in the model from Table 6.3 (Grechka et al., 2000). Each symbol corresponds to a fixed source-receiver azimuth, as indicated by the legend.

z (km)	V_{P0} (km/s)	V_{S0} (km/s)	$\epsilon^{(1)}$	$\epsilon^{(2)}$	$\delta^{(1)}$	$\delta^{(2)}$	$\gamma^{(1)}$	$\gamma^{(2)}$	$\zeta^{(1)}$	$\zeta^{(2)}$	$\zeta^{(3)}$
1.0	2.0	1.0	0.10	0.20	0.10	0.15	−0.10	0.15	−0.02	−0.03	0.05
1.5	2.5	1.1	0.05	0.12	0.03	0.10	0.05	0.25	0.02	−0.01	−0.07
2.5	3.0	1.4	0.07	0.10	0.05	0.07	0.03	0.12	−0.02	−0.01	0.03

Table 6.3: Parameters of the three-layer monoclinic model used in the inversion; z is the depth of the bottom of each layer (Grechka et al., 2000).

$\epsilon^{(1)}$	$\epsilon^{(2)}$	$\delta^{(1)}$	$\delta^{(2)}$	$\gamma^{(1)}$	$\gamma^{(2)}$	$\zeta^{(1)}$	$\zeta^{(2)}$	$\zeta^{(3)}$
0.11	0.20	0.10	0.15	−0.09	0.17	−0.02	−0.03	0.05
0.04	0.12	0.03	0.10	0.02	0.22	0.02	0.00	−0.06
0.07	0.10	0.05	0.07	0.02	0.11	−0.02	−0.01	0.03

Table 6.4: Estimated anisotropy parameters for the model from Table 6.3 (Grechka et al., 2000).

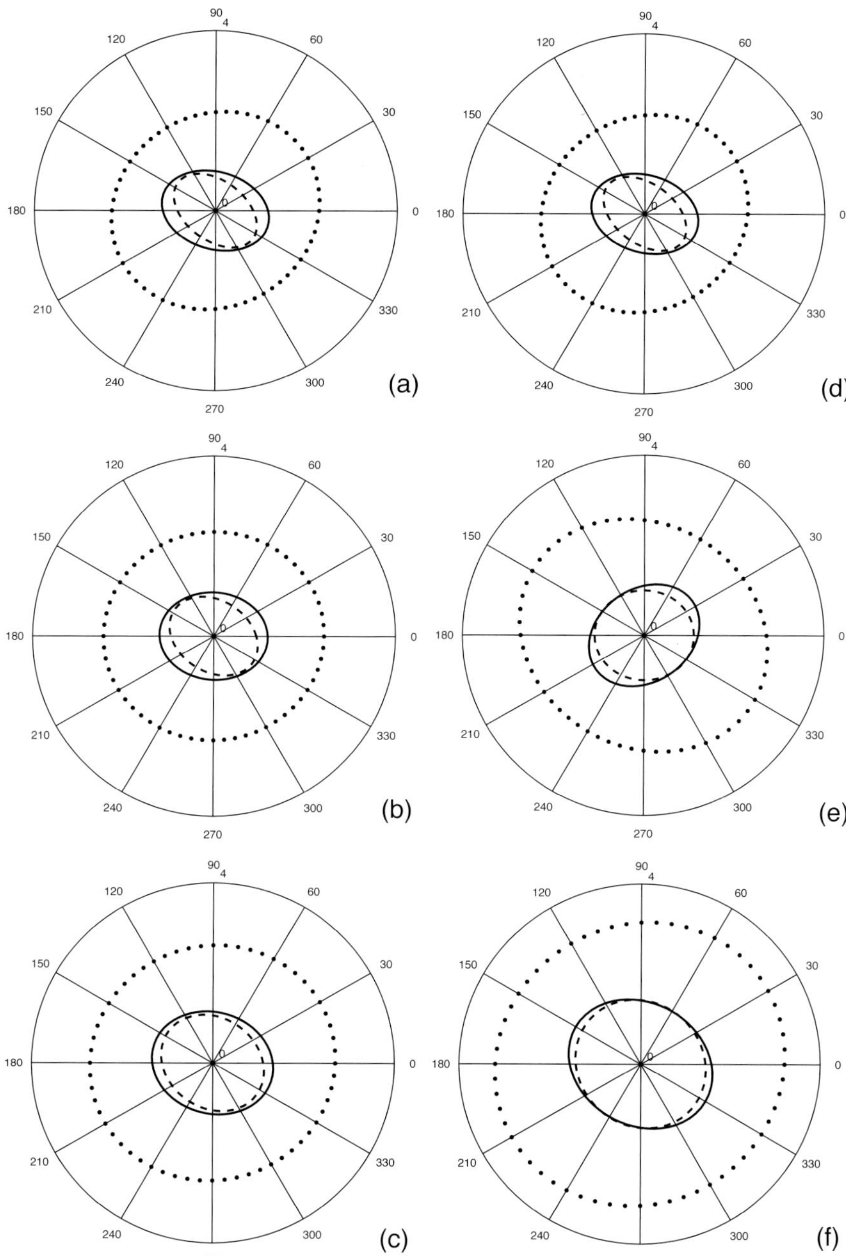

Figure 6.18: Effective NMO ellipses obtained from ray-traced traveltimes for buried receivers at depths (a) z=1.0 km, (b) 1.5 km, and (c) 2.5 km (Grechka et al., 2000). The interval NMO ellipses computed from the generalized Dix equation for the (d) first (top) layer (it coincides with the effective ellipse), (e) second layer, and (f) third layer. The P-wave ellipses are marked by dotted lines, S_1-wave ellipses by solid lines, and S_2-wave ellipses by dashed lines.

derived in section 1.6, which helps compute the effective NMO ellipses of pure SS reflections from PP and PS data. Alternatively, wide-azimuth SS (ΨS) traveltimes can be generated using the 3D version of the PP+PS=SS method introduced in Chapter 4. Then the interval NMO ellipses of PP, S_1S_1, and S_2S_2 reflections are found by applying the generalized Dix differentiation or velocity-independent layer stripping (VILS). Finally, the interval symmetry-plane NMO velocities of the PP- and SS-waves are inverted for the anisotropy parameters under the assumption that the layer thickness (or one of the vertical velocities) is known.

The method was tested on physical-modeling reflection data acquired over a block of orthorhombic Phenolic material. The NMO ellipses and zero-offset traveltimes of PP-, PS_1-, and PS_2-waves were estimated from semblance analysis on several differently oriented CMP lines and inverted for the orientation of the vertical symmetry planes and eight (out of nine) Tsvankin's parameters. It should be mentioned that extrapolation of the NMO ellipses yielded the moveout velocities of the physically *nonexistent* converted PSH-waves in the vertical symmetry planes. The parameter $\delta^{(3)}$, which is not constrained by conventional-spread reflection traveltimes, was found from the group velocity of the direct P-wave traveling in the horizontal symmetry plane. The high accuracy of the obtained orthorhombic model was confirmed in several ways, including a comparison of the inverted and measured horizontal velocities of P- and S-waves. One of the main issues in implementing this algorithm for field data is reliable separation of the split converted waves PS_1 and PS_2, especially for models with depth-varying azimuths of the symmetry planes.

In the third section we extended joint inversion of wide-azimuth PP and PS traveltimes to multiple orthorhombic layers separated by dipping or curved interfaces. Following the approach described in Chapters 4 and 5, mode-converted waves were replaced in velocity analysis with the pure shear reflections produced by the PP+PS=SS method. The NMO ellipses, zero-offset traveltimes, and reflection time slopes of the PP-, S_1S_1-, and S_2S_2-waves served as the input to multicomponent stacking-velocity inversion (*tomography*) generalized for orthorhombic symmetry. The feasibility of depth-domain interval parameter estimation strongly depends on the available reflector dips and orientations. For a plane reflector beneath a single orthorhombic layer, the inversion is nonunique when the dip plane is close to one of the vertical symmetry planes of the medium. Stable parameter estimation becomes possible if the dip direction deviates by more than $10°$ from the nearest symmetry plane and the dip reaches $15° - 20°$. In principle, the algorithm can reconstruct a layered orthorhombic model in depth if the dip and orientation of each interface satisfy the conditions established for a single layer. However, numerical tests indicate that it is necessary to put tight constraints on the interval vertical velocities to avoid instability in the inversion of noise-contaminated reflection data.

Finally, we discussed moveout inversion of wide-azimuth, multicomponent (PP and PS, or PP and SS) data for monoclinic models with a horizontal symmetry plane. The inversion procedure was facilitated by introducing a Thomsen-style notation designed to capture the combinations of the stiffness elements responsible for the kinematic signatures of reflected waves. In addition to the seven anisotropy parameters defined by Tsvankin (1997a) for orthorhombic symmetry, this notation includes the parameters $\zeta^{(1,2,3)}$ responsible for the rotation of the pure-mode NMO ellipses with respect to the polarization directions of the vertically traveling shear waves. If the thickness of a horizontal monoclinic layer is known, the shear-wave polarizations at vertical incidence and the NMO ellipses of PP-, S_1S_1-, and S_2S_2-waves provide sufficient information for obtaining eleven (out of twelve; the only exception is $\delta^{(3)}$) medium parameters. To carry out interval parameter estimation for stratified monoclinic models, the single-layer inversion has to be preceded by shear-wave polarization analysis and either generalized Dix differentiation or velocity-independent layer stripping (VILS).

Appendices for Chapter 6

6A Coordinate frame for monoclinic media

The fourth-rank stiffness tensor $c_{ijk\ell}^{(\mathrm{mnc})}$ for a monoclinic medium with a horizontal symmetry plane can be written in the form of a 6×6 stiffness matrix using the Voigt notation (see equation 6.25). Rotating the tensor $c_{ijk\ell}^{(\mathrm{mnc})}$ by an arbitrary angle θ around the vertical axis x_3 yields a tensor

$$\tilde{c}_{i'j'k'\ell'}^{(\mathrm{mnc})} = c_{ijk\ell}^{(\mathrm{mnc})}\, r_{i'i}\, r_{j'j}\, r_{k'k}\, r_{\ell'\ell}\,, \tag{6.58}$$

which has the same structure as $c_{ijk\ell}^{(\mathrm{mnc})}$; summation over repeated indices from 1 to 3 is implied. The elements $r_{m'm}$ in equation 6.58 form the rotation matrix $\mathbf{r}$:

$$\mathbf{r} = \begin{pmatrix} \cos\theta & \sin\theta & 0 \\ -\sin\theta & \cos\theta & 0 \\ 0 & 0 & 1 \end{pmatrix}. \tag{6.59}$$

If the tensor $\mathbf{c}^{(\mathrm{mnc})}$ is rotated by the angle

$$\theta = \frac{1}{2}\tan^{-1}\left(\frac{2c_{45}}{c_{55}-c_{44}}\right), \tag{6.60}$$

the element $\tilde{c}_{45}$ goes to zero, as discussed in more detail by Slawinski (2010, section 5.6.3), Helbig (1994, section 3.6), and Mensch and Rasolofosaon (1997). The horizontal axes x_1 and x_2 in the coordinate frame for which $\tilde{c}_{45}=0$ are parallel to the polarization vectors of the vertically propagating shear waves. This result follows from the Christoffel equation:

$$[G_{ik} - \rho\,\delta_{ik}]\,U_k = 0\,, \tag{6.61}$$

where ρ is the density, δ_{ik} is Kronecker's symbolic δ, $\mathbf{U}$ is the unit polarization vector of a plane wave, and $\mathbf{G}$ is the Christoffel matrix dependent on the slowness vector $\mathbf{p}$:

$$G_{ik} = c_{ijk\ell}\, p_j\, p_\ell\,. \tag{6.62}$$

For vertical propagation (i.e., for $\mathbf{p}=[0,0,p_3]$) in the coordinate frame for which $c_{45}=0$, equation 6.61 reduces to

$$\begin{bmatrix} c_{55}\,p_3^2-\rho & 0 & 0 \\ 0 & c_{44}\,p_3^2-\rho & 0 \\ 0 & 0 & c_{33}\,p_3^2-\rho \end{bmatrix} \begin{bmatrix} U_1 \\ U_2 \\ U_3 \end{bmatrix} = 0\,. \tag{6.63}$$

The polarization vectors $\mathbf{U}$, which represent the eigenvectors of the Christoffel matrix, are aligned with the axes of this coordinate frame. The third row of the matrix in equation 6.63 describes the vertically polarized P-wave with the slowness

$$q^P \equiv p_3^P = \sqrt{\frac{\rho}{c_{33}}}\,, \tag{6.64}$$

while the first and the second rows define the slownesses and polarizations of the split shear waves. Assuming that the fast wave S_1 is polarized in the x_1-direction (i.e., $c_{55} > c_{44}$), the shear-wave slownesses are

$$q^{S1} \equiv p_3^{S1} = \sqrt{\frac{\rho}{c_{55}}} \quad \text{and} \quad q^{S2} \equiv p_3^{S2} = \sqrt{\frac{\rho}{c_{44}}}\,. \tag{6.65}$$

6B NMO ellipses in a horizontal monoclinic layer

Equation 6.33 of the main text can be used to model pure-mode NMO ellipses in a horizontal, homogeneous, arbitrarily anisotropic layer. For monoclinic media, the vertical slowness q of the zero-offset ray is given for P-, S_1-, and S_2-waves in equations 6.64 and 6.65. Since the slowness vector of the zero-offset ray is perpendicular to the reflector, the horizontal slownesses p_1 and p_2 vanish. As discussed in Chapter 1, the derivatives $q_{,ij} \equiv \partial^2 q/(\partial p_i \partial p_j)$ can be computed from the Christoffel equation 6.61 treated as an implicit relation between the slowness components, $F(p_1, p_2, q(p_1, p_2)) = 0$ (equations 1.9 and 1.10). Substitution of the vertical slownesses and their derivatives into equation 6.33 yields the matrices $\mathbf{W}$ responsible for the exact NMO ellipses. The expressions listed below are obtained in the coordinate system associated with the S-wave polarization directions at vertical incidence (i.e., in the system for which $c_{45} = 0$).

- PP-wave:

$$W_{11}^P = \frac{1}{f^P}\Big[(c_{33}-c_{55})(c_{23}^2 + 2c_{23}c_{44} + c_{33}c_{44}) + c_{36}^2(c_{33}-c_{44})\Big], \tag{6.66}$$

$$W_{12}^P = -\frac{c_{36}}{f^P}\Big[c_{33}(c_{44}+c_{55}+c_{13}+c_{23}) - c_{44}(c_{55}+c_{13}) - c_{55}(c_{44}+c_{23})\Big], \tag{6.67}$$

$$W_{22}^P = \frac{1}{f^P}\Big[(c_{33}-c_{44})(c_{13}^2 + 2c_{13}c_{55} + c_{33}c_{55}) + c_{36}^2(c_{33}-c_{55})\Big], \tag{6.68}$$

where

$$\begin{aligned} f^P = {} & \frac{1}{\rho}\Big\{(c_{13}^2 + c_{33}c_{55})(c_{23}^2 + 2c_{23}c_{44} + c_{33}c_{44}) \\ & + 2c_{13}\big[c_{23}^2 c_{55} + c_{44}c_{55}(c_{33}+2c_{23}) - c_{36}^2(c_{23}+c_{44})\big] \\ & + c_{36}^2\big[c_{33}(c_{44}+c_{55}) - 2c_{55}(2c_{44}+c_{23}) + c_{36}^2\big]\Big\}. \end{aligned} \tag{6.69}$$

- S_1S_1-wave:

$$W_{11}^{S1} = \frac{c_{66}(c_{55}-c_{33}) + c_{36}^2}{f^{S1}}, \tag{6.70}$$

$$W_{12}^{S1} = \frac{c_{16}(c_{33}-c_{55}) - c_{36}(c_{13}+c_{55})}{f^{S1}}, \tag{6.71}$$

$$W_{22}^{S1} = \frac{c_{13}^2 + 2c_{13}c_{55} + c_{55}^2 + c_{11}(c_{55}-c_{33})}{f^{S1}}, \tag{6.72}$$

where

$$f^{S1} = \frac{1}{\rho}\Big[c_{66}(c_{13}+c_{55})^2 + (c_{33}-c_{55})(c_{16}^2 - c_{11}c_{66}) \\ - 2c_{16}c_{36}(c_{13}+c_{55}) + c_{11}c_{36}^2\Big]. \tag{6.73}$$

- S_2S_2-wave:

$$W_{11}^{S2} = \frac{c_{23}^2 + 2c_{23}c_{44} + c_{44}^2 + c_{22}(c_{44}-c_{33})}{f^{S2}}, \tag{6.74}$$

$$W_{12}^{S2} = \frac{c_{26}(c_{33}-c_{44}) - c_{36}(c_{23}+c_{44})}{f^{S2}}, \tag{6.75}$$

$$W_{22}^{S2} = \frac{c_{66}(c_{44}-c_{33}) + c_{36}^2}{f^{S2}}, \tag{6.76}$$

where

$$f^{S2} = \frac{1}{\rho}\Big[c_{66}(c_{23}+c_{44})^2 + (c_{33}-c_{44})(c_{26}^2 - c_{22}c_{66}) \\ - 2c_{26}c_{36}(c_{23}+c_{44}) + c_{22}c_{36}^2\Big]. \tag{6.77}$$

Chapter 7

Estimation of anisotropy from VSP data

The previous chapters were devoted to estimation of anisotropy from the kinematics of reflected waves. Extending Thomsen notation to symmetries lower than transverse isotropy made it possible to build realistic models of the azimuthally anisotropic subsurface. The success of the traveltime inversion was ensured by our ability to identify the combinations of the stiffness coefficients that control such key signatures measured from wide-azimuth data as the NMO ellipse and nonhyperbolic moveout. Still, because reflection traveltimes are insensitive to velocity and anisotropy variations on a fine scale, moveout inversion produces models that have relatively low spatial resolution.

In principle, there are two options for improving the resolution of subsurface velocity fields. One is to include information about seismic amplitudes either directly (see Chapter 8) or via full-waveform inversion of reflection data (e.g., Tarantola, 1987; Virieux and Operto, 2009). Realizing the potential of full-waveform inversion, however, largely belongs to the future because only preliminary (albeit encouraging) results for anisotropic media are currently available (Chang and McMechan, 2009). The second option is to make use of acquisition geometries that differ from conventional surface recording. Here, we discuss application of vertical seismic profiling (VSP) geometries in anisotropic parameter estimation and demonstrate their ability to resolve in situ anisotropy with the spatial resolution close to the dominant seismic wavelength.

The term "VSP" refers to observations of elastic waves excited by sources located at or near the earth's surface and recorded by geophones placed in a borehole. A string of such geophones tracks the evolution of the seismic wavefield as it propagates through the medium. The fundamentals and early applications of the VSP method, which originated in the Soviet Union in the late 1950s, are covered in the book by Gal'perin (1971). Western geophysicists became aware of potential applications of VSP technology to reservoir exploration and development after the English translation of Gal'perin's book was published by the SEG in 1974.

This chapter is focused on the inversion of borehole slowness and polarization measurements for local anisotropy parameters. For more details on processing of shear waves in VSP geometries, the reader is referred to MacBeth (2002). We will show that, similarly to surface seismic, a sufficiently wide angular aperture is the main prerequisite for successful VSP inversion. Also, as is the case for reflection data, supplementing P-waves with shear-wave data is generally beneficial for constraining the anisotropy parameters, although the feasibility of employing multicomponent data depends on the structural complexity of the overburden.

A laterally homogeneous overburden represents the best-case scenario for resolving local anisotropy near downhole geophones. Preservation of the horizontal slowness along the ray then helps obtain the most general (triclinic) stiffness tensor from wide-azimuth, multicomponent VSP data. This tensor can then be analyzed to determine if the medium has a higher symmetry, such as orthotropy or transverse isotropy. Hence, VSP surveys can provide an exciting opportunity to establish the symmetry from the data as opposed to assuming it a priori, as we often had to do in the previous chapters.

Estimating the symmetry of the medium along with the pertinent anisotropy parameters is no longer feasible when the overburden is characterized by strong lateral heterogeneity caused, for instance, by the presence of salt bodies. Such overburden complexity prevents us from reconstructing the slowness surface at the receiver location and usually renders shear-wave arrivals too noisy to be used in the inversion. Nonetheless, valuable information about anisotropy can be obtained solely from P-wave traveltimes and polarizations. This type of inversion is facilitated by introducing Thomsen-style anisotropy parameters specifically tailored to P-wave signatures recorded in VSP geometries.

7.1 Stratified models

The first-break times measured by a single downhole geophone in a walkaway VSP survey, such as those displayed in Figure 7.1, provide the most direct way of estimating the effective anisotropy above the receiver. For a vertical borehole, the vertical velocity is found directly from the zero-offset time and the known geophone depth. Fitting P-wave traveltimes for a range of source-receiver offsets with models of a chosen symmetry yields the δ- and ϵ-type anisotropy parameters. The accuracy of estimating the effective values of ϵ in vertically heterogeneous VTI media and of $\epsilon^{(1)}$, $\epsilon^{(2)}$, and $\delta^{(3)}$ (see equations 1.80, 1.83, and 1.86 in Appendix 1B) in orthorhombic media is governed by the spatial distribution of sources, the ratio of maximum offset to geophone depth, and the velocity gradients. A comparable accuracy in the ϵ- and δ-parameters is typically achieved when the offset-to-depth ratio varies between unity and 1.5 (the corresponding ratio for surface reflection data is between two and three). An example of such data set is given in Figure 7.1. More details on the inversion of the effective parameters of VTI media from P-wave VSP data can be found in Grechka and Tsvankin (1998b) and Tsvankin (2005; see his Figure 7.25 and the corresponding discussion).

Although the benefits of estimating the effective anisotropy can be questioned on the grounds that unaccounted velocity heterogeneity tends to bias the obtained ani-

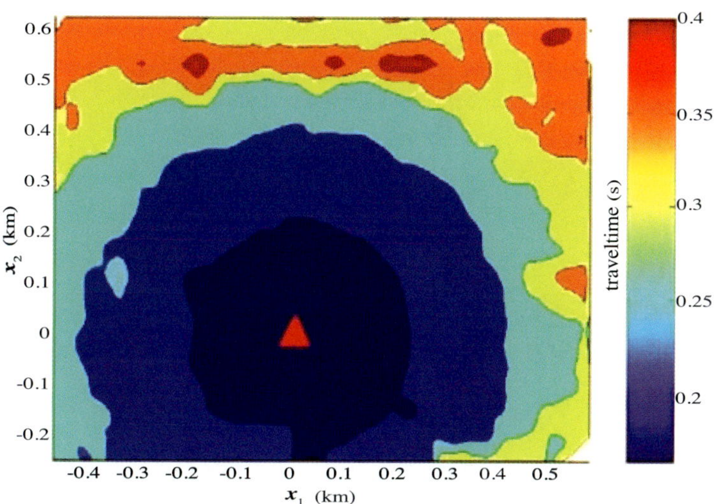

Figure 7.1: First-break P-wave times recorded at a depth of 304.8 m at Vacuum field, New Mexico, USA. The red triangle at the coordinate origin indicates the location of the vertical borehole. The coordinates x_1 and x_2 specify the source locations with respect to the wellhead. The data were acquired by the Reservoir Characterization Project at Colorado School of Mines.

sotropy parameters (Grechka and Tsvankin, 2002b), the effective values are sufficient at least for time processing. Also, if our goal is just to image seismic reflection data, the traveltime surface in Figure 7.1 is useful by itself because it provides the exact depth-migration operator at the geophone location. Whether or not this operator can be applied away from the borehole depends on the type and magnitude of spatial velocity variations.

7.1.1 Slowness method

The influence of heterogeneity on anisotropy estimates from VSP data has been recognized and at least partially addressed since the earliest attempts to perform anisotropic VSP processing. Historically, the first assumption made about the overburden was that it is composed of horizontal, laterally homogeneous layers. This assumption led to the technique known as the *slowness method* (Gaiser, 1990; Miller and Spencer, 1994; Jílek et al., 2003).

The idea of the slowness method is explained in Figure 7.2. Suppose that we measure the traveltimes t of waves excited at the surface and recorded by a string of geophones placed in a vertical borehole with depth increment dh. If the VSP data are sorted into common-shot gathers, the vertical slowness $q \equiv p_3$ at each geophone location is

$$q \equiv dt/dh\,. \tag{7.1}$$

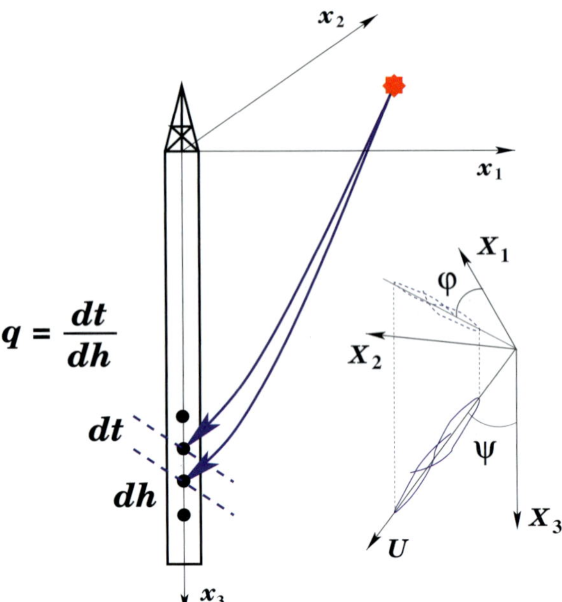

Figure 7.2: Measurements used for estimating anisotropy in VSP geometry (Grechka et al., 2007). The traveltime difference dt between the geophones (dots) separated by the distance dh along a borehole defines the apparent slowness $q{=}dt/dh$. Three-component traces recorded by each downhole geophone yield the direction of particle motion $\mathbf{U}$ described by the polarization polar angle ψ and azimuth φ. The azimuths of the horizontal components X_1 and X_2 of the geophones usually vary along the tool string and should be treated as unknown, unless independently measured.

Equation 7.1 follows from the general definition of slowness as the traveltime gradient. If the recorded traces are resorted into common-receiver gathers, we can calculate the derivatives of the same traveltime t with respect to the source coordinates x_i $(i = 1,\, 2)$ and obtain the horizontal slowness components at the earth's surface:

$$p_i \equiv \partial t / \partial x_i, \quad (i = 1,\, 2)\,. \tag{7.2}$$

Based on the assumption of lateral homogeneity above the receiver, the horizontal slownesses p_i given by equation 7.2 can be used at each geophone depth. Indeed, according to Snell's law, the slownesses p_1 and p_2 remain unchanged along a ray propagating in a laterally homogeneous medium. Hence, by computing the traveltime gradients from multiazimuth walkaway VSP data, one can obtain a portion of the slowness surface $q(p_1,\, p_2)$ (or a portion of the slowness curve in 2D) at the downhole geophone locations. Once the surface $q(p_1,\, p_2) \equiv p_3(p_1,\, p_2)$ has been constructed (see Figures 7.4d-f below for examples of the slowness surfaces estimated from field data), the density-normalized stiffness tensor $\mathbf{c}$ for a chosen anisotropic symmetry can be computed by solving the Christoffel equation written as

$$\det\left(c_{ijkl}\, p_j\, p_k \;-\; \delta_{il}\right) = 0\,, \tag{7.3}$$

where δ_{il} is Kronecker's symbolic delta. Each source-receiver pair provides one nonlinear equation 7.3 for the stiffness components c_{ijkl}. Application of the slowness method thus results in a nonlinear inverse problem whose solution depends on the angle coverage of the VSP data.

7.1.2 Slowness-polarization method

The direction of particle motion (polarization) of waves recorded by downhole geophones represents an inherent part of VSP data. When heterogeneity in the vicinity of the borehole is not extreme and different modes can be separated on VSP seismograms, it is usually possible to measure the body-wave polarization vectors $\mathbf{U}$ (inset in Figure 7.2), at least for the direct arrivals:

$$\mathbf{U} = \left[\sin\psi\cos\varphi,\; \sin\psi\sin\varphi,\; \cos\psi\right]. \tag{7.4}$$

Because both the slowness and polarization vectors pertain to the same geophone locations, they can be combined to estimate the local anisotropy in a rock volume near the borehole. The linear size of this volume is approximately equal to the dominant seismic wavelength.

Joint inversion of the vectors $\mathbf{p}$ and $\mathbf{U}$ leads to the so-called *slowness-polarization* technique for estimating the pertinent anisotropy parameters. Using this method, the density-normalized stiffness tensor $\mathbf{c}$ can be found directly from the Christoffel equation:

$$\Gamma_i^{(\mathrm{Q})}(\mathbf{c}) \equiv c_{ijkl}\, p_j^{(\mathrm{Q})}\, p_k^{(\mathrm{Q})}\, U_l^{(\mathrm{Q})} \;-\; U_i^{(\mathrm{Q})} = 0\,, \quad (i = 1,\, 2,\, 3;\; \mathrm{Q} = \mathrm{P},\, \mathrm{S}_1,\, \mathrm{S}_2)\,. \tag{7.5}$$

The stiffness coefficients c_{ijkl} are obtained by solving a linear inverse problem of minimizing the vectors $\mathbf{\Gamma}^{(Q)}$ for the available source-receiver pairs. If all three body waves (P, S_1, and S_2) were recorded, and the data aperture at geophone locations approaches a full hemisphere, the local triclinic stiffness tensor can be obtained from equations 7.5 without any assumptions about the medium symmetry (Dewangan and Grechka, 2003).

As an example, Figure 7.3 displays the components of the triclinic stiffness tensor $\mathbf{c}$ inverted from the noise-contaminated slowness and polarization vectors of P-, S_1-, and S_2-waves computed for 144 wavefront normal vectors:

$$\mathbf{n} = \left[\sin\theta_1\cos\theta_2,\; \sin\theta_1\sin\theta_2,\; \cos\theta_1\right],$$

where

$$(\theta_1 = 10°,\, 20°,\, \ldots,\, 80°;\; \theta_2 = 20°,\, 40°,\, \ldots,\, 360°)\,.$$

All coefficients c_{ij}, including those with near-zero values (that is, when the medium is close to orthorhombic), are accurately recovered. The high stability of the inverse

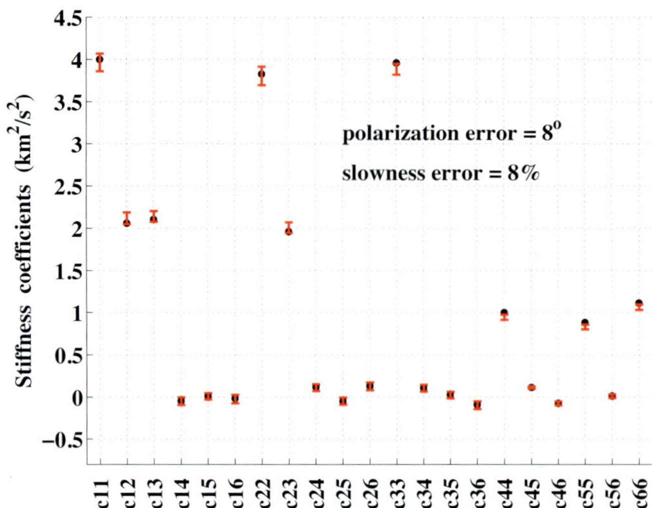

Figure 7.3: Inversion of the noise-contaminated slowness and polarization vectors of P-, S_1-, and S_2-waves for the stiffnesses of a triclinic medium. The black dots indicate the model values, and the red bars are the 95% confidence intervals. The input data were contaminated by Gaussian noise with the standard deviations equal to 8° for the polarization vectors and 8% for the slowness components.

problem is corroborated by the singular-value decomposition of the Fréchet-derivative matrix $\partial \mathbf{\Gamma} / \partial \mathbf{c}$, where $\mathbf{\Gamma}$ is given by equation 7.5. The ratio of the minimum and maximum singular values of that matrix is equal to 0.1, which indicates that the slowness-polarization data tightly constrain all 21 stiffness elements c_{ij}.

The relationships between phase and group velocities as well as phase and group angles make it possible to formulate inverse problems in two different contexts. The slowness used in this section represents the inverse of phase velocity. A similar method that operates with group velocity and polarization was developed by Bóna and Slawinski (2008) and Bóna et al. (2008), who also evaluated the stability of algorithms based on phase and group quantities.

7.1.3 Case study from Vacuum field

Dewangan and Grechka (2003) applied the slowness-polarization method to a 3D multicomponent walkaway VSP data set acquired by the Reservoir Characterization Project (Colorado School of Mines) at Vacuum field, New Mexico, USA. The data were recorded by a string of 10 three-component (3C) geophones placed in a vertical well between 304.8 m and 439.8 m depth with a 15-m increment. Vertical and horizontal vibrators were used to excite P- and S-waves (respectively) at 250 locations uniformly distributed around the borehole with offsets reaching 510 m. Details of data acquisition and initial processing can be found in Michaud (2001), who applied the technique described by DiSeina et al. (1984) to estimate the P-wave polarization vectors. Also, using Alford's (1986) rotation of the sources and receivers, Michaud (2001) separated the fast and slow shear waves and determined their polarization directions (Figures 7.4a-c).

To compute the slowness components, it is necessary to pick P- and S-wave traveltimes. Figure 7.1 shows the first-break P-wave times at the shallowest receiver in this survey. To improve the stability of traveltime differentiation (equations 7.1 and 7.2), the time picks were smoothed with a quartic polynomial in the source-geophone offset. The results of the slowness calculation in Figures 7.4d-f show that the P-wave slowness contours (Figure 7.4d) noticeably deviate from circles. Hence, an azimuthally anisotropic model is required to fit the measurements. Also, the polar coverage of the data is extremely good (the maximum polar angle reaches 75°–80°), so we can expect comparable accuracy in the stiffness coefficients that govern wave propagation in all directions.

The data in Figure 7.4 were inverted for the stiffness coefficients of triclinic media (Figure 7.5). To obtain a suite of inversion results rather than a single model, the polarizations and slownesses were contaminated with Gaussian noise. According to the results of Michaud (2001), the errors in the polarization direction and slowness components are 12° and 5%, respectively. These values were used as the corresponding standard deviations.

The estimated stiffnesses are accurately described by an orthorhombic medium with a nearly horizontal symmetry plane. Indeed, the local x_3-axis of the best-fit orthorhombic model has a tilt of approximately 4°, which means that one of the symmetry planes is practically horizontal. The azimuth of the $[x_1, x_3]$ symmetry plane is $25° \pm 15°$SE; Tsvankin's anisotropy parameters are shown in Figure 7.6. The model correctly reproduces the main features of the slowness plots (Figures 7.4d-f). In particular, the shear-wave splitting parameter $\gamma^{(S)} \approx \gamma^{(1)} - \gamma^{(2)} \approx 0.10$ (defined by equation 1.87) matches that computed from Figures 7.4e,f. Also the elongation of the P-wave slowness contours in the approximately southeast-northwest (x_1) direction (Figure 7.4d) indicates that $\delta^{(2)} < \delta^{(1)}$, which agrees with the inversion results in Figure 7.6.

It should be emphasized that the orthorhombic symmetry of the formation near the borehole was *inferred* from the data rather than assumed a priori. The measured slownesses and polarizations were inverted for the most general, triclinic stiffness tensor, whose symmetry proved to be close to orthorhombic. These results are in agreement with those from several other studies done at Vacuum field. For instance, Mattocks (1998) examined polarizations and energy focusing of shear waves in the same area and concluded that the overburden (shallow 300 m) has orthorhombic symmetry. In addition, the azimuth of the borehole breakouts is close to 32°SE (Scuta, 1997).

The problem of inferring the symmetry class and estimating the orientation of symmetry planes and axes was explored by Kochetov and Slawinski (2009a, 2009b, 2009c). In particular, they considered the concept of distance in the space spanned by

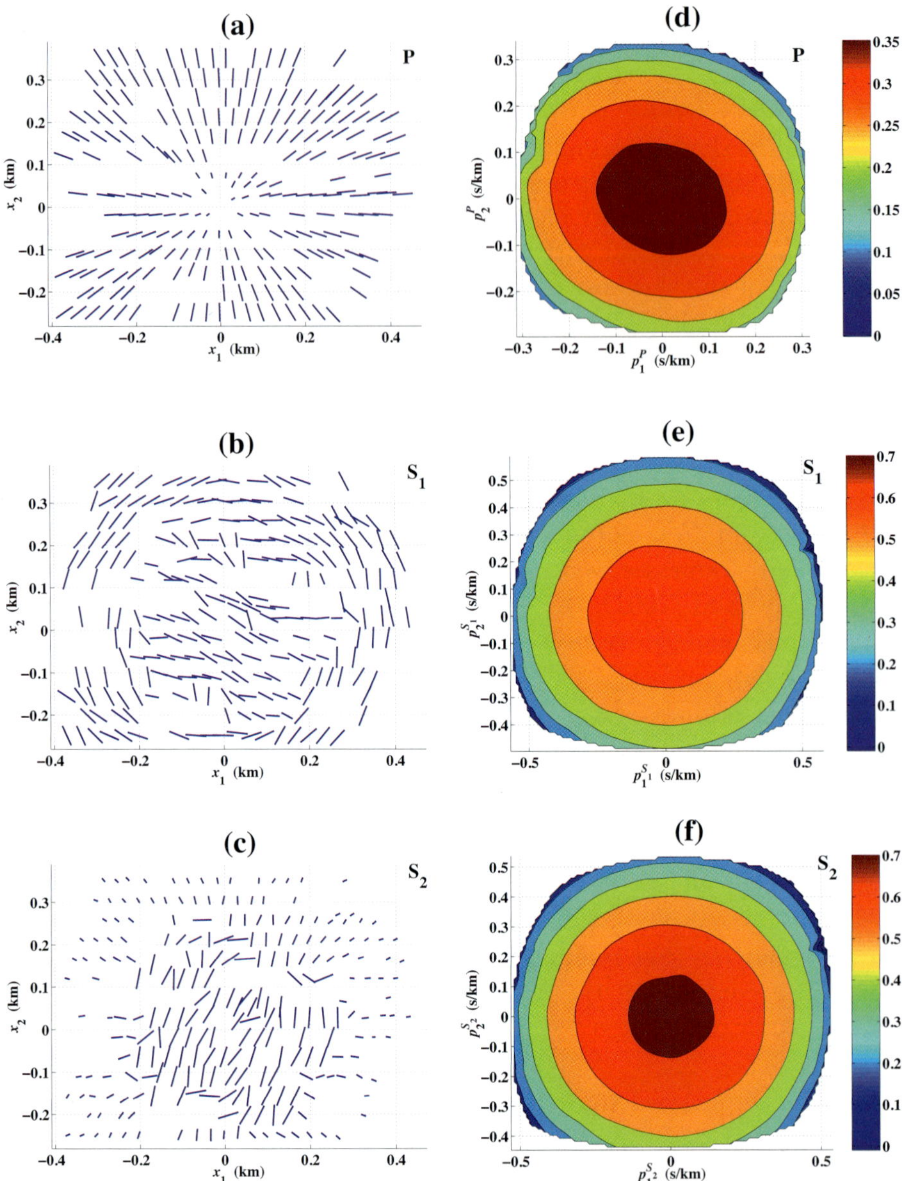

Figure 7.4: Slownesses and polarizations at Vacuum field for the downhole receiver at a depth of 304.8 m (Dewangan and Grechka, 2003). (a, b, c) The horizontal projections of the polarization vectors (shorter ticks correspond to waves polarized closer to the vertical) plotted at the source locations; and (d, e, f) the vertical slowness components $p_3^{(Q)}(p_1^{(Q)}, p_2^{(Q)})$ (in s/km). The top row corresponds to P-waves, middle row to S_1-waves, and bottom row to S_2-waves. The borehole is located at the coordinate origin.

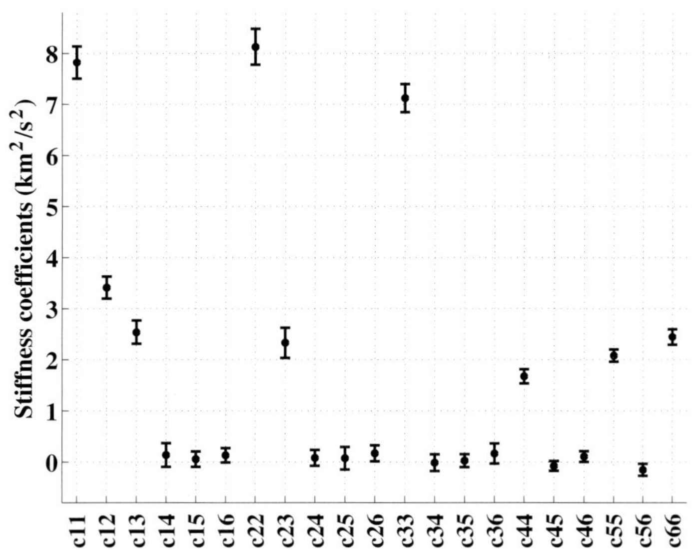

Figure 7.5: Stiffness coefficients of triclinic media inverted from the polarization and slowness vectors in Figure 7.4 (Dewangan and Grechka, 2003). The dots indicate the mean values, the bars correspond to the 95% confidence intervals.

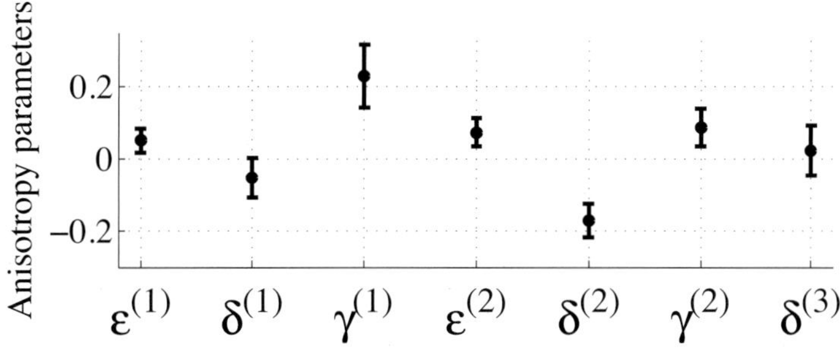

Figure 7.6: Anisotropy parameters of the inverted orthorhombic model. As in Figure 7.5, the dots and bars mark the mean values and the 95% confidence intervals, respectively. The estimated vertical velocities (not shown) are V_{P0}=2.66±0.05 km/s and V_{S0}=1.44±0.04 km/s. This is a corrected version of Figure 10 in Dewangan and Grechka (2003).

the stiffness coefficients and the question of "being close to" rather than "belonging to" a given symmetry class. Such a formulation helps identify the symmetry that best represents the real material for given experimental data. Kochetov and Slawinski (2009a) applied this methodology to the data set from Vacuum field described above and obtained results consistent with those of Dewangan and Grechka (2003).

7.2 Models with laterally heterogeneous overburden

If the elastic properties above downhole geophones vary laterally, the horizontal slowness components p_1 and p_2 calculated at the earth's surface are no longer preserved along rays and cannot be used at the geophone locations. Treating the horizontal slownesses as unknowns makes the inversion based on the Christoffel equation 7.5 nonlinear and causes significant deterioration in the accuracy of the stiffness coefficients (Dewangan and Grechka, 2003).

Also, traveltimes of P- or S-waves recorded by a downhole geophone are not guaranteed to produce a plausible effective anisotropic model because they are distorted by the lateral heterogeneity of the overburden. As an example, the first-break traveltime contours in Figure 7.7 exhibit not only a shift of the traveltime minimum to the southeast from the geophone location (star), but also a sharp time change at about 2000 ft (610 m) to the west from the borehole. This traveltime discontinuity is caused by a fault, which is also recognizable in reflection data (Franco et al., 2007; Grechka et al., 2007).

Although the shift of the traveltime minimum should be attributed to lateral heterogeneity, it also can be reproduced using a homogeneous anisotropic model that does not have a horizontal symmetry plane. No homogeneous model, however, yields a sufficiently rapid velocity change to explain the traveltime behavior due to the velocity contrast across the fault. Hence, Figure 7.7 demonstrates the importance of taking into account subsurface heterogeneity in anisotropic analysis of VSP data. Failure to do so would produce erroneous or even implausible effective anisotropy parameters.

An alternative approach to infer local anisotropy from VSP data acquired beneath a laterally heterogeneous overburden was proposed by White et al. (1983) and further developed for VTI media by de Parscau (1991) and Hsu et al. (1991). The idea is to combine two types of local quantities shown in Figure 7.2 – the apparent slownesses q obtained by differentiating the arrival times along a vertical borehole and the polarization vectors $\mathbf{U}$ measured by three-component (3C) geophones. Thus, the horizontal slowness components, which might vary laterally, never enter the inversion, making this slowness-polarization technique theoretically applicable to subsurface of any complexity. Indeed, the only assumption behind this method is that the wavefronts crossing the borehole are locally planar, which is the standard geometrical-seismics approximation.

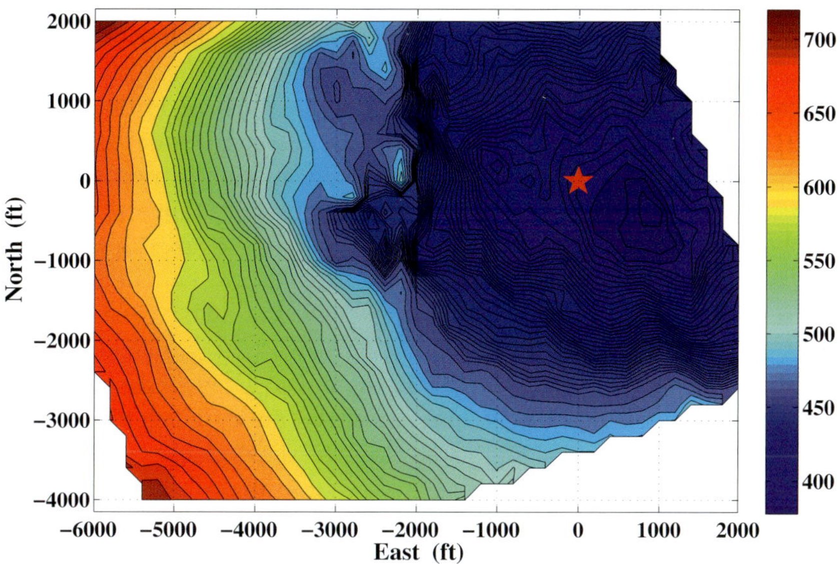

Figure 7.7: First-break P-wave traveltimes (in ms) recorded at a depth of 4509 ft (1374 m) in a vertical borehole (star) at Rulison field (Grechka et al., 2007). The data were acquired by the Reservoir Characterization Project at Colorado School of Mines.

De Parscau (1991) and Hsu et al. (1991) discussed which parameters of VTI media are constrained by the vertical slownesses and polarizations. They concluded that the density-normalized stiffness coefficients c_{11}, c_{13}, c_{33}, and c_{55} or, equivalently, the Thomsen parameters V_{P0}, V_{S0}, ϵ, and δ, can be estimated from the combination of P- and SV-wave VSP data when the propagation angles range from 0° to about 80° from the vertical. In addition, de Parscau (1991) and Hsu et al. (1991) note that neither P nor SV data alone can be inverted for all four pertinent stiffnesses or Thomsen parameters. This theoretical conclusion was confirmed by Horne and Leaney (2000).

The need to operate with SV-waves to ensure the uniqueness of parameter estimation is often inconvenient in practice because it requires identifying SV arrivals and separating them from SH-waves. Typically, the level of noise for shear modes is much higher than for P-waves, especially if the overburden is complex. Also, separation of the two shear modes is problematic when the geophone orientations in a vertical borehole are not measured independently (Figure 7.2) and have to be estimated from the data. Yet another issue arises when the local symmetry is lower than TI, and shear waves cannot be treated as simply SV and SH. Enforcing the SV-wave properties on either S_1- or S_2-waves, as would be required by the inversion algorithm, seriously biases the shear-wave slownesses and polarizations. Thus, it is important to find out what information about anisotropy can be inferred solely from P-wave VSP data acquired beneath complicated structures (e.g., subsalt).

7.2.1 Governing parameters for P-wave VSPs in VTI media

To perform inversion based on P-waves alone, we need to identify the parameter combinations constrained by the dependence of the vertical slowness on the polarization direction. The first attempt to address this question was made by Williamson and Maocec (2001), who noticed that the ratio of the vertical velocities V_{S0}/V_{P0} had to be known a priori to resolve the parameters ϵ and δ. Confirming their conclusion, Grechka and Mateeva (2007) proposed to reparameterize the inverse problem in such a way that the ratio V_{S0}/V_{P0} is absorbed by the well-constrained quantities.

The P-wave slowness-of-polarization dependence $\mathbf{p}(\mathbf{U})$ can be simplified under the assumption of weak anisotropy (Appendix 7A). The most general result of Appendix 7A is equation 7.27 for the function $\mathbf{p}(\mathbf{U})$, which is applicable to arbitrarily anisotropic (triclinic) media. In a typical VSP experiment, we measure the projection of the local slowness vector $\mathbf{p}$ onto the borehole direction $\boldsymbol{\ell}$, which yields the apparent slowness $p_\ell = \mathbf{p} \cdot \boldsymbol{\ell}$. Equation 7.27 indicates that the dependence $p_\ell(\mathbf{U})$ constrains up to 15 combinations of 21 stiffness coefficients for an arbitrary direction $\boldsymbol{\ell}$ (Zheng and Pšenčík, 2002).

In this section, we are interested in a particular special case of equation 7.27 corresponding to a vertical borehole in a VTI medium. Hence, the receiver array measures the vertical slowness component $p_3 \equiv q$ described by the vector $\boldsymbol{\ell} = [0,\ 0,\ 1]$. Because of the rotational symmetry around the vertical, q depends just on the polar polarization angle ψ. The weak-anisotropy approximation for $q(\psi)$ is given by equations 7.32, 7.36, and 7.37 in Appendix 7A:

$$q(\psi) = \frac{\cos\psi}{V_{P0}} \left(1 + \delta_{\mathrm{VSP}} \sin^2\psi + \eta_{\mathrm{VSP}} \sin^4\psi\right), \tag{7.6}$$

where the newly introduced anisotropy parameters δ_{VSP} and η_{VSP} are defined as

$$\delta_{\mathrm{VSP}} \equiv (f_0 - 1)\,\delta \tag{7.7}$$

and

$$\eta_{\mathrm{VSP}} \equiv (2f_0 - 1)\,\eta\,; \tag{7.8}$$

with

$$f_0 = \frac{1}{1 - V_{S0}^2/V_{P0}^2}\,. \tag{7.9}$$

The term $\cos\psi/V_{P0}$ in equation 7.6 describes the P-wave vertical slowness $q(\psi)$ in a purely isotropic medium with the velocity V_{P0} and the term in parentheses represents the contribution of anisotropy. This anisotropic term in equation 7.6 is strikingly similar to the one that appears in the linearized P-wave phase velocity V_P as a function of the phase angle θ (Thomsen, 1986; Tsvankin, 2005) rewritten in terms of δ and η:

$$V_P(\theta) = V_{P0}\left(1 + \delta \sin^2\theta + \eta \sin^4\theta\right). \tag{7.10}$$

Based on this similarity, we expect that the pairs $\{\delta_{\mathrm{VSP}},\ \eta_{\mathrm{VSP}}\}$ and $\{\delta,\ \eta\}$ play comparable roles for VSP data acquired in vertical boreholes and for surface reflection data, respectively. Indeed, δ_{VSP} is responsible for the near-vertical variation of $q(\psi)$, while η_{VSP} governs the vertical slowness at larger polarization angles. The parameters δ_{VSP} and η_{VSP} absorb the shear-wave vertical velocity V_{S0}, rendering its value unnecessary for fitting the P-wave slowness-of-polarization functions.

To analyze the sensitivity of $q(\psi)$ to δ and η at different polarization angles ψ, it is necessary to specify the ratio g_0 of the squared vertical velocities:

$$g_0 \equiv \frac{V_{S0}^2}{V_{P0}^2} \,. \tag{7.11}$$

For small ratios g_0, the factor f_0 (equation 7.9) can be well approximated by

$$f_0 = \frac{1}{1 - g_0} \approx 1 + g_0 \,. \tag{7.12}$$

Substituting equation 7.12 into equations 7.6 – 7.8 yields

$$q(\psi) = \frac{\cos\psi}{V_{P0}} \left[1 + g_0 \delta \sin^2\psi + (1 + 2\, g_0)\, \eta \sin^4\psi \right] \,. \tag{7.13}$$

For the typical velocity ratio $V_{S0}/V_{P0} = 0.5$, the parameters δ and η are multiplied with 0.25 and 1.5, respectively. Therefore, the η-term dominates the δ-term in equation 7.13 for polarization angles $\psi \geq 25°$ with the vertical. At $\psi < 25°$, the influence of δ on $q(\psi)$ is still relatively minor because the product $g_0 \sin^2\psi$ is much smaller than unity, and $q(\psi)$ is determined mostly by the vertical velocity V_{P0}.

Although approximation 7.10 is limited to weak anisotropy, P-wave kinematics even in strongly anisotropic VTI media are governed by the same parameters – V_{P0}, ϵ, and δ (no influence of V_{S0}). Likewise, linearized equation 7.6 suggests that a reduced set of quantities (V_{P0}, δ_{VSP}, and η_{VSP}) might be sufficient to fit P-wave VSP data. Indeed, Figure 7.8 confirms that the exact $q(\psi)$-curves (computed by solving the quadratic equation 7.43 in Appendix 7B, which is valid for arbitrary strength of anisotropy) nearly overlap for a family of VTI models with fixed values of V_{P0}, δ_{VSP}, and η_{VSP}. The Thomsen parameters for these models vary from moderate $\epsilon = 0.19$ and $\delta = 0.03$ to uncommonly large $\epsilon = 6.40$ and $\delta = 3.30$ (not typos). Clearly, these vastly different models cannot be distinguished using the slowness-of-polarization dependence without independent information about the V_{S0}/V_{P0} ratio. Even though the weak-anisotropy approximation (dashed line in Figure 7.8) slightly deviates from the exact curve, it has allowed us to identify the parameter combinations responsible for the function $q(\psi)$.

7.2.2 Case study from the Gulf of Mexico

Grechka and Mateeva (2007) applied the described methodology to a 2D P-wave walkaway VSP data set acquired in the deepwater Gulf of Mexico (Figure 7.9). The data comprise 677 shot locations placed at 100 ft (30.5 m) intervals along a line, with a maximum offset of 36,660 ft (11,170 m) from the wellhead. A receiver tool,

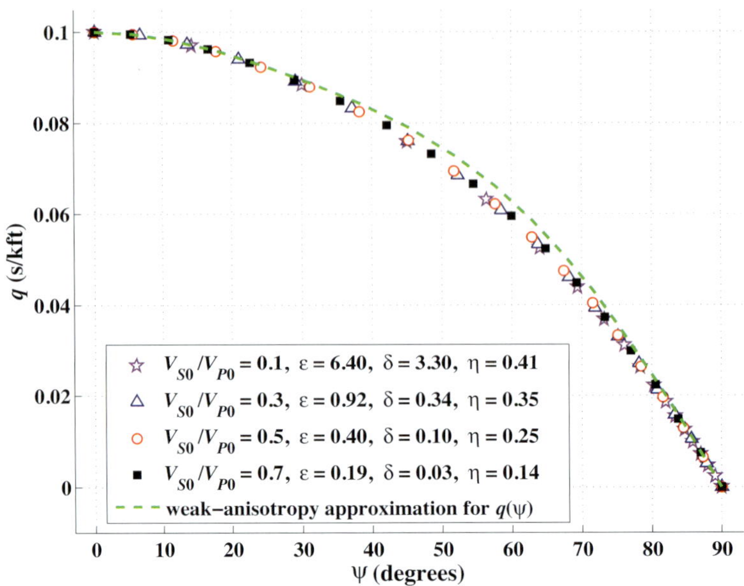

Figure 7.8: Exact function $q(\psi)$ for different V_{S0}/V_{P0} ratios but fixed parameters V_{P0}=10 kft/s (3 km/s), δ_{VSP}=0.033, and η_{VSP}=0.417 (Grechka and Mateeva, 2007). The dashed line is the weak-anisotropy approximation 7.6 for $q(\psi)$.

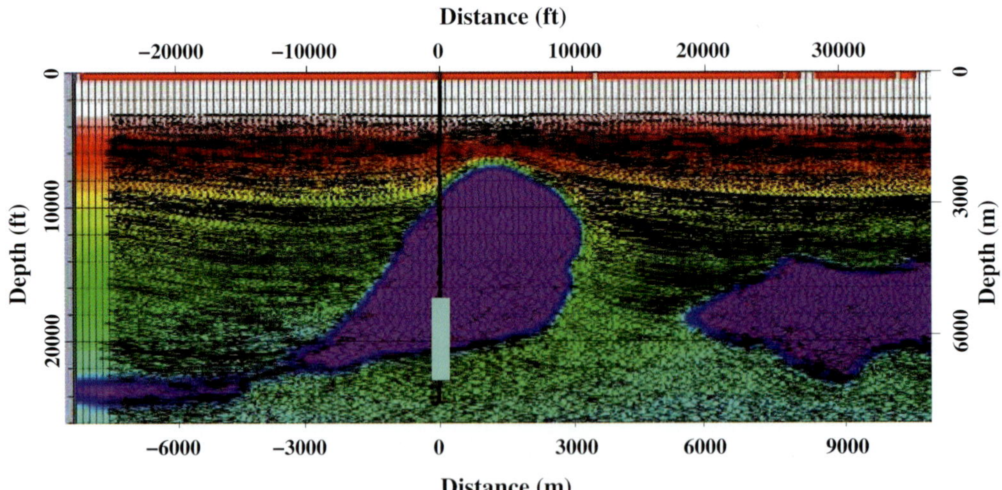

Figure 7.9: Geometry of a walkaway VSP survey (Grechka and Mateeva, 2007). Surface shot locations are shown in red, the positions of 3C receivers in the borehole are marked in cyan. Depth-migrated surface P-wave data are displayed against the background of the isotropic velocity model (white color corresponds to the water velocity, and magenta to the velocity in the salt).

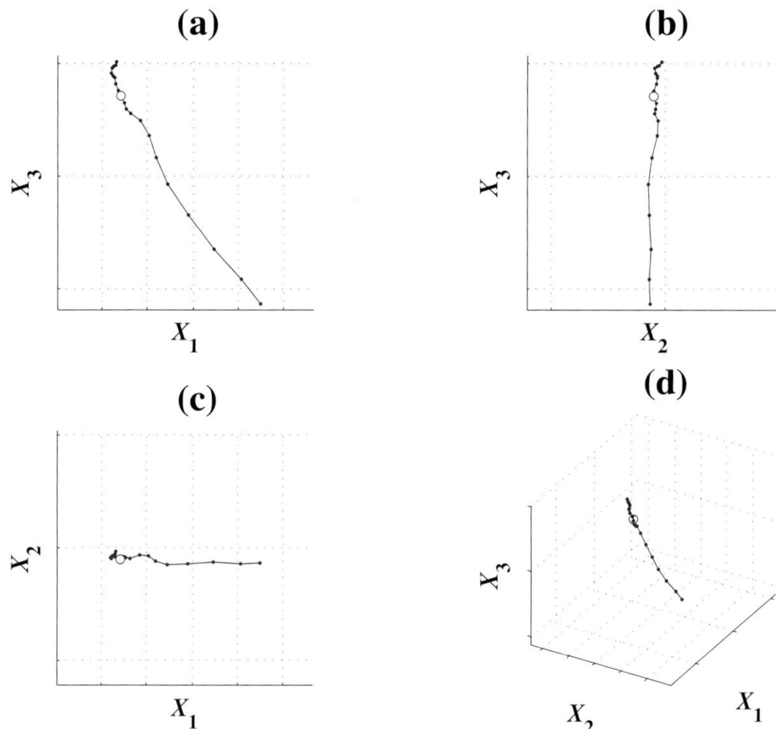

Figure 7.10: Particle-motion hodogram for a typical source-receiver pair used in the inversion (Grechka and Mateeva, 2007). The geophone axis X_3 is vertical; the axes X_1 and X_2 are horizontal, but their azimuths are unknown. The open circles are the picked first-break times, the dots mark the particle motion at 2-ms time increments. The hodogram corresponds to approximately one-quarter of the dominant P-wave period.

which included 24 3C geophones 100 ft (30.5 m) apart, covered the 17,500 – 22,250 ft (5330 – 6780 m) depth interval. The geophones were placed both in and beneath the salt body (Figure 7.9). Because the geophone azimuths in the borehole were not measured and the sources were located along a single line, azimuthal anisotropy could not be resolved and the medium was assumed to have VTI symmetry.

The slowness and polarization picks made from the data were subjected to the criteria below to ensure that the wavefronts crossing the borehole can be treated as locally planar. In other words, it was verified that the wavefield could be described by the geometrical-seismics approximation. Each slowness-polarization pair used in the inversion had to possess the following features:

1. Linearity of particle motion. — P-waves typically have quasi-linear polarization at sufficiently large distances from seismic sources (i.e., in the far-field). The linearity of particle motion can be estimated directly from the recorded waveforms (Figure 7.10).

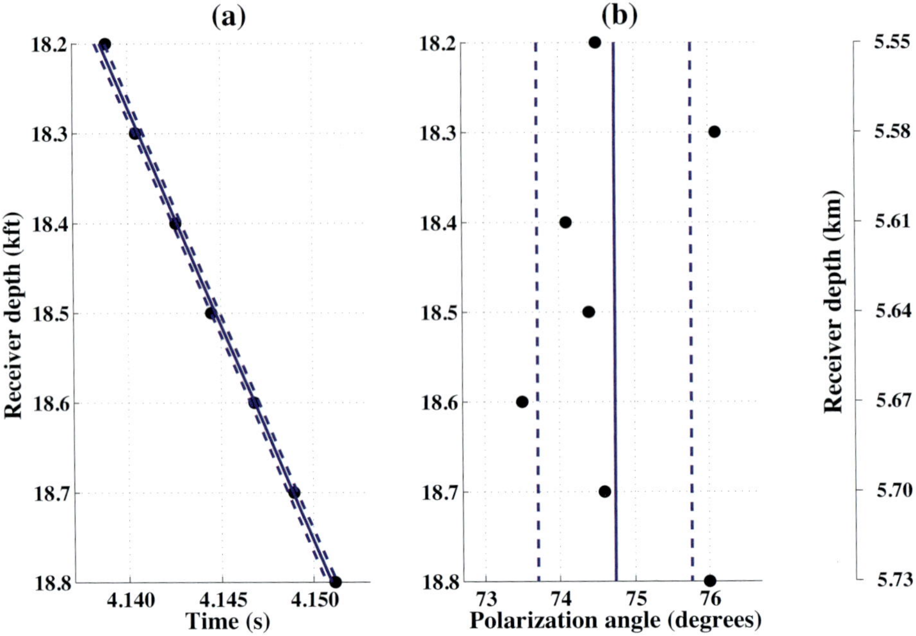

Figure 7.11: (a) Traveltimes and (b) polarization angles of P-waves measured in a borehole (Grechka and Mateeva, 2007). The dots mark the quantities picked from several adjacent traces in a common-shot gather. The solid lines are (a) the linear traveltime fit; and (b) the mean polarization angle, which give one slowness-polarization pair $q(\psi)$ for the inversion. The dashed lines correspond to $\pm$ one standard deviation (0.25 ms for the traveltimes and 1° for the polarization angles) from the best-fit lines.

2. Locally linear moveout. — For a plane wave, the orientation of the slowness vector does not change along the wavefront. Consequently, the traveltimes recorded over a small depth interval in the borehole should be linear in a common-shot gather. This property was used both to calculate the apparent slowness q and to reject time picks exhibiting too much scatter around a straight line (Figure 7.11a).

3. Small scatter of polarization angles. — If the wavefront is close to being planar, the polarization angles ψ for a fixed shot location should not vary significantly in the vicinity of a geophone (Figure 7.11b). Although the observed angles ψ are never constant because of noise, their standard deviation is only about 1°.

Typical parameter-estimation results both in the salt and below it are shown in Figure 7.12. The salt body turned out to be nearly isotropic, with the corresponding data points in Figure 7.12a tightly clustered around the isotropic curve described by $q_{\text{iso}}(\psi) = \cos\psi / V_{P0}$. It should be emphasized that the salt parameters were estimated with high confidence stemming from an almost ideal data coverage. As seen in Figure 7.12a, the polarization angle ψ reaches 75° from the vertical, leading to precise estimates of both δ_{VSP} and η_{VSP}. The deviation of the best-fit VTI slowness-

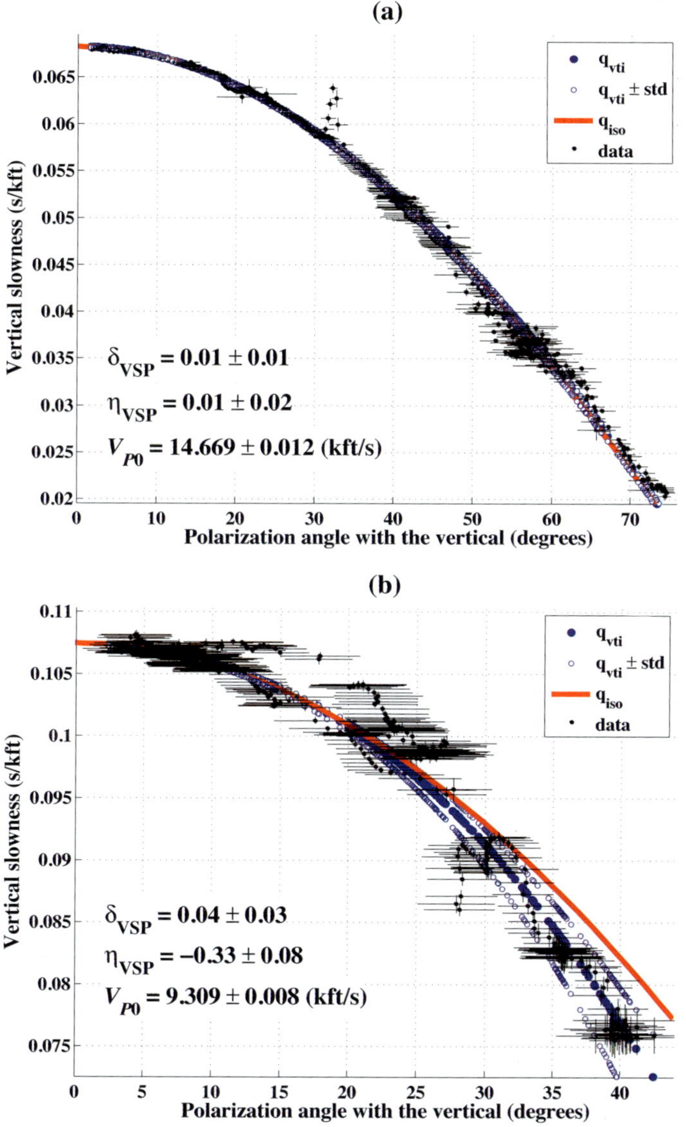

Figure 7.12: Slowness-polarization data and estimated parameters (a) in the salt at a depth of 18,500 ft (5640 m) and (b) beneath the salt at a depth of 21,750 ft (6630 m) (Grechka and Mateeva, 2007). The crosses associated with each data point (black dots) mark the standard deviations in the picked ψ and q values. The solid circles correspond to the best-fit $q(\psi)$ VTI function and open circles to that function $\pm$ one standard deviation. The obtained P-wave vertical velocities are (a) V_{P0}=4471$\pm$4 m/s and (b) V_{P0}=2837$\pm$3 m/s. The red lines show the isotropic function $q(\psi)$ for the estimated V_{P0}. The inversion employed the exact equations from Appendix 7B.

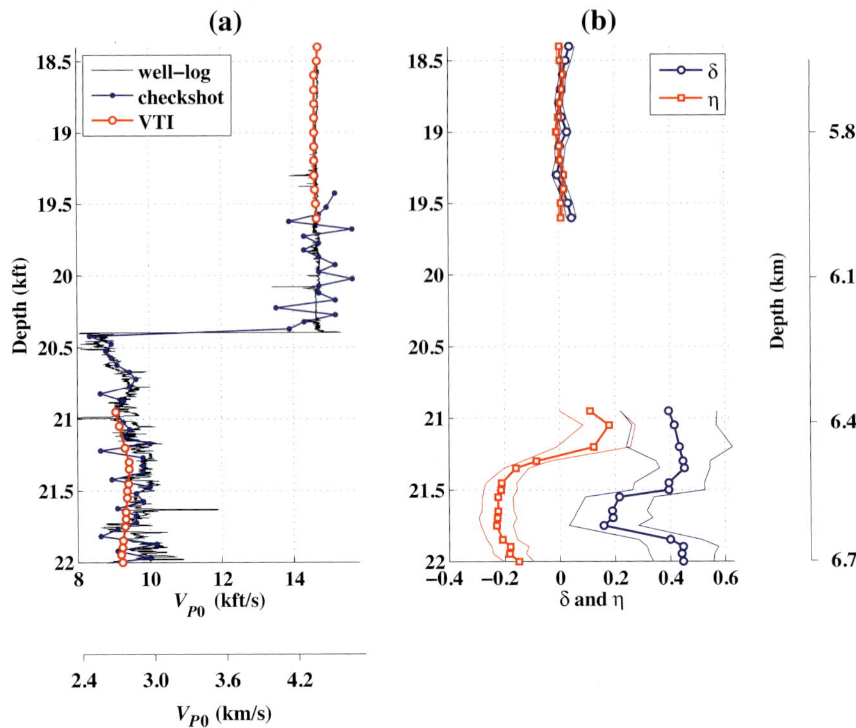

Figure 7.13: Inverted (a) vertical velocities and (b) anisotropy parameters (Grechka and Mateeva, 2007). The thin solid lines in (b) are the standard deviations of δ and η computed from the uncertainties in the picked values of ψ and q.

polarization curve in the salt from the isotropic function $q_{\text{iso}}(\psi)$ is smaller than the measurement uncertainties.

In contrast, the subsalt sediments are noticeably anisotropic (Figure 7.12b) with a negative value of η_{VSP} (the corresponding $\eta = -0.23 \pm 0.06$). At polarization angles $\psi > 30°$, all data points clearly fall below the isotropic curve $q_{\text{iso}}(\psi)$. The angular aperture of the subsalt data in Figure 7.12b is reduced by the strong velocity contrast at the base of salt (compare the values of V_{P0} in Figures 7.12a,b), which bends seismic rays toward the vertical when they exit the salt body. As a result, the slowness-polarization inversion yields larger standard deviations of the anisotropy parameters δ_{VSP} and η_{VSP} compared to those in the salt.

Figure 7.13 shows the parameter-estimation results for the entire data set. The anisotropy is represented in terms of the parameters δ and η computed by supplementing the obtained values of δ_{VSP} and η_{VSP} with the ratio V_{S0}/V_{P0}. The vertical velocities obtained from sonic logs were averaged over depth intervals of 600 ft (180 m) to infer anisotropy from the VSP data. The parameter values near the base of the salt could not be determined because of interference with reflected and transmitted waves, which invalidates the key assumption of a single, locally plane body wave.

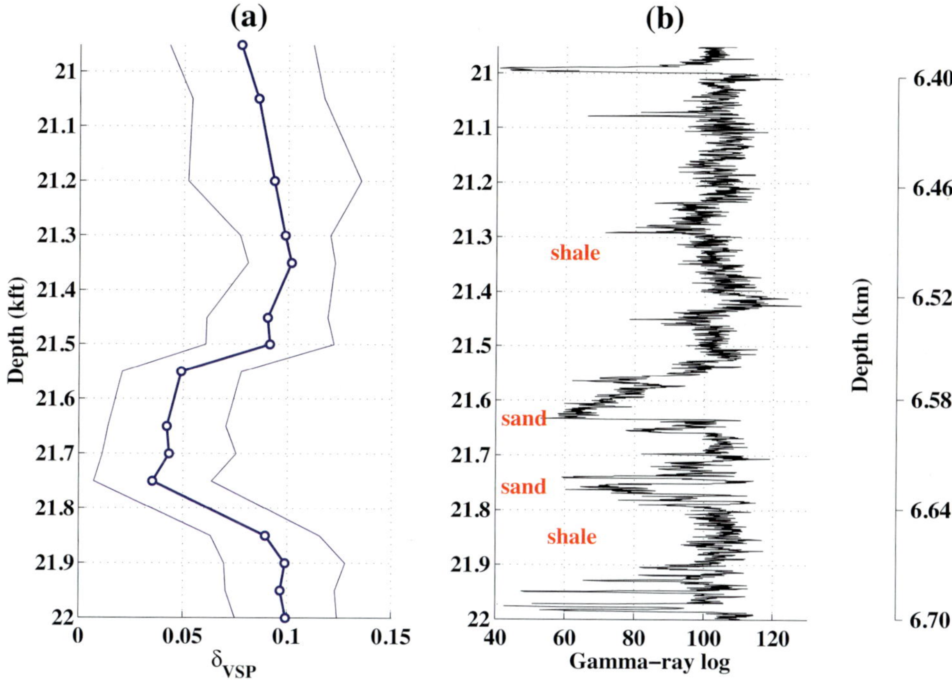

Figure 7.14: Comparison of (a) the inverted anisotropy parameter $\delta_{\rm VSP}$ with (b) a recorded gamma-ray log for subsalt sediments (Grechka and Mateeva, 2007). The thin lines on plot (a) mark the standard deviation of $\delta_{\rm VSP}$.

The velocity and anisotropy estimates in Figure 7.13 differ distinctly for the salt and the sedimentary section. As mentioned above, the salt turns out to be nearly isotropic on seismic scale (Figure 7.13b), which is consistent with the conventional assumption made in velocity analysis of reflection data in many subsalt plays. The absence of measurable anisotropy in salt bodies is noteworthy because individual salt (NaCl) crystals are known to be anisotropic and have cubic symmetry (e.g., Sun, 1994). Most likely, the orientations of salt crystals are random, and the effective medium (see Chapter 9) on the scale of seismic wavelength (about 300 m in the salt) is isotropic.

It is interesting that the anellipticity parameter η in the subsalt sediments is mostly negative and averages about -0.2 (Figure 7.13b). Negative η-values have sometimes been measured on rock samples (e.g., more than a dozen shale and sandstone cores with $\eta < 0$ are cited in Thomsen, 1986) but they have seldom been observed in seismic data. In contrast, the behavior of the parameter δ in Figure 7.13b is much easier to explain. First, the depth variation of δ essentially reproduces that of the inverted parameter $\delta_{\rm VSP}$ (Figure 7.14a). Second, δ and $\delta_{\rm VSP}$ are clearly correlated with lithology (marked on the gamma-ray log in Figure 7.14b), which confirms the well-established fact that shales typically possess stronger anisotropy than do sands.

7.2.3 Governing parameters for orthorhombic symmetry

Just as for VTI symmetry, general weak-anisotropy approximation 7.27 for the P-wave function $\mathbf{p}(\mathbf{U})$ can also be simplified for orthorhombic symmetry (equations 7.28–7.30). If the vertical-velocity ratio V_{S0}/V_{P0} is unknown, it can be incorporated into the "effective" VSP anisotropy parameters, as was done above for VTI media. The weak-anisotropy approximation for the vertical slowness q as a function of the P-wave polar (ψ) and azimuthal (φ) polarization angles is given by equations 7.30 and 7.32–7.34:

$$q(\psi,\,\varphi) = \frac{\cos\psi}{V_{P0}}\left[1 + \left(\delta^{(1)}_{\text{VSP}}\sin^2\varphi + \delta^{(2)}_{\text{VSP}}\cos^2\varphi\right)\sin^2\psi \right. \\ \left. + \left(\eta^{(1)}_{\text{VSP}}\sin^2\varphi + \eta^{(2)}_{\text{VSP}}\cos^2\varphi - \eta^{(3)}_{\text{VSP}}\sin^2\varphi\cos^2\varphi\right)\sin^4\psi\right], \tag{7.14}$$

where

$$\delta^{(i)}_{\text{VSP}} \equiv (f_0 - 1)\,\delta^{(i)}, \quad (i = 1,\,2) \tag{7.15}$$

and

$$\eta^{(i)}_{\text{VSP}} \equiv (2\,f_0 - 1)\,\eta^{(i)}, \quad (i = 1,\,2,\,3)\,. \tag{7.16}$$

The parameters $\delta^{(i)}$ are defined in equations 1.81 and 1.84, $\eta^{(i)}$ in equations 1.88–1.90, and f_0 in equation 7.9. The δ-term multiplied with $\sin^2\psi$ represents the azimuthally varying parameter δ (responsible for the P-wave NMO ellipse) scaled by the factor $(f_0 - 1)$. Similarly, the η-term multiplied with $\sin^4\psi$ is proportional to the azimuthally varying parameter η, which governs the nonhyperbolic moveout of P-waves (equation 3.26).

Inspection of equations 7.14–7.16 reveals the following:

1. The function $q(\psi,\,\varphi)$ constrains up to five anisotropy parameters ($\delta^{(1)}_{\text{VSP}}$, $\delta^{(2)}_{\text{VSP}}$, $\eta^{(1)}_{\text{VSP}}$, $\eta^{(2)}_{\text{VSP}}$, and $\eta^{(3)}_{\text{VSP}}$) depending on the polar and azimuthal angular coverage.
2. The velocity ratio V_{S0}/V_{P0} is required only to make the transition from the VSP parameters to those needed for reflection data processing ($\delta^{(1,2)}$ and $\eta^{(1,2,3)}$).

7.2.4 Case study from Rulison field

Here, following Grechka et al. (2007), we test the presented theoretical findings on a VSP data set acquired at tight-gas Rulison field (see section 3.4) by the Reservoir Characterization Project at Colorado School of Mines. Figure 7.7, which was discussed earlier in conjunction with analysis of the influence of lateral heterogeneity on VSP inversion, shows the first-break P-wave traveltimes picked at one of the downhole receivers.

The Rulison VSP survey differs from the Gulf-of-Mexico data set described above in two important respects. First, the data were acquired for a wide range of azimuths (Figure 7.7); second, the orientations of 3C borehole geophones were calculated from gyro and well-deviation surveys. The acquisition geometry and the possibility of measuring the azimuthal polarization angle make these data suitable for estimating the local parameters of orthorhombic media.

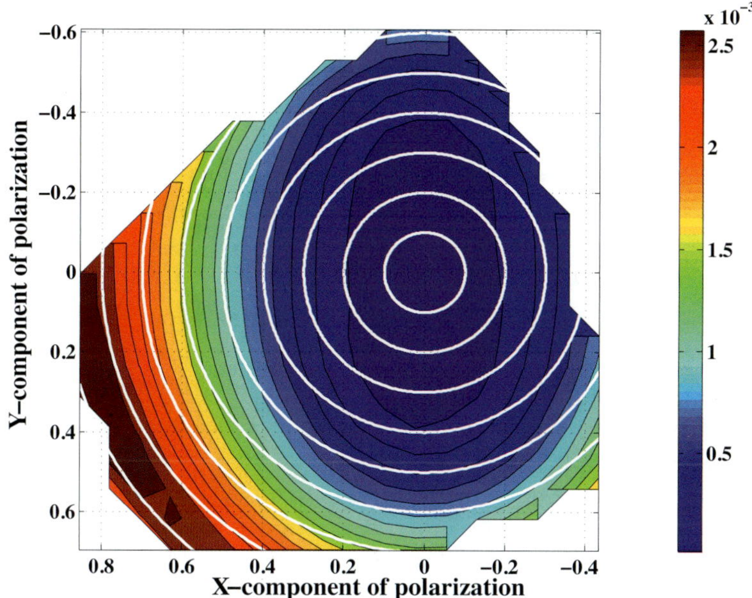

Figure 7.15: Azimuthally varying P-wave vertical slowness q (in s/kft) expressed through the horizontal polarization components in a borehole at Rulison field (Grechka et al., 2007). The function $q(\psi, \varphi)$ is computed for the best-fit orthorhombic model in the depth range 4510 ft – 4910 ft (1375 m – 1497 m) and corrected for the isotropic dependence $q_{\text{iso}}(\psi) = \cos\psi / V_{P0}$. The white circles mark the vertical slowness for azimuthally isotropic media.

Figure 7.15 displays the vertical slowness for the best-fit orthorhombic model corrected for isotropy by subtracting the isotropic slowness-of-polarization function $q_{\text{iso}}(\psi)$. The removal of the isotropic component emphasizes the azimuthal variation of q, which is substantial for near-vertical propagation (or for small horizontal polarization components) but becomes less apparent away from the vertical. The parameters of the obtained orthorhombic model and their standard deviations are as follows: $V_{P0} = 14,527 \pm 48$ ft/s (4428 ± 15 m/s), $\delta^{(1)} = -0.01 \pm 0.06$, $\delta^{(2)} = 0.12 \pm 0.05$, $\eta^{(1)} = 0.07 \pm 0.03$, $\eta^{(2)} = 0.04 \pm 0.02$, and $\eta^{(3)} = -0.02 \pm 0.03$. These parameters were calculated by supplementing VSP estimates of $\delta^{(1,2)}_{\text{VSP}}$ and $\eta^{(1,2,3)}_{\text{VSP}}$ with the velocity ratio V_{S0}/V_{P0} measured from sonic logs.

It is instructive to note a significant difference between $\delta^{(1)}$ and $\delta^{(2)}$, which stands out despite the sizeable uncertainties in both parameters. In contrast, the parameters $\eta^{(1)}$ and $\eta^{(2)}$ are close to each other, while $\eta^{(3)} \approx 0$ within the error bars. Therefore, the VSP inversion produces an azimuthally varying parameter δ, whereas η is practically independent of azimuth. This type of orthorhombic symmetry is consistent with the presence of gas-filled vertical fractures in a VTI host rock (Grechka, 2007) – a plausible model for Rulison field.

7.3 Summary

It has long been recognized that seismic data recorded in VSP geometries contain valuable information for estimating in situ anisotropy. While sufficiently wide angular data coverage is the main condition for meaningful anisotropic inversion, the parameter set that can be obtained from a particular survey depends on the type of heterogeneity of the overburden and on the availability of shear-wave data.

Type of heterogeneity of the overburden. — If the overburden is laterally homogeneous, the horizontal slowness components $p_{1,2}$ estimated at the surface remain constant along the raypath. Combined with the vertical slownesses $q \equiv p_3$ measured in the borehole, the values of $p_{1,2}$ make is possible to reconstruct slowness curves for 2D walkaway surveys or slowness surfaces for 3D wide-azimuth data. The inversion of these slowness functions is performed in terms of the Thomsen (or Thomsen-style for azimuthal anisotropy) parameters widely used in processing of reflection data. A more comprehensive characterization of local anisotropy near the borehole can be achieved by combining slowness and polarization measurements, which results in linear inversion for the stiffness coefficients.

This methodology is no longer applicable when lateral heterogeneity in the overburden distorts the horizontal slownesses. Then the anisotropic inversion can operate with the dependence of the vertical slowness q on the polarization direction $\mathbf{U}$. If only P-wave data are available, the function $q(\mathbf{U})$ constrains up to 15 combinations of the 21 stiffness coefficients. Restricting the azimuthal aperture or adding more unknowns (for example, geophone orientations when the polarization azimuths are essential) reduces the number of recoverable stiffnesses or anisotropy parameters. It is convenient to represent the dependence $q(\mathbf{U})$ for P-waves through the modified Thomsen-style anisotropy parameters specifically tailored to VSP applications. We introduced these parameters for VTI and orthorhombic models and showed on two case studies that they can be resolved in stable fashion from P-wave walkaway VSP data.

Availability of shear waves. — If S-waves have acceptable quality, it is beneficial to combine them with P-waves to constrain a more complete set of the anisotropy parameters. In the most favorable scenario, the direct P and S arrivals are recorded beneath a laterally homogeneous overburden by geophones whose azimuthal orientations are independently measured. Then the slowness surfaces and polarizations of P-waves and split S-waves can be inverted for the most general (triclinic) symmetry, provided the sources are deployed for a sufficiently wide range of offsets and azimuths. We demonstrated the feasibility of this algorithm by estimating the triclinic stiffness tensor (which proved to be close to orthorhombic) from wide-azimuth, multicomponent VSP data acquired at Vacuum field in New Mexico, USA. To our knowledge, no other type of seismic data can provide such a comprehensive characterization of in situ anisotropy.

Shear waves, however, are not excited in the majority of VSP surveys. Instead, they are recorded in the form of PS mode conversions, which are often too noisy for stable anisotropic inversion. Also, the quality of shear-wave data deteriorates for a laterally heterogeneous overburden. Then VSP inversion has to operate with P-wave slownesses and polarizations, as discussed above.

One of the most important conclusions suggested by the case studies examined in this chapter is that the anisotropy parameters estimated from VSP data are consistent with independent information about lithology and fractures. Clearly, linking the measured anisotropy to its plausible physical causes might prove critical for reservoir exploitation and development. One such link – the relationship between natural fracture sets and the effective anisotropy of fractured media – is examined more closely in Chapter 9.

Appendices for Chapter 7

7A Weak-anisotropy approximation for the function p(U)

Following Grechka and Mateeva (2007), here we present the weak-anisotropy approximation for the P-wave slowness vector $\mathbf{p}$ in terms of the P-wave unit polarization vector $\mathbf{U}$. The first-order perturbation theory is used to derive the approximate function $\mathbf{p}(\mathbf{U})$ for triclinic media, which is then specialized for higher symmetries.

The reference (background) medium is isotropic with the density-normalized stiffness tensor

$$c_{ijkl}^{\text{iso}} = (V_P^2 - 2V_S^2)\,\delta_{ij}\,\delta_{kl} + V_S^2\,(\delta_{ik}\,\delta_{jl} + \delta_{il}\,\delta_{jk})\,, \quad (i,\,j,\,k,\,l = 1,\,2,\,3)\,, \tag{7.17}$$

where V_P and V_S are the P- and S-wave velocities and δ_{ij} is Kronecker's delta. The P-wave slowness vector in the background is

$$p_m^{\text{iso}} = \frac{n_m}{V_P}\,, \quad (m = 1,\,2,\,3)\,, \tag{7.18}$$

where $\mathbf{n}$ is a unit vector in the slowness (wavefront normal) direction. Because for isotropic media the P-wave polarization vector $\mathbf{U}^{\text{iso}} = \mathbf{n}$, equation 7.18 can be rewritten as

$$p_m^{\text{iso}} = \frac{U_m^{\text{iso}}}{V_P}\,, \quad (m = 1,\,2,\,3)\,, \tag{7.19}$$

yielding the well-known isotropic relationship between the P-wave slowness and polarization.

Next, we consider an anisotropic perturbation $\Delta\mathbf{c}$ of the reference stiffness tensor $\mathbf{c}^{\text{iso}}$ (equation 7.17). This perturbation results in generally different P-wave slowness and polarization vectors for the same slowness direction $\mathbf{n}$ (that is, $\mathbf{p} \neq \mathbf{p}^{\text{iso}}$, $\mathbf{U} \neq \mathbf{U}^{\text{iso}}$, and the perturbed vectors $\mathbf{p}$ and $\mathbf{U}$ are no longer parallel). To obtain an explicit weak-anisotropy approximation for $\mathbf{p}(\mathbf{U})$, we assume that the stiffness perturbation is small, and the norms of the tensors $\Delta\mathbf{c}$ and $\mathbf{c}^{\text{iso}}$ are related as

$$||\Delta\mathbf{c}|| \ll ||\mathbf{c}^{\text{iso}}||\,. \tag{7.20}$$

The magnitudes of all perturbation components Δc_{ijkl} are assumed to be of the same order. Then the approximate dependence $\mathbf{p}(\mathbf{U})$ can be derived by linearizing all pertinent quantities in $\Delta\mathbf{c}$.

7A.1 Triclinic media

We begin with a triclinic perturbation $\Delta\mathbf{c}$ from the isotropic background. Using the result of Backus (1965) for P-wave phase velocity leads to the following weak-anisotropy approximation for the perturbed slowness vector:

$$p_m = \frac{n_m}{V_P}\left(1 - \frac{\Delta c_{ijkl}}{2V_P^2}\, n_i n_j n_k n_l\right), \quad (m = 1,\, 2,\, 3)\,, \tag{7.21}$$

where summation from 1 to 3 with respect to repeated indices is implied.

To relate the original wavefront normal $\mathbf{n}$ to the perturbed P-wave polarization vector $\mathbf{U}$, we use the following expression (e.g., Pšenčík and Gajewski, 1998; Farra, 2001):

$$n_m = U_m - f\left(n_m^{(2)} n_i^{(2)} + n_m^{(3)} n_i^{(3)}\right) \frac{\Delta c_{ijkl}}{V_P^2}\, n_j n_k U_l\,, \quad (m = 1,\, 2,\, 3)\,, \tag{7.22}$$

where

$$f = \frac{1}{1 - V_S^2/V_P^2}\,, \tag{7.23}$$

and $\mathbf{n}^{(2)}$, $\mathbf{n}^{(3)}$ are unit vectors orthogonal to $\mathbf{n} \equiv \mathbf{n}^{(1)}$ and to each other,

$$n_i^{(r)} n_i^{(t)} = \delta_{rt}\,, \quad (r,\, t = 1,\, 2,\, 3)\,. \tag{7.24}$$

We also introduce two mutually orthogonal unit vectors, $\mathbf{U}^{(2)}$ and $\mathbf{U}^{(3)}$, that lie in the plane normal to the perturbed polarization vector, $\mathbf{U} \equiv \mathbf{U}^{(1)}$. The vectors $\mathbf{U}^{(1,2,3)}$ satisfy the orthogonality relationships,

$$U_i^{(r)} U_i^{(t)} = \delta_{rt}\,, \quad (r,\, t = 1,\, 2,\, 3)\,, \tag{7.25}$$

which can be used to define the vectors $\mathbf{U}^{(2)}$ and $\mathbf{U}^{(3)}$.

The difference between the vectors $\mathbf{n}$ and $\mathbf{U}$ in equation 7.22 is linear in $\Delta\mathbf{c}$. Therefore, within the framework of the linearized approximation the vectors $\mathbf{n}^{(r)}$ in the term proportional to $\Delta\mathbf{c}$ can be replaced with $\mathbf{U}^{(r)}$:

$$n_m = U_m - f\left(U_m^{(2)} U_i^{(2)} + U_m^{(3)} U_i^{(3)}\right) \frac{\Delta c_{ijkl}}{V_P^2}\, U_j U_k U_l\,, \quad (m = 1,\, 2,\, 3)\,. \tag{7.26}$$

Finally, substituting equation 7.26 into equation 7.21 and dropping the second- and higher-order terms in $\Delta\mathbf{c}$ yields

$$p_m = \frac{U_m}{V_P} - \left[\frac{U_m U_i}{2} + f\left(U_m^{(2)} U_i^{(2)} + U_m^{(3)} U_i^{(3)}\right)\right] \frac{\Delta c_{ijkl}}{V_P^3}\, U_j U_k U_l\,, \quad (m = 1,\, 2,\, 3)\,. \tag{7.27}$$

Equation 7.27, which represents the explicit weak-anisotropy approximation for $\mathbf{p}(\mathbf{U})$ for triclinic media, has the following properties:

1. The first term coincides with the isotropic function $\mathbf{p}(\mathbf{U})$ (equation 7.19), while the second is the anisotropic correction.

2. It can be easily verified that the term involving the vectors $\mathbf{U}^{(2)}$ and $\mathbf{U}^{(3)}$ does not depend on their orientation in the plane orthogonal to $\mathbf{U}$. This property allows us to choose the vectors $\mathbf{U}^{(2)}$ and $\mathbf{U}^{(3)}$ based on computational convenience.

In particular, they do not have to coincide with the polarizations of the split shear waves propagating in the direction $\mathbf{p}/|\mathbf{p}|$ in the perturbed anisotropic medium.

3. Given the reference isotropic velocities V_P and V_S, equation 7.27 can be viewed as a linear system to be solved for the unknown coefficients Δc_{ijkl} using the (presumably measured) slownesses and polarizations. In accordance with the findings of Zheng and Pšenčík (2002), the polarization dependence $p_m(\mathbf{U})$ of any slowness component in equation 7.27 constrains up to 15 combinations of the 21 stiffness coefficients. Those combinations (in the Voigt notation) are: Δc_{11}, $\Delta c_{12} + 2\,\Delta c_{66}$, $\Delta c_{13} + 2\,\Delta c_{55}$, $\Delta c_{14} + 2\,\Delta c_{56}$, Δc_{15}, Δc_{16}, Δc_{22}, $\Delta c_{23} + 2\,\Delta c_{44}$, Δc_{24}, $\Delta c_{25} + 2\,\Delta c_{46}$, Δc_{26}, Δc_{33}, Δc_{34}, Δc_{35}, and $\Delta c_{36} + 2\,\Delta c_{45}$.

4. Knowledge of the reference shear-wave velocity V_S is important for estimation of the stiffness perturbations. Equation 7.27 suggests that the P-wave velocity V_P can be recovered simultaneously with Δc_{ijkl}, if f (equation 7.23) is known. However, the factor f (or the velocity ratio V_S/V_P) itself cannot be inferred from $\mathbf{p}(\mathbf{U})$ because it is multiplied with the stiffness perturbations. This issue will be clarified later in this appendix after deriving the approximate dependence $\mathbf{p}(\mathbf{U})$ for higher-symmetry models.

7A.2 Orthorhombic media

For orthorhombic models it is convenient to specify the P-wave slowness and polarization vectors in the coordinate frame whose axes coincide with the symmetry directions (intersections of the symmetry planes) of the medium (see section 2.5). Then the stiffness perturbations Δc_{ijkl} can be expressed through the anisotropy parameters and substituted into equation 7.27. The reference isotropic velocities are chosen to be equal to the vertical velocities V_{P0} and V_{S0} given by equations 1.78 and 1.79. The weak-anisotropy approximation of the function $\mathbf{p}(\mathbf{U})$ takes the following form:

$$
\begin{aligned}
p_1 = \frac{\sin\psi\cos\phi}{V_{P0}} \Big\{ 1 &- f_0\,\delta^{(2)} + (f_0 - 1)\left(\delta^{(1)}\sin^2\phi + \delta^{(2)}\cos^2\phi\right)\sin^2\psi \\
&+ \eta^{(1)}\sin^2\psi\,\sin^2\phi\left[(f_0 - 1)\sin^2\psi - f_0\cos^2\psi\right] \\
&- \eta^{(2)}\sin^2\psi\left[\sin^2\psi\cos^2\phi + f_0(2\cos^2\psi - \cos 2\psi\,\sin^2\phi)\right] \\
&+ \eta^{(3)}\sin^2\psi\,\sin^2\phi\left[\,f_0\cos^2\psi + (\cos^2\phi - f_0\cos 2\phi)\,\sin^2\psi\right]\Big\},
\end{aligned}
\tag{7.28}
$$

$$
\begin{aligned}
p_2 = \frac{\sin\psi\sin\phi}{V_{P0}} \Big\{ 1 &- f_0\,\delta^{(1)} + (f_0 - 1)\left(\delta^{(2)}\cos^2\phi + \delta^{(1)}\sin^2\phi\right)\sin^2\psi \\
&+ \eta^{(1)}\sin^2\psi\,\left[(2f_0 - 1)\sin^2\psi\,\sin^2\phi - f_0\,(1 + \sin^2\phi)\right] \\
&+ \eta^{(2)}\sin^2\psi\,\cos^2\phi\left[(f_0 - 1)\sin^2\psi - f_0\cos^2\psi\right] \\
&+ \eta^{(3)}\sin^2\psi\,\cos^2\phi\left[f_0 - (2f_0 - 1)\sin^2\psi\,\sin^2\phi\right]\Big\},
\end{aligned}
\tag{7.29}
$$

$$p_3 = \frac{\cos\psi}{V_{P0}} \Big[1 + (f_0 - 1)\left(\delta^{(1)} \sin^2\phi + \delta^{(2)} \cos^2\phi\right) \sin^2\psi + (2f_0 - 1)\left(\eta^{(1)} \sin^2\phi + \eta^{(2)} \cos^2\phi - \eta^{(3)} \sin^2\phi \cos^2\phi\right) \sin^4\psi \Big], \quad (7.30)$$

where ψ and ϕ are the polar and azimuthal polarization angles,

$$\mathbf{U} = [\sin\psi \cos\phi, \ \sin\psi \sin\phi, \ \cos\psi], \quad (7.31)$$

and

$$f_0 = \frac{1}{1 - V_{S0}^2/V_{P0}^2}. \quad (7.32)$$

According to inequality 7.20, all anisotropy parameters are assumed to be much smaller than unity. Therefore, equations 7.28 – 7.32 could have been derived without the intermediate linearization (equation 7.27) of $\mathbf{p}(\mathbf{U})$ in terms of the stiffness perturbations.

The analytic results (equations 7.28 – 7.32) allow us to draw several important conclusions:

1. The function $\mathbf{p}(\mathbf{U})$ in orthorhombic media constrains up to five anisotropy parameters: $\delta^{(1)}$, $\delta^{(2)}$, $\eta^{(1)}$, $\eta^{(2)}$, and $\eta^{(3)}$.

2. The ratio of vertical velocities V_{S0}/V_{P0} (or the factor f_0, equation 7.32) has to be known a priori in order to estimate the parameters $\delta^{(1,2)}$ and $\eta^{(1,2,3)}$.

3. The vertical slowness p_3 measured from wide-azimuth P-wave VSP data recorded in a vertical well is described by equation 7.30. Anisotropy contributes to p_3 through the following "effective" parameters, which absorb the factor f_0:

$$\delta^{(i)}_{\text{VSP}} = (f_0 - 1)\,\delta^{(i)}, \quad (i = 1, 2) \quad (7.33)$$

and

$$\eta^{(i)}_{\text{VSP}} = (2f_0 - 1)\,\eta^{(i)}, \quad (i = 1, 2, 3). \quad (7.34)$$

7A.3 VTI media

Vertical transverse isotropy represents a special case of orthorhombic symmetry with $\delta^{(1)} = \delta^{(2)} = \delta$, $\eta^{(1)} = \eta^{(2)} = \eta$, and $\eta^{(3)} = 0$ (Tsvankin, 1997a, 2005). Making these substitutions in equations 7.28 – 7.30 yields

$$p_i = \frac{U_i}{V_{P0}} \left\{ 1 + \delta\left[(f_0 - 1)\sin^2\psi - f_0\right] + \eta \sin^2\psi \left[(2f_0 - 1)\sin^2\psi - 2f_0\right] \right\}, \quad (i = 1, 2), \quad (7.35)$$

$$p_3 = \frac{U_3}{V_{P0}} \left[1 + \delta\,(f_0 - 1)\sin^2\psi + \eta\,(2f_0 - 1)\sin^4\psi \right], \quad (7.36)$$

where U_i and f_0 are defined in equations 7.31 and 7.32, respectively. The influence of anisotropy on P-wave VSP data recorded in a vertical well (equation 7.36) is governed by the effective parameters

$$\delta_{\text{VSP}} = (f_0 - 1)\,\delta \quad \text{and} \quad \eta_{\text{VSP}} = (2f_0 - 1)\,\eta. \quad (7.37)$$

The linearized relationship between the P-wave slowness and polarization vectors in VTI media can be also obtained from the weak-anisotropy approximation for the polarization angle ψ (Tsvankin, 2005, his equation 1.78):

$$\psi = \theta + f_0 \left(\delta + 2\eta \sin^2\theta\right) \sin\theta \cos\theta \,, \tag{7.38}$$

where θ is the polar phase angle. The P-wave slowness vector is then approximately given by

$$\begin{bmatrix} p_1 \\ p_2 \\ p_3 \end{bmatrix} = \frac{1 - \delta \sin^2\theta - \eta \sin^4\theta}{V_{P0}} \begin{bmatrix} \sin\theta \, \cos\phi \\ \sin\theta \, \sin\phi \\ \cos\theta \end{bmatrix} . \tag{7.39}$$

Combining equations 7.38 and 7.39 and applying further linearization in the anisotropy parameters represents an alternative way of deriving equations 7.35 and 7.36.

7B Exact function p(ψ) for VTI media

The simplicity of the VTI model makes it possible to obtain an exact equation for computing the P-wave phase angle θ and then the slowness vector $\mathbf{p}$ from the polarization angle ψ (Grechka and Mateeva, 2007). Assuming that the wave propagates in the $[x_1, x_3]$-plane, the vectors $\mathbf{U}$ and $\mathbf{p}$ can be written as

$$\mathbf{U} = \begin{bmatrix} \sin\psi \\ 0 \\ \cos\psi \end{bmatrix} \tag{7.40}$$

and

$$\mathbf{p} = \frac{1}{V(\theta)} \begin{bmatrix} \sin\theta \\ 0 \\ \cos\theta \end{bmatrix} , \tag{7.41}$$

where $V(\theta)$ is the P-wave phase velocity. The angles θ and ψ are related by the Christoffel equation (e.g., Tsvankin, 2005, his equation 1.35):

$$\begin{bmatrix} c_{11} \sin^2\theta + c_{55} \cos^2\theta - V^2 & (c_{13} + c_{55}) \sin\theta \cos\theta \\ (c_{13} + c_{55}) \sin\theta \cos\theta & c_{55} \sin^2\theta + c_{33} \cos^2\theta - V^2 \end{bmatrix} \begin{bmatrix} \sin\psi \\ \cos\psi \end{bmatrix} = 0 \,. \tag{7.42}$$

If $c_{13} + c_{55} \neq 0$, the phase velocity V can be eliminated by solving one of the linear equations 7.42 for V^2 and substituting the obtained expression into the other equation. This results in a quadratic equation for $\tan\theta$:

$$(c_{11} - c_{55}) \tan^2\theta + 2(c_{13} + c_{55}) \cot 2\psi \tan\theta - (c_{33} - c_{55}) = 0 \,. \tag{7.43}$$

For typical (i.e., not anomalously anisotropic) solids with $c_{11} > c_{55}$ and $c_{33} > c_{55}$, equation 7.43 always has two real roots. These two solutions correspond to the phase angles θ of P- and SV-waves whose polarization directions are defined by the angle ψ

(the P-wave solution θ is closer to ψ than the SV-wave solution). Once the angle θ is found, the P-wave phase velocity V can be obtained as the square root of the larger eigenvalue of the Christoffel matrix in equation 7.42. Then the exact slowness vector $\mathbf{p}(\psi)$ is calculated from its definition 7.41.

Equation 7.43 was derived under the assumption that $c_{13} \neq -c_{55}$. For media with $c_{13} + c_{55} = 0$, the Christoffel matrix in equation 7.42 becomes diagonal, and the polarizations of the two waves (which can no longer be classified as "P" and "SV") are parallel and perpendicular to the symmetry axis for all angles θ (Helbig, 1994). Because the polarization angle ψ for each wave is constant, the dependence $\theta(\psi)$ ceases to exist. Such a model, however, is not likely to be encountered in practice because it has uncommon negative values of the stiffness c_{13} (c_{55} is always positive). Note that when $c_{13} = -c_{55}$, the parameter δ reaches its theoretical minimum (Tsvankin, 2005):

$$\delta_{\min} = \frac{1}{2}\left(\frac{c_{55}}{c_{33}} - 1\right). \tag{7.44}$$

Chapter 8

Prestack amplitude analysis of wide-azimuth data

In Chapters 2, 3, 5, and 6 we discussed moveout inversion of wide-azimuth reflection data for the parameters of transversely isotropic, orthorhombic, and monoclinic media. Essential information for anisotropic parameter estimation and fracture characterization can also be obtained from azimuthally dependent reflection coefficients. Analysis of prestack amplitude variation with offset and azimuth (often called *azimuthal AVO analysis* or *AVAZ*) is one of the most effective tools for seismic characterization of fractures and in situ stress field (e.g., Mallick et al., 1998; Lynn et al., 1999b; Bakulin et al., 2000a,b). The main advantage of amplitude methods compared to traveltime inversion is their superior vertical resolution, which makes AVO analysis suitable for relatively thin reservoirs. Also, body-wave amplitudes are highly sensitive to azimuthal velocity variations associated with vertical fracture systems and nonhydrostatic stresses.

We start by reviewing the reflection/transmission problem and plane-wave reflection coefficients for a boundary between azimuthally anisotropic halfspaces. For orthorhombic models with the same orientation of the vertical symmetry planes above and below the reflector, the symmetry-plane reflection coefficients can be directly adapted from the corresponding VTI equations. Application of the weak-contrast, weak-anisotropy approximation shows that the azimuthal variation of the P-wave AVO gradient in orthorhombic and HTI media is controlled by the shear-wave splitting coefficient and the parameter responsible for the eccentricity of the NMO ellipse.

Because AVO algorithms aim to operate with the reflection coefficient at the target horizon, an important element of AVO processing is a correction for the source/receiver directivity and propagation factors, most notably for the geometrical spreading in the overburden. Geometrical spreading of PP- and PS-waves in laterally homogeneous, arbitrarily anisotropic media can be represented as a function of the spatial derivatives of reflection traveltime. This formalism, combined with the generalized Alkhalifah-Tsvankin nonhyperbolic moveout equation (see Chapter 3), helps devise the moveout-based anisotropic spreading correction (MASC) for wide-azimuth P-wave data from layered orthorhombic media.

To illustrate the performance of P-wave azimuthal moveout and AVO analyses for a typical tight fractured reservoir, we present a case study from gas-sand Rulison field in Colorado, USA. The advanced processing sequence includes 3D nonhyperbolic moveout inversion, computation of the effective and interval NMO ellipses, anisotropic geometrical-spreading correction (MASC), and estimation of the azimuthally varying AVO gradient. Reservoir heterogeneity significantly complicates analysis and correlation of the kinematic and dynamic attributes, but azimuthal AVO anomalies still provide insight into the orientation and spatial distribution of natural fractures.

Whereas conventional AVO analysis is restricted to relatively small, subcritical incidence angles, long-offset reflection data are particularly sensitive to anisotropy. A promising way to use wide-angle reflection events in anisotropic parameter estimation is by measuring and inverting the critical angle for reflections from high-velocity layers (*critical-angle reflectometry*). We give a concise description of the critical angles in VTI and orthorhombic media and show how the critical offset can be found from the AVO response.

The chapter concludes with a section devoted to analysis of P- and S-wave attenuation coefficients in the presence of both velocity and attenuation anisotropy. After defining the quality-factor matrix Q_{ij} through the ratios of the real and imaginary parts of the stiffness coefficients, we introduce Thomsen-style parameters responsible for angle-dependent attenuation in TI and orthorhombic media. Approximate attenuation coefficients expressed through these parameters provide the foundation for anisotropic inversion of field attenuation measurements.

8.1 Basics of azimuthal AVO analysis

Reflection and transmission coefficients contain information about the local medium properties near the interface averaged on the scale of seismic wavelength. Here, we describe the reflection/transmission problem for azimuthally anisotropic media with the main focus on P-wave reflectivity in the symmetry planes of orthorhombic models.

8.1.1 Reflection/transmission coefficients in anisotropic media

Derivation of plane-wave reflection and transmission coefficients in the presence of anisotropy follows the same main steps as in isotropic media, reviewed below. For a more comprehensive treatment of the reflection/transmission problem and AVO analysis in anisotropic media, we refer the reader to the book by Rüger (2002) and thesis by Jílek (2002b).

Reflection/transmission problem

According to Snell's law, the projection of the slowness vector onto the reflector (here assumed to be horizontal) should be identical for the incident, reflected, and transmitted waves (Figure 8.1). The conservation of the horizontal slowness can be used to obtain the vertical slownesses $q = p_3$ and takeoff phase angles for the scattered (i.e., reflected and transmitted) waves. Whereas this operation is straightforward for

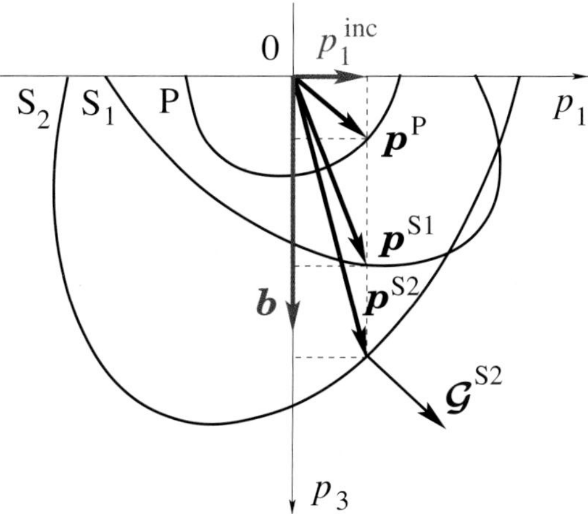

Figure 8.1: Illustration of Snell's law for horizontal interfaces in anisotropic media. The plot shows the vertical cross-sections of the slowness surfaces of P-, S_1-, and S_2-waves in one of the halfspaces. The horizontal projections of the slowness vectors $\mathbf{p}^{\rm P}$, $\mathbf{p}^{\rm S1}$, and $\mathbf{p}^{\rm S2}$ are equal to the horizontal slowness $p_1^{\rm inc}$ of the incident wave. The corresponding vertical slownesses $q{=}p_3$ of the three modes should be found from the Christoffel equation. The group-velocity vectors (e.g., $\mathcal{G}^{\rm S2}$), which define seismic rays, are orthogonal to the slowness surfaces.

isotropic media, evaluation of q in the presence of anisotropy involves solving the Christoffel equation formulated in terms of the slowness vector.

In the most general case, the vertical slownesses in each halfspace represent the roots of a sixth-order polynomial, with the solutions corresponding to three (P, S_1, S_2) upgoing and three downgoing modes. If the vertical incidence plane coincides with a plane of symmetry in both halfspaces, transversely polarized SH-waves are decoupled, which reduces the order of the polynomial for P- and SV-waves to four. For models with a horizontal symmetry plane, the slowness surface is symmetric with respect to the reflector, and the squared vertical slownesses q^2 can be found analytically from a cubic equation. If both symmetry conditions are satisfied simultaneously (e.g., in VTI media), q^2 can be obtained from a quadratic equation for P- and SV-waves and a linear equation for SH-waves. In addition to the vertical slownesses, the Christoffel equation also yields the polarization vectors needed to solve the reflection/transmission problem.

The boundary conditions for two halfspaces in welded contact consist of the continuity of the three-component displacement and traction vectors across the interface. The displacements and tractions of the scattered waves depend on their slowness and polarization vectors determined from the Christoffel equation and are proportional to the corresponding reflection/transmission coefficients. In general, the boundary conditions provide six linear equations for the six reflection and transmission coefficients of P-, S_1-, and S_2-waves. However, for VTI media and the vertical symmetry planes of HTI and orthorhombic media, the number of equations reduces to four for P- and

SV-waves and to two for SH-waves. Exact expressions for the reflection coefficients in VTI models can be found in Daley and Hron (1977), Graebner (1992), and Rüger (1997, 2002).

Weak-contrast, weak-anisotropy approximations

Although computation of reflection/transmission coefficients is relatively straightforward even for low-symmetry models, it is difficult to analyze the contributions of multiple anisotropy parameters solely by performing numerical simulations. Unfortunately, exact expressions for reflection/transmission coefficients are complicated even for isotropic media and cannot even be obtained in closed form outside symmetry planes of anisotropic models (e.g., Fryer and Frazer, 1984; Jílek, 2002a,b). Therefore, the solutions of the reflection/transmission problem are often studied by means of weak-contrast, weak-anisotropy approximations, which have a much simpler form and reduce the number of independent parameters. These approximations are derived by linearizing the boundary conditions and then the reflection/transmission coefficients in the small quantities, which include the relative jumps in the elastic parameters across the interface and the dimensionless anisotropy parameters. For example, Shuey (1985) employed the weak-contrast assumption to derive the P-wave reflection coefficient in isotropic media as a function of the contrasts in the P- and S-wave velocities and density. Shuey's approximation is widely used in conventional (isotropic) AVO analysis.

The P-wave reflection coefficient for a boundary between VTI media depends on the P- and S-wave vertical velocities (V_{P0} and V_{S0}), density ρ, and the parameters δ and ϵ (Banik, 1987; Thomsen, 1993; Rüger, 1997; Tsvankin, 2005). Following Rüger (1997, 2002), the weak-contrast, weak-anisotropy approximation for that coefficient can be represented as

$$\begin{aligned} R^{(\mathrm{VTI})} &= \frac{1}{2}\frac{\triangle Z}{\bar{Z}} + \frac{1}{2}\left[\frac{\triangle V_{P0}}{\bar{V}_{P0}} - \left(\frac{2\bar{V}_{S0}}{\bar{V}_{P0}}\right)^2 \frac{\triangle G}{\bar{G}} + \triangle\delta\right]\sin^2\theta \\ &+ \frac{1}{2}\left[\frac{\triangle V_{P0}}{\bar{V}_{P0}} + \triangle\epsilon\right]\sin^2\theta\,\tan^2\theta\,, \end{aligned} \tag{8.1}$$

where θ is the incidence phase angle, $Z = \rho V_{P0}$ is the P-wave vertical impedance, and $G = \rho V_{S0}^2$ is the S-wave vertical rigidity modulus. By $\triangle B = B_2 - B_1$ we denote the difference between the values of the parameter B below and above the reflector, and $\bar{B} = (B_1 + B_2)/2$ is the average value.

The normal-incidence reflection coefficient, which is equal to the fractional difference between the impedances of the reflecting and incidence media, is often called the *AVO intercept.* The term multiplied with $\sin^2\theta$ (*AVO gradient*) is responsible for the angle variation of $R^{(\mathrm{VTI})}$ near the vertical, while the contribution of the "curvature term," proportional to the product ($\sin^2\theta\,\tan^2\theta$), becomes significant at larger incidence angles.

8.1.2 Symmetry-plane reflection coefficients for orthorhombic media

The equivalence between the symmetry planes of TI and orthorhombic models helps adapt the VTI reflection/transmission coefficients for symmetry-plane propagation in orthorhombic media. Suppose a horizontal boundary separates two orthorhombic halfspaces with the same orientation of the vertical symmetry planes (the other symmetry plane is assumed to be horizontal). Because of the identical form of the Christoffel equation, the symmetry-plane velocities and polarization vectors on both sides of the reflector are described by the same expressions as in VTI media (see section 2.5). Therefore, the boundary conditions in the symmetry planes have the VTI form, and the reflection/transmission coefficients can be found directly from the corresponding VTI equations (Rüger, 2002; Tsvankin, 2005).

Although this "equivalence principle" is valid for exact symmetry-plane reflection/transmission coefficients, we restrict ourselves to weak-contrast, weak-anisotropy approximations. Linearized P-wave reflection coefficients for the full range of azimuths in arbitrarily anisotropic media can be found in Vavryčuk and Pšenčík (1998) and are not reproduced here.

Equation 8.1 can be adapted for both vertical symmetry planes of orthorhombic media by simply making appropriate parameter substitutions (see Appendix 1B). For the $[x_1, x_3]$-plane, which is taken to have the same azimuth above and below the reflector, the parameter δ is replaced with $\delta^{(2)}$ and ϵ with $\epsilon^{(2)}$. Likewise, in the $[x_2, x_3]$-plane the substitutions are $\delta \rightarrow \delta^{(1)}$ and $\epsilon \rightarrow \epsilon^{(1)}$.

One important subtlety in this adaptation is the difference between the vertical velocities of the split shear waves. As discussed in section 6.1, P-waves in the symmetry planes $[x_1, x_3]$ and $[x_2, x_3]$ are coupled to two different shear modes. For example, suppose that the fast mode S_1 at vertical incidence is polarized in the x_2-direction, and the slow mode S_2 in the x_1-direction. Then the S_1-wave has in-plane (SV) polarization in the $[x_2, x_3]$-plane, and S_2 plays the role of the SV-wave in the $[x_1, x_3]$-plane. Because the P-wave is coupled to the in-plane polarized (SV) shear mode, the vertical velocity V_{S0} in equation 8.1 corresponds to the S_1-wave in the $[x_2, x_3]$-plane and to the S_2-wave in the $[x_1, x_3]$-plane.

According to the definitions in Appendix 1B, V_{S0} is the vertical velocity of the S-wave polarized in the x_1-direction (here assumed to be the S_2-wave) and, therefore, should be used in the expression for the reflection coefficient in the $[x_1, x_3]$-plane. For the $[x_2, x_3]$-plane, V_{S0} must be replaced with the vertical velocity of the S_1-wave given by equation 6.14:

$$V_{S1} = V_{S0} \sqrt{\frac{1 + 2\gamma^{(1)}}{1 + 2\gamma^{(2)}}} \, . \tag{8.2}$$

The S-wave vertical rigidity modulus in the $[x_2, x_3]$-plane becomes $G_{S1} = \rho V_{S1}^2$. Using equation 8.2 and applying the weak-anisotropy approximation, we find

$$G_{S1} = \rho V_{S0}^2 \left(1 + 2\gamma^{(1)} - 2\gamma^{(2)}\right) = G \left(1 + 2\gamma^{(S)}\right) , \tag{8.3}$$

where $\gamma^{(S)}$ is the linearized version of the shear-wave splitting coefficient defined in equation 1.87.

Then the approximate P-wave reflection coefficients in the symmetry planes take the following form:

$$R^{[x_1,\,x_3]} = \frac{1}{2}\frac{\triangle Z}{\bar{Z}} + \frac{1}{2}\left[\frac{\triangle V_{P0}}{\bar{V}_{P0}} - \left(\frac{2\bar{V}_{S0}}{\bar{V}_{P0}}\right)^2 \frac{\triangle G}{\bar{G}} + \triangle\delta^{(2)}\right]\sin^2\theta + \frac{1}{2}\left[\frac{\triangle V_{P0}}{\bar{V}_{P0}} + \triangle\epsilon^{(2)}\right]\sin^2\theta\,\tan^2\theta\,, \tag{8.4}$$

$$R^{[x_2,\,x_3]} = \frac{1}{2}\frac{\triangle Z}{\bar{Z}} + \frac{1}{2}\left[\frac{\triangle V_{P0}}{\bar{V}_{P0}} - \left(\frac{2\bar{V}_{S1}}{\bar{V}_{P0}}\right)^2 \frac{\triangle G_{S1}}{\bar{G}_{S1}} + \triangle\delta^{(1)}\right]\sin^2\theta + \frac{1}{2}\left[\frac{\triangle V_{P0}}{\bar{V}_{P0}} + \triangle\epsilon^{(1)}\right]\sin^2\theta\,\tan^2\theta\,. \tag{8.5}$$

Rüger (2002) gives equations 8.4 and 8.5 in a slightly different notation. Substituting equation 8.3 into equation 8.5 and linearizing the result in the small parameters, we rewrite the reflection coefficient in the $[x_2,\, x_3]$-plane as

$$R^{[x_2,\,x_3]} = \frac{1}{2}\frac{\triangle Z}{\bar{Z}} + \frac{1}{2}\left[\frac{\triangle V_{P0}}{\bar{V}_{P0}} - \left(\frac{2\bar{V}_{S0}}{\bar{V}_{P0}}\right)^2 \left(\frac{\triangle G}{\bar{G}} + 2\triangle\gamma^{(S)}\right) + \triangle\delta^{(1)}\right]\sin^2\theta + \frac{1}{2}\left[\frac{\triangle V_{P0}}{\bar{V}_{P0}} + \triangle\epsilon^{(1)}\right]\sin^2\theta\,\tan^2\theta\,. \tag{8.6}$$

The behavior of the P-wave reflection coefficient between the symmetry planes can be analyzed using the results of Vavryčuk and Pšenčík (1998). As in VTI media, the anisotropy parameters make a substantial contribution to the AVO response, which is evident from the weighting factors. In particular, for typical vertical-velocity ratios the multiplier of the coefficient $\gamma^{(S)}$ in the AVO gradient in equation 8.6 can be twice as large as that of the jump in V_{P0}.

Equations 8.4 and 8.6 contain the same terms involving Z, V_{P0}, and G and, therefore, are well-suited for evaluating the azimuthal variation of the P-wave reflection coefficient between the symmetry planes:

$$R^{[x_1,x_3]} - R^{[x_2,x_3]} = \left[\left(\frac{2\bar{V}_{S0}}{\bar{V}_{P0}}\right)^2 \triangle\gamma^{(S)} + \frac{1}{2}\left(\triangle\delta^{(2)} - \triangle\delta^{(1)}\right)\right]\sin^2\theta + \frac{1}{2}\left[\triangle\epsilon^{(2)} - \triangle\epsilon^{(1)}\right]\sin^2\theta\,\tan^2\theta\,. \tag{8.7}$$

If the incidence medium is isotropic, equation 8.7 includes only the splitting parameter

$\gamma^{(S)}$ and the differences $(\delta^{(2)} - \delta^{(1)})$ and $(\epsilon^{(2)} - \epsilon^{(1)})$ in the reflecting halfspace:

$$R^{[x_1,x_3]} - R^{[x_2,x_3]} = \left[\left(\frac{2\bar{V}_{S0}}{\bar{V}_{P0}}\right)^2 \gamma^{(S)} + \frac{1}{2}\left(\delta^{(2)} - \delta^{(1)}\right)\right] \sin^2\theta + \frac{1}{2}\left[\epsilon^{(2)} - \epsilon^{(1)}\right] \sin^2\theta \, \tan^2\theta \,. \tag{8.8}$$

Because the P-wave is coupled to two different shear modes in the vertical symmetry planes, the azimuthal variation of the P-wave AVO gradient is strongly dependent on the S-wave splitting parameter $\gamma^{(S)}$. In fact, for the common velocity ratio $\bar{V}_{P0}/\bar{V}_{S0} = 2$, the weighting factor of $\gamma^{(S)}$ in equation 8.8 is twice as large as that of $(\delta^{(2)} - \delta^{(1)})$. Therefore, azimuthal AVO analysis of P-wave data provides information about shear-wave splitting, often used as a direct indicator of fracturing. The difference $(\delta^{(2)} - \delta^{(1)})$ controls the eccentricity of the P-wave NMO ellipse (equation 2.62) and, for a sufficiently thick reflecting layer, can be estimated from azimuthal moveout analysis (see Chapter 2).

The results of this section can be readily modified for horizontal transverse isotropy, which represents a special case of orthorhombic symmetry. If the symmetry axis of the reflecting HTI halfspace points in the x_1-direction, the parameters $\delta^{(1)}$ and $\epsilon^{(1)}$ vanish, and equation 8.8 simplifies to

$$R^{[x_1,x_3]} - R^{[x_2,x_3]} = \left[\left(\frac{2\bar{V}_{S0}}{\bar{V}_{P0}}\right)^2 \gamma^{(S)} + \frac{\delta^{(2)}}{2}\right] \sin^2\theta + \frac{\epsilon^{(2)}}{2} \sin^2\theta \, \tan^2\theta \,. \tag{8.9}$$

The linearized P-wave reflection coefficient for out-of-plane propagation in HTI media is analyzed in Rüger (2002). HTI symmetry is typically caused by a system of parallel, vertical, penny-shaped cracks in isotropic host rock. For such a model, the coefficient $\gamma^{(S)}$ is close to the crack density, while the parameters $\delta^{(2)}$ and $\epsilon^{(2)}$ (often denoted by $\delta^{(V)}$ and $\epsilon^{(V)}$) are proportional to the crack density and also depend on fluid infill (Rüger, 2002). Therefore, the azimuthally varying P-wave AVO response is sensitive to the crack density and saturation, although the inversion of the P-wave AVO gradient for the fracture parameters is generally nonunique even for HTI media (see Chapter 9).

If azimuthal anisotropy is caused by a single fracture set (the model can be HTI or orthorhombic), the maximum AVO gradient can be either parallel or perpendicular to the fractures. This 90°-uncertainty in the fracture azimuth can be resolved using a priori information or other signatures, such as the P-wave NMO ellipse or the S-wave polarization directions. Application of P-wave azimuthal AVO analysis to fracture characterization is discussed in more detail in section 8.3, where we present a case study from Rulison field in Colorado, USA (that data set was used in section 3.4).

The symmetry-plane reflection coefficients of pure SS-waves and mode-converted PS-waves can also be adapted from the corresponding VTI equations (Rüger, 2002; Jílek, 2002a,b). To perform this adaptation correctly, one has to take into account the shear-wave polarization properties in orthorhombic media described above. Indeed, each split S-wave has in-plane (SV) polarization in one vertical symmetry plane and transverse (SH) polarization in the other. Therefore, if the S_1-wave at vertical incidence is polarized in the x_2-direction, its reflection coefficient in the $[x_2, x_3]$-plane should be obtained from the VTI equation for the SV-wave. The S_1-wave reflection coefficient in the $[x_1, x_3]$-plane, however, has to be adapted from the VTI equation for the SH-wave; the same logic applies to the S_2-wave. Note that this approach was used in section 6.1 to derive the NMO ellipses of the split S-waves in an orthorhombic layer. For a detailed analysis of reflection coefficients of mode-converted waves in anisotropic media we refer the reader to Jílek (2002b), who also developed joint AVO inversion of wide-azimuth PP and PS reflections for TI and orthorhombic models.

8.2 Geometrical spreading in azimuthally anisotropic media

In addition to the dependence on the reflection coefficient, the amplitude signature of reflected waves is influenced by the source/receiver directivity and such propagation factors as the geometrical spreading, attenuation, and transmission coefficients along the raypath (Martinez, 1993; Maultzsch et al., 2003b). The most significant contribution to the amplitude distortions in the overburden typically is made by geometrical spreading. If the medium above the reflector is anisotropic, geometrical spreading can focus or defocus seismic energy like a 3D optical lens and, therefore, significantly change the amplitude distribution along the wavefronts of the incident and reflected waves. As discussed in detail by Tsvankin (1995b, 2005), the influence of geometrical spreading on the AVO signature for VTI media can be comparable to that of the anisotropic term in the reflection coefficient. Therefore, the angle-dependent reflection coefficient beneath anisotropic layers, which is needed for AVO analysis, cannot be reconstructed without an appropriate geometrical-spreading correction.

Because plane waves in 1D models are not subject to geometrical spreading, such a correction can be implemented in the $\tau-p$ domain. Wang and McCowan (1989), Dunne and Beresford (1998), and van der Baan (2004) employed plane-wave decomposition to remove the geometrical-spreading factor from the amplitudes of primary and multiple reflections (including mode conversions) in horizontally layered media. The discussion here, however, is limited to the more common implementation of geometrical-spreading correction in the time-offset domain.

The most straightforward way to compute anisotropic geometrical spreading is by performing dynamic ray tracing (e.g., Gajewski and Pšenčík, 1990; Červený, 2001). For simple homogeneous models it is possible to use analytic approximations of the Green's function, such as those for P- and SV-waves in a TI layer presented by Tsvankin (1995b, 2005). Modeling methods, however, require accurate information about the medium parameters, which is seldom available in practice.

An alternative approach, more suitable for purposes of AVO processing, is based on expressing geometrical spreading through the traveltimes of reflection events. Geometrical spreading in the time-offset domain is related to the convergence or divergence of ray beams (Gajewski and Pšenčík, 1987) and can be obtained directly from the spatial traveltime derivatives. For example, an algorithm to compute geometrical spreading from coarsely-gridded traveltime tables was introduced by Vanelle and Gajewski (2003). Ursin and Hokstad (2003) represented the geometrical-spreading factor of P-waves in horizontally layered VTI media in terms of reflection traveltime described by the Tsvankin-Thomsen nonhyperbolic moveout equation 3.3. A simpler geometrical-spreading approximation for VTI media can be obtained using the Alkhalifah-Tsvankin moveout equation 3.36 parameterized by the NMO velocity and the anellipticity parameter η.

Distortions caused by geometrical spreading in reflection amplitudes become especially pronounced for azimuthally anisotropic media. Below we follow the work of Xu et al. (2005) to discuss a traveltime-based equation for geometrical spreading of reflected waves recorded over horizontally layered, arbitrarily anisotropic media. This equation provides the foundation for moveout-based anisotropic spreading correction for wide-azimuth data developed by Xu and Tsvankin (2006a, 2008).

8.2.1 Analytic description of geometrical spreading

Geometrical spreading describes the amplitude decay of seismic wave caused by the expansion of its wavefront away from the source. The relative geometrical spreading $L(R, S)$ between the source S and receiver R is an essential part of the ray-theory Green's function (e.g., Červený, 2001):

$$G_{in}(R, T, S, T_0) = \frac{U_n(S)\, U_i(R)\, e^{i\, T^G(R,\, S)}}{4\pi\, L(R, S)\, \sqrt{V(S)\, V(R)\, \rho(S)\, \rho(R)}}\, R^C\, \delta\big[T - T_0 - t(R, S)\big]\,, \quad (8.10)$$

where T is the recording time, T_0 is the excitation time, $\mathbf{U}(S)$ and $\mathbf{U}(R)$ are the polarization vectors at the source and receiver, respectively, $V(S)$ and $\rho(S)$ are the phase velocity and density at the source location, $V(R)$ and $\rho(R)$ are the same quantities at the receiver, $T^G(R, S)$ is the complete phase shift, R^C is the product of the reflection/transmission coefficients along the raypath, $\delta(t)$ is the temporal delta function, and $t(R, S)$ is the traveltime between the source and receiver.

Relative geometrical spreading can be expressed through the spatial derivatives of the traveltime around the raypath. For sources and receivers confined to a horizontal plane, the factor L is given by (Goldin, 1986; Červený, 2001, his equations 4.10.46 and 4.10.50):

$$L(R, S) = \sqrt{\frac{\cos\psi^s \cos\psi^r}{|\det \mathbf{M}^{\text{mix}}(R, S)|}}\,, \quad (8.11)$$

where ψ^s is the ray angle (i.e., the angle between the ray and the vertical) at the source, ψ^r is the ray angle at the receiver, and the matrix $\mathbf{M}^{\text{mix}}$ is given by

$$\mathbf{M}^{\text{mix}} = \begin{pmatrix} \dfrac{\partial^2 t}{\partial x_1^s \partial x_1^r} & \dfrac{\partial^2 t}{\partial x_1^s \partial x_2^r} \\ \dfrac{\partial^2 t}{\partial x_2^s \partial x_1^r} & \dfrac{\partial^2 t}{\partial x_2^s \partial x_2^r} \end{pmatrix} ; \tag{8.12}$$

$(x_1^s,\, x_2^s)$ and $(x_1^s,\, x_2^s)$ are the Cartesian coordinates of the source and receiver.

If the subsurface is laterally homogeneous, signatures of pure (nonconverted) modes depend just on the distance x between the source and receiver and the azimuth α of the source-receiver line. Then equation 8.11 reduces to the following function of the reflection traveltime t (see the derivation in Appendix 8A):

$$L(x,\alpha) = (\cos\psi^s \cos\psi^r)^{1/2}\, x \left[\frac{\partial^2 t}{\partial x^2} \left(\frac{\partial^2 t}{\partial \alpha^2} + x \frac{\partial t}{\partial x} \right) - \left(\frac{\partial^2 t}{\partial x \partial \alpha} - \frac{1}{x} \frac{\partial t}{\partial \alpha} \right)^2 \right]^{-1/2} . \tag{8.13}$$

Equation 8.13 is valid for vertically heterogeneous (e.g., horizontally layered) media of arbitrary anisotropic symmetry. It should be mentioned that the corresponding equation 4 in Xu et al. (2005) is missing the term $[\partial^2 t/(\partial x \partial \alpha)]$, which contributes to the spreading factor outside vertical symmetry planes of the medium.

8.2.2 Applications of the moveout-based spreading equation

The representation of the factor $L(x,\alpha)$ in terms of reflection traveltime helps gain insight into the influence of both polar and azimuthal velocity variations on geometrical spreading. In vertical symmetry planes the derivative $\partial t/\partial \alpha$ goes to zero, and equation 8.13 takes the form

$$L(x) = (\cos\psi^s \cos\psi^r)^{1/2}\, x \left[\frac{\partial^2 t}{\partial x^2} \left(\frac{\partial^2 t}{\partial \alpha^2} + x \frac{\partial t}{\partial x} \right) \right]^{-1/2} . \tag{8.14}$$

Equation 8.14 confirms the conclusion of Tsvankin (1997a, 2005) that the equivalence between the symmetry planes of orthorhombic and VTI media cannot be extended to geometrical spreading. Specifically, the derivative $\partial^2 t/\partial \alpha^2$, which generally does not vanish even in symmetry planes, reflects the influence of azimuthal velocity variations on symmetry-plane amplitudes. This 3D character of geometrical spreading is explained by the dependence of the wavefront curvature on both in-plane and out-of-plane (azimuthal) velocity variations. Note that if the medium is azimuthally anisotropic, even the traveltime derivatives with respect to offset depend on the orientation of the source-receiver line.

For isotropic and VTI media, where $\partial^2 t/\partial \alpha^2 = 0$ and $\psi^s = \psi^r = \psi$, equation 8.14 further simplifies to (Ursin and Hokstad, 2003)

$$L(x) = \cos\psi \left[\frac{1}{x} \frac{\partial^2 t}{\partial x^2} \frac{\partial t}{\partial x} \right]^{-1/2} . \tag{8.15}$$

The main practical importance of equation 8.13 is in providing the analytic foundation for geometrical-spreading correction of wide-azimuth reflection data. As shown in Chapter 3, long-spread P-wave moveout in layered orthorhombic media is accurately described by the generalized Alkhalifah-Tsvankin equation 3.37. Substitution of equation 3.37 into equation 8.13 yields the geometrical-spreading factor as a function of the best-fit moveout coefficients. This approach was used by Xu and Tsvankin (2006a) to devise the moveout-based anisotropic spreading correction (MASC) for wide-azimuth P-wave data. MASC employs the 3D nonhyperbolic semblance algorithm introduced in Chapter 3 to estimate the parameters of equation 3.37, which serve as the input to the geometrical-spreading computation. Because the correction involves only the moveout function and the group angles at the source/receiver locations, it does not require knowledge of the velocity field beneath the uppermost layer.

Synthetic tests of Xu and Tsvankin (2006a,b) demonstrated the importance of anisotropic spreading correction in azimuthal AVO analysis for typical othorhombic models. Application of MASC becomes necessary when the azimuthal variation of the geometrical spreading reaches at least 1/3 that of the reflection coefficient. A case study of azimuthal AVO analysis with anisotropic spreading correction is described in the next section.

The derivation in Appendix 8A assumes reflection moveout to be symmetric (i.e., independent of the sign of offset), but the final result is valid for mode-converted data (Xu and Tsvankin, 2008). Reflection traveltimes of converted waves can be asymmetric even in the absence of lateral heterogeneity, if the medium does not have a horizontal symmetry plane (see Chapters 4 and 5). However, the only modification required to account for the possible moveout asymmetry of mode conversions is extension of the range of azimuths to 2π, which is formally accomplished by changing definition 8.118:

$$\alpha = \tan^{-1}\left[\frac{x_2^r - x_2^s}{x_1^r - x_1^s}\right] \qquad (x_1^r - x_1^s > 0)\,, \tag{8.16}$$

$$\alpha = \tan^{-1}\left[\frac{x_2^r - x_2^s}{x_1^r - x_1^s}\right] + \pi \qquad (x_1^r - x_1^s < 0)\,. \tag{8.17}$$

Spreading computation for events with asymmetric moveout has to be performed separately for "reciprocal" source-receiver pairs with azimuths $\alpha \pm \pi$. Hence, geometrical-spreading equation 8.13 can be used for any reflected wave (pure or converted) in laterally homogeneous, arbitrarily anisotropic media.

8.3 Case study of azimuthal AVO and moveout analysis

This section is based on the work by Xu and Tsvankin (2007) who carried out azimuthal AVO and moveout analysis of wide-azimuth P-wave data acquired above a fractured reservoir at tight-gas Rulison field in Colorado (Figure 8.2). Careful amplitude

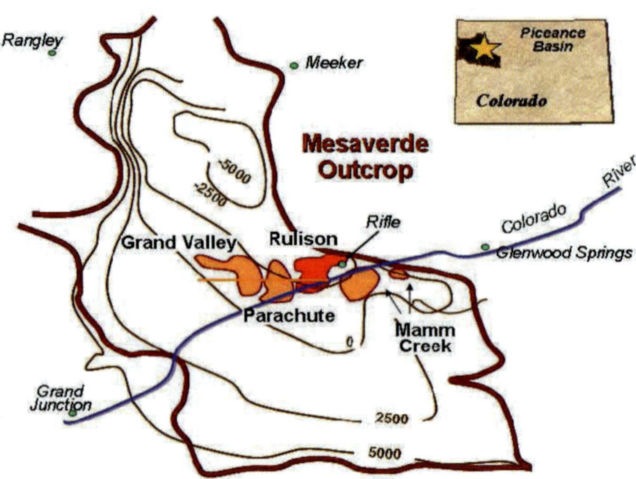

Figure 8.2: Location of Rulison field in the South Piceance Basin, Colorado, USA (Xu and Tsvankin, 2007).

processing allowed them to obtain accurate AVO gradients at the top and bottom of the reservoir, which provided important insights into the orientation and spatial distribution of fractures. P-wave data from Rulison field were used in Chapter 3 (section 3.4) to test the 3D velocity-independent layer-stripping algorithm (VILS).

In many respects, the Rulison case study is typical for tight (low-porosity) fractured gas reservoirs, which are common in North America. Geomechanical properties and the flow of hydrocarbons in tight reservoirs are largely governed by natural fracture networks. Hence reliable estimation of fracture density and orientation is important for cost-effective hydrocarbon production and hydraulic completion. While direct information about fracturing can be obtained using borehole methods such as image log analysis, which provide estimates of fracture counts and orientations on various scales, borehole measurements are sensitive only to rock properties in the immediate vicinity of the well.

Section 8.1 outlines the analytic basis for employing azimuthally varying P-wave reflection coefficients in fracture characterization. AVO analysis of wide-azimuth P-wave data has been successfully used to estimate the dominant fracture and fault orientation and, in some cases, identify "soft spots" of intense fracturing (e.g., Mallick et al., 1998; Gray et al., 2002; Neves et al., 2003; Gray and Todorovic-Marinic, 2004). For instance, Hall and Kendall (2003) showed that the azimuth of the minimum P-wave AVO gradient at Valhall field in the North Sea is generally aligned with faults mapped from coherency analysis of surface data. The link between the AVO signature and fracture properties is explored in more detail in Chapter 9.

Azimuthal AVO analysis has both advantages and drawbacks compared with moveout inversion. While the AVO response can give high-resolution estimates of the elastic properties at the top or bottom of the reservoir, moveout attributes (e.g., the NMO ellipse) depend on the average parameters of the reservoir layer. When combined together, azimuthal AVO and moveout attributes might offer improved understanding of the spatial distribution and physical properties of fractures.

Estimates of azimuthally varying moveout usually are more robust and less distorted by standard preprocessing algorithms than amplitude measurements. In contrast, reflection amplitudes are more sensitive to anisotropy and can be used for reservoirs whose thickness is small compared to their depth (but not compared to the predominant wavelength). Both moveout and amplitude methods face difficulties caused by the near surface (e.g., statics errors and coupling problems) and have to account for propagation phenomena in the overburden. Interval moveout analysis, based on Dix-type equations or layer stripping of reflection traveltimes (see sections 2.1 and 3.4), becomes unstable if the target layer has a relatively small thickness-to-depth ratio. In such a case, for purposes of moveout inversion the target horizon can, at the expense of some loss in resolution, be combined with a layer above or below it. It is less common in seismic fracture characterization to properly compensate for amplitude distortions in the overburden caused, in particular, by anisotropic geometrical spreading. As discussed in the previous section, the geometrical-spreading factor can be removed from the amplitude response using the moveout-based anisotropic spreading correction (MASC), an essential step in this case study.

8.3.1 Geologic background

Gas production at Rulison field comes primarily from the Williams Fork formation, which consists of channel sand lenses embedded in fine-grained levee deposition (Figure 3.28). The reservoir is capped by the UMV shale (used as the target layer in section 3.4), while the Cameo coal beneath the reservoir is believed to provide the source for the gas accumulation. The unconformity at the top of the Mesaverde group underlies a massive shale formation.

The reservoir lithology is classified as tight sand with the matrix permeability on the order of micro-Darcies and porosity of $6\% - 12\%$. The top several hundred feet of the reservoir formation are saturated with water, which is replaced by gas in the lower part of the reservoir (Cumella and Ostby, 2003). The pay section is relatively thick (about 1200 ft, or 366 m) and is considered to be normally pressured or slightly overpressured. Because the matrix permeability is low, characterization of natural fracture networks has vital importance for cost-effective development of the field.

8.3.2 Data acquisition and processing

The wide-azimuth P-wave data used here are part of a 3D multicomponent seismic survey acquired by the Reservoir Characterization Project (RCP) at CSM over a 2.2×2.5 km area of Rulison field. The orthogonal acquisition geometry (Figure 8.3) was designed to reach optimal balance between the uniformity of the azimuthal distribution and the layout economy. The data coverage is especially dense near the center

Bin size	55×55 ft
Number of receiver locations	1500
Number of source locations	770
Receiver grid	110-ft inline spacing, 330 ft between lines
Source grid	110-ft inline spacing, 660 ft between lines
Receiver array	1, 3C VectorSeis System Four
Source array	Mertz 18
P-wave sweep range	6 – 120 Hz

Table 8.1: Acquisition parameters of the RCP survey at Rulison field (Xu and Tsvankin, 2007).

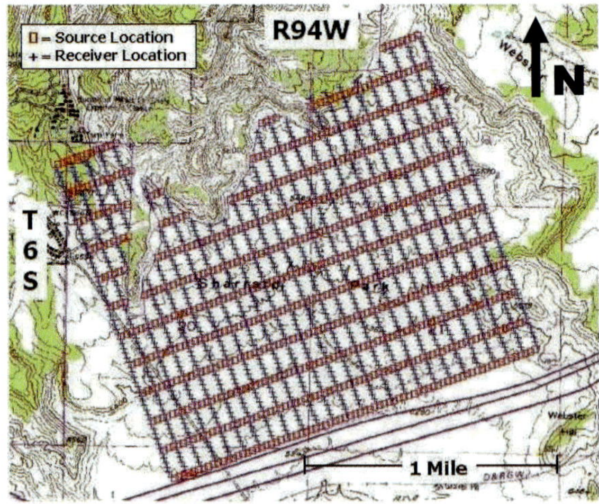

Figure 8.3: Seismic acquisition grid for the Rulison survey (Xu and Tsvankin, 2007).

of the survey area, with the highest fold reaching 225 for a small bin size of 55×55 ft (17×17 m; see Table 8.1 and Figure 3.30). This data set served as the baseline survey for a time-lapse monitoring study conducted by RCP.

To improve offset and azimuthal coverage, traces with close midpoints were collected into superbins. The choice of superbin size is nontrivial and required conducting a number of tests. Relatively small superbins suffer from nonuniformity of the distribution of offsets and azimuths; on the other hand, using large superbins increases heterogeneity-induced distortions. After experimenting with several bin sizes, 5×5 superbins were chosen. Further increase in size reduces semblance values in 3D nonhyperbolic moveout analysis, likely because of the influence of lateral heterogeneity. CMP locations used in this study occupy a square area in the center of the RCP survey (Figure 3.30) where the fold per superbin varies from 1500 to 5000.

Although typical CMP supergathers look somewhat noisy (Figure 8.4a), the data

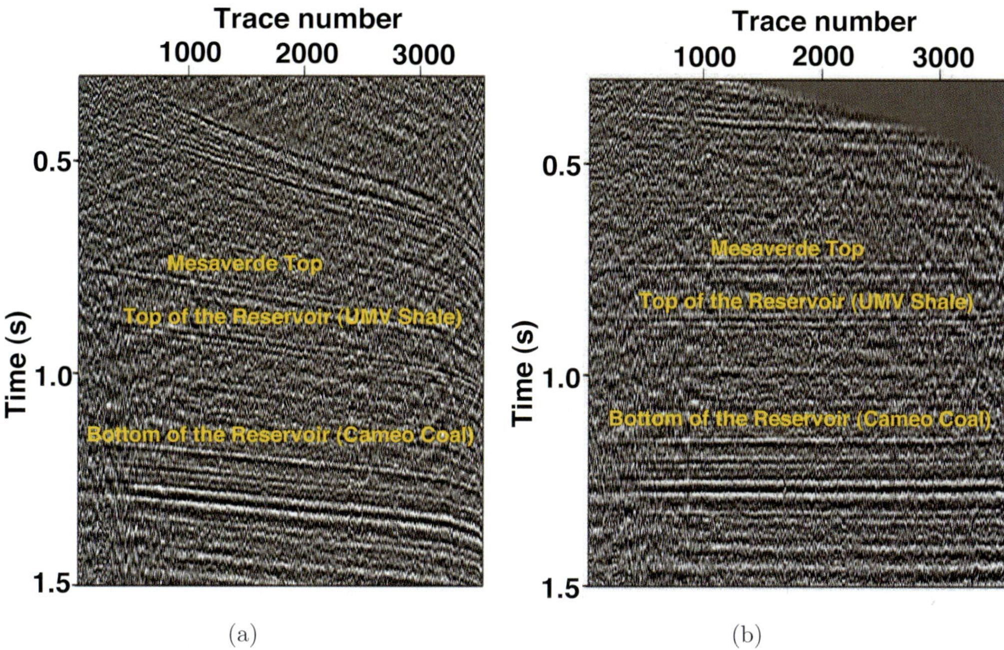

Figure 8.4: Preprocessed CMP supergather in the upper left corner of the study area (a) before and (b) after application of azimuthally varying hyperbolic moveout correction (Xu and Tsvankin, 2007). The maximum offset is 7700 ft (2347 m), which corresponds to an offset-to-depth ratio of 1.5 for the reservoir top (UMV shale).

quality is above average for land surveys. The statics correction was performed by the RCP contractor (Veritas; currently CGGVeritas). As suggested by Vasconcelos and Grechka (2007), ground roll was suppressed using a slope filter to minimize azimuthal distortions. Hyperbolic moveout correction with the best-fit NMO ellipse does not completely flatten relatively shallow reflection events, for which the maximum offset-to-depth ratio is much larger than unity (Figure 8.4b). For example, the bending at far offsets (i.e., a "hockey stick") of the UMV shale reflection indicates the presence of nonhyperbolic moveout generated in the overburden (see section 3.4 for more details).

Similarly to the NMO ellipse, the azimuthal variation of the AVO gradient can be approximated by an elliptical curve, unless the gradient changes sign with azimuth (section 8.1). The subsurface structure at Rulison field is close to layer-cake (see Figure 3.29), which simplifies application of azimuthal moveout and AVO inversion. For laterally homogeneous formations with fracture-induced azimuthal anisotropy, the orientation and eccentricity of the NMO and AVO ellipses depend primarily on the direction, density, and fluid saturation of subsurface fracture systems.

Because dividing 3D data into azimuthal sectors could introduce bias caused by uneven distribution of offsets and azimuths, Xu and Tsvankin (2007) adopted a more robust "global" approach that honors the azimuth of each trace in moveout and AVO analysis. Wide-azimuth CMP supergathers were processed using the nonhyperbolic moveout-inversion method for layered orthorhombic media described in section 3.3. This semblance-based algorithm is designed to reconstruct the effective NMO ellipse and to obtain the azimuthally varying anellipticity $\eta(\alpha)$ defined by the parameters $\eta^{(1,2,3)}$. Note that for models with depth-varying azimuths of the vertical symmetry planes, the principal directions of $\eta(\alpha)$ deviate from the axes of the NMO ellipse. As discussed in section 8.2, the estimated effective moveout parameters also serve as the input to the moveout-based anisotropic spreading correction.

To carry out azimuthal AVO analysis, raw amplitudes of the major reflection events were picked along the reconstructed nonhyperbolic traveltime surfaces and corrected for geometrical spreading using MASC. The AVO intercept and gradient for each supergather were estimated by fitting the AVO ellipse to the corrected amplitudes. To assess the influence of the anisotropic spreading correction on the azimuthal AVO response, the amplitude processing was repeated with conventional t^2-gain correction instead of MASC.

The offset and azimuthal coverage in CMP superbins increases from the corners of the study area (Figure 8.5) toward the center. Uniform azimuthal coverage, which is critical for fracture characterization, extends to offsets of about 5000 ft (1524 m), implying that the NMO and AVO ellipses for the Mesaverde Top and the top of the reservoir should be largely free from acquisition-related bias. The results for the bottom of the reservoir (Cameo coal; depth 7000 ft or 2134 m), however, might bear some acquisition footprint, particularly at the edges of the study area.

8.3.3 Results of azimuthal analysis

The major reflectors discussed here include the top of the Mesaverde group, the top of the reservoir (UMV shale), and the bottom of the reservoir (Cameo coal). To minimize possible edge distortions, some of the processed CMP superbins included source/receiver locations outside the study area. Both NMO and AVO ellipses are represented by their eccentricity and the orientation (azimuth) of the major axis computed for each common midpoint.

Mesaverde Top

The AVO ellipses for the Mesaverde Top exhibit a distinctive anomaly near the east boundary of the study area (Figures 8.6a,b). At the center of the anomaly, the AVO gradient in one principal azimuthal direction is more than twice as large as in the orthogonal direction. The axes of the AVO ellipses in the area of the anomaly have azimuths close to 45° and 135°. Because the reflection coefficient responds to local changes of rock properties across the interface, the azimuthal AVO anomaly in Figure 8.6 could be associated with an intensely fractured zone near the Mesaverde Top. The AVO-gradient map offers potentially valuable information for the operating

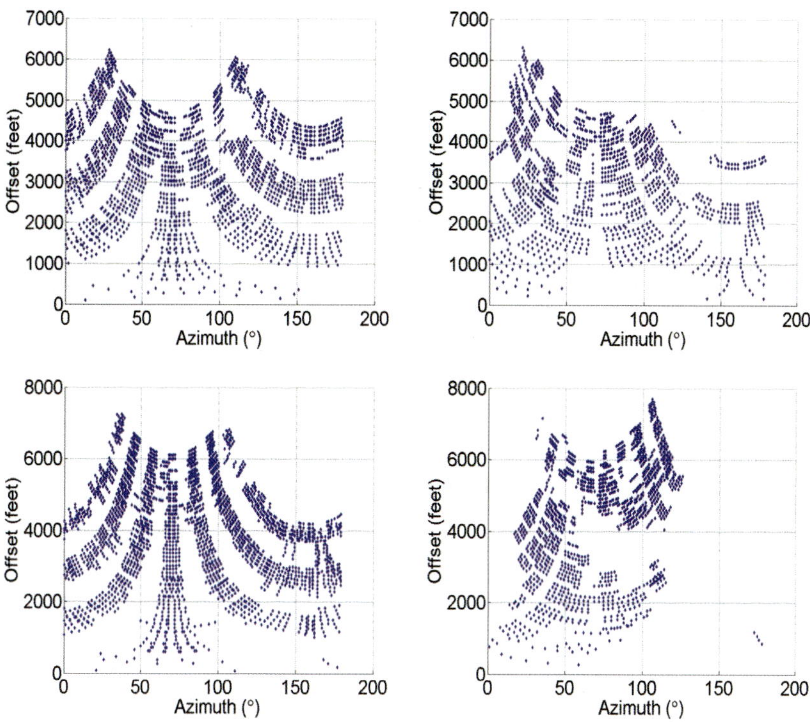

Figure 8.5: Distribution of offsets and azimuths for CMP superbins in the four corners of the study area (Xu and Tsvankin, 2007). Full azimuthal coverage is achieved for offsets up to about 5000 ft (1524 m).

company, which is interested in using formations above the Mesaverde Top to store production water.

In contrast, the NMO ellipticity for this reflection is negligible (Figure 8.6c). Although the parameter η estimated from nonhyperbolic moveout inversion is substantial (0.15 on average), it is almost the same in both vertical symmetry planes of the medium. On the whole, the reflection traveltime (governed by the effective properties of the entire overburden) and, consequently, geometrical spreading for the Mesaverde Top is weakly dependent on azimuth. Comparison of the first two columns of Figure 8.6 confirms that MASC has almost no influence on the azimuthal AVO response.

Reservoir top (UMV shale)

As for the Mesaverde Top, the only pronounced AVO-gradient anomaly at the top of the reservoir is located near the east boundary of the study area (Figure 8.7a). The magnitude of this anomaly, however, is about 30% higher than that for the Mesaverde Top, and the location of the maximum ellipticity is somewhat shifted. The right column in Figure 8.7 shows the interval NMO ellipses in the UMV shale computed

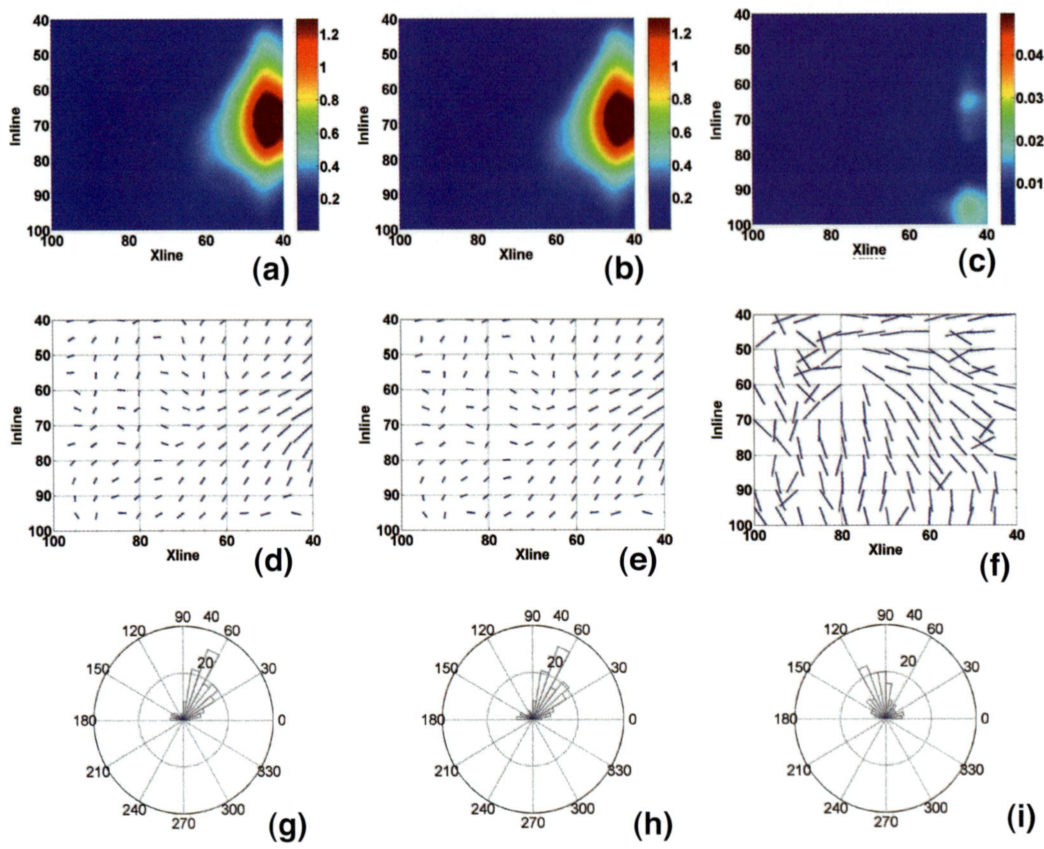

Figure 8.6: AVO and NMO ellipses estimated for the reflection from the Mesaverde Top (Xu and Tsvankin, 2007). The first two columns display the AVO ellipses computed using MASC (left) and conventional t^2-gain correction (center). The right column shows the effective NMO ellipses. In the top row [panels (a), (b), and (c)] are maps of the eccentricity of the ellipses calculated by subtracting unity from the ratio of the major and minor axes; the major axis of the AVO ellipse corresponds to the larger absolute value of the AVO gradient. The middle row [panels (d), (e), and (f)] shows the azimuth of the major axis of the ellipse (the north-south axis is vertical); the length of the ticks is proportional to the eccentricity. The longest ticks on panels (d) and (e) correspond to an eccentricity of 1.25 and on panel (f) to an eccentricity of 0.03. In the bottom row [panels (g), (h), and (i)] are the rose diagrams of the azimuths from panels (d), (e), and (f), respectively.

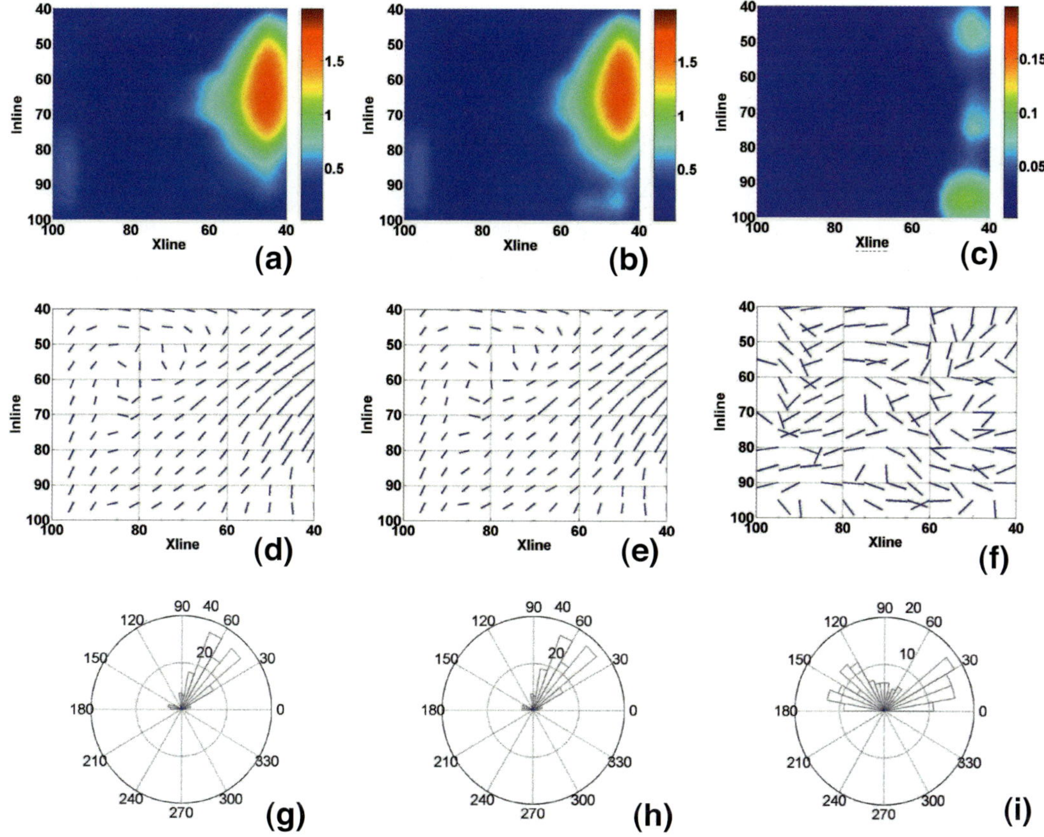

Figure 8.7: AVO ellipses for the top of the reservoir (left and center columns) and the interval NMO ellipses in the UMV shale (right column); same display as in Figure 8.6 (Xu and Tsvankin, 2007). The longest ticks on panels (d) and (e) correspond to an eccentricity of 1.55 and on panel (f) to an eccentricity of 0.12.

from the generalized Dix equation 1.32. Even in the areas of higher eccentricity, the orientations of the ellipses are almost random, which suggests (in agreement with the results of section 3.4) that azimuthal anisotropy in the shale is weak. For that reason, the anisotropic spreading correction leaves the azimuthal AVO response practically unchanged. The AVO anomaly in Figure 8.7 is likely caused by a "soft spot" of high fracture density in the upper part of the reservoir.

Reservoir bottom (Cameo coal)

The azimuthal seismic attributes for the bottom of the reservoir are shown in Figure 8.8. Two significant AVO-gradient anomalies appear in the upper right and lower left corners of the study area (Figures 8.8a,b). The magnitude of both anomalies is

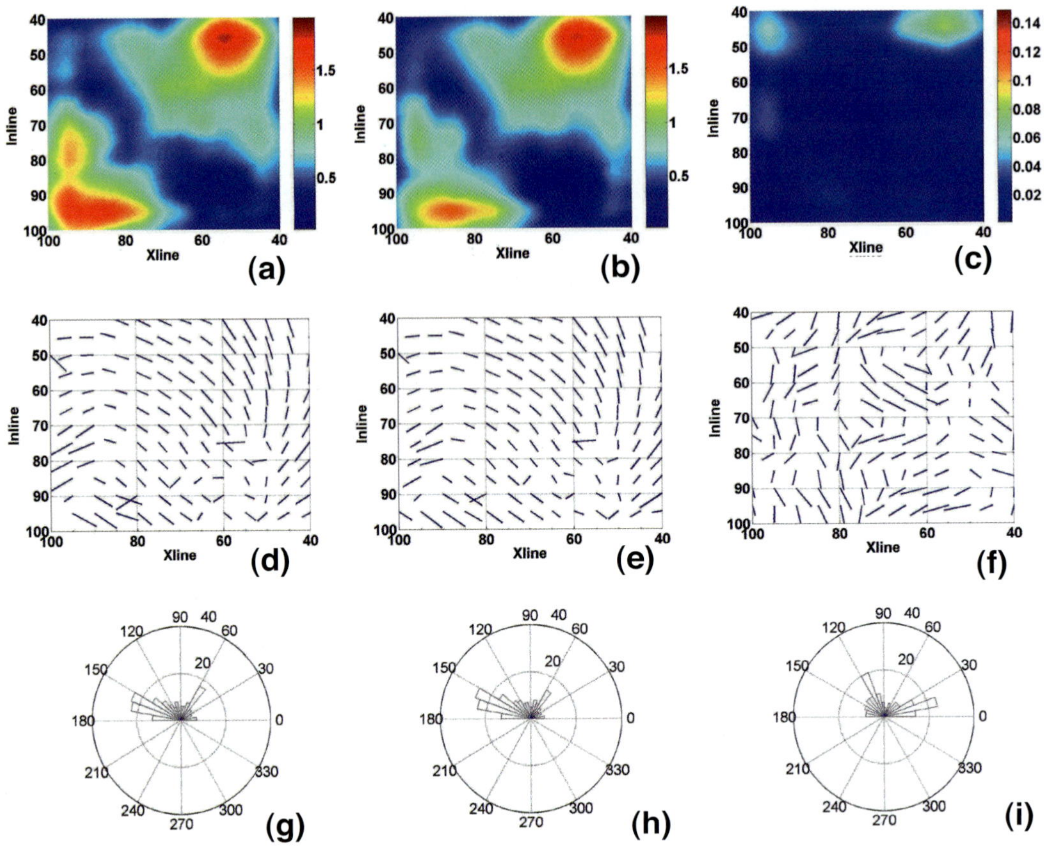

Figure 8.8: AVO ellipses for the bottom of the reservoir (left and center columns) and the interval NMO ellipses in the reservoir (right column); same display as in Figure 8.6 (Xu and Tsvankin, 2007). The longest ticks on panels (d) and (e) correspond to an eccentricity of 1.6 and on panel (f) to an eccentricity of 0.09.

close to 1.6, which means that the AVO gradient changes by a factor of 2.6 between the axes of the AVO ellipse. The azimuth of the major axis (Figures 8.8d,e) exhibits a strikingly regular pattern, which might be related to the geomechanical processes that produce wrenching faults in the area (Jansen, 2005). According to the orientation of the AVO ellipse (Figures 8.8g,h), the average fracture azimuth at the bottom of the reservoir should be close to N70W.

The large thickness of the reservoir ensures stable computation of the interval NMO ellipses (the right column in Figure 8.8). The only noticeable azimuthal NMO anomaly is located in the upper right corner of the area and partially overlaps with one of the azimuthal AVO anomalies. The maximum NMO ellipticity is close to 8%, which translates into a difference of about 0.08 between the interval anisotropy parameters $\delta^{(1)}$ and $\delta^{(2)}$ in the vertical symmetry planes.

In contrast to the results for shallower reflectors, the anisotropic spreading correction (MASC) has a larger impact on the azimuthal AVO response for the base of the reservoir. The AVO anomaly in the lower left corner of the study area becomes more pronounced and spatially coherent after application of MASC (compare Figures 8.8a and 8.8b).

Most existing case studies of azimuthal AVO analysis are conducted for the tops of reservoir formations (e.g., Neves et al. 2003). Using synthetic modeling for fractured gas sands, Sayers and Rickett (1997) concluded that the reservoir bottom often produces a stronger azimuthal AVO anomaly. However, since Sayers and Rickett (1997) did not apply an anisotropic spreading correction, their modeled amplitudes were likely influenced by both the reflection coefficient and the azimuthally varying geometrical spreading inside the reservoir. The above results for Rulison field demonstrate the importance of the anisotropic spreading correction for reflectors beneath fractured formations.

8.3.4 Discussion of processing results

Acquisition footprint

Because full azimuthal coverage is achieved for offsets up to approximately 5000 ft (1524 m), the signatures for the reservoir base (Cameo coal; depth 7000 ft or 2134 m) might be biased towards the dominant acquisition directions from 40° to 100°. The orientations of neither AVO nor NMO ellipses for the bottom of the reservoir, however, exhibit any noticeable bias (Figures 8.8d,e,f). In particular, the azimuths of the AVO ellipses are practically random in the lower right corner of the area where the AVO eccentricity is small (Figures 8.8d,e). The absence of the acquisition footprint can be explained by the orthogonality of the acquisition layout, which ensures that 80% of all traces fall into the offset range with complete azimuthal coverage.

Error analysis

A possible problem in moveout analysis is related to the bias caused by varying superbin size. The NMO ellipticity systematically increases over the area when the superbin size changes from 5×5 to 9×9. Because this increase in ellipticity is accompanied by lower semblance values, the larger superbins should be more influenced by lateral heterogeneity. On average, the semblance value for the top of the reservoir decreases from around 0.6 for 5×5 superbins to 0.45 for 9×9 superbins, while the effective NMO ellipticity increases by 0.04. Apparently, then, 5×5 superbins produce more reliable azimuthal seismic attributes.

Because the offset-to-depth-ratio for the bottom of the reservoir reaches only 1.6 in the center of the study area, the anellipticity parameters $\eta^{(1,2,3)}$ for that interface are not tightly constrained. Nonhyperbolic moveout inversion for the reservoir base influenced the AVO anomalies only in the lower left corner of the area by producing relatively large values (exceeding 0.2) of the difference between the effective parameters $\eta^{(1)}$ and $\eta^{(2)}$. This difference caused a pronounced azimuthal variation of the geometrical-spreading factor, which distorted the corresponding AVO anomaly and required application of MASC (Figures 8.8a,b). The spatial consistency and magnitude of the estimated value of $\eta^{(1)} - \eta^{(2)}$ indicates that the contribution of MASC is unlikely to be a processing artifact.

The robustness of the AVO ellipticity can be evaluated from the correlation between the magnitude of the AVO anomalies and the regularity of the ellipse orientation (Figures 8.6, 8.7, and 8.8). When the ellipticity is smaller than 0.3, the azimuths of the maximum AVO gradient are random, which is particularly clear in the upper left quarter of Figure 8.7d and the lower right quarter of Figure 8.8d. The AVO azimuths show a more regular pattern for ellipticities exceeding 0.3. Because AVO gradients are estimated independently at each CMP location with no data overlap between adjacent gathers and no smoothing, the confidence level for AVO ellipticity is greater than 0.3. The magnitude of the major azimuthal AVO anomalies at the bottom of the reservoir is five times this confidence level.

Correlation between the NMO and AVO ellipses

In principle, combining the NMO ellipse with the azimuthally varying AVO gradient (see Chapter 9) can help constrain the anisotropic velocity model and fracture parameters (Rüger and Tsvankin, 1997; Bakulin et al., 2000a). This approach is feasible when the reservoir is thick enough for reliable estimation of the interval NMO ellipses and the variation of major fracture properties (orientation, density, fluid saturation) with depth is mild. In the presence of strong vertical heterogeneity, however, the difference in resolution between amplitude and traveltime methods complicates joint analysis of NMO and AVO responses.

Although the thickness of the reservoir formation at Rulison is sufficient for moveout inversion, there is no clear correlation between the azimuthal NMO and AVO attributes. The only area of increased interval NMO ellipticity in the reservoir does overlap with an azimuthal AVO anomaly, but NMO ellipses inside the second AVO anomaly are quasi-circular. Most likely, the vertical and lateral heterogeneity of the Williams Fork formation strongly influences the interval moveout, which reflects the average parameters of the reservoir. In contrast, the azimuthal AVO response after spreading correction depends on the local medium properties above and below the reflector.

Comparison with fault distribution and EMI logs

Because both fractures and faults respond to subsurface stress fields, enhanced fracture zones are often associated with fault locations. It is, therefore, instructive to compare the AVO results with the fault distribution at Rulison field. Cumella and Ostby (2003) suggested that faults in the area follow a wrenching pattern. Employing the wrenching fault model, Jansen (2005) carried out fault mapping by applying automated curvature measurements to poststack P-wave images. The primary

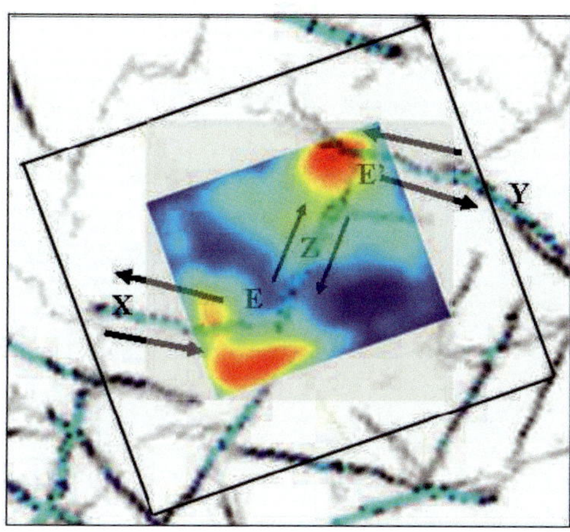

Figure 8.9: Comparison of the eccenticity of the AVO ellipses (Figure 8.8a) and the fault system for the bottom of the reservoir (Xu and Tsvankin, 2007). The faults (solid lines with arrows) were mapped by Jansen (2005) using poststack P-wave images; the arrows indicate the slip movement. The black rectangle marks the RCP survey area.

fault system at the bottom of the reservoir is aligned with the average azimuth of the major axis of the AVO ellipse (N70W), while secondary step-over faults trend along N30E (Figure 8.9). Interestingly, the AVO-gradient anomalies are located at the intersections of the two wrenching fault systems (marked by E and E′), where stress concentration is likely to induce intense fracturing. Also, the orientation of the AVO ellipses (Figures 8.8d,e) exhibits a rotation pattern, which seems to support the wrenching fault model.

An Electrical MicroImager (EMI) log was available in well RWF 542-20, located between the left-corner AVO anomaly and the center of the study area. The dominant fracture directions obtained from the EMI log and from the azimuthal AVO analysis for the reservoir base differ by less than 10° (Figure 8.10).

Quantitative AVO inversion

The AVO gradient was estimated by expressing the reflection coefficient as a quadratic function of the source and receiver coordinates. While this representation is justified for the NMO ellipse, it is inadequate for quantitative inversion of the AVO response. Indeed, the plane-wave reflection coefficient has to be treated as a function of the incidence phase angle or horizontal slowness. Accurate computation of the incidence angle at the reflector, however, requires knowledge of the interval anisotropy parameters in the overburden. Also, the reconstructed reflection coefficient has to be calibrated using borehole information (well logs). Additional complications in AVO inversion at Rulison field are caused by the presence of multiple fracture sets in parts of the reservoir (see section 9.6) and intense fracturing in the Cameo coal.

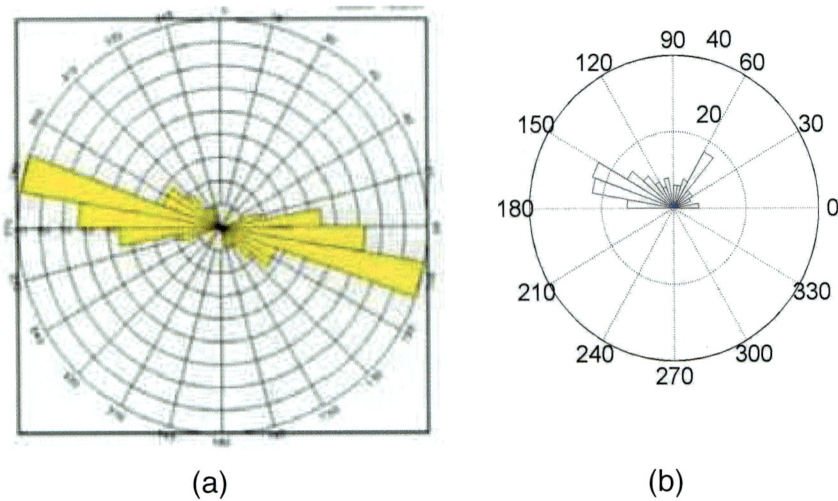

Figure 8.10: Rose diagrams of the fracture orientation obtained from an EMI log and azimuthal AVO analysis (Xu and Tsvankin, 2007). The fracture directions were (a) counted in a well within the reservoir; (b) estimated from the AVO ellipses at the bottom of the reservoir (Figure 8.8g).

8.4 Anisotropic critical-angle reflectometry

Most existing AVO algorithms, including the one described in the previous section, operate with the AVO gradient estimated from conventional-spread reflection data. Despite the increased offset range of modern seismic surveys, wide-angle AVO analysis (Pankhurst et al., 2002) is not common in practice. Including longer offsets can potentially reduce the ambiguity of amplitude inversion, especially for multicomponent data (Jílek, 2002b). The lower data quality and phase changes at large incidence angles, however, cause serious complications for accurate amplitude picking and AVO processing.

This section is based on the results of Landrø and Tsvankin (2007), who suggested an alternative method (called *critical-angle reflectometry*) of employing long-offset data in anisotropic parameter estimation. When the P-wave velocity in a reservoir is higher than that in the cap rock, it is possible to record PP reflections incident upon the reservoir at critical and postcritical angles. For low-velocity reservoirs, critical reflections are generated at the reservoir base. Landrø et al. (2004) proposed to use critical-angle measurements in time-lapse seismics because the shift in the critical angle or offset with time helps monitor reservoir production.

If either the overburden or reservoir itself is azimuthally anisotropic, the critical angle and the corresponding critical offset vary with azimuth. For 3D data with good coverage in offset and azimuth, the azimuthal variation of the critical angle can be used to identify the vertical symmetry planes of the model and constrain the anisotropy parameters (in particular, for fracture-characterization applications). In contrast to AVO analysis, this method does not require accurate amplitude picking for a wide range of offsets because the critical angle can be estimated from the point of the fastest increase of the reflection amplitude (see below). It should be mentioned that ultrasonic critical-angle reflectometry is an established technology for measuring and interpreting the critical angle in human bones, some composite materials, etc. (Antich and Mehta, 1997).

Although postcritical reflections are well documented in seismological literature, implementation of critical-angle reflectometry is not straightforward. First, the method requires a substantial velocity increase at the top or bottom of the target layer. (Most carbonate and consolidated sand reservoirs have a higher velocity than that in the overburden.) Second, the azimuthally varying critical angle should be measured from wide-azimuth, long-offset data routinely acquired only in ocean-bottom-cable (OBC) surveys. Multiazimuth land data seldom include sufficiently long offsets, whereas modern wide-azimuth streamer surveys are typically acquired for purposes of subsalt imaging rather than fracture characterization. Third, exploration targets are often located at significant depths and overlain by a sequence of sedimentary layers. Multiples and mode conversions in the overburden can create interference with the critical-angle event from the reservoir. Also, thin layering at the target level can change the properties of the critical-angle reflection and complicate its detection.

Still, under favorable circumstances the critical angle can be estimated from seismic reflection data. Figure 8.11 displays a raw seismic gather that contains an interpretable reflection (marked by the arrow) from a boundary with a large positive velocity contrast. Despite the interference with noise and overburden events, this reflection remains visible for offsets up to about 4.5 km. The interfering arrivals were successfully attenuated by standard f-k filtering (Figure 8.12). To make noise suppression near the critical angle more effective, the filtering was applied with a narrow range of apparent velocities, which attenuated the target event at near offsets.

Amplitude analysis of the target reflection (Figure 8.13) was used to identify the point of the fastest amplitude increase (the inflection point), which is presumed to correspond to the critical offset $x_{\rm cr}$ (see below). The value of $x_{\rm cr} \approx 4$ km estimated from the AVO curve is in good agreement with an independent prediction made using borehole data (well logs) from the area. Note that despite the generally high level of noise at far offsets, the sharp amplitude increase near the critical angle should help in detecting the critical offset. This case study supports the feasibility of critical-angle reflectometry in the presence of a significant velocity contrast at the target level.

8.4.1 Critical angle for VTI and orthorhombic media

We consider a plane P- or S-wave incident upon a horizontal interface that separates two anisotropic halfspaces. If the transmitted wave (again, it can be P or S) has a higher velocity, for a certain incidence angle its group-velocity vector becomes horizontal. This angle, which is usually called critical, depends on the phase-velocity

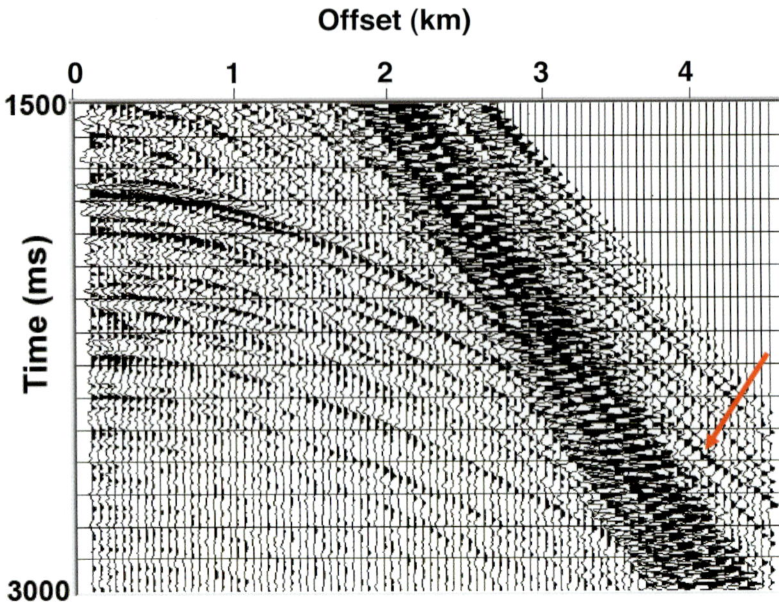

Figure 8.11: Common-receiver field gather of the vertical displacement component (Landrø and Tsvankin, 2007). The arrow marks a reflection from the top of a high-velocity layer (target event).

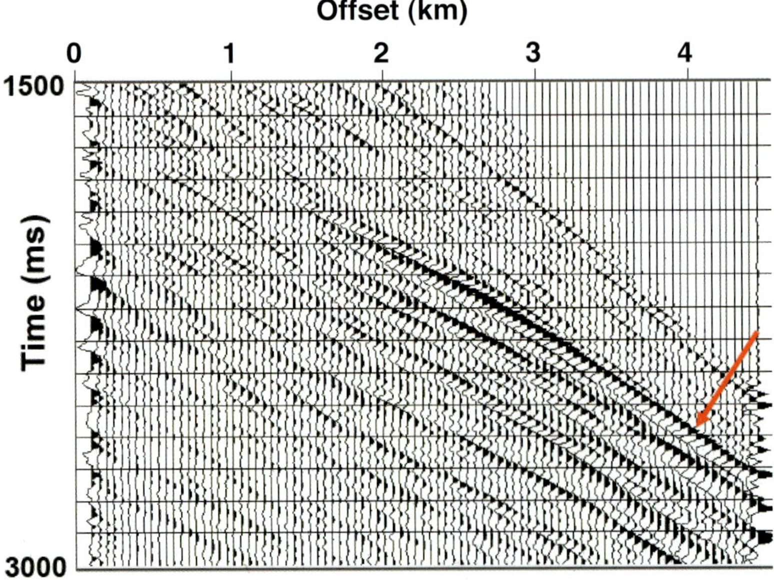

Figure 8.12: Gather from Figure 8.11 after application of f-k filtering designed to attenuate the overburden noise (Landrø and Tsvankin, 2007). The target event is clearly visible at offsets larger than 1500 m.

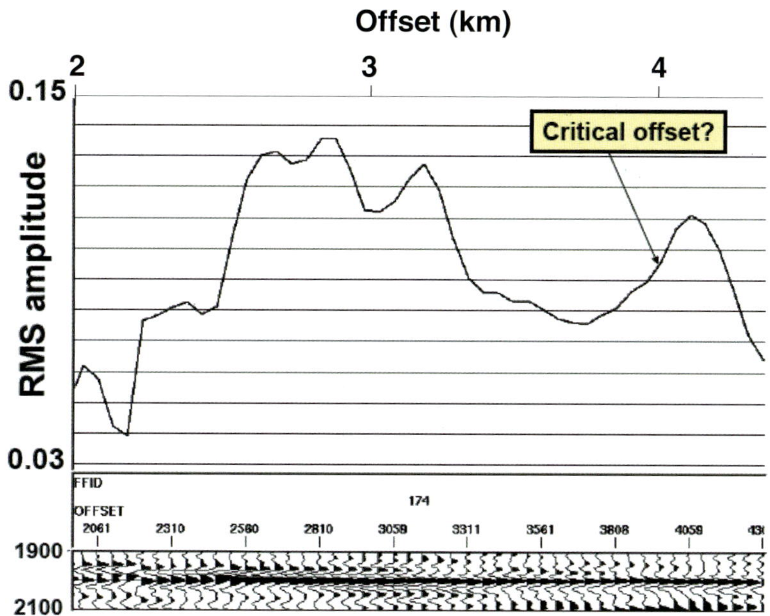

Figure 8.13: Estimated rms amplitude of the target reflection from Figure 8.12 computed in a 50 ms window (Landrø and Tsvankin, 2007). The offset of the fastest amplitude increase (assumed to correspond to the critical-angle reflection) is slightly smaller than 4 km. The windowed section on the bottom shows the flattened target event.

function and, therefore, is influenced by anisotropy. For simplicity, here the reflector is assumed to coincide with a symmetry plane in the reflecting medium, which implies that at the critical angle the phase-velocity vector of the transmitted wave is also horizontal. According to Snell's law, the horizontal slowness of all reflected and transmitted waves should be equal to that of the incident wave, which yields

$$\frac{\sin\theta_{\rm cr}}{V_1(\theta_{\rm cr})} = \frac{1}{V_{\rm hor,2}}\,, \tag{8.18}$$

where $\theta_{\rm cr}$ is the critical phase angle with the vertical, $V_1(\theta)$ is the phase velocity in the incidence medium, and $V_{\rm hor,2}$ is the horizontal phase velocity in the reflecting medium. Hereafter in this section, the subscripts "1" and "2" denote the incident and reflecting media, respectively. All quantities in equation 8.18 are computed in the vertical plane that contains the slowness vector of the incident wave. If the wave propagates outside symmetry planes in azimuthally anisotropic media, the corresponding critical ray might not lie in the same vertical plane.

The critical angle for a particular wave mode (e.g., PP, SS, or SP) can be found by solving equation 8.18 with the appropriate phase-velocity function $V_1(\theta)$ and the horizontal velocity $V_{\rm hor,2}$. It is also possible to compute the critical angle from the

Christoffel equation in the incidence medium expressed in terms of the slowness components (this approach is used in deriving reflection/transmission coefficients). After setting the horizontal slowness to $(1/V_{\text{hor},2})$, the Christoffel equation can be solved for the vertical slowness of the incident wave and, therefore, for the angle θ_{cr}. If the wavefield is excited by a point source, the segment of the incident wavefront orthogonal to the critical phase direction θ_{cr} is responsible for generating the head wave, which interferes with the reflected wave in the postcritical domain (for more details, see below).

VTI media

The velocity function and the critical angle for vertical transverse isotropy are azimuthally invariant. The P-wave horizontal velocity in the reflecting medium is given in terms of the vertical velocity V_{P0} and parameter ϵ by

$$V_{\text{hor},2} = V_{P0,2}\sqrt{1+2\epsilon_2}\,. \tag{8.19}$$

If the incidence medium is isotropic with the velocity $V_{P0,1}$, the exact P-wave critical angle can be easily found from equation 8.18:

$$\sin\theta_{\text{cr}} = \frac{V_{P0,1}}{V_{P0,2}\sqrt{1+2\epsilon_2}}\,. \tag{8.20}$$

The existence of the critical angle requires that $V_{P0,1} < \left(V_{P0,2}\sqrt{1+2\epsilon_2}\right)$. Because the parameter ϵ is typically nonnegative, the isotropic condition $V_{P0,1} < V_{P0,2}$ is sufficient for most VTI models.

To express the critical angle for a VTI/VTI interface through the anisotropy parameters, it is convenient to use the weak-anisotropy approximation for the phase velocity in the incidence medium (Thomsen, 1986):

$$V_1(\theta) = V_{P0,1}\left(1+\delta_1\sin^2\theta\cos^2\theta+\epsilon_1\sin^4\theta\right). \tag{8.21}$$

The velocity $V_{\text{hor},2}$ (equation 8.19) can be linearized in ϵ_2:

$$V_{\text{hor},2} = V_{P0,2}\left(1+\epsilon_2\right). \tag{8.22}$$

Substituting equations 8.21 and 8.22 into equation 8.18, we find

$$\sin\theta_{\text{cr}} = \frac{V_{P0,1}\left(1+\delta_1\sin^2\theta_{\text{cr}}\cos^2\theta_{\text{cr}}+\epsilon_1\sin^4\theta_{\text{cr}}\right)}{V_{P0,2}\left(1+\epsilon_2\right)}\,. \tag{8.23}$$

Within the framework of the linearized approximation, the angle θ_{cr} in the terms involving ϵ and δ can be replaced by its isotropic value $\theta_{\text{cr,is}}$:

$$\sin\theta_{\text{cr,is}} = \frac{V_{P0,1}}{V_{P0,2}} = n\,; \tag{8.24}$$

the velocity ratio n is sometimes called the *refraction index.* Then the linearized right-hand side of equation 8.23 no longer contains the unknown critical angle:

$$\sin\theta_{\text{cr}} = n\left[1 - \epsilon_2 + \delta_1 n^2 + (\epsilon_1 - \delta_1)\, n^4\right] . \tag{8.25}$$

Because the vertical-velocity ratio $n < 1$, the critical angle is particularly sensitive to the parameter ϵ_2 responsible for the P-wave horizontal velocity in the reflecting medium.

The critical offset on reflection data is determined by the critical group (ray) angle ψ_{cr}, which can be computed from θ_{cr} using the known group-velocity equations for VTI media. In the weak-anisotropy approximation (Tsvankin, 2005),

$$\psi_{\text{cr}} = \theta_{\text{cr}} + \left[\delta_1 + 2(\epsilon_1 - \delta_1)\sin^2\theta_{\text{cr}}\right]\sin 2\theta_{\text{cr}} . \tag{8.26}$$

Orthorhombic media

Suppose the reflector separates two orthorhombic halfspaces with a horizontal symmetry plane and the same orientation of the vertical symmetry planes. As discussed by Tsvankin (1997a, 2005), all kinematic signatures in the symmetry planes of orthorhombic media (assumed to coincide with the coordinate planes) can be adapted from the corresponding VTI equations by using the appropriate anisotropy parameters defined in Appendix 1B. For example, if the incidence medium is isotropic, the exact critical angle in the $[x_1, x_3]$-plane is found by replacing the parameter ϵ in equation 8.20 with $\epsilon^{(2)}$:

$$\sin\theta_{\text{cr}}\,(\alpha = 0^\circ) = \frac{V_{P0,1}}{V_{P0,2}\sqrt{1 + 2\epsilon_2^{(2)}}} ; \tag{8.27}$$

α is the azimuth with respect to the x_1-axis.

When both media are orthorhombic, the weak-anisotropy approximation for the critical angle in the $[x_1, x_3]$-plane is adapted from equation 8.25:

$$\sin\theta_{\text{cr}}\,(\alpha = 0^\circ) = n\left[1 - \epsilon_2^{(2)} + \delta_1^{(2)}\, n^2 + (\epsilon_1^{(2)} - \delta_1^{(2)})\, n^4\right] ; \tag{8.28}$$

$n = V_{P0,1}/V_{P0,2}$. To obtain the critical angles in the $[x_2, x_3]$-plane, the parameters $\epsilon^{(2)}$ and $\delta^{(2)}$ in equations 8.27 and 8.28 have to be replaced by $\epsilon^{(1)}$ and $\delta^{(1)}$, respectively.

In the weak-anisotropy limit, the kinematic analogy between orthorhombic and VTI media remains valid for 2D P-wave propagation even outside the symmetry planes if the parameters ϵ and δ are expressed as the following functions of azimuth (Tsvankin, 1997a, 2005):

$$\delta(\alpha) = \delta^{(1)}\sin^2\alpha + \delta^{(2)}\cos^2\alpha , \tag{8.29}$$

$$\epsilon(\alpha) = \epsilon^{(1)}\sin^4\alpha + \epsilon^{(2)}\cos^4\alpha + (2\epsilon^{(2)} + \delta^{(3)})\sin^2\alpha\cos^2\alpha . \tag{8.30}$$

Substituting $\delta(\alpha)$ and $\epsilon(\alpha)$ from equations 8.29 and 8.30 into equation 8.25 yields the linearized P-wave critical angle for an arbitrary azimuth α:

$$\sin\theta_{\rm cr}(\alpha) = n\left\{1 - \epsilon_2(\alpha) + \delta_1(\alpha)\, n^2 + \left[\epsilon_1(\alpha) - \delta_1(\alpha)\right] n^4\right\}. \qquad (8.31)$$

As in VTI media, the critical offset is determined by the group angle that corresponds to the critical phase angle $\theta_{\rm cr}$ (see equation 8.26).

Numerical examples

The magnitude of the azimuthal variation of the critical angle for typical orthorhombic models (e.g., Bakulin et al., 2000b) and the accuracy of the weak-anisotropy approximation are illustrated in Figures 8.14–8.16. When the incidence medium is isotropic, the azimuthal dependence of $\theta_{\rm cr}$ is controlled by the P-wave velocity in the horizontal plane of the reflecting orthorhombic medium (equation 8.18). For weak anisotropy, the horizontal velocity in the reflecting medium is described by the parameter $\epsilon_2(\alpha)$ (equation 8.30).

The difference between the horizontal velocities in the symmetry planes is approximately proportional to $\epsilon_2^{(1)} - \epsilon_2^{(2)}$. For the model in Figure 8.14, this difference is equal to 0.15, which translates into a change in $\theta_{\rm cr}$ between the symmetry planes that exceeds 5°. The extrema of the function $\theta_{\rm cr}(\alpha)$, however, do not necessarily coincide with the symmetry-plane directions. The maximum of $\theta_{\rm cr}$ at an azimuth close to 30° in Figure 8.14 is associated with a decrease in the P-wave horizontal velocity away from the $[x_1, x_3]$-plane caused by the negative value of $\delta_2^{(3)}$. After reaching a minimum at $\alpha \approx 30°$, the velocity increases toward the symmetry plane $[x_2, x_3]$, which reduces the critical angle for azimuths $\alpha > 30°$.

The model in Figure 8.15 has a smaller value (0.05) of the difference between $\epsilon_2^{(1)}$ and $\epsilon_2^{(2)}$, and the maximum azimuthal variation of the critical angle is limited to about 3°. For both models, the weak-anisotropy approximation does not deviate far from the exact solution and correctly reproduces the trend of $\theta_{\rm cr}(\alpha)$.

When the incidence medium is orthorhombic (Figure 8.16), the critical angle becomes a function of all five anisotropy parameters responsible for the P-wave velocity above the reflector ($\epsilon_1^{(1)}$, $\epsilon_1^{(2)}$, $\delta_1^{(1)}$, $\delta_1^{(2)}$, and $\delta_1^{(3)}$). Because the reflecting halfspace in Figure 8.16 is isotropic, the azimuthal variation of $\theta_{\rm cr}$ replicates that of the phase velocity at the critical angle. As was the case in Figures 8.14 and 8.15, the azimuthal dependence of the critical angle in Figure 8.16 is more pronounced for the model with a larger difference between the parameters $\epsilon^{(1)}$ and $\epsilon^{(2)}$.

Variations of the critical angle on the order of 5–6° would change the critical offset for deep interfaces by hundreds of meters, which should be detectable on AVO plots similar to that in Figure 8.13. Estimation of $\theta_{\rm cr}$ requires acquisition of long-offset, wide-azimuth data with a sufficiently high signal-to-noise ratio at offsets close to critical. Furthermore, because the function $\theta_{\rm cr}(\alpha)$ often has extrema between the symmetry planes, dense azimuthal coverage likely is essential for the success of critical-angle reflectometry.

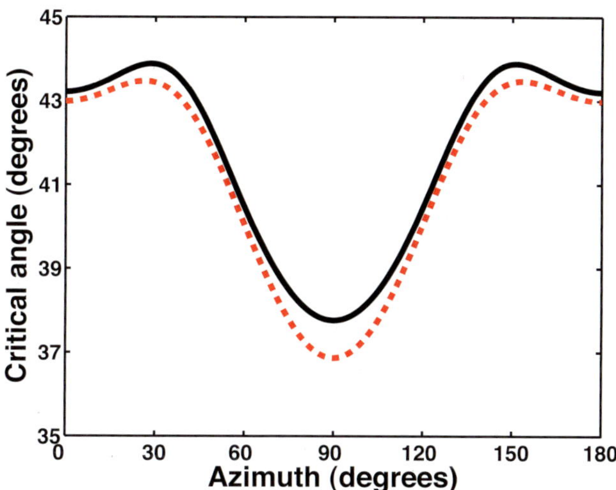

Figure 8.14: P-wave critical angle as a function of azimuth for an interface between isotropic (incidence) and orthorhombic (reflecting) media (Landrø and Tsvankin, 2007). The azimuth is measured from the symmetry plane $[x_1, x_3]$. The solid curve is the exact angle θ_{cr} computed from equation 8.18 using the phase-velocity function given in Tsvankin (2005); the dashed curve is the weak-anisotropy approximation 8.31. The P-wave velocity in the isotropic medium is $V_{P0,1}$=2100 m/s; the relevant parameters of the orthorhombic medium are $V_{P0,2}$=2800 m/s, $\epsilon_2^{(1)}$=0.25, $\epsilon_2^{(2)}$=0.1, and $\delta_2^{(3)}=-0.1$.

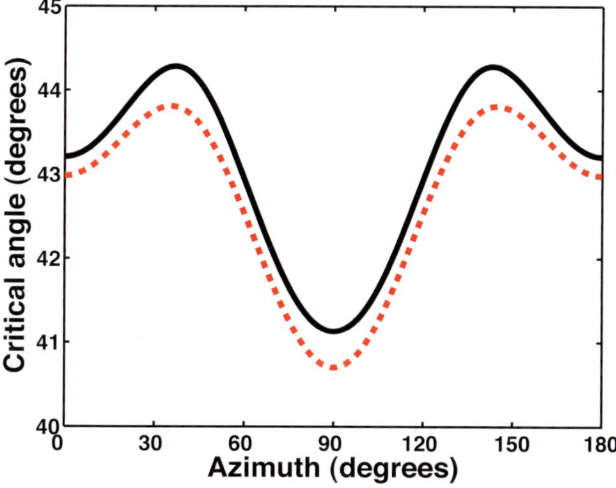

Figure 8.15: P-wave critical angle for the same model as that in Figure 8.14, except for $\epsilon_2^{(1)}$=0.15 (Landrø and Tsvankin, 2007).

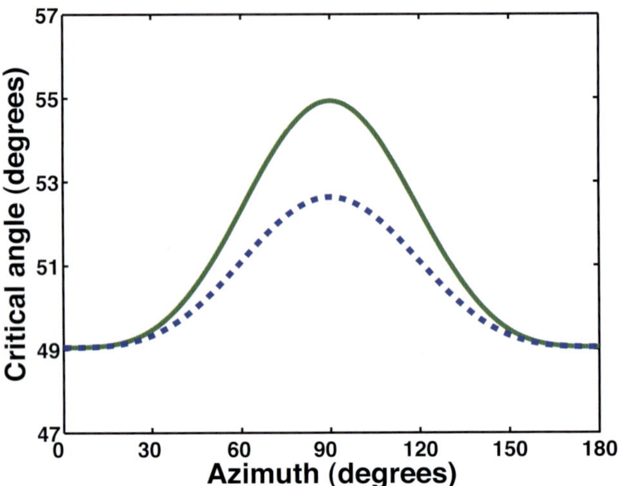

Figure 8.16: P-wave critical angle obtained from approximation 8.31 for an interface between orthorhombic (incidence) and isotropic (reflecting) media (Landrø and Tsvankin, 2007). The P-wave velocity in the isotropic halfspace is $V_{P0,2}$=2800 m/s. The solid curve is computed for the following parameters of the orthorhombic medium: $V_{P0,1}$=2100 m/s, $\epsilon_1^{(1)}$=0.25, $\epsilon_1^{(2)}$=0.1, $\delta_1^{(1)}$=0.05, $\delta_1^{(2)}$=−0.1, and $\delta_1^{(3)}$=−0.1. For the dashed curve, all parameters are the same, except for $\epsilon_1^{(1)}$=0.15.

Critical angle for converted and shear waves

Analysis of multicomponent data here is restricted to VTI media and symmetry planes of orthorhombic media, where P-waves are coupled with in-plane polarized SV-waves. Analytic description of the split shear waves outside symmetry planes of azimuthally anisotropic media is much more complicated and cannot be based on the analogy with vertical transverse isotropy. Equation 8.25 can be adapted for mode-converted and shear waves in VTI media by applying the "substitution rule" described in Tsvankin (2005). In the weak-anisotropy limit, any kinematic signature of SV-waves can be obtained from the corresponding P-wave approximation by replacing V_{P0} with the shear-wave vertical velocity V_{S0}, δ with the SV-wave parameter $\sigma \equiv (V_{P0}^2/V_{S0}^2)(\epsilon - \delta)$, and setting the parameter ϵ in the P-wave equation to zero.

Using this recipe, the critical angle for the SS (i.e., SVSV) transmission into a high-velocity medium can be found from equation 8.25 as

$$\sin \theta_{\text{cr,SS}} = n_{\text{SS}} \left[1 + \sigma_1 n_{\text{SS}}^2 (1 - n_{\text{SS}}^2) \right] , \tag{8.32}$$

where $n_{\text{SS}} = V_{S0,1}/V_{S0,2}$. Both the vertical and horizontal velocities of SV-waves are equal to V_{S0}, which explains why $\theta_{\text{cr,SS}}$ is independent of anisotropy in the reflecting halfspace.

Similarly, for the S-to-P (i.e., SV-to-P) transmission we have

$$\sin \theta_{\text{cr,SP2}} = n_{\text{SP2}} \left[1 - \epsilon_2 + \sigma_1 n_{\text{SP2}}^2 (1 - n_{\text{SP2}}^2) \right] ; \tag{8.33}$$

$n_{\text{SP2}} = V_{S0,1}/V_{P0,2}$.

For the S-to-P reflection in the incidence medium,

$$\sin\theta_{\text{cr,SP1}} = n_{\text{SP1}}\left[1 - \epsilon_1 + \sigma_1 n^2_{\text{SP1}}(1 - n^2_{\text{SP1}})\right]; \tag{8.34}$$

$n_{\text{SP1}} = V_{S0,1}/V_{P0,1}$.

In the atypical case of an extremely strong velocity contrast, the SV-wave velocity in the reflecting layer ($V_{S0,2}$) can be higher than the P-wave velocity in the incidence medium ($V_{P0,1}$). The approximate critical angle for the P-to-S transmission is

$$\sin\theta_{\text{cr,PS}} = n_{\text{PS}}\left[1 + \delta_1 n^2_{\text{PS}} + (\epsilon_1 - \delta_1)\, n^4_{\text{PS}}\right]; \tag{8.35}$$

$n_{\text{PS}} = V_{P0,1}/V_{S0,2}$.

With the appropriate parameter substitutions, equations 8.32 – 8.35 remain valid in the vertical symmetry planes of orthorhombic media. This adaptation of the VTI equations to orthorhombic models should take into account that SV-waves in the symmetry planes $[x_1, x_3]$ and $[x_2, x_3]$ represent two different modes (S_1 and S_2; see sections 6.1 and 8.1). It should be mentioned that the accuracy of weak-anisotropy approximations for SV-waves usually is much lower than for P-waves because of the relatively large values of the parameter σ (Tsvankin, 2005).

The critical angles of pure shear and mode-converted waves can be potentially estimated from the AVO response of the reflected SS-wave. Also, the PS-wave reflection coefficient has a peak corresponding to the critical angle of PP-waves (Mehdizadeh et al., 2005), and the offset of that peak is smaller the critical offset for the PP reflection. Therefore, amplitude analysis of mode conversions might help in estimating the PP-wave critical angle.

8.4.2 Inversion of P-wave critical-angle measurements

The analytic results described above indicate that estimates of the critical angle can be used in anisotropic inversion. Here, the parameter-estimation issues are explored for both VTI and orthorhombic models.

VTI media

If the vertical-velocity ratio n has been found, for example, from borehole data, an estimate of the P-wave critical angle can be used to constrain the parameters ϵ and δ. For a VTI layer beneath a purely isotropic overburden, the angle θ_{cr} depends on just ϵ_2 – the parameter responsible for the horizontal velocity in the reflecting medium (equation 8.20). In the weak-anisotropy approximation, ϵ_2 can be found from equation 8.25:

$$\epsilon_2 = 1 - \frac{\sin\theta_{\text{cr}}}{n}. \tag{8.36}$$

When the reflecting medium is isotropic, while the overburden is VTI, equation 8.25 for θ_{cr} has two unknowns (ϵ_1 and δ_1), and this ambiguity cannot be resolved without additional information. For the general VTI/VTI model, measurement of the critical angle provides a potentially useful relationship between the parameters ϵ_1, δ_1, and ϵ_2. Because equation 8.25 represents a linearized approximation, more accurate estimates of the anisotropy parameters can be obtained by solving equation 8.18 with the exact velocity function.

Orthorhombic media

In azimuthal AVO analysis, the best-constrained parameter is the difference between the AVO gradients in the vertical symmetry planes. Likewise, critical-angle reflectometry can provide an estimate of the difference between the symmetry-plane critical angles. As follows from equation 8.31,

$$\begin{aligned}\Delta(\sin\theta_{\mathrm{cr}}) &\equiv \sin\theta_{\mathrm{cr}}(0^\circ) \;-\; \sin\theta_{\mathrm{cr}}(90^\circ) \\ &= -\,(\Delta\epsilon_2)\,n + (\Delta\delta_1)\,n^3 + (\Delta\epsilon_1 - \Delta\delta_1)\,n^5\,,\end{aligned} \tag{8.37}$$

where $\Delta\delta_1 = \delta_1^{(2)} - \delta_1^{(1)}$, $\Delta\epsilon_1 = \epsilon_1^{(2)} - \epsilon_1^{(1)}$, and $\Delta\epsilon_2 = \epsilon_2^{(2)} - \epsilon_2^{(1)}$. If both media are orthorhombic, equation 8.37 contains three independent anisotropic parameter combinations, which cannot be resolved without additional information.

For the simplest case of a purely isotropic overburden, however, the critical angle is influenced by just $\epsilon_2(\alpha)$. Then the difference $\Delta(\sin\theta_{\mathrm{cr}})$ can be used to find

$$\Delta\epsilon_2 = \epsilon_2^{(2)} - \epsilon_2^{(1)} = -\frac{\Delta(\sin\theta_{\mathrm{cr}})}{n}\,. \tag{8.38}$$

If reliable critical-angle measurements in both symmetry planes are available and the velocity ratio n is known, the parameters $\epsilon_2^{(1)}$ and $\epsilon_2^{(2)}$ can be resolved individually using VTI equations 8.20 or 8.36.

As discussed above, the azimuthally varying critical angle can have maxima or minima between the symmetry planes. The azimuth and magnitude of these extrema are quite sensitive to the parameter $\delta_2^{(3)}$, which influences the horizontal velocity outside the symmetry planes. Therefore, if the incidence medium is isotropic (or if its parameters have been estimated), the ratio n is known, and the data have dense azimuthal coverage, the function $\theta_{\mathrm{cr}}(\alpha)$ can be inverted for all three relevant anisotropy parameters of the reflecting medium – $\epsilon_2^{(1)}$, $\epsilon_2^{(2)}$, and $\delta_2^{(3)}$.

8.4.3 Estimation of the critical angle from reflection data

Near-critical reflection amplitude

To identify the critical angle on surface reflection data, it is necessary to analyze the amplitude of the reflected wave at near-critical offsets. The behavior of the reflection coefficient near the critical angle is seldom analyzed in exploration seismology because conventional processing techniques operate with subcritical reflection data. It is well known from both analytic and numerical results for isotropic media that the reflection

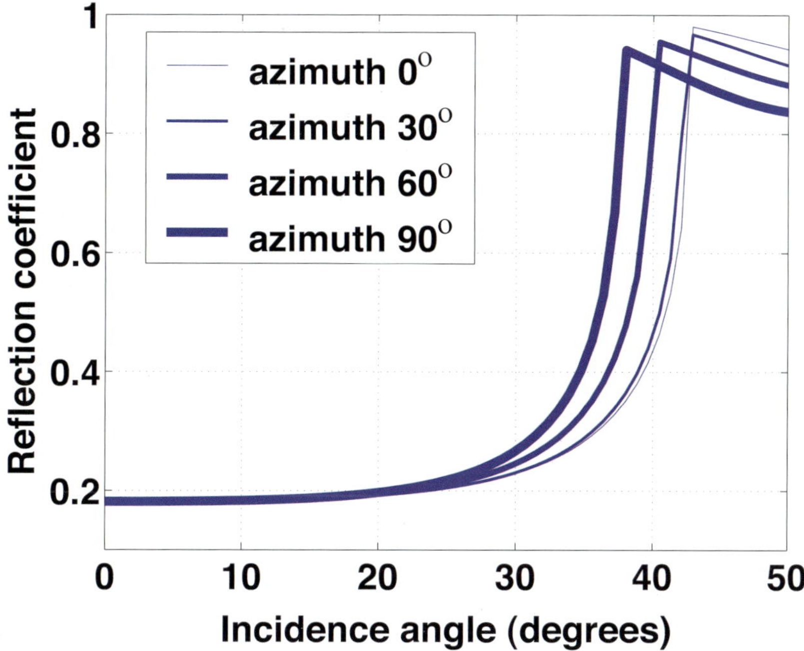

Figure 8.17: Real part of the exact PP-wave reflection coefficient computed in four azimuthal directions for an interface between isotropic (incidence) and orthorhombic (reflecting) media (Landrø and Tsvankin, 2007). The velocities and density in the incidence medium are $V_{P0,1}$=2100 m/s, $V_{S0,1}$=1200 m/s, and ρ_1=2.4 g/cm^3. For the reflecting medium, $V_{P0,2}$=2800 m/s, $V_{S0,2}$=1200 m/s, ρ_2=2.6 g/cm^3, $\epsilon_2^{(1)}$=0.25, $\epsilon_2^{(2)}$=0.1, $\delta_2^{(1)}$=0.05, $\delta_2^{(2)}$ = −0.1, $\delta_2^{(3)}$=−0.1, $\gamma_2^{(1)}$=0.28, and $\gamma_2^{(2)}$=0.15. The azimuthally varying critical angle for this model is plotted in Figure 8.14.

coefficient rapidly increases toward the critical angle, where its derivative becomes infinite (Aki and Richards, 2002; Tsvankin, 1995c). The P-wave reflection coefficient for an interface between isotropic and orthorhombic media varies with azimuth, but its shape near the azimuthally dependent critical angle $\theta_{\text{cr}}(\alpha)$ is practically the same (Figure 8.17). After a sharp peak at θ_{cr}, the real part of the reflection coefficient slowly decreases in the postcritical domain. Note that the critical angles for all four azimuths in Figure 8.17 are in excellent agreement with the results in Figure 8.14.

The AVO curve of the reflected wave at near-critical incidence, however, is not determined exclusively by the plane-wave reflection coefficient. First, because the amplitude of the reflected wave (as opposed to the reflection coefficient) has a continuous derivative, it increases beyond θ_{cr}. Second, the reflected wave in the postcritical domain undergoes rapid phase changes and interferes with the head wave. For a range of postcritical offsets, the reflected and head waves form a single arrival that cannot be described by the geometrical-seismics approximation based on the plane-wave reflection coefficient (Červený, 1962; Tsvankin, 1995c).

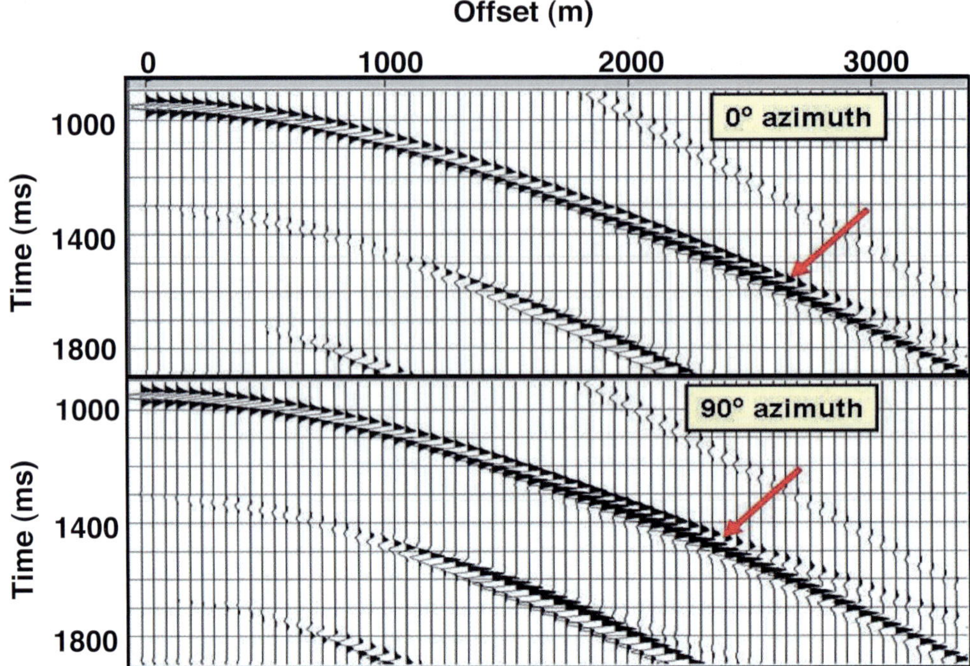

Figure 8.18: Synthetic shot gathers of the vertical displacement for the reflection from an orthorhombic layer (Landrø and Tsvankin, 2007). The sources and receivers are located in the incident isotropic halfspace (the free surface was removed); the model parameters are the same as in Figure 8.17. The gathers are computed in the symmetry planes α=0° (top) and α=90° (bottom). The zero-offset traveltime for the PP-wave reflection from the layer's top is about 950 ms. The head wave splits off from the reflected wave at offsets (marked by the arrows) between 2000 m and 3000 m.

With increasing separation between the two waves, their interference produces a frequency-dependent AVO curve that does not follow the angle dependence of the plane-wave reflection coefficient. In particular, the amplitude maximum of this interference arrival is shifted toward *postcritical* offsets, while the critical angle is generally close to the point of the fastest amplitude increase (Červený, 1961, 1962). Although Červený's analysis is limited to isotropic models, his conclusion agrees with the exact modeling results for an isotropic/orthorhombic interface discussed below. It should be emphasized, however, that the shift of the amplitude maximum relative to the critical offset x_{cr} varies with frequency, azimuth, and model parameters.

Synthetic example using reflectivity modeling

Critical-angle reflectometry was tested on a model that includes an orthorhombic layer beneath an isotropic halfspace. The seismograms in Figure 8.18 were generated with the anisotropic reflectivity method (Chin et al., 1984; Fryer and Frazer, 1984;

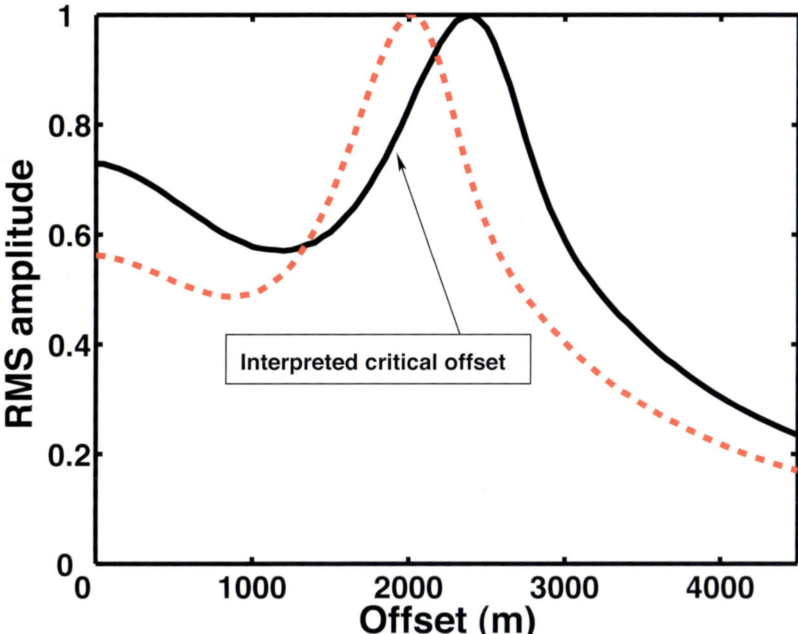

Figure 8.19: Amplitude of the reflected PP-wave from Figure 8.18 as a function of offset for the azimuths α=0° (solid curve) and α=90° (dashed) (Landrø and Tsvankin, 2007). The arrow marks the point of the fastest amplitude increase (i.e., of the largest slope of the AVO curve) for α=0°.

Tsvankin and Chesnokov, 1990), which produces exact 3D wavefields for horizontally layered media. The goal was to estimate the critical angle for the PP reflection from the top of the orthorhombic layer.

In both vertical symmetry planes a weak, but clearly visible head wave splits off from the target PP reflection. The separation of the head wave occurs at a smaller offset for azimuth $\alpha = 90°$, which is indicative of a decrease in the critical angle. The critical offset was assumed to correspond to the point of the fastest amplitude increase (i.e., the inflection point) on the AVO curves (Figure 8.19). The critical angle, computed from the critical offset using the known velocity in the incidence medium, changes from 43° for $\alpha = 0°$ to 38.5° for $\alpha = 90°$. These estimates of $\theta_{\rm cr}$, as well as those for off-symmetry directions, are close to the exact values (Figure 8.20).

Although the maximum critical angle is observed at $\alpha \approx 30°$, the symmetry-plane azimuths can be found from extrema of the function $\theta_{\rm cr}(\alpha)$ that are separated by 90°. If the vertical-velocity ratio n is known, the measurements of $\theta_{\rm cr}$ can be inverted for all three relevant parameters of the orthorhombic medium: $\epsilon_2^{(1)}$, $\epsilon_2^{(2)}$, and $\delta_2^{(3)}$. The linearized approximation 8.38 for this model gives an error in the difference $(\epsilon_2^{(1)} - \epsilon_2^{(2)})$ of about 0.05.

For laterally homogeneous media the slope (dx/dt) of the linear moveout of the

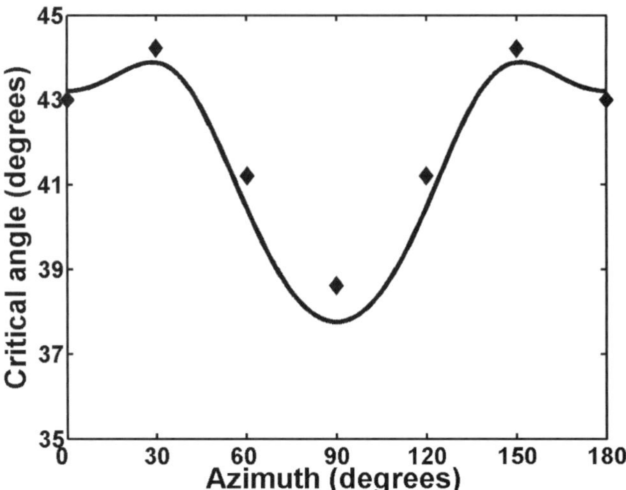

Figure 8.20: Variation of the critical angle with azimuth for the model from Figure 8.18 (Landrø and Tsvankin, 2007). The diamonds mark the estimates obtained from the AVO curves; the solid line is the exact angle θ_{cr}.

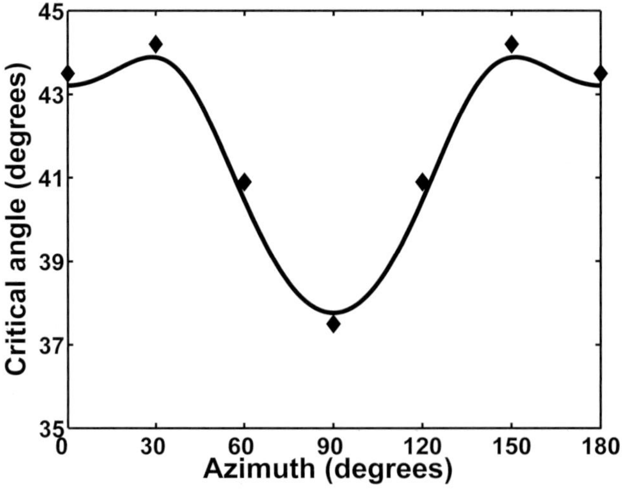

Figure 8.21: Same as Figure 8.20 but the diamonds mark the critical angle estimated from the horizontal velocity of the head wave using equation 8.18 (Landrø and Tsvankin, 2007).

head wave is equal to the horizontal velocity in the reflecting layer. Therefore, the critical angle can be computed from equation 8.18, in which the slope of the head-wave moveout for each azimuth is substituted for the horizontal phase velocity $V_{\text{hor},2}$. The resulting function $\theta_{\text{cr}}(\alpha)$ in Figure 8.21 is even more accurate than the one obtained from the AVO curve, which supports the feasibility of using the moveout of head waves in critical-angle reflectometry. (As discussed below, if the vertical velocity in the reflecting layer is known, $V_{\text{hor},2}$ can be inverted directly for the parameters $\epsilon_2^{(1)}$, $\epsilon_2^{(2)}$, and $\delta_2^{(3)}$ without computing the critical angle.) In practice, however, it might be difficult to operate with head waves because their amplitudes are relatively small, and the long offsets required for their detection are seldom available.

8.4.4 Implementation issues

Several practical issues related to the potential application of critical-angle reflectometry to field data were discussed in the introductory part of this section. Also, quantitative analysis of critical-angle measurements faces a number of serious challenges. One of them is related to the estimation of the critical offset x_{cr} from AVO curves. Although the point of the fastest amplitude increase (the inflection point) in the synthetic test proved to be close to x_{cr}, the shift of the critical offset with respect to the amplitude maximum varies with frequency. Therefore, it is desirable to verify the obtained value of x_{cr} by computing the AVO curve for the velocity model built after application of critical-angle reflectometry (possibly in combination with other methods). The critical angle can be also estimated by picking the offset of the amplitude maximum and applying a model-dependent offset shift.

If the overburden is anisotropic, amplitudes of reflected waves can be distorted by angle-dependent geometrical spreading (see sections 8.2 and 8.3). However, in the absence of strong heterogeneity in the overburden, geometrical spreading of P-waves varies slowly with offset and is unlikely to have a major impact in the narrow range of near-critical offsets used in the method. If necessary, the geometrical-spreading factor can be computed directly from reflection traveltimes and removed from the AVO response. Another factor that might change the shape of the near-critical amplitude function is vertical velocity gradient below the reflector, which transforms the head wave into a diving (refracted) mode.

Even if the critical offset has been accurately estimated from the reflection amplitudes, computation of the corresponding critical angle at the target horizon requires knowledge of the overburden velocity model. An apparent azimuthal variation of the critical offset can be created by reflector dip, although for mild dips of several degrees such offset distortions are insignificant. For anisotropic media it is also necessary to account for the difference between the group (ray) angle responsible for the critical offset and the phase angle of the critical ray analyzed here. While conventional AVO analysis has to deal with the same problem, the critical-angle reflectometry is more sensitive to errors in the offset-to-angle conversion because it relies on a single angle measurement for a given azimuth.

The transition from the critical offset to the critical angle can be avoided for laterally homogeneous models, in which the ray parameter of the critical reflection at the surface is equal to the slowness of the horizontal ray in the reflecting medium. Therefore, the slope of the traveltime function measured at the critical offset in two orthogonal directions gives a direct estimate of the horizontal slowness vector $\mathbf{p}$ beneath the reflector. For high-quality data sets with uncommonly long offsets, the horizontal velocity in the reflecting layer can also be obtained from the linear moveout of the head wave (see Figure 8.21). For laterally homogeneous models it may be advantageous to identify the critical-angle reflection by applying the $\tau - p$ transform (van der Baan, 2004, 2005).

The azimuthally varying horizontal velocity in the reflecting orthorhombic layer is governed by the parameter $\epsilon(\alpha)$ (equation 8.30) and can be inverted for $\delta^{(3)}$ and for the difference between $\epsilon^{(1)}$ and $\epsilon^{(2)}$ without knowledge of the overburden velocity model. If the vertical velocity in the reflecting medium is known, the parameters $\epsilon^{(1)}$ and $\epsilon^{(2)}$ can be resolved individually. Furthermore, the symmetry-plane horizontal velocities can be combined with the corresponding interval NMO velocities in the reflecting layer to compute the interval anellipticity parameters $\eta^{(1)}$ and $\eta^{(2)}$ responsible for time processing of P-wave data in orthorhombic media (see section 3.5). This approach might provide more accurate η-parameters than does nonhyperbolic moveout inversion, which is hampered by the trade-offs between the quadratic and quartic moveout coefficients. If the reflecting layer is VTI, the P-wave horizontal and NMO velocities yield the time-processing parameter η.

On the whole, critical-angle reflectometry can hardly be regarded as a reliable stand-alone method for anisotropic parameter estimation. Rather, it should be used in combination with other techniques that operate with azimuthally varying traveltimes and amplitudes.

8.5 Elements of anisotropic attenuation analysis

Attenuation (absorption) dissipates the energy of seismic waves and changes their amplitude and frequency content. Physical-modeling experiments have shown that attenuation in anisotropic rocks can vary with direction, and attenuation anisotropy often is much stronger than velocity anisotropy (Hosten et al., 1987; Zhu et al., 2007b; Chichinina et al., 2009). The same conclusions were drawn by Carcione (2000) who computed wavefields propagating through attenuative shale layers. Among possible physical causes of attenuation anisotropy are small-scale aligned fractures (e.g., Mavko and Nur, 1979; Rathore et al., 1995; Chichinina et al., 2006), interbedding of thin attenuative layers (e.g., Zhu et al., 2007a), stress-induced phenomena (e.g., Liu et al., 1993; Prasad and Nur, 2003), and azimuthal scattering (Willis et al., 2004).

Here we describe the macroscopic behavior of anisotropic anelastic media, but do not address the physical mechanisms that make attenuation angle-dependent. Unless attenuation is unusually high, its contribution to plane-wave reflection and transmission coefficients (see section 8.1) is relatively small (Stovas and Ursin, 2003; Behura and Tsvankin, 2009b). Amplitudes of reflected waves are more influenced by the attenuation along the raypath, which is described by the attenuation coefficient related to the imaginary part of the wave vector. Variation of attenuation coefficients with direction might cause distortions of the AVO response (in particular, of the AVO gradient). On the other hand, estimation of anisotropic attenuation can provide sensitive attributes for characterization of fractured reservoirs and lithology discrimination.

A detailed discussion of wave propagation in anisotropic attenuative media in terms of the complex stiffness coefficients can be found in Carcione (2001). This section is based on the results of Zhu and Tsvankin (2006, 2007), who showed that attenuation analysis in TI and orthorhombic media can be substantially facilitated by introducing dimensionless anisotropy parameters responsible for the attenuation coefficients of P- and S-waves.

8.5.1 Analytic background

Christoffel equation for anisotropic attenuative media

The equation of motion for anisotropic attenuative media has the same form as that for purely elastic models (e.g., Helbig, 1994; Carcione, 2001):

$$\rho \frac{\partial^2 u_m}{\partial t^2} - \tilde{c}_{mjkl} \frac{\partial^2 u_k}{\partial x_j \partial x_l} = f_m , \tag{8.39}$$

where $\mathbf{u}$ is the displacement vector, ρ is the mass density, t is the time, $\mathbf{f}$ is the vector of body forces, and $\tilde{c}_{mjkl}$ is the stiffness tensor which becomes complex in the presence of attenuation. Summation over repeated indices (each index changes from 1 to 3) is implied, and the notation $\tilde{c}$ denotes a complex quantity. Equation 8.39 is valid for linearly elastic, homogeneous (or weakly heterogeneous) media with arbitrary symmetry.

To solve equation 8.39 for plane-wave propagation, the source term $\mathbf{f}$ has to be set to zero. The displacement of a harmonic plane wave in attenuative media is

$$\tilde{\mathbf{u}} = \tilde{\mathbf{U}} e^{i(\omega t - \tilde{\mathbf{k}} \cdot \mathbf{x})}, \tag{8.40}$$

where $\tilde{\mathbf{U}}$ is the complex polarization vector, ω is the angular frequency, and $\tilde{\mathbf{k}}$ is the complex wave vector: $\tilde{\mathbf{k}} = \mathbf{k} - i\mathbf{k}^I$. The real part $\mathbf{k}$ of the wave vector is responsible for the phase velocity and defines planes of constant phase. The imaginary term $\mathbf{k}^I$ is sometimes called the *attenuation vector* because it controls attenuation-induced amplitude decay (Figure 8.22). In general, $\mathbf{k}^I$ is not parallel to $\mathbf{k}$, which means that the planes of constant phase and constant amplitude do not coincide and the direction of the fastest amplitude decay deviates from the phase-velocity vector (e.g., Krebes and Slawinski, 1991; Krebes and Le, 1994; Červený and Pšenčík, 2005a, 2005b). The angle between $\mathbf{k}^I$ and $\mathbf{k}$ is often called the *inhomogeneity angle*, denoted here by ξ.

In plane-wave propagation, the inhomogeneity angle represents a free parameter, except for certain *forbidden directions* of $\mathbf{k}^I$ where solutions of the wave equation

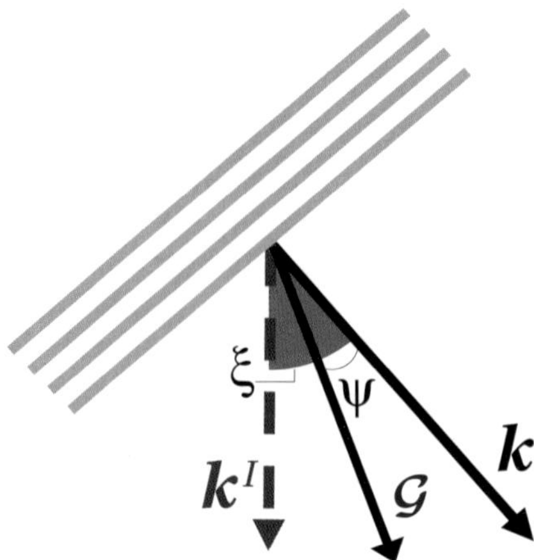

Figure 8.22: Plane wave with a nonzero inhomogeneity angle ξ between the real ($\mathbf{k}$) and imaginary ($\mathbf{k}^I$) parts of the wave vector. The wave propagates in the direction $\mathbf{k}$ (perpendicular to the planes of constant phase) and attenuates most rapidly in the direction $\mathbf{k}^I$. The group-velocity vector $\boldsymbol{\mathcal{G}}$ makes the angle ψ with the phase direction.

do not exist. In contrast, for wavefields excited by point sources, the value of ξ is determined by the model parameters. If the source is located in a weakly attenuative homogeneous medium, the angle ξ is usually small, and the rate of attenuation is highest near the propagation direction (Ben-Menahem and Singh, 1981; Vavryčuk, 2007). The inhomogeneity angle, however, can become relatively large in stratified media with significant attenuation contrasts across layer boundaries. Still, as will be discussed below, the angle ξ has almost no influence on the group attenuation coefficient estimated from seismic data (Behura and Tsvankin, 2009a).

The Christoffel equation describes the velocities and polarizations of plane waves in anisotropic media. Substitution of the plane-wave displacement (equation 8.40) into the wave equation 8.39 leads to the general Christoffel equation that accounts for both anisotropy and attenuation:

$$\left[\tilde{c}_{mjkl}\,\tilde{k}_j\,\tilde{k}_l - \rho\,\omega^2\delta_{mk}\right]\tilde{U}_k = 0\,; \tag{8.41}$$

δ_{mk} is Kronecker's symbolic δ.

If the vectors $\mathbf{k}$ and $\mathbf{k}^I$ are parallel to one another (i.e., the angle $\xi = 0°$), the wave vector can be represented as $\tilde{\mathbf{k}} = \mathbf{n}\,(k - i k^I)$, where $\mathbf{n}$ is the unit vector in the phase direction, $k = |\mathbf{k}|$, and $k^I = |\mathbf{k}^I|$. Then equation 8.41 simplifies to

$$\left[\tilde{G}_{mk} - \frac{\rho\,\omega^2}{(k - i k^I)^2}\,\delta_{mk}\right]\tilde{U}_k = 0\,, \tag{8.42}$$

where $\tilde{G}_{mk} = \tilde{c}_{mjkl}\,n_j\,n_l$ is the complex Christoffel matrix. The ratio $\omega/(k - ik^I)$ can be treated as the complex phase velocity $\tilde{V}$ (Carcione, 2001), which allows us to rewrite equation 8.42 in the same form as in nonattenuative media (e.g., see equation 1.10 in Tsvankin, 2005):

$$\left[\tilde{G}_{mk} - \rho \tilde{V}^2 \delta_{mk}\right] \tilde{U}_k = 0 . \tag{8.43}$$

The three equations (corresponding to $m = 1, 2, 3$) making up system 8.43 are complex and can be solved for both the real (k) and imaginary (k^I) parts of the wave vector. Before analyzing these solutions for TI and orthorhombic media, we review the relationship between group and phase attenuation and the influence of the inhomogeneity angle on attenuation coefficients measured from seismic data.

Group attenuation and the influence of the inhomogeneity angle

The ratio k^I/k, which an be obtained from the Christoffel equation, determines the rate of amplitude decay per wavelength and is called the normalized phase attenuation coefficient $\mathcal{A}$:

$$\mathcal{A} \equiv \frac{k^I}{k} . \tag{8.44}$$

For a nonzero inhomogeneity angle ξ, the coefficient $\mathcal{A}$ is a measure of attenuation along the vector $\mathbf{k}^I$ rather than $\mathbf{k}$. In seismic data processing, however, the attenuation is evaluated along the group-velocity direction (raypath), which generally deviates from both $\mathbf{k}$ and $\mathbf{k}^I$. For example, attenuation coefficients are often estimated using the spectral-ratio method (e.g., Johnston and Toksöz, 1981), which has been extended to anisotropic media (Zhu et al., 2007b). In particular, Behura and Tsvankin (2009c) combined the spectral-ratio method with velocity-independent layer stripping (VILS, see section 3.4) to estimate interval attenuation in layered media from reflection seismic data.

Under the assumption that reflection/transmission coefficients, source/receiver directivity patterns, and geometrical spreading are independent of frequency, the spectral-ratio method produces the attenuation along the raypath normalized by the angular frequency ω (Zhu et al., 2007b; Behura and Tsvankin, 2009c):

$$S_a = \frac{k_g^I \, d}{\omega} , \tag{8.45}$$

where k_g^I is the average group attenuation coefficient and d is the path length. If the the ray segment used in equation 8.45 lies in a homogeneous medium, the length d is equal to the product of the group velocity $\mathcal{G}$ along the ray and the traveltime t:

$$S_a = \frac{k_g^I \, \mathcal{G} \, t}{\omega} = \mathcal{A}_g \, t ; \tag{8.46}$$

$\mathcal{A}_g = k_g^I/(\omega/\mathcal{G})$ is the normalized group attenuation coefficient. Because the traveltime t can be obtained from kinematic analysis, the coefficient $\mathcal{A}_g$ is the measure of attenuation provided by seismic data.

When the medium is anisotropic (or isotropic, but with a large inhomogeneity angle), the group-velocity vector deviates from the phase direction defined by the real part $\mathbf{k}$ of the wave vector. The group attenuation coefficient k_g^I can be found by projecting the phase attenuation vector $\mathbf{k}^I$ onto the group direction (Figure 8.22). For simplicity, we assume that the group-velocity vector lies in the plane formed by the vectors $\mathbf{k}$ and $\mathbf{k}^I$.[1] If the angle between the phase- and group-velocity vectors is denoted by ψ, the coefficient k_g^I is given by

$$k_g^I = k^I \cos(\xi - \psi)\,, \tag{8.47}$$

and

$$\mathcal{A}_g = \frac{k_g^I\,\mathcal{G}}{\omega} = \frac{k^I \cos(\xi - \psi)\,\mathcal{G}}{\omega}\,. \tag{8.48}$$

The phase velocity $V = \omega/k$ represents the projection of the group-velocity vector onto the phase direction, and

$$\mathcal{G} = \frac{V}{\cos\psi} = \frac{\omega}{k\cos\psi}\,. \tag{8.49}$$

Substituting equation 8.49 into equation 8.48, we find

$$\mathcal{A}_g = \frac{k^I}{k}\,\frac{\cos(\xi - \psi)}{\cos\psi}\,. \tag{8.50}$$

In general, k^I and $\mathcal{A}_g$ should be treated as functions of the inhomogeneity angle (typically the influence of ξ on k and ψ is weak). Taking into account the definition of the phase attenuation coefficient (equation 8.44), equation 8.50 can be rewritten as

$$\mathcal{A}_g(\xi) = \frac{k^I(\xi)}{k}\,\frac{\cos(\xi - \psi)}{\cos\psi} = \mathcal{A}(\xi)\,\frac{\cos(\xi - \psi)}{\cos\psi}\,. \tag{8.51}$$

For a zero inhomogeneity angle, equation 8.51 reduces to

$$\mathcal{A}_g(\xi = 0^\circ) = \mathcal{A}(\xi = 0^\circ). \tag{8.52}$$

Thus, although the group and phase directions (and velocities) differ, for $\xi = 0^\circ$ the group and phase attenuation coefficients are identical.

Behura and Tsvankin (2009a) demonstrated that equation 8.52 is a special case of a much more general result valid for arbitrarily anisotropic, attenuative media. Although k^I varies with ξ, the product $k^I(\xi)\cos(\xi - \psi)$ and, therefore, the group attenuation coefficient $\mathcal{A}_g$ as a whole is insensitive to the inhomogeneity angle. Even for uncommonly high attenuation and strong velocity and attenuation anisotropy, $\mathcal{A}_g(\xi)$ practically coincides with the phase attenuation coefficient for a *zero* inhomogeneity angle (Behura and Tsvankin, 2009a):

$$\mathcal{A}_g(\xi) \simeq \mathcal{A}(\xi = 0^\circ). \tag{8.53}$$

[1] A more general analysis that leads to the same result can be found in Behura and Tsvankin (2009a).

The contribution of ξ to $\mathcal{A}_g$ becomes significant only for large inhomogeneity angles approaching the forbidden directions.

Result 8.53 has fundamental importance in attenuation analysis for both isotropic and anisotropic media. Indeed, the inhomogeneity angle is extremely difficult to estimate from seismic data, and its potential influence on group attenuation could greatly complicate interpretation of attenuation coefficients. According to equation 8.53, however, seismic attenuation analysis yields the phase attenuation coefficient for $\xi = 0°$, which represents a direct measure of the intrinsic attenuation of the material. In fact, the coefficient $\mathcal{A}(\xi = 0°)$ is proportional to the inverse quality factor in the phase direction. Behura and Tsvankin (2009c) confirmed the accuracy of equation 8.53 by processing synthetic data from layered anisotropic media with strong attenuation. Therefore, it is sufficient for us to examine the solutions of the simpler Christoffel equations 8.42 or 8.43 derived for a zero inhomogeneity angle.

8.5.2 Attenuation coefficients for TI media

Our choice of terminology is designed to draw a distinction between the anisotropy of velocity and attenuation. To ensure consistency with existing literature on purely elastic anisotropic media, terms such as *anisotropic medium* or *transversely isotropic (TI) medium* refer to velocity anisotropy. When discussing attenuative models, we explicitly specify the character of attenuation. For example, the term *TI medium with isotropic attenuation* means that the model is transversely isotropic with respect to the velocity function but the attenuation is isotropic (i.e., independent of direction).

Here, we follow the work of Zhu and Tsvankin (2006), who studied plane-wave propagation in transversely isotropic media with isotropic and TI attenuation. The symmetry axis is assumed to be vertical (VTI), but because all results are derived for a homogeneous medium, they can be readily adapted to models with any symmetry-axis orientation.

Anisotropic quality-factor matrix

The most common measure of attenuation used for isotropic media is the quality factor Q, which is inversely proportional to the attenuation coefficient. When the medium is anisotropic and attenuation is directionally dependent, a single value of Q is insufficient. To ensure consistency in describing both isotropic and anisotropic attenuation coefficients, we follow Carcione (2001) in introducing the *quality-factor matrix* as

$$Q_{ij} \equiv \frac{c_{ij}}{c^I_{ij}}, \tag{8.54}$$

where c_{ij} and c^I_{ij} are the real and the imaginary parts of the stiffness coefficient $\tilde{c}_{ij}$, and there is no summation over i or j. It is assumed that $c^I_{ij} \neq 0$; otherwise, the medium is purely elastic, and the quality factor is infinite.

In many cases, the physical reasons for attenuation and velocity anisotropy in TI media are expected to be similar (e.g., preferential orientation of clay platelets in shales). For such models, the matrices c_{ij} and c^I_{ij} have the same symmetry, and the quality-factor matrix for a vertical symmetry axis takes the form

$$\mathbf{Q} = \begin{bmatrix} Q_{11} & Q_{12} & Q_{13} & 0 & 0 & 0 \\ Q_{12} & Q_{11} & Q_{13} & 0 & 0 & 0 \\ Q_{13} & Q_{13} & Q_{33} & 0 & 0 & 0 \\ 0 & 0 & 0 & Q_{55} & 0 & 0 \\ 0 & 0 & 0 & 0 & Q_{55} & 0 \\ 0 & 0 & 0 & 0 & 0 & Q_{66} \end{bmatrix}. \tag{8.55}$$

As follows from the rotational symmetry of the model,

$$Q_{12} = Q_{11} \frac{c_{11} - 2c_{66}}{c_{11} - 2c_{66}\dfrac{Q_{11}}{Q_{66}}}. \tag{8.56}$$

When both the real and imaginary parts of the stiffness matrix have isotropic structure, the quality factor is described by only two independent parameters, Q_{33} and Q_{55}:

$$\mathbf{Q} = \begin{bmatrix} Q_{33} & Q_{13} & Q_{13} & 0 & 0 & 0 \\ Q_{13} & Q_{33} & Q_{13} & 0 & 0 & 0 \\ Q_{13} & Q_{13} & Q_{33} & 0 & 0 & 0 \\ 0 & 0 & 0 & Q_{55} & 0 & 0 \\ 0 & 0 & 0 & 0 & Q_{55} & 0 \\ 0 & 0 & 0 & 0 & 0 & Q_{55} \end{bmatrix}. \tag{8.57}$$

The element $Q_{13} = Q_{12}$ can be obtained from Q_{33} and Q_{55} as

$$Q_{13} = Q_{33} \frac{c_{33} - 2c_{55}}{c_{33} - 2c_{55}\dfrac{Q_{33}}{Q_{55}}}. \tag{8.58}$$

P-wave attenuation in isotropic media is controlled by Q_{33}, while Q_{55} is responsible for attenuation of shear waves. Laboratory measurements (Gautam et al., 2003) indicate that the Q-factor for P-waves can be either larger or smaller than that for SV-waves (e.g., depending on the mobility of fluids in the rock). When $Q_{33} = Q_{55}$, all nonzero components of the **Q**-matrix are identical:

$$Q_{ij} = Q\,. \tag{8.59}$$

As discussed below, if the quality-factor matrix is given by equation 8.59, the attenuation for both P- and S-waves does not vary with direction (i.e., is isotropic), even in the presence of velocity anisotropy.

We assume Q to be independent of frequency, which is often a good approximation for the seismic frequency band. In a more rigorous description of attenuation, the complex stiffness components and the quality factor vary with frequency, as does the velocity. Our results, however, can still be applied for any given frequency, and the definitions of the Thomsen-style attenuation-anisotropy parameters given below do not change (although the parameters become frequency-dependent).

SH-wave attenuation

For waves propagating in the $[x_1, x_3]$-plane of VTI media, the Christoffel equation splits into an equation for SH-waves polarized in the x_2-direction and two coupled equations for in-plane polarized P- and SV-waves. We start with the SH-wave solution, which has a relatively simple form.

As follows from the Christoffel equation 8.42, the SH-wavenumbers k and k^I satisfy the following equation:

$$(k - i k^I)^2 \, (\tilde{c}_{66} \sin^2\theta + \tilde{c}_{55} \cos^2\theta) - \rho\,\omega^2 = 0\,; \tag{8.60}$$

θ is the phase angle with the symmetry axis.

The solutions of equation 8.60 are analyzed in Appendix 8B. The assumption of isotropic Q for SH-waves implies $Q_{55} = Q_{66} = Q$ and does not involve the condition $Q_{33} = Q_{55}$. For media with $Q_{55} = Q_{66}$, the imaginary part of equation 8.60 gives a simple expression for the attenuation coefficient (equation 8.146; also see equation 2.122 in Carcione, 2001):

$$\mathcal{A}_{SH} = \frac{k^I}{k} = \sqrt{1 + Q^2} \; - Q\,. \tag{8.61}$$

Hence, for isotropic Q the coefficient $\mathcal{A}_{SH}$ is independent of angle and is given by the familiar expression for isotropic media. When attenuation is weak (i.e., $1/Q \ll 1$), equation 8.61 reduces to

$$\mathcal{A}_{SH} = \frac{1}{2Q}\,. \tag{8.62}$$

The weak-attenuation approximation 8.62 is close to the exact attenuation coefficient for the practically important range $Q > 10$ and breaks down only for strongly attenuative media (Figure 8.23).

Next, we analyze SH-wave propagation in VTI media with VTI attenuation. As shown in Appendix 8B, the imaginary part of the Christoffel equation 8.60 provides the following solution for the attenuation coefficient (equation 8.139):

$$\mathcal{A}_{SH} \equiv \frac{k^I}{k} = \sqrt{1 + (Q_{55}\, b)^2} \; - Q_{55}\, b\,, \tag{8.63}$$

where

$$b \equiv \frac{(1+2\gamma)\sin^2\theta + \cos^2\theta}{(1+2\gamma)\,\dfrac{Q_{55}}{Q_{66}}\,\sin^2\theta + \cos^2\theta}\,; \tag{8.64}$$

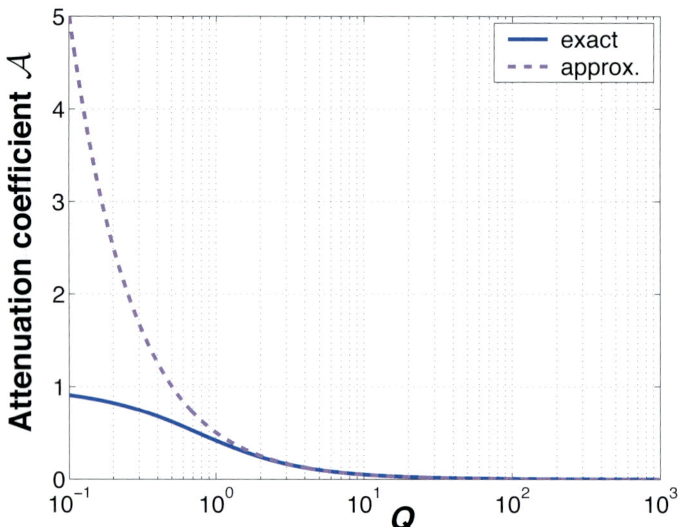

Figure 8.23: Normalized attenuation coefficient $\mathcal{A}$ for SH-waves as a function of the Q-factor for a medium with $Q_{55} = Q_{66} = Q$ (Zhu and Tsvankin, 2006). The solid line is the exact $\mathcal{A}$ from equation 8.61; the dashed line is the weak-attenuation approximation 8.62.

γ is Thomsen's velocity-anisotropy parameter for SH-waves. Therefore, the directional dependence of the attenuation coefficient is influenced by the velocity anisotropy, which can be expected from the definition of $\mathcal{A}_{SH}$ and from the coupling of the real and imaginary parts of the wave vector in the Christoffel equation. In the weak-attenuation limit, equation 8.63 becomes

$$\mathcal{A}_{SH} = \frac{1}{2Q_{55}b}. \tag{8.65}$$

According to equation 8.65, Q_{55} is multiplied with the directionally dependent parameter b to form the effective quality factor for the SH-wave, $Q_{55}^{\text{eff}} = Q_{55}b$. At vertical incidence ($\theta = 0°$), $b = 1$ and $\mathcal{A}_{SH} = 1/(2Q_{55})$. In the horizontal direction ($\theta = 90°$), $b = Q_{66}/Q_{55}$ and $\mathcal{A}_{SH} = 1/(2Q_{66})$. For intermediate propagation directions, b reflects the coupling between the SH-wave velocity-anisotropy parameter γ and the ratio of the quality-factor elements Q_{55} and Q_{66}. The contribution of the ratio Q_{55}/Q_{66} in equation 8.64 is used below to define an attenuation-anisotropy parameter analogous to γ.

The real part of the Christoffel equation 8.60 can be used to obtain the SH-wave phase velocity (equation 8.141):

$$V_{SH}(\theta) = \zeta_Q V_{SH}^{\text{elast}}(\theta), \tag{8.66}$$

where V_{SH}^{elast} is the phase velocity in the reference nonattenuative medium (equation 8.142) and ζ_Q is defined in equation 8.143. For a wide range of Q-values, the phase velocity is well-described by the weak-attenuation approximation (equation 8.144):

$$V_{SH}(\theta) = \left[1 + \frac{1}{2(Q_{55}b)^2}\right] V_{SH}^{\text{elast}}(\theta). \tag{8.67}$$

Equation 8.67 shows that for a realistic magnitude of Q_{ij} the influence of attenuation on the phase-velocity function can be ignored. For models with moderate attenuation anisotropy, b does not significantly deviate from unity ($b=1$ for $Q_{55}=Q_{66}$). Even for media with a low $Q_{55}=10$, the remaining attenuation term $1/(2Q_{55}^2)$ in equation 8.67 changes the velocity V_{SH} only by 0.5%. This analysis does not take into account velocity dispersion, which is usually weak in the frequency band typical for reflection seismology.

Thomsen-style notation for VTI attenuation

Because of the coupling between P- and SV-waves, equations governing their velocity and attenuation are more complicated than those for SH-waves. Zhu and Tsvankin (2006) obtained closed-form solutions (their equations 19 – 21) for the attenuation coefficients of P- and SV-waves under the assumption that both attenuation anisotropy and attenuation itself are weak. Those expressions, derived in terms of the stiffness coefficients and quality-factor elements, are quite accurate but cumbersome and do not provide sufficient insight into the behavior of the attenuation coefficients.

The description of TI attenuation can be significantly simplified by extending the principle of Thomsen notation to directionally dependent attenuation coefficients (Zhu and Tsvankin, 2006). The Q_{ij}-matrix for VTI media with VTI attenuation contains five independent elements, which can be replaced by two reference (isotropic) quantities and three dimensionless parameters (ϵ_Q, δ_Q, and γ_Q) responsible for attenuation anisotropy. To maintain close similarity with Thomsen notation for velocity anisotropy and optimize this parameterization for reflection data, it is convenient to choose the P- and S-wave attenuation coefficients in the symmetry (vertical) direction as the reference values:

$$\mathcal{A}_{P0} = Q_{33}\left(\sqrt{1+\frac{1}{Q_{33}^2}} - 1\right) \approx \frac{1}{2Q_{33}}, \tag{8.68}$$

$$\mathcal{A}_{S0} = Q_{55}\left(\sqrt{1+\frac{1}{Q_{55}^2}} - 1\right) \approx \frac{1}{2Q_{55}}. \tag{8.69}$$

The coefficient $\mathcal{A}_{S0}$ is also responsible for the SV-wave attenuation in the isotropy plane. The linearization of the square-roots in equations 8.68 and 8.69 produces approximations accurate to the first order in $1/Q_{ij}$.

The attenuation-anisotropy parameter γ_Q is defined as the fractional difference between the SH-wave attenuation coefficients in the horizontal and vertical directions (see equation 8.65):

$$\gamma_Q \equiv \frac{1/Q_{66} - 1/Q_{55}}{1/Q_{55}} = \frac{Q_{55} - Q_{66}}{Q_{66}}. \tag{8.70}$$

This definition is analogous to that of the Thomsen parameter γ, which is close to the fractional difference between the horizontal and vertical velocities of SH-waves. The parameter γ_Q controls the magnitude of the SH-wave attenuation anisotropy; for isotropic Q ($Q_{55} = Q_{66}$), $\gamma_Q = 0$.

Substitution of γ_Q into equation 8.64 for the parameter b results in

$$b = \frac{(1 + 2\gamma)\sin^2\theta + \cos^2\theta}{(1 + 2\gamma)(1 + \gamma_Q)\sin^2\theta + \cos^2\theta}. \tag{8.71}$$

When both γ and γ_Q are small ($|\gamma| \ll 1$, $|\gamma_Q| \ll 1$), b can be linearized in these parameters:

$$b = 1 - \gamma_Q \sin^2\theta. \tag{8.72}$$

Then the SH-wave attenuation coefficient from equation 8.65 becomes independent of γ. Further linearization of equation 8.65 in γ_Q yields

$$\mathcal{A}_{SH} = \mathcal{A}_{S0}\,(1 + \gamma_Q \sin^2\theta)\,, \tag{8.73}$$

where $\mathcal{A}_{S0}$ is defined in equation 8.69. Equation 8.73 has the same form as the phase velocity of SH-waves linearized in the parameter γ (Thomsen, 1986). Note, however, that the exact phase velocity, unlike the exact attenuation coefficient, is described by a simple function of γ that corresponds to elliptical anisotropy.

From equation 8.73, γ_Q determines the sign and rate of the variation of $\mathcal{A}_{SH}(\theta)$ away from the vertical direction. When $\gamma_Q > 0$, the factor b decreases with the phase angle θ, which causes an increase in the attenuation coefficient (Figure 8.24). If the magnitude of the SH-wave velocity anisotropy is small (i.e., $|\gamma| \ll 1$), approximation 8.73 gives an accurate estimate of the attenuation coefficient even for relatively large absolute values of γ_Q reaching 0.4 (Figure 8.24b).

Another attenuation-anisotropy parameter, ϵ_Q, can be defined by analogy with the Thomsen parameter ϵ as the fractional difference between the P-wave attenuation coefficients in the horizontal and vertical directions:

$$\epsilon_Q \equiv \frac{1/Q_{11} - 1/Q_{33}}{1/Q_{33}} = \frac{Q_{33} - Q_{11}}{Q_{11}}. \tag{8.74}$$

To complete the description of TI attenuation, we need to introduce a parameter similar to Thomsen's δ that involves the quality-factor element Q_{13}. The parameter δ has proved to be useful in describing signatures of reflected P-waves in VTI media because it determines the second derivative of the P-wave phase-velocity function in the vertical direction (the first derivative goes to zero). Therefore, the idea of Thomsen notation can be preserved by defining δ_Q through the second derivative of the P-wave attenuation coefficient at $\theta = 0$:

$$\delta_Q \equiv \frac{1}{2\mathcal{A}_{P0}} \left.\frac{d^2\mathcal{A}_P}{d\theta^2}\right|_{\theta=0^\circ}. \tag{8.75}$$

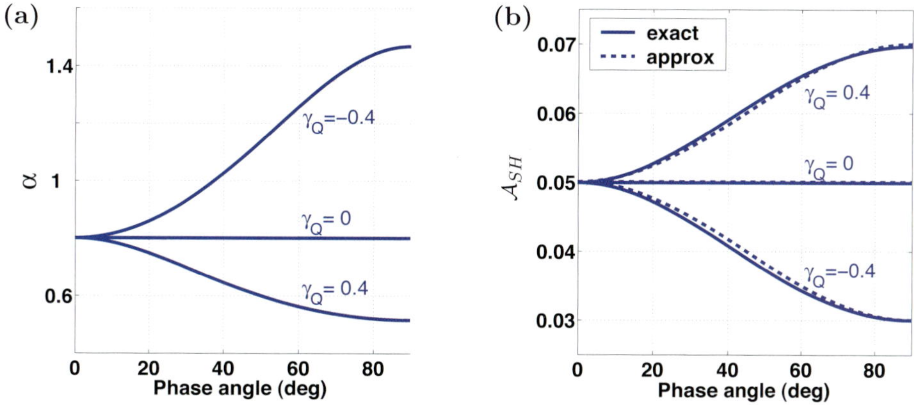

Figure 8.24: (a) Factor b and (b) the exact and approximate SH-wave attenuation coefficients for a medium with γ=0.1, Q_{55}=10, and γ_Q varying from −0.4 to 0.4 (Zhu and Tsvankin, 2006). The exact $\mathcal{A}_{SH}$ (solid line) is computed from equation 8.63, and the approximate $\mathcal{A}_{SH}$ (dashed) from equation 8.73.

In other words, the parameter δ_Q controls the curvature of the coefficient $\mathcal{A}_P(\theta)$ in the vertical direction. Thus, the role of δ_Q in describing the P-wave attenuation anisotropy is similar to that of δ in the P-wave phase-velocity equation (Thomsen, 1986; Tsvankin, 2005). Because the first derivative of $\mathcal{A}_P$ for $\theta = 0$ is equal to zero, δ_Q is responsible for the angular variation of the P-wave attenuation coefficient near the vertical direction.

Assuming that both attenuation and attenuation anisotropy are weak, Zhu and Tsvankin (2006) found the following expression for δ_Q:

$$\delta_Q \equiv \frac{\dfrac{Q_{33} - Q_{55}}{Q_{55}} c_{55} \dfrac{(c_{13} + c_{33})^2}{c_{33} - c_{55}} + 2 \dfrac{Q_{33} - Q_{13}}{Q_{13}} c_{13} (c_{13} + c_{55})}{c_{33} (c_{33} - c_{55})} . \tag{8.76}$$

The accuracy of equation 8.76 in describing P-wave attenuation coefficients is verified by the numerical tests below. Unless attenuation is uncommonly strong, the phase velocities of P- and SV-waves are close to those in the reference nonattenuative medium and do not vary with the parameters ϵ_Q and δ_Q. The presence of the real-valued stiffnesses in equation 8.76, however, indicates that δ_Q might be influenced by velocity anisotropy. Indeed, the stiffness coefficient c_{13} for small $|\delta|$ can be approximated as (Tsvankin, 2005, p. 20)

$$c_{13} = c_{33} (1 + \delta) - 2c_{55} . \tag{8.77}$$

Substitution of equation 8.77 into equation 8.76 shows that δ_Q depends on the parameter δ. Linearizing equation 8.76 in the squared vertical-velocity ratio,

$$g \equiv \frac{c_{55}}{c_{33}} = \frac{V_{S0}^2}{V_{P0}^2} , \tag{8.78}$$

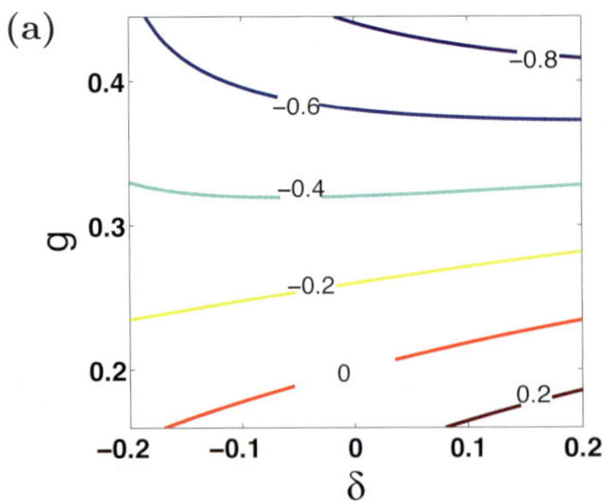

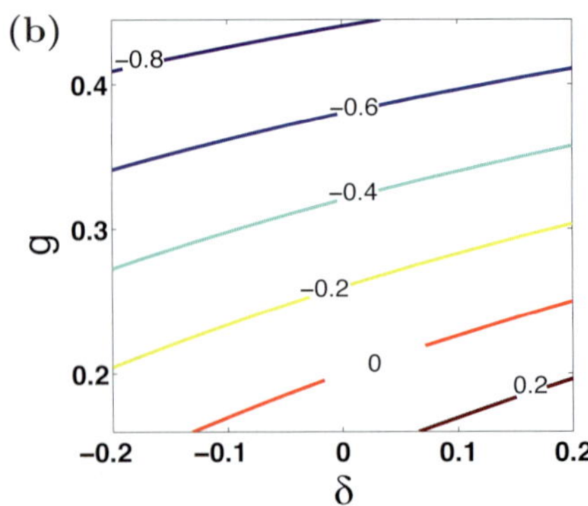

Figure 8.25: Contour plot of the parameter δ_Q as a function of δ and $g = V_{S0}^2/V_{P0}^2$ computed from (a) equation 8.76 and (b) equation 8.79 (Zhu and Tsvankin, 2006). The other parameters are Q_{33}=4, Q_{55}=8, and Q_{13}=3. The range of g-values corresponds to $1.5 < V_{P0}/V_{S0} < 2.5$.

leads to a simplified definition of δ_Q:

$$\delta_Q = 4\frac{Q_{33} - Q_{55}}{Q_{55}}\,g + 2\frac{Q_{33} - Q_{13}}{Q_{13}}\Big(1 + 2\delta - 2g\Big). \tag{8.79}$$

The values of δ_Q for a wide range of the parameters δ and g are shown in Figure 8.25. The elements Q_{33} and Q_{55} are taken from the experimental results of Gautam et al. (2003) for Rim sandstone at a frequency of 25 Hz. With this choice of the quality-factor matrix, the parameter δ_Q is quite sensitive to the squared velocity ratio g and reaches large negative values for hard rocks with $g > 0.35$. While the simpler equation 8.79 for δ_Q is sufficiently accurate for small magnitudes of both δ and g ($-0.1 < \delta < 0.2$ and $g < 0.2$), it produces a significant error for $g > 0.25$.

In combination with the Thomsen parameters for the velocity function (V_{P0}, V_{S0}, ϵ, δ, and γ), the parameters $\mathcal{A}_{P0}$, $\mathcal{A}_{S0}$, ϵ_Q, δ_Q, and γ_Q fully characterize the attenuation of P-, SV-, and SH-waves.

Approximate attenuation coefficients for P- and SV-waves

Zhu and Tsvankin (2006) derived approximate coefficients $\mathcal{A}$ of P- and SV-waves in terms of the attenuation-anisotropy parameters by assuming simultaneously:

1. Weak attenuation ($1/Q_{ij} \ll 1$).
2. Weak attenuation anisotropy ($|\epsilon_Q| \ll 1$, $|\delta_Q| \ll 1$).
3. Weak velocity anisotropy ($|\epsilon| \ll 1$, $|\delta| \ll 1$).

For P-waves, the attenuation coefficient linearized in the anisotropy parameters ϵ, δ, ϵ_Q, and δ_Q is given by

$$\mathcal{A}_P = \mathcal{A}_{P0}\left(1 + \delta_Q \sin^2\theta\cos^2\theta + \epsilon_Q \sin^4\theta\right), \tag{8.80}$$

where $\mathcal{A}_{P0}$ is defined in equation 8.68. The angle dependence of the approximate $\mathcal{A}_P$ is governed by just the attenuation-anisotropy parameters ϵ_Q and δ_Q, although δ_Q itself is influenced by velocity anisotropy. The parameter δ_Q is responsible for the attenuation coefficient in near-vertical directions, while ϵ_Q controls $\mathcal{A}_P$ near the horizontal plane. If both ϵ_Q and δ_Q go to zero, the approximate coefficient $\mathcal{A}_P$ becomes isotropic.

As is the case for SH-waves, the form of equation 8.80 coincides with that of the weak-anisotropy approximation for P-wave phase velocity (Thomsen, 1986):

$$V_P = V_{P0}\left(1 + \delta \sin^2\theta \cos^2\theta + \epsilon \sin^4\theta\right). \tag{8.81}$$

To obtain equation 8.81 from the attenuation coefficient in equation 8.80, we need only replace $\mathcal{A}_{P0}$ with V_{P0}, ϵ_Q with ϵ, and δ_Q with δ.

The SV-wave attenuation coefficient can be approximately expressed in terms of the anisotropy parameters as

$$\mathcal{A}_{SV} = \mathcal{A}_{S0}\,\frac{1 + \left(\dfrac{2\sigma}{g_Q} + \dfrac{\epsilon_Q - \delta_Q}{g\,g_Q}\right)\sin^2\theta\cos^2\theta}{1 + 2\sigma \sin^2\theta\cos^2\theta}, \tag{8.82}$$

where $\mathcal{A}_{S0}$ and g are defined in equations 8.69 and 8.78, respectively, σ is the SV-wave velocity-anisotropy parameter $[\sigma \equiv (V_{P0}^2/V_{S0}^2)(\epsilon - \delta)]$, and

$$g_Q \equiv \frac{Q_{33}}{Q_{55}}. \tag{8.83}$$

To increase accuracy, equation 8.82 is not fully linearized in the parameter σ. If $|\sigma| \ll 1$ and $|2\sigma/g_Q + (\epsilon_Q - \delta_Q)/(g\,g_Q)| \ll 1$, equation 8.82 can be simplified to

$$\mathcal{A}_{SV} = \mathcal{A}_{S0}\left(1 + \sigma_Q \sin^2\theta\cos^2\theta\right); \tag{8.84}$$

$$\sigma_Q \equiv \frac{1}{g_Q}\left[2\sigma(1 - g_Q) + \frac{\epsilon_Q - \delta_Q}{g}\right]. \tag{8.85}$$

Hence, the parameter σ_Q controls SV-wave attenuation anisotropy, including the curvature of the coefficient $\mathcal{A}_{SV}$ in the symmetry direction. Equation 8.84 can be converted into Thomsen's (1986) approximation for the SV-wave phase velocity by replacing $\mathcal{A}_{S0}$ with V_{S0} and σ_Q with σ. It should be emphasized, however, that σ_Q is a function of both attenuation-anisotropy and velocity-anisotropy parameters. Depending on the sign of σ_Q, the coefficient $\mathcal{A}_{SV}$ has either a maximum or a minimum at $\theta = 45°$.

According to approximate expression 8.80, the P-wave attenuation coefficient is isotropic (i.e., independent of angle) if

$$\epsilon_Q = \delta_Q = 0. \tag{8.86}$$

Similarly, the coefficient $\mathcal{A}_{SV}$ for SV-waves (equation 8.84) is isotropic if $\sigma_Q = 0$; that is (see equation 8.85), when

$$\epsilon_Q - \delta_Q = -2(1 - g_Q)(\epsilon - \delta) . \tag{8.87}$$

The coefficients $\mathcal{A}_P$ and $\mathcal{A}_{SV}$ are both isotropic when $\epsilon_Q = \delta_Q = 0$ and either $g_Q = 1$ or $\epsilon = \delta$ (elliptical anisotropy). If $g_Q = 1$ and $\epsilon_Q = \delta_Q = 0$, all pertinent quality-factor elements are identical:

$$Q_{11} = Q_{33} = Q_{13} = Q_{55} , \tag{8.88}$$

which makes the exact P- and SV-wave attenuation coefficients isotropic even in the presence of arbitrary velocity anisotropy. The conditions $\epsilon = \delta$ and $\epsilon_Q = \delta_Q = 0$, however, guarantee that only the approximate attenuation coefficients are directionally invariant, unless $\epsilon = \delta = 0$.

The linearized P-wave attenuation coefficient in equation 8.80 does not contain the shear-wave vertical coefficient $\mathcal{A}_{S0}$. Although this approximation becomes inaccurate with increasing magnitude of the anisotropy parameters, $\mathcal{A}_P$ remains independent of $\mathcal{A}_{S0}$ even for models with strong attenuation and pronounced velocity and attenuation anisotropy (Zhu and Tsvankin, 2006). Therefore, P-wave attenuation in VTI media is governed primarily by a reduced set of parameters: $\mathcal{A}_{P0}$, ϵ_Q, and δ_Q. A similar result is valid for the P-wave phase-velocity function, which is insensitive to the shear-wave vertical velocity V_{S0}. In contrast, SV-wave attenuation is strongly influenced by the vertical P-wave attenuation coefficient $\mathcal{A}_{P0}$ through the parameter σ_Q.

Numerical examples

The accuracy of approximate solutions 8.80, 8.82, and 8.84 is illustrated by the numerical tests in Figures 8.26, 8.28, and 8.29. The P-wave attenuation coefficient in Figure 8.26 has a maximum at an angle slightly smaller than 45° because ϵ_Q and δ_Q have opposite signs. If the signs of ϵ_Q and δ_Q coincide, $\mathcal{A}_P$ varies monotonically between the vertical and horizontal directions. The polar $\mathcal{A}_{SV}$-curve has a concave shape because σ_Q in equation 8.84 is negative and large in absolute value. Approximations 8.82 and 8.84 predict a minimum of the SV-wave attenuation coefficient at $\theta = 45°$. The extrema of the exact coefficients $\mathcal{A}$ for both P- and SV-waves in Figure 8.26, however, are somewhat shifted from their approximate positions toward the vertical axis.

The linearized expressions for the attenuation coefficients give satisfactory results for near-vertical propagation directions with angles θ up to about 30°. The error becomes noticeable for intermediate angles $30° < \theta < 75°$ and then decreases again near the horizontal plane. Note that velocity anisotropy for the model from Figure 8.26 (see Figure 8.27) is significant; the concave slowness surface of SV-waves indicates that the SV-wavefront develops a cusp centered at $\theta = 45°$ because σ reaches 0.75. The magnitude of the attenuation-anisotropy parameters is large, with δ_Q close to unity

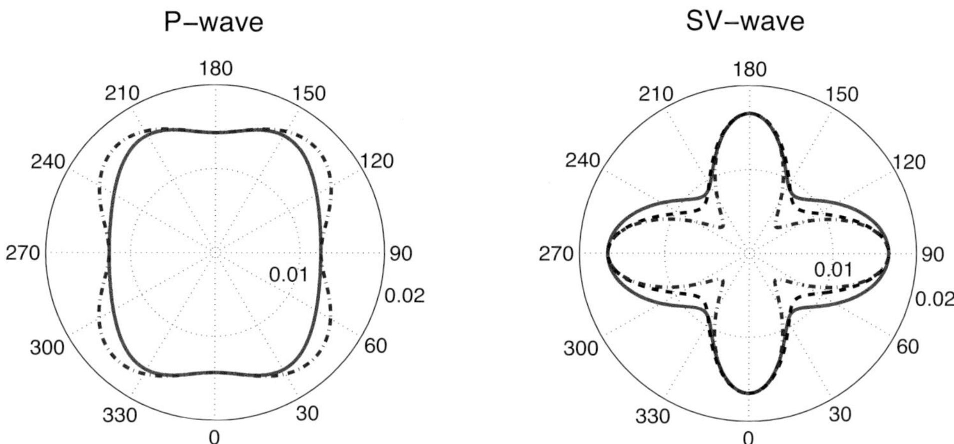

Figure 8.26: Attenuation coefficients of P-waves (left) and SV-waves (right) as functions of the phase angle, which is marked on the perimeter (Zhu and Tsvankin, 2006). The solid curves are the exact coefficients $\mathcal{A}$; the dash-dotted curves are approximations 8.80 for P-waves and 8.84 for SV-waves; the dashed curve on the right plot is the approximate SV-wave coefficient from equation 8.82. The model parameters are V_{P0}=2.42 km/s, V_{S0}=1.4 km/s, ϵ=0.4, δ=0.15, Q_{33}=35, Q_{55}=30, ϵ_Q=−0.125, and δ_Q=0.94 (δ_Q is computed from equation 8.76).

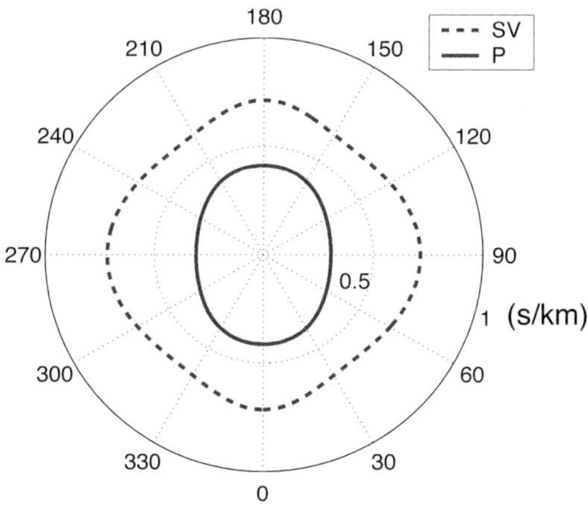

Figure 8.27: Slownesses of P- and SV-waves for the model from Figure 8.26 (Zhu and Tsvankin, 2006).

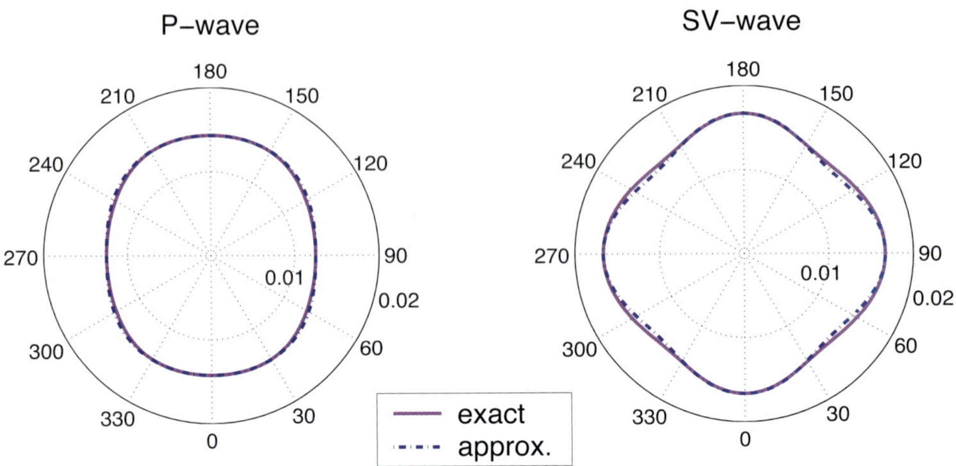

Figure 8.28: P- and SV-wave attenuation coefficients for a model with V_{P0}=2.42 km/s, V_{S0}=1.4 km/s, ϵ=0.125, δ=−0.05, Q_{33}=35, Q_{55}=30, ϵ_Q=−0.125, and δ_Q=0.05 (Zhu and Tsvankin, 2006). The solid curves are the exact coefficients $\mathcal{A}$; the dash-dotted curves are approximations 8.80 and 8.84.

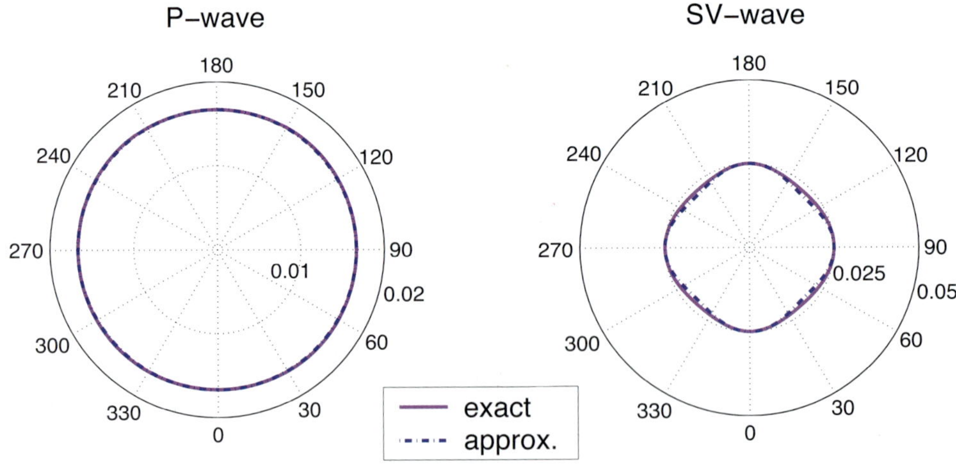

Figure 8.29: P- and SV-wave attenuation coefficients for a model with ϵ=0.4, δ=0.15, and ϵ_Q=δ_Q=0; the other parameters are the same as those in Figure 8.28 (Zhu and Tsvankin, 2006). The solid curves are the exact coefficients $\mathcal{A}$; the dash-dotted curves are approximations 8.80 and 8.84.

and $\sigma_Q = -2.93$. Equation 8.82 provides a better approximation for the SV-wave attenuation coefficient near the minimum of $\mathcal{A}_{SV}$ than does equation 8.84. Except for the vicinity of the attenuation minimum, however, the accuracy of the two equations is comparable.

For models with smaller magnitudes of the anisotropy parameters (Figure 8.28), equations 8.80 and 8.84 become sufficiently accurate over the full range of phase angles. Numerical testing shows that the error of approximate solutions 8.80, 8.82, and 8.84 is controlled primarily by the strength of *velocity anisotropy*, even if the magnitude of attenuation anisotropy is much larger. Apparently, higher-order anisotropic terms dropped from the linearized equations for $\mathcal{A}$ depend mostly on ϵ and δ. Therefore, the influence of velocity anisotropy (for fixed attenuation-anisotropy parameters) on exact attenuation coefficients might not be negligible, as further illustrated below by synthetic examples for orthorhombic media.

When $\epsilon_Q = \delta_Q = 0$ (Figure 8.29), the approximate P-wave attenuation computed from equation 8.80 is isotropic. The exact coefficient $\mathcal{A}_P$ deviates from a circle only slightly because of the influence of quadratic and higher-order terms in the anisotropy parameters. The parameter σ_Q (equation 8.85) for this model does not vanish, so the SV-wave attenuation coefficient in Figure 8.29b varies with angle.

8.5.3 Attenuation coefficients for orthorhombic media

Polar and azimuthal velocity variations in fractured orthorhombic formations are likely to be accompanied by directionally dependent attenuation. Indeed, as mentioned above, systems of aligned fractures or pores are among the most common physical reasons for attenuation anisotropy. Discussion of the physical mechanisms (in particular, wave-induced fluid flow) responsible for effective attenuation anisotropy of fractured porous media can be found in MacBeth (1999), Pointer et al. (2000), Chapman (2003, 2009), and Brajanovski et al. (2005).

The high sensitivity of P- and S-wave attenuation coefficients to anisotropy makes them potentially useful fracture-detection attributes, which can supplement azimuthally varying kinematic signatures and AVO gradients. Physical modeling indicates that the P-wave attenuation coefficient in a medium with a single system of aligned fractures or pores is highest in the direction perpendicular to the fractures (Akbar et al., 1993; Maultzsch et al., 2003a). Similar results were obtained by Hosten et al. (1987) for an orthorhombic sample made of composite material. Furthermore, Vasconcelos and Jenner (2005) and Clark et al. (2009) found substantial azimuthal variation of P-wave attenuation estimated from reflection field data. The correlation of azimuthally dependent attenuation of reflection events recorded over a fractured reservoir with horizontal permeability was discussed by Lynn et al. (1999a).

The main challenge in describing attenuation anisotropy in orthorhombic materials is the large number of independent parameters that control the attenuation coefficients. Because of the coupling between velocity and attenuation anisotropy, the phase attenuation coefficient $\mathcal{A}$ (equation 8.44) depends on nine real-valued stiffnesses and nine elements of the quality-factor matrix (for a fixed orientation of the symmetry planes). Here, we follow the work of Zhu and Tsvankin (2007) who extended Tsvankin's (1997a, 2005) notation to attenuative orthorhombic media and obtained concise, physically intuitive expressions for the attenuation coefficients.

Equivalence between attenuative orthorhombic and VTI media

Basic features of orthorhombic symmetry and Tsvankin's velocity-anisotropy parameters are described in section 2.5 and Appendix 1B. It is convenient to choose a Cartesian coordinate system aligned with the natural coordinate frame so that the coordinate planes coincide with the three mutually orthogonal symmetry planes of the medium (Figure 2.36). We assume that both the real (c_{ij}) and imaginary (c^I_{ij}) parts of the complex stiffness matrix have orthorhombic symmetry with the same orientation of the symmetry planes. Then the quality-factor matrix Q_{ij} defined in equation 8.54 has the same structure as that of the real-valued matrix c_{ij}, which governs phase velocity. Note that the physical-modeling results of Hosten et al. (1987) indicate that the symmetry of the attenuation coefficient in an orthorhombic sample closely resembles that of the velocity function.

The general Christoffel equation 8.41 takes the following form in the $[x_1, x_3]$ symmetry plane of attenuative orthorhombic media:

$$\begin{bmatrix} \tilde{c}_{11}\tilde{k}_1^2 + \tilde{c}_{55}\tilde{k}_3^2 - \rho\,\omega^2 & 0 & (\tilde{c}_{13}+\tilde{c}_{55})\tilde{k}_1\tilde{k}_3 \\ 0 & \tilde{c}_{66}\tilde{k}_1^2 + \tilde{c}_{44}\tilde{k}_3^2 - \rho\,\omega^2 & 0 \\ (\tilde{c}_{13}+\tilde{c}_{55})\tilde{k}_1\tilde{k}_3 & 0 & \tilde{c}_{55}\tilde{k}_1^2 + \tilde{c}_{33}\tilde{k}_3^2 - \rho\,\omega^2 \end{bmatrix} \begin{bmatrix} \tilde{U}_1 \\ \tilde{U}_2 \\ \tilde{U}_3 \end{bmatrix} = 0\,. \quad (8.89)$$

Equation 8.89 is valid for arbitrary orientations of the real and imaginary parts of the complex wave vector $\tilde{\mathbf{k}} = \mathbf{k} - i\,\mathbf{k}^I$ in the $[x_1, x_3]$-plane (i.e., for any inhomogeneity angle). As is the case for nonattenuative media, equation 8.89 is almost identical to the Christoffel equation for VTI media with VTI attenuation. The only difference between the two equations is that while for VTI models $\tilde{c}_{44} = \tilde{c}_{55}$, that generally does not hold for orthorhombic symmetry. The stiffness $\tilde{c}_{44}$, however, influences only the decoupled shear (SH) mode polarized perpendicular to the propagation plane $[x_1, x_3]$, while $\tilde{c}_{55}$ contributes to the velocity and attenuation of the in-plane polarized waves (P and SV).

The Christoffel equation in the symmetry planes $[x_2, x_3]$ and $[x_1, x_2]$ can be obtained from equation 8.89 by making the simple substitutions in the subscripts listed in Tsvankin (1997a, 2005). Therefore, the equivalence between the Christoffel equation in VTI media and symmetry planes of orthorhombic media holds for attenuative models as long as the real and imaginary parts of the stiffness matrix have the same symmetry. This equivalence was used by Tsvankin (1997a, 2005) to develop Thomsen-style notation for the velocity functions of P- S_1-, and S_2-waves in orthorhombic media. As shown throughout this volume, Tsvankin's parameterization significantly facilitates analytic description of seismic signatures and provides a basis for inversion and processing of wide-azimuth reflection and VSP data. Zhu and Tsvankin (2007) introduced a similar notation for attenuative orthorhombic media with the main goal of defining the parameter combinations that control the directionally dependent attenuation coefficients.

Hereafter, the real ($\mathbf{k}$) and imaginary ($\mathbf{k}^I$) parts of the wave vector are assumed to be parallel to one another, so the inhomogeneity angle is set to zero. The same assumption was made in the analysis of attenuation coefficients in VTI media. As discussed above, the inhomogeneity angle has almost no influence on the group attenuation coefficient $\mathcal{A}_g$ estimated from seismic data (Behura and Tsvankin, 2009a). According to equation 8.53, for a wide range of inhomogeneity angles $\mathcal{A}_g$ is practically equal to the normalized phase attenuation coefficient $\mathcal{A} \equiv k^I/k$ ($k = |\mathbf{k}|$; $k^I = |\mathbf{k}^I|$) evaluated for a zero inhomogeneity angle. Therefore, the projections of the wave vector in equation 8.89 can be written as

$$\tilde{k}_1 = (k - ik^I)\sin\theta \quad \text{and} \quad \tilde{k}_3 = (k - ik^I)\cos\theta\,, \tag{8.90}$$

where θ is the phase angle with the vertical.

Attenuation-anisotropy parameters

To make their notation suitable for reflection data, Zhu and Tsvankin (2007) chose the P- and S-wave vertical attenuation coefficients ($\mathcal{A}_{P0}$ and $\mathcal{A}_{S0}$) as the reference isotropic quantities. The coefficient $\mathcal{A}_{S0}$ corresponds to the shear wave polarized in the x_1-direction; it can be either the fast or slow mode depending on the relationship between the stiffnesses c_{44} and c_{55}. The coefficients $\mathcal{A}_{P0}$ and $\mathcal{A}_{S0}$ accurate to the second order in $1/Q$ are given by

$$\mathcal{A}_{P0} \equiv \frac{1}{2Q_{33}}\,, \tag{8.91}$$

$$\mathcal{A}_{S0} \equiv \frac{1}{2Q_{55}}\,. \tag{8.92}$$

To characterize the attenuation of waves propagating in the $[x_1, x_3]$-plane, it is convenient to define three attenuation-anisotropy parameters analogous to the VTI parameters ϵ_Q, δ_Q, and γ_Q introduced above. The parameter $\epsilon_Q^{(2)}$ (the superscript "2" stands for the x_2-axis perpendicular to the $[x_1, x_3]$-plane) quantifies the fractional difference between the P-wave attenuation coefficients in the x_1- and x_3-directions. The same measure of SH-wave attenuation anisotropy is denoted by $\gamma_Q^{(2)}$. The parameter $\delta_Q^{(2)}$ determines the curvature of the P-wave attenuation coefficient in the $[x_1, x_3]$-plane for $\theta = 0$.

All three attenuation-anisotropy parameters are defined below by the corresponding VTI equations 8.70, 8.74, 8.75, 8.76, and 8.79. The only change needs to be made in equation 8.70 because $\gamma_Q^{(2)}$ is expressed through Q_{44}, not Q_{55} (for VTI media, $Q_{44} = Q_{55}$).

$$\epsilon_Q^{(2)} \equiv \frac{Q_{33} - Q_{11}}{Q_{11}}\,, \tag{8.93}$$

$$\delta_Q^{(2)} \equiv \frac{1}{2\mathcal{A}_{P0}} \left. \frac{d^2 \mathcal{A}_P^{(2)}(\theta)}{d\theta^2} \right|_{\theta=0}$$

$$\simeq \frac{\dfrac{Q_{33}-Q_{55}}{Q_{55}} c_{55} \dfrac{(c_{13}+c_{33})^2}{c_{33}-c_{55}} + 2\,\dfrac{Q_{33}-Q_{13}}{Q_{13}} c_{13}(c_{13}+c_{55})}{c_{33}(c_{33}-c_{55})} \tag{8.94}$$

$$\approx 4\,\frac{Q_{33}-Q_{55}}{Q_{55}}\, g^{(2)} + 2\,\frac{Q_{33}-Q_{13}}{Q_{13}} \left(1 + 2\delta^{(2)} - 2g^{(2)}\right), \tag{8.95}$$

$$\gamma_Q^{(2)} \equiv \frac{Q_{44}-Q_{66}}{Q_{66}}, \tag{8.96}$$

where $\mathcal{A}_P^{(2)}$ is the P-wave attenuation coefficient in the $[x_1, x_3]$-plane, $g^{(2)} \equiv c_{55}/c_{33}$, and $\delta^{(2)}$ is the velocity-anisotropy parameter defined in equation 1.84. Equation 8.95 is simplified by dropping quadratic and higher-order terms in $g^{(2)}$ and $\delta^{(2)}$. According to its definition, the parameter $\delta_Q^{(2)}$ controls P-wave attenuation for near-vertical propagation in the $[x_1, x_3]$-plane. Although $\epsilon_Q^{(2)}$, $\delta_Q^{(2)}$, and $\gamma_Q^{(2)}$ are equivalent to the corresponding VTI parameters, for orthorhombic media the stiffnesses and quality-factor elements with the subscripts "55" and "44" generally differ, and cannot be interchanged.

By making the same substitutions in the subscripts ($11 \rightarrow 22$, $13 \rightarrow 23$, $55 \rightarrow 44$, and $44 \rightarrow 55$) as those in Tsvankin (1997a, 2005), equations 8.93–8.96 can be adapted to introduce the attenuation-anisotropy parameters in the $[x_2, x_3]$-plane:

$$\epsilon_Q^{(1)} \equiv \frac{Q_{33}-Q_{22}}{Q_{22}}, \tag{8.97}$$

$$\delta_Q^{(1)} \equiv \frac{1}{2\mathcal{A}_{P0}} \left. \frac{d^2 \mathcal{A}_P^{(1)}(\theta)}{d\theta^2} \right|_{\theta=0}$$

$$\simeq \frac{\dfrac{Q_{33}-Q_{44}}{Q_{44}} c_{44} \dfrac{(c_{23}+c_{33})^2}{c_{33}-c_{44}} + 2\,\dfrac{Q_{33}-Q_{23}}{Q_{23}} c_{23}(c_{23}+c_{44})}{c_{33}(c_{33}-c_{44})} \tag{8.98}$$

$$\approx 4\,\frac{Q_{33}-Q_{44}}{Q_{44}}\, g^{(1)} + 2\,\frac{Q_{33}-Q_{23}}{Q_{23}} \left(1 + 2\delta^{(1)} - 2g^{(1)}\right), \tag{8.99}$$

$$\gamma_Q^{(1)} \equiv \frac{Q_{55}-Q_{66}}{Q_{66}}, \tag{8.100}$$

where $\mathcal{A}_P^{(1)}$ is measured in the $[x_2, x_3]$-plane, $g^{(1)} \equiv c_{44}/c_{33}$, and $\delta^{(1)}$ is the velocity-anisotropy parameter defined in equation 1.81.

Because attenuation coefficients should be positive (otherwise, amplitude will increase with distance), the diagonal components of the Q_{ij}-matrix must be positive as well. This constraint implies that the parameters $\epsilon_Q^{(1)}$, $\epsilon_Q^{(2)}$, $\gamma_Q^{(1)}$, and $\gamma_Q^{(2)}$ are always larger than -1.

The quality-factor element Q_{12}, which is not involved in the above definitions, can be used to introduce a δ_Q-parameter in the $[x_1, x_2]$-plane (x_1 is treated as the symmetry axis of the equivalent VTI model):

$$\delta_Q^{(3)} \equiv \frac{1}{2\mathcal{A}_P^{(3)}(\theta=0)} \left. \frac{d^2\mathcal{A}_P^{(3)}(\theta)}{d\theta^2} \right|_{\theta=0}$$

$$\simeq \frac{\dfrac{Q_{11}-Q_{66}}{Q_{66}} c_{66} \dfrac{(c_{11}+c_{12})^2}{c_{11}-c_{66}} + 2\dfrac{Q_{11}-Q_{12}}{Q_{12}} c_{12}(c_{12}+c_{66})}{c_{11}(c_{11}-c_{66})} \tag{8.101}$$

$$\approx 4\frac{Q_{11}-Q_{66}}{Q_{66}} g^{(3)} + 2\frac{Q_{11}-Q_{12}}{Q_{12}} \left(1 + 2\delta^{(3)} - 2g^{(3)}\right). \tag{8.102}$$

The coefficient $\mathcal{A}_P^{(3)}$ is measured in the $[x_1, x_2]$-plane as a function of the phase angle θ with the x_1-axis, $g^{(3)} \equiv c_{66}/c_{11}$, and $\delta^{(3)}$ is the velocity-anisotropy parameter defined in equation 1.86. Although it is possible to add the parameters $\epsilon_Q^{(3)}$ and $\gamma_Q^{(3)}$ in the $[x_1, x_2]$-plane, they would be redundant.

The nine attenuation-anisotropy parameters defined in equations 8.91–8.102, combined with the nine velocity-anisotropy parameters (Appendix 1B), are sufficient to fully characterize both velocity and attenuation in orthorhombic media. An additional parameter of practical importance, which is responsible for the differential attenuation of the split S-waves in the vertical direction, is discussed below.

Approximate attenuation coefficients in the symmetry planes

The equivalence with VTI media means that, just as for velocity anisotropy, the symmetry-plane attenuation coefficients of all three modes can be adapted from the corresponding VTI equations. The approximate SH-wave attenuation coefficient in the $[x_1, x_3]$-plane can be found from equation 8.73 as

$$\mathcal{A}_{SH}^{(2)} = \bar{\mathcal{A}}_{S0}\left(1 + \gamma_Q^{(2)} \sin^2\theta\right), \tag{8.103}$$

where

$$\bar{\mathcal{A}}_{S0} = \frac{1}{2Q_{44}} = \mathcal{A}_{S0}\frac{1+\gamma_Q^{(1)}}{1+\gamma_Q^{(2)}} \tag{8.104}$$

is the vertical attenuation coefficient for the S-wave polarized in the x_2-direction. Equation 8.103, which is linearized in the anisotropy parameters, is obtained by replacing the parameter γ_Q in equation 8.73 by $\gamma_Q^{(2)}$ and using the appropriate vertical coefficient $\bar{\mathcal{A}}_{S0}$. Likewise, the linearized SH-wave attenuation coefficient in the $[x_2, x_3]$-plane is

$$\mathcal{A}_{SH}^{(1)} = \mathcal{A}_{S0}\left(1 + \gamma_Q^{(1)} \sin^2\theta\right). \tag{8.105}$$

It should be emphasized that the term "SH-wave" refers to two different shear modes in the vertical symmetry planes (see section 6.1). For example, if $c_{44} > c_{55}$, the fast shear wave S_1 represents an SH-wave in the $[x_1, x_3]$-plane where it is polarized in the x_2-direction. For propagation in the $[x_2, x_3]$-plane, however, the S_1-wave becomes an SV mode with in-plane polarization.

The difference between the attenuation coefficients of the vertically traveling split shear waves can be quantified by the *attenuation splitting parameter* $\gamma_Q^{(S)}$:

$$\gamma_Q^{(S)} \equiv \left| \frac{\bar{\mathcal{A}}_{S0} - \mathcal{A}_{S0}}{\mathcal{A}_{S0}} \right| = \frac{|\gamma_Q^{(1)} - \gamma_Q^{(2)}|}{1 + \gamma_Q^{(2)}} \approx |\gamma_Q^{(1)} - \gamma_Q^{(2)}|\,. \tag{8.106}$$

The definition 8.106 is analogous to that of the widely used S-wave velocity splitting parameter $\gamma^{(S)}$ (equation 1.87). The parameter $\gamma_Q^{(S)}$ is expected to play an important role in shear-wave attenuation analysis and seismic fracture characterization.

Substituting the attenuation-anisotropy parameters $\epsilon_Q^{(2)}$ and $\delta_Q^{(2)}$ into the VTI equations 8.80 and 8.84 obtained in the weak-attenuation, weak-anisotropy approximation yields the P- and SV-wave attenuation coefficients in the $[x_1, x_3]$-plane:

$$\mathcal{A}_P^{(2)} = \mathcal{A}_{P0}\left(1 + \delta_Q^{(2)} \sin^2\theta \cos^2\theta + \epsilon_Q^{(2)} \sin^4\theta\right), \tag{8.107}$$

$$\mathcal{A}_{SV}^{(2)} = \mathcal{A}_{S0}\left(1 + \sigma_Q^{(2)} \sin^2\theta \cos^2\theta\right), \tag{8.108}$$

where

$$\sigma_Q^{(2)} \equiv \frac{1}{g_Q^{(2)}}\left[2\sigma^{(2)}(1 - g_Q^{(2)}) + \frac{\epsilon_Q^{(2)} - \delta_Q^{(2)}}{g^{(2)}}\right]; \tag{8.109}$$

$g_Q^{(2)} \equiv Q_{33}/Q_{55} = \mathcal{A}_{S0}/\mathcal{A}_{P0}$, and $\sigma^{(2)} \equiv (\epsilon^{(2)} - \delta^{(2)})/g^{(2)}$. Equations 8.107 and 8.108 have the same form as the corresponding linearized phase-velocity expressions, but the dependence of the attenuation-anisotropy parameter $\delta_Q^{(2)}$ (equation 8.95) on $\delta^{(2)}$ and $\sigma_Q^{(2)}$ (equation 8.109) on $\sigma^{(2)}$ reflects the coupling between attenuation and velocity anisotropy. The linearized coefficients $\mathcal{A}_P^{(1)}$ and $\mathcal{A}_{SV}^{(1)}$ in the $[x_2, x_3]$-plane are also adapted from the VTI equations by using the attenuation-anisotropy parameters $\epsilon_Q^{(1)}$ and $\delta_Q^{(1)}$.

As in VTI media, phase velocity is practically independent of attenuation even for orthorhombic models with high attenuation and strong attenuation anisotropy (Zhu and Tsvankin, 2007). The influence of attenuation on velocity cannot be ignored only for large inhomogeneity angles. Therefore, in the absence of pronounced velocity dispersion, velocity analysis for typical orthorhombic models can be carried out without taking attenuation into account. Inversion for the attenuation-anisotropy parameters, however, requires some velocity information because the attenuation coefficient $\mathcal{A}$ estimated for a given ray (group) direction must be computed for the corresponding phase angle. Still, for models with moderate velocity anisotropy the difference between the group and phase directions does not represent a significant source of error in attenuation analysis (Behura and Tsvankin, 2009c).

Azimuthal variation of P-wave attenuation

Because of the difficulties in linearizing shear-wave velocities and attenuation coefficients for out-of-plane phase directions, here we consider only azimuthal dependence of P-wave attenuation. While the attenuation coefficients of the split shear waves can be studied for all azimuths by solving the Christoffel equation numerically, the range of validity of such plane-wave solutions in describing radiation from seismic sources is significantly reduced by distortions associated with S-wave point singularities (e.g., Crampin, 1991).

Zhu and Tsvankin (2007) obtained the following linearized P-wave attenuation coefficient outside the symmetry planes:

$$\mathcal{A}_P(\theta, \alpha) = \mathcal{A}_{P0}\left[1 + \delta_Q(\alpha)\sin^2\theta\cos^2\theta + \epsilon_Q(\alpha)\sin^4\theta\right], \tag{8.110}$$

where θ is the phase angle with the vertical (polar angle), α is the phase angle with the x_1-axis (azimuthal angle), and

$$\delta_Q(\alpha) = \delta_Q^{(1)}\sin^2\alpha + \delta_Q^{(2)}\cos^2\alpha\,, \tag{8.111}$$

$$\epsilon_Q(\alpha) = \epsilon_Q^{(1)}\sin^4\alpha + \epsilon_Q^{(2)}\cos^4\alpha + (2\epsilon_Q^{(2)} + \delta_Q^{(3)})\sin^2\alpha\cos^2\alpha\,. \tag{8.112}$$

Hence, the approximate attenuation coefficient in any vertical plane $\alpha = \text{const}$ is described by the VTI equation 8.80 with the azimuthally varying parameters $\epsilon_Q(\alpha)$ and $\delta_Q(\alpha)$. For wave propagation in the $[x_1, x_3]$-plane ($\alpha = 0°$), $\epsilon_Q = \epsilon_Q^{(2)}$, $\delta_Q = \delta_Q^{(2)}$, and equation 8.110 reduces to equation 8.107. Similarly, for the $[x_2, x_3]$-plane ($\alpha = 90°$), $\epsilon_Q = \epsilon_Q^{(1)}$ and $\delta_Q = \delta_Q^{(1)}$.

Equations 8.110 – 8.112 become identical to the linearized P-wave phase-velocity equations 1.107 – 1.109 in Tsvankin (2005) (also see our equations 8.29 and 8.30) if $\mathcal{A}_{P0}$ is replaced with V_{P0} and each attenuation-anisotropy parameter is replaced with the corresponding velocity parameter (i.e., $\delta_Q^{(i)}$ with $\delta^{(i)}$, and $\epsilon_Q^{(i)}$ with $\epsilon^{(i)}$). This similarity is explained by the same symmetry imposed on the real and imaginary parts of the stiffness matrix and by the assumption of a zero inhomogeneity angle. An important difference between the coefficient $\mathcal{A}_P$ and the phase-velocity function, however, is that the parameters $\delta_Q^{(1)}$, $\delta_Q^{(2)}$, and $\delta_Q^{(3)}$ include a contribution of velocity anisotropy. Also, as shown below, the exact coefficient $\mathcal{A}_P$ is influenced by the velocity-anisotropy parameters even for fixed values of $\delta_Q^{(1,2,3)}$.

Attenuative TI models with a vertical and horizontal symmetry axis can be treated as special cases of orthorhombic media. For VTI media with VTI attenuation, all vertical planes are identical, and there is no variation in velocity or attenuation in the horizontal (isotropy) plane. Therefore,

$$\epsilon_Q^{(1)} = \epsilon_Q^{(2)} = \epsilon_Q\,, \quad \delta_Q^{(1)} = \delta_Q^{(2)} = \delta_Q\,, \quad \delta_Q^{(3)} = 0\,. \tag{8.113}$$

Then $\epsilon_Q(\alpha) = \epsilon_Q$, $\delta_Q(\alpha) = \delta_Q$, and equation 8.110 reduces to equation 8.80.

For HTI media with the symmetry axis (for both velocity and attenuation) pointing in the x_1-direction, there are no property variations in the $[x_2, x_3]$-plane, and

$\epsilon_Q^{(1)} = \delta_Q^{(1)} = \gamma_Q^{(1)} = 0$. Also, the parameters $\delta^{(3)}$ and $\delta_Q^{(3)}$ are not independent because the $[x_1, x_2]$-plane is equivalent to the $[x_1, x_3]$-plane; in the weak-anisotropy approximation, $\delta_Q^{(3)} = \delta_Q^{(2)} - 2\epsilon_Q^{(2)}$. The azimuthally dependent attenuation-anisotropy parameters (equations 8.111 and 8.112) for HTI models become

$$\delta_Q^{\mathrm{HTI}}(\alpha) = \delta_Q^{(2)} \cos^2 \alpha \,, \tag{8.114}$$

$$\epsilon_Q^{\mathrm{HTI}}(\alpha) = \epsilon_Q^{(2)} \cos^4 \alpha + \delta_Q^{(2)} \sin^2 \alpha \cos^2 \alpha \,. \tag{8.115}$$

The linearized P-wave attenuation coefficient in equation 8.110 is independent of the parameters $\mathcal{A}_{S0}$, $\gamma_Q^{(1)}$, and $\gamma_Q^{(2)}$. Numerical tests show that this conclusion holds for the *exact* coefficient $\mathcal{A}_P$ in models with strong attenuation and pronounced velocity and attenuation anisotropy. Therefore, for a fixed orientation of the symmetry planes, P-wave attenuation is controlled primarily by the vertical coefficient $\mathcal{A}_{P0}$ and five attenuation-anisotropy parameters ($\epsilon_Q^{(1,2)}$ and $\delta_Q^{(1,2,3)}$). An equivalent result is valid for the phase-velocity function of P-waves in orthorhombic media, which is governed by the vertical velocity V_{P0} and the parameters $\epsilon^{(1,2)}$ and $\delta^{(1,2,3)}$. However, even if all attenuation-anisotropy parameters are held constant, the exact P-wave attenuation coefficient depends somewhat on $\epsilon^{(1,2)}$ and $\delta^{(1,2,3)}$ (i.e., on velocity anisotropy), as discussed below.

In Figure 8.30, the linearized equation 8.110 is compared with the exact solution for a medium with pronounced orthorhombic attenuation. Similar to the results for VTI media, the weak-anisotropy approximation is sufficiently accurate for near-vertical propagation directions with polar angles up to about 30°, but the error increases for $30° < \theta < 75°$. When the incidence plane is close to either vertical symmetry plane (i.e., the azimuth α approaches 0° or 90°), the approximate solution also gives an accurate estimate of $\mathcal{A}_P$ near the horizontal direction. The error of equation 8.110 for the full range of polar and azimuthal angles is less than 15%. Note that while the magnitude of the velocity-anisotropy parameters for this model is moderate (both $\epsilon^{(1)}$ and $\epsilon^{(2)}$ are about 0.3), the attenuation-anisotropy parameters are twice as large.

To identify the source of errors in the weak-anisotropy approximation, the test was repeated with a purely isotropic velocity model (Figure 8.31). Although the attenuation parameters were kept the same, setting the velocity-anisotropy parameters to zero practically eliminated the error of equation 8.110 in Figure 8.31. Evidently, while the linearized equation for $\mathcal{A}_P$ includes just the attenuation parameters, the exact attenuation coefficient is somewhat influenced by velocity anisotropy (even for fixed parameters $\delta_Q^{(1,2,3)}$). Furthermore, as in VTI media, the magnitude of the velocity-anisotropy parameters is almost entirely responsible for the accuracy of the weak-anisotropy approximation for $\mathcal{A}_P$. Overall, equation 8.110 gives an adequate qualitative description of P-wave attenuation for models with weak or moderate velocity anisotropy, even if attenuation anisotropy is much stronger.

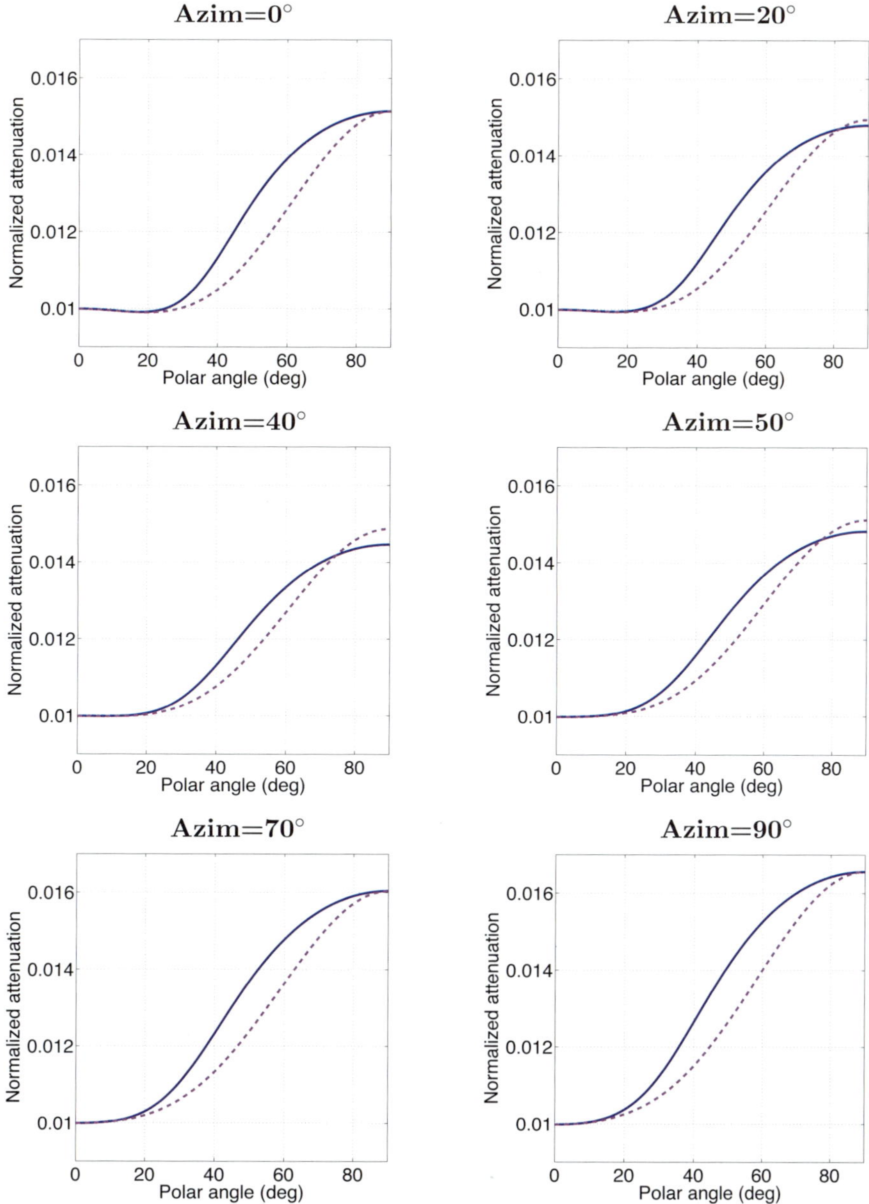

Figure 8.30: Comparison of the exact P-wave attenuation coefficient $\mathcal{A}_P$ (solid curves) computed from the Christoffel equation with the linearized approximation 8.110 (dashed) for a model with orthorhombic velocity and attenuation (Zhu and Tsvankin, 2007). The azimuth α (shown on top) is fixed for each plot. The relevant velocity parameters correspond to a system of vertical cracks embedded in a VTI background (Schoenberg and Helbig, 1997): $\epsilon^{(1)}$=0.329, $\epsilon^{(2)}$=0.258, $\delta^{(1)}$=0.083, $\delta^{(2)}$=−0.078, and $\delta^{(3)}$=−0.106. The coefficient $\mathcal{A}_{P0}$=0.01 (Q_{33}=50), and each relevant attenuation-anisotropy parameter is twice the corresponding velocity-anisotropy parameter: $\epsilon_Q^{(1)}$=0.658, $\epsilon_Q^{(2)}$=0.516, $\delta_Q^{(1)}$=0.166, $\delta_Q^{(2)}$=−0.156, and $\delta_Q^{(3)}$=−0.212.

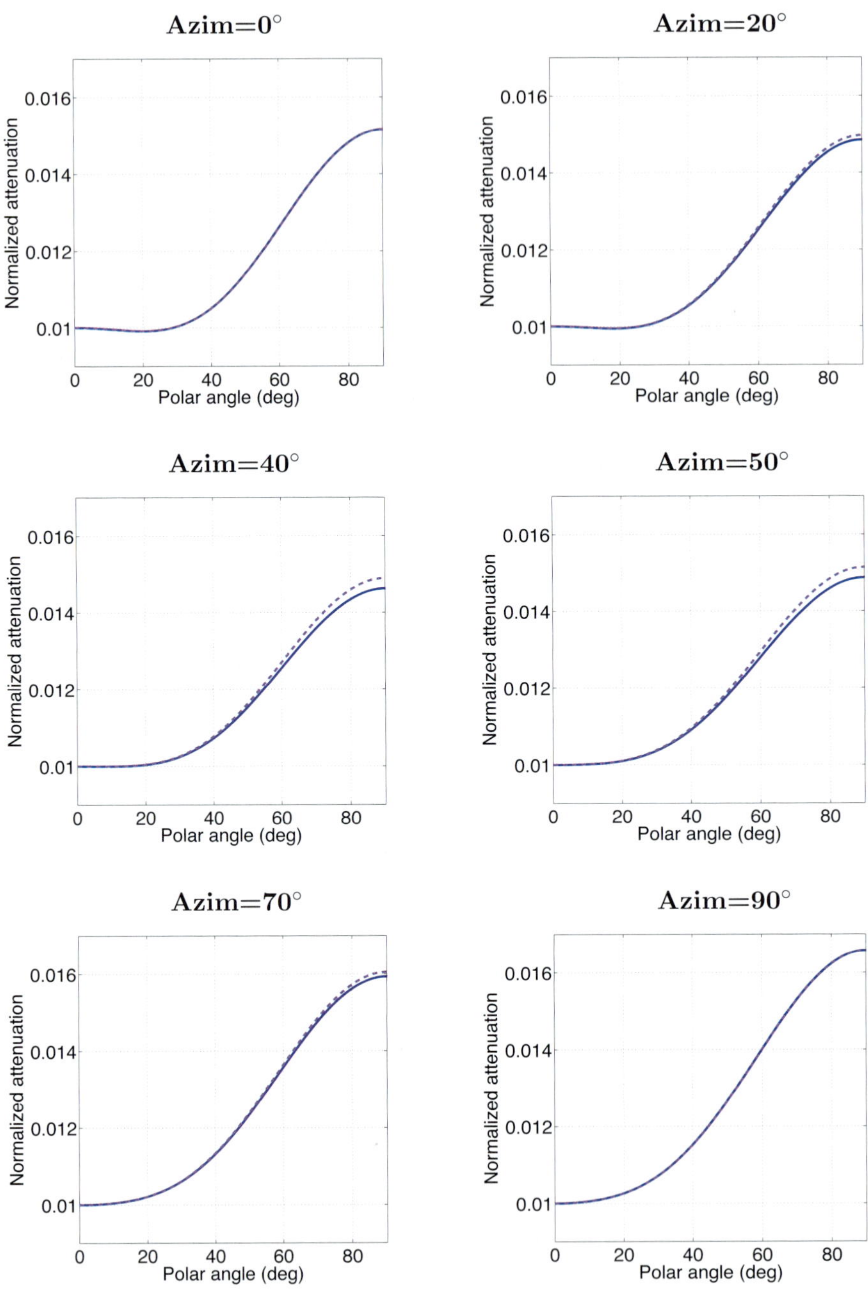

Figure 8.31: Comparison of the exact coefficient $\mathcal{A}_P$ (solid curves) with the linearized approximation 8.110 (dashed) for a model with orthorhombic attenuation but a purely isotropic velocity function (Zhu and Tsvankin, 2007). The attenuation parameters (and, therefore, the linearized coefficient $\mathcal{A}_P$) are the same as those in Figure 8.30, but all velocity-anisotropy parameters are set to zero.

8.6 Summary

In this chapter we addressed several essential aspects of amplitude processing and inversion of wide-azimuth reflection data. P-wave azimuthal AVO analysis has proved to be an efficient fracture-characterization technique because of its superior (compared to traveltime methods) vertical resolution and the high sensitivity of reflection amplitudes to the pertinent anisotropy parameters. To explain the behavior of azimuthally varying plane-wave reflection coefficients, it is convenient to apply linearized weak-contrast, weak-anisotropy approximations. For a boundary between two orthorhombic halfspaces with the same orientation of the vertical symmetry planes, the symmetry-plane reflection coefficients can be obtained by simply adapting the corresponding VTI equations. The difference between the P-wave AVO gradients in the symmetry planes strongly depends on the shear-wave splitting parameter $\gamma^{(S)}$, which makes the azimuthal AVO response a sensitive indicator of fracturing. The azimuthal variation of the AVO gradient in both orthorhombic and HTI media is also influenced by a parameter responsible for the eccentricity of the NMO ellipse. For a more detailed discussion of reflection coefficients and AVO response in azimuthally anisotropic media, we refer the reader to the comprehensive monograph by Rüger (2002). Analysis of the Green's function and properties of point-source radiation in TI and orthorhombic media can be found in Tsvankin (2005).

One of serious challenges in the implementation of anisotropic AVO analysis is reliable recovery of the reflection coefficient at the target horizon from surface data. It is particularly important to properly correct for the geometrical spreading in the overburden because variations of spreading with offset and azimuth can significantly distort the azimuthal AVO signature. The relative geometrical-spreading factor for reflected waves recorded above a horizontally layered medium can be expressed through the traveltime derivatives with respect to offset and azimuth. For stratified orthorhombic media, the derivatives of P-wave traveltime can be obtained from the Alkhalifah-Tsvankin nonhyperbolic moveout equation discussed in Chapter 3. Then the offset- and azimuth-dependent geometrical spreading is found as a function of the effective moveout parameters. This moveout-based anisotropic spreading correction (MASC) removes the influence of an anisotropic overburden on the AVO response without knowledge of the interval velocity model.

Seismic methods play an important role in characterization of tight fractured reservoirs with substantial spatial variability of fracture density. We described the case study of Xu and Tsvankin (2007), who carried out moveout and amplitude analysis of wide-azimuth P-wave data for a fractured gas-sand formation at Rulison field, Colorado. The processing sequence, designed for layered orthorhombic media, fully honored the azimuthal variation of long-offset traveltimes and amplitudes without sectoring the data. The AVO ellipse, obtained by computing the azimuthally varying AVO gradient on common-midpoint supergathers, proved to be the most sensitive fracture-detection attribute. In particular, two areas of extremely high AVO ellipticity were identified at the bottom of the reservoir. These two anomalies coincide with the intersections of wrenching fault systems, where one can expect concentration of stress. Also, the dominant fracture direction (N70W) determined from the AVO ellipses is aligned with one of the fault systems and agrees with EMI logs. This geologic evidence suggests that the AVO anomalies indeed correspond to "soft spots" of high fracture density.

The AVO response can also be used to estimate the critical angle of long-offset reflections from high-velocity layers (we call this method *seismic critical-angle reflectometry*). The main diagnostic features of the critical reflection are the rapid amplitude increase at the critical angle and the subsequent separation of the head wave. For a high-velocity orthorhombic layer beneath an isotropic overburden, the azimuthally dependent P-wave critical angle can be inverted for the symmetry-plane directions and the anisotropy parameters $\epsilon^{(1)}$, $\epsilon^{(2)}$, and $\delta^{(3)}$. Computation of the critical phase angle at the target reflector from surface data requires knowledge of the velocity field in the overburden. This complication can be avoided, however, for laterally homogeneous media by calculating the horizontal velocity in the reflecting layer directly from the ray parameter at the critical offset. For orthorhombic models, the azimuthal variation of the P-wave horizontal velocity constrains the difference $(\epsilon^{(1)} - \epsilon^{(2)})$ and the parameter $\delta^{(3)}$. Also, the combination of the symmetry-plane horizontal and NMO velocities yields the time-processing parameters $\eta^{(1)}$ and $\eta^{(2)}$ without using nonhyperbolic moveout. Synthetic testing confirmed that critical-angle reflectometry may provide valuable complementary information for seismic fracture characterization.

We concluded the chapter by presenting an analytic framework for describing and inverting the attenuation coefficients of P- and S-waves in anelastic TI and orthorhombic media. Attenuation-related amplitude decay along seismic rays is governed by the normalized phase attenuation coefficient $\mathcal{A}$ computed for a zero inhomogeneity angle (i.e., for the attenuation vector parallel to the phase direction). Analysis of the attenuation coefficient can be greatly simplified by introducing dimensionless attenuation-anisotropy parameters based on the same principle as that of Thomsen's notation. In the weak-attenuation, weak-anisotropy approximation, the coefficients $\mathcal{A}$ represented through the attenuation-anisotropy parameters have exactly the same form as the corresponding linearized phase-velocity functions. P-wave attenuation in orthorhombic models is controlled primarily by a reduced parameter set that includes the vertical attenuation coefficient $\mathcal{A}_{P0}$ and the attenuation-anisotropy parameters $\epsilon_Q^{(1,2)}$ and $\delta_Q^{(1,2,3)}$ defined by analogy with the velocity parameters $\epsilon^{(1,2)}$ and $\delta^{(1,2,3)}$. While the influence of attenuation on velocity typically is weak, the exact attenuation coefficient $\mathcal{A}_P$ for fixed values of the attenuation parameters is somewhat dependent on P-wave velocity anisotropy (i.e., on the parameters $\epsilon^{(1,2)}$ and $\delta^{(1,2,3)}$). The feasibility of estimating the attenuation-anisotropy parameters from wide-angle amplitude measurements was demonstrated by Zhu et al. (2007b) and Chichinina et al. (2009) on physical-modeling data.

Appendices for Chapter 8

8A Relative geometrical spreading as a function of reflection traveltime

The relative geometrical spreading $L(R, S)$ can be obtained in terms of the mixed second-order traveltime derivatives with respect to the source and receiver coordinates (equations 8.11 and 8.12). Here, the factor $L(R, S)$ is expressed through the azimuthally varying reflection traveltime of a pure (nonconverted) mode recorded over a laterally homogeneous medium. The final result (with a minor modification) is also valid for converted waves, as discussed in the main text.

The traveltime-derivative matrix $\mathbf{M}^{\text{mix}}$ needed to obtain $L(R, S)$ is defined in equation 8.12:

$$\mathbf{M}^{\text{mix}} = \begin{pmatrix} \dfrac{\partial^2 t}{\partial x_1^s \partial x_1^r} & \dfrac{\partial^2 t}{\partial x_1^s \partial x_2^r} \\ \dfrac{\partial^2 t}{\partial x_2^s \partial x_1^r} & \dfrac{\partial^2 t}{\partial x_2^s \partial x_2^r} \end{pmatrix}, \tag{8.116}$$

where x_1^s and x_2^s are the horizontal Cartesian coordinates of the source, and x_1^r and x_2^r are the receiver coordinates. Both sources and receivers are assumed to be located at the earth's surface (horizontal plane).

In general, $\mathbf{M}^{\text{mix}}$ is a function of four independent variables – x_1^s, x_2^s, x_1^r, and x_2^r. For laterally homogeneous media, however, pure-mode traveltime in the horizontal plane depends only on the distance x between the source and receiver and the azimuth α of the source-receiver line with respect to the x_1-axis:

$$x = \sqrt{(x_1^r - x_1^s)^2 + (x_2^r - x_2^s)^2}\,, \tag{8.117}$$

$$\alpha = \tan^{-1}\left(\frac{x_2^r - x_2^s}{x_1^r - x_1^s}\right). \tag{8.118}$$

It follows from equations 8.117 and 8.118 that in the absence of lateral heterogeneity,

$$\frac{\partial t}{\partial x_i^s} = -\frac{\partial t}{\partial x_i^r}, \quad (i = 1, 2)\,. \tag{8.119}$$

Taking equation 8.119 into account, the matrix $\mathbf{M}^{\text{mix}}$ in equation 8.116 can be rewritten as

$$\mathbf{M}^{\text{mix}} = -\begin{pmatrix} \dfrac{\partial^2 t}{\partial (x_1^r)^2} & \dfrac{\partial^2 t}{\partial x_1^r \partial x_2^r} \\ \dfrac{\partial^2 t}{\partial x_1^r \partial x_2^r} & \dfrac{\partial^2 t}{\partial (x_2^r)^2} \end{pmatrix}. \tag{8.120}$$

Since the traveltime in equation 8.120 is differentiated only with respect to the receiver coordinates, with no loss in generality the source can be placed at the origin of the coordinate system (i.e., $x_1^s = x_2^s = 0$), and equations 8.117 and 8.118 reduce to

$$x = \sqrt{(x_1^r)^2 + (x_2^r)^2}\,, \tag{8.121}$$

$$\alpha = \tan^{-1}\left(\frac{x_2^r}{x_1^r}\right). \tag{8.122}$$

For brevity, hereafter the superscript "r" is omitted.

If the traveltime t is expressed as a function of x and α, the derivatives in equation 8.120 become

$$\begin{aligned}\frac{\partial^2 t}{\partial x_i \partial x_j} &= \frac{\partial^2 t}{\partial x^2}\frac{\partial x}{\partial x_i}\frac{\partial x}{\partial x_j} + \frac{\partial t}{\partial x}\frac{\partial^2 x}{\partial x_i \partial x_j} + \frac{\partial^2 t}{\partial x \partial \alpha}\left(\frac{\partial \alpha}{\partial x_i}\frac{\partial x}{\partial x_j} + \frac{\partial \alpha}{\partial x_j}\frac{\partial x}{\partial x_i}\right) \\ &+ \frac{\partial^2 t}{\partial \alpha^2}\frac{\partial \alpha}{\partial x_i}\frac{\partial \alpha}{\partial x_j} + \frac{\partial t}{\partial \alpha}\frac{\partial^2 \alpha}{\partial x_i \partial x_j}\,, \qquad (i, j = 1, 2)\,.\end{aligned} \tag{8.123}$$

No summation over i and j in equation 8.123 is implied. The derivatives of x and α with respect to the receiver coordinates can be obtained from equations 8.121 and 8.122:

$$\frac{\partial x}{\partial x_1} = \frac{x_1}{x}\,, \quad \frac{\partial x}{\partial x_2} = \frac{x_2}{x}\,, \tag{8.124}$$

$$\frac{\partial \alpha}{\partial x_1} = -\frac{x_2}{x^2}\,, \quad \frac{\partial \alpha}{\partial x_2} = \frac{x_1}{x^2}\,, \tag{8.125}$$

$$\frac{\partial^2 x}{\partial x_1^2} = \frac{x_2^2}{x^3}\,, \quad \frac{\partial^2 x}{\partial x_2^2} = \frac{x_1^2}{x^3}\,, \quad \frac{\partial^2 x}{\partial x_1 \partial x_2} = -\frac{x_1 x_2}{x^3}\,, \tag{8.126}$$

$$\frac{\partial^2 \alpha}{\partial x_1^2} = \frac{2x_1 x_2}{x^4}\,, \quad \frac{\partial^2 \alpha}{\partial x_2^2} = -\frac{2x_1 x_2}{x^4}\,, \quad \frac{\partial^2 \alpha}{\partial x_1 \partial x_2} = \frac{x_2^2 - x_1^2}{x^4}\,. \tag{8.127}$$

It is convenient to align the source-receiver direction with the x_1-axis, which does not change the form of the final result below. Setting $x_2 = 0$ and $x_1 = x$ (assuming $x_1 > 0$) simplifies equations 8.124 – 8.127:

$$\frac{\partial x}{\partial x_1} = 1\,, \quad \frac{\partial x}{\partial x_2} = 0\,, \tag{8.128}$$

$$\frac{\partial \alpha}{\partial x_1} = 0\,, \quad \frac{\partial \alpha}{\partial x_2} = \frac{1}{x}\,, \tag{8.129}$$

$$\frac{\partial^2 x}{\partial x_1^2} = 0\,, \quad \frac{\partial^2 x}{\partial x_2^2} = \frac{1}{x}\,, \quad \frac{\partial^2 x}{\partial x_1 \partial x_2} = 0\,, \tag{8.130}$$

$$\frac{\partial^2 \alpha}{\partial x_1^2} = 0\,, \quad \frac{\partial^2 \alpha}{\partial x_2^2} = 0\,, \quad \frac{\partial^2 \alpha}{\partial x_1 \partial x_2} = -\frac{1}{x^2}\,. \tag{8.131}$$

Substituting equations 8.128 – 8.131 into equation 8.123, we obtain

$$\frac{\partial^2 t}{\partial x_1^2} = \frac{\partial^2 t}{\partial x^2}, \tag{8.132}$$

$$\frac{\partial^2 t}{\partial x_2^2} = \frac{1}{x^2}\frac{\partial^2 t}{\partial \alpha^2} + \frac{1}{x}\frac{\partial t}{\partial x}, \tag{8.133}$$

and

$$\frac{\partial^2 t}{\partial x_1 \partial x_2} = \frac{1}{x}\frac{\partial^2 t}{\partial x \partial \alpha} - \frac{1}{x^2}\frac{\partial t}{\partial \alpha}. \tag{8.134}$$

Note that $\partial x/\partial x_1$ and $\partial \alpha/\partial x_2$ in equations 8.128 and 8.129 change sign for $x_1 < 0$ (see equations 8.124 and 8.125). Nevertheless, as is straightforward to verify, equations 8.132 – 8.134 are independent of the sign of x_1 provided that traveltime is reciprocal with respect to the source and receiver positions (which is the case for pure modes).

Using equations 8.132 – 8.134, the determinant of the matrix $\mathbf{M}^{\text{mix}}$ is found from equation 8.120 as

$$\det \mathbf{M}^{\text{mix}} = \frac{1}{x^2}\frac{\partial^2 t}{\partial x^2}\left(\frac{\partial^2 t}{\partial \alpha^2} + x\,\frac{\partial t}{\partial x}\right) - \frac{1}{x^2}\left(\frac{\partial^2 t}{\partial x \partial \alpha} - \frac{1}{x}\frac{\partial t}{\partial \alpha}\right)^2. \tag{8.135}$$

Equation 8.135 makes it possible to express the relative geometrical spreading from equation 8.11 through the traveltime derivatives with respect to offset and azimuth:

$$L(x,\alpha) = (\cos\psi^s \cos\psi^r)^{1/2}\, x \left[\frac{\partial^2 t}{\partial x^2}\left(\frac{\partial^2 t}{\partial \alpha^2} + x\,\frac{\partial t}{\partial x}\right) - \left(\frac{\partial^2 t}{\partial x \partial \alpha} - \frac{1}{x}\frac{\partial t}{\partial \alpha}\right)^2\right]^{-1/2}. \tag{8.136}$$

The azimuth α contributes to the spreading factor only through $\partial t/\partial \alpha$ and $\partial^2 t/\partial \alpha^2$, which are invariant under rotation of the horizontal coordinate axes. Therefore, although equation 8.136 was derived for $\alpha = 0$, it requires no modification for arbitrary azimuth of the source-receiver line.

8B SH-wave attenuation in VTI media

Propagation of SH-waves in an attenuative VTI medium is described by the Christoffel equation 8.60, which is valid for a zero inhomogeneity angle. By introducing Thomsen's parameter γ, the imaginary part of equation 8.60 can be written as

$$k^2 - (k^I)^2 - 2Q_{55}\, b k k^I = 0, \tag{8.137}$$

where

$$b \equiv \frac{(1+2\gamma)\sin^2\theta + \cos^2\theta}{(1+2\gamma)\,\dfrac{Q_{55}}{Q_{66}}\sin^2\theta + \cos^2\theta}. \tag{8.138}$$

The physically meaningful (positive) solution of equation 8.137 is

$$\frac{k^I}{k} = \sqrt{1 + (Q_{55}\, b)^2} \; - \; Q_{55}\, b \,. \tag{8.139}$$

The real part of equation 8.60 expressed through b takes the form

$$(c_{66} \sin^2\theta + c_{55} \cos^2\theta) \left[k^2 - (k^I)^2 + \frac{2 k k^I}{Q_{55}\, b} \right] - \rho\, \omega^2 = 0 \,. \tag{8.140}$$

Substituting k^I from equation 8.139 into equation 8.140, we obtain the phase velocity of SH-waves:

$$V_{SH} = \frac{\omega}{k} = \zeta_{_Q} V_{SH}^{\text{elast}} \,, \tag{8.141}$$

where V_{SH}^{elast} is the phase velocity in a purely elastic VTI medium,

$$V_{SH}^{\text{elast}} = \sqrt{\frac{c_{66} \sin^2\theta + c_{55} \cos^2\theta}{\rho}} \,, \tag{8.142}$$

and $\zeta_{_Q}$ is the factor responsible for the influence of attenuation:

$$\zeta_{_Q} \equiv \sqrt{\frac{2 \left[\sqrt{1 + (Q_{55}\, b)^2} \; - Q_{55}\, b \right] \left[1 + (Q_{55}\, b)^2 \right]}{Q_{55}\, b}} \,. \tag{8.143}$$

For weak attenuation,

$$\zeta_{_Q} = 1 + \frac{1}{2 (Q_{55}\, b)^2} \,. \tag{8.144}$$

When $Q_{55} = Q_{66} = Q$, the quality-factor matrix is isotropic for SH-wave propagation. Then $b = 1$, and equations 8.137 and 8.139 reduce to

$$k^2 - (k^I)^2 - 2 Q k k^I = 0 \tag{8.145}$$

and

$$\frac{k^I}{k} = \sqrt{1 + Q^2} \; - \; Q \,. \tag{8.146}$$

The phase-velocity factor $\zeta_{_Q}$ (equation 8.143) for $Q_{55} = Q_{66}$ depends just on Q:

$$\zeta_{_Q} = \sqrt{\frac{2 \left(\sqrt{1 + Q^2} \; - Q \right) \left(1 + Q^2 \right)}{Q}} \,. \tag{8.147}$$

Chapter 9

Seismic characterization of natural fractures

This chapter is devoted to analysis of seismic signatures of naturally fractured rocks. Unlike the previous chapters, where we discussed estimation of seismic anisotropy without addressing its underlying physical causes in detail, here we concentrate on anisotropy induced by the presence of aligned fractures. Our motivation for emphasizing this particular reason for effective anisotropy is purely practical: fractured formations contain substantial oil and gas reserves. Production of hydrocarbons from these formations is facilitated by networks of natural fractures that might provide highly permeable pathways for fluid flow.

Current state of the art of seismic characterization of fractured reservoirs can be described as follows:

1. Understanding of the influence of small-scale fractures on seismic signatures is based on so-called effective media theories designed to replace a microheterogeneous rock volume containing fractures with a homogeneous one that has exactly the same overall (or effective) elastic properties in the regime of static deformation.

2. The existing industry methodologies for seismic characterization of fractured rocks are largely limited to the simplest model of a single set of parallel, penny-shaped, vertical cracks embedded in an otherwise isotropic background. The anisotropy parameters of this effective medium, which is transversely isotropic with a horizontal symmetry axis (HTI), can be estimated from reflection data and related to the fractures. Seismic signatures that can be inverted for the HTI parameters include shear-wave splitting (e.g., Gaiser and Van Dok, 2005), P- and S-wave NMO ellipses (Tsvankin, 1997b; Bakulin et al., 2000a; Berthet et al., 2004), and azimuthal AVO response (Bakulin et al., 2000a; del Monte et al., 2004; Gray and Todorovic-Marinic, 2004; Minsley et al., 2004; Todorovic-Marinic et al., 2004; Bachrach et al., 2009).

The choice of static effective media schemes for the theoretical framework reflects the fact of the significant difference in size between typical seismic wavelengths (10^{+1} to 10^{+2} m) and the lengths of fractures (10^{-2} to 10^{-1} m) that contribute to production. The assumption of a single fracture set responsible for fluid flow, however, is often inadequate because numerous observations around the globe reveal the presence of multiple open fracture systems (e.g, Laubach et al., 2004; Narr et al., 2006).

In this chapter, following Grechka and Kachanov (2006b) and Grechka (2009), we address the main aspects of seismic fracture characterization. We begin with an overview of effective media theories for solids containing multiple, differently oriented fracture sets. The overview will naturally lead us to:

1. Analysis of the accuracy of two popular effective media schemes proposed by Schoenberg (1980) and Hudson (1980).
2. Realization that some geometric features of fractures cannot be inferred from the effective properties.
3. Identification of the fracture properties constrained by wide-azimuth seismic reflection data.

We conclude the chapter with a case study of seismic characterization of multiple fracture sets using multicomponent, wide-azimuth reflection data.

9.1 Effective media theories for fractured models

9.1.1 Basic concepts and definitions

A standard point of departure for the discussion of mechanics of microheterogeneous media (e.g., Nemat-Nasser and Hori, 1999; Milton, 2002) is introduction of the representative volume element (RVE). On an intuitive level, the RVE can be understood as follows.

Consider a piece of rock with linear size L and volume $\mathcal{V}$. Any rock volume usually contains small-scale heterogeneities such as pores, cracks, and individual grains, whose characteristic lengths are denoted by ℓ. The elastic properties of the medium can be described by the spatially variable stiffness tensor $\mathbf{c}(\mathbf{x})$, where $\mathbf{x}$ is the Cartesian coordinate vector. Given the many heterogeneities in the examined rock volume, we shall assume that $\ell \ll L$.

The rock itself, however, appears statistically homogeneous (see Ostoja-Starzewski, 2008, for the definition and detailed discussion of this term) in the sense that the displacement field of a long seismic wave (i.e., the wavelength $\Lambda \gg L$) is nearly constant inside $\mathcal{V}$. Furthermore, if our microheterogeneous rock volume is RVE, this large-scale displacement field remains unchanged when we replace this particular volume with a different one that has the same (constant) average elastic properties or, more precisely, the same effective stiffness tensor $\mathbf{c}_e$. Thus, we can say that $\mathbf{c}_e$ "homogenizes" the original stiffness tensor $\mathbf{c}(\mathbf{x})$. In addition, the effective stiffness tensor $\mathbf{c}_e$ of the RVE is guaranteed to be independent of physical measurements that have to be performed on $\mathbf{c}(\mathbf{x})$ to obtain $\mathbf{c}_e$ (Huet, 1990).

Consequently, the first goal of homogenization or effective media theories is to derive the tensor $\mathbf{c}_e$ from $\mathbf{c}(\mathbf{x})$. The second, more practical goal is to determine what information about $\mathbf{c}(\mathbf{x})$ can be inferred from either the full tensor $\mathbf{c}_e$ or certain combinations of its components, which can be estimated, for example, from seismic data. Both issues are discussed below.

The effective stiffness tensor $\mathbf{c}_e$ of a heterogeneous solid relates the stress $\boldsymbol{\tau}$ and the strain $\boldsymbol{\varepsilon}$ tensors averaged over the RVE via Hooke's law,

$$\tau_{ij} = c_{e,ijkl}\,\varepsilon_{kl}\,, \quad (i,\, j = 1,\, 2,\, 3)\,. \tag{9.1}$$

Although Hooke's law is applicable to both static and dynamic elastic processes, here we use it only in the static regime. This regime corresponds to probing the RVE with infinitely long waves, that is, with $\Lambda = \infty$.

Equation 9.1 can be represented in an equivalent form,

$$\varepsilon_{ij} = s_{e,ijkl}\,\tau_{kl}\,, \quad (i,\, j = 1,\, 2,\, 3)\,, \tag{9.2}$$

where $\mathbf{s}_e$ is the compliance tensor defined as the inverse of the stiffness tensor:

$$\mathbf{s}_e \equiv (\mathbf{c}_e)^{-1}\,. \tag{9.3}$$

Formulation 9.2 turns out to be more appropriate for fractured solids because cracks represent sources of extra strain. This can be made explicit by splitting the tensor $\mathbf{s}_e$ into the background compliance $\mathbf{s}_b$ and the fracture-related compliance $\Delta\mathbf{s}$:

$$\mathbf{s}_e \equiv \mathbf{s}_b + \Delta\mathbf{s}\,. \tag{9.4}$$

Then equation 9.2 can be rewritten as

$$\varepsilon_{ij} = s_{b,ijkl}\,\tau_{kl} + \Delta s_{ijkl}\,\tau_{kl} = s_{b,ijkl}\,\tau_{kl} + \Delta\varepsilon_{ij}\,, \quad (i,\, j = 1,\, 2,\, 3)\,, \tag{9.5}$$

where

$$\Delta\varepsilon_{ij} = \Delta s_{ijkl}\,\tau_{kl}\,, \quad (i,\, j = 1,\, 2,\, 3)\,, \tag{9.6}$$

is the extra strain induced by the cracks.

The strain $\Delta\boldsymbol{\varepsilon}$ can be expressed through the displacement discontinuities across the crack surfaces that have areas S. For flat cracks, this results in (Vavakin and Salganik, 1975)

$$\Delta\varepsilon_{ij} = \frac{1}{2\mathcal{V}} \sum_{(k)} \left[(n_i B_j + B_i n_j)\, S\right]^{(k)}, \quad (i,\, j = 1,\, 2,\, 3)\,, \tag{9.7}$$

where $\mathbf{B}$ is the displacement-discontinuity vector averaged over the crack area, $\mathbf{n}$ is the unit normal to the crack face, and the sum is taken over all cracks (k) in volume $\mathcal{V}$. Thus, obtaining the effective compliance is reduced to finding the vectors $\mathbf{B}^{(k)}$.

9.1.2 Noninteraction approximation

Up to this point, our discussion has been fairly general. It involved defining a crack as a flat inhomogeneity and relying on the linearity of Hooke's law. To proceed further, however, we need to make certain assumptions. Indeed, the displacement-discontinuity vector $\mathbf{B}^{(k)}$ at the kth fracture is a function of the local tractions that depend on the positions and orientations of all adjacent cracks because their presence might create complicated local stress fields (see, for example, Figures 9.3 and 9.4 below). Therefore, strictly speaking, complete microstructural information is required to account for the contribution of each individual fracture to the effective elasticity. Such detailed information, however, is unavailable in most practical applications. To circumvent this difficulty, we make an important assumption that interactions of different fractures in the stress fields can be ignored so that each crack "senses" only the far-field stress $\boldsymbol{\tau}$. This assumption, first applied to fractured solids by Bristow (1960), leads to the so-called *noninteraction approximation* (NIA).

Several other approximations, reviewed by Grechka and Kachanov (2006b), have been proposed to account for fracture interactions. They typically assume noninteracting cracks to be placed into either the effective elastic matrix or the effective stress field. Here we concentrate on the NIA because, being the simplest scheme, it accurately describes the influence of thin fractures (but not round pores) on the effective elasticity (e.g., see Figures 9.5 – 9.7 below). In the NIA, the displacement-discontinuity vector $\mathbf{B}$ is given in terms of the symmetric, second-rank excess fracture-compliance tensor $\mathbf{Z}$ (Schoenberg, 1980; Kachanov, 1992). This tensor relates the vector $\mathbf{B}$ to the uniform traction $\mathbf{T} = \mathbf{n} \cdot \boldsymbol{\tau}$ induced at the crack face by the remotely applied stress $\boldsymbol{\tau}$,

$$B_i = n_l \, \tau_{lj} \, Z_{ji} \, , \quad (i = 1, 2, 3) \, . \tag{9.8}$$

The eigenvectors of the symmetric, positive definite tensor $\mathbf{Z}$ are the principal compliance directions of a flat crack that has arbitrary shape. For a purely isotropic host material, one of the eigenvectors coincides with the crack normal $\mathbf{n}$, while the other two lie in the crack plane (Sevostianov and Kachanov, 2002). The eigenvalue Z_N of the tensor $\mathbf{Z}$ corresponding to the eigenvector $\mathbf{n}$ is called the *normal* excess crack compliance; the other two eigenvalues Z_{T1} and Z_{T2} are known as the *tangential* or *shear* excess crack compliances.

9.1.3 Fracture types

Scalar cracks

The simplest and yet most important special case is when the $\mathbf{Z}$-tensor is diagonal,

$$Z_{ij} = \mathcal{Z} \, \delta_{ij} \, , \quad (i, j = 1, 2, 3) \, , \tag{9.9}$$

where $\mathcal{Z}$ is a positive scalar and δ_{ij} is Kronecker's delta. Schoenberg and Sayers (1995) call such cracks *scalar*. All excess compliances of scalar cracks are equal, $Z_N = Z_{T1} = Z_{T2} = \mathcal{Z}$.

The scalar nature of the excess fracture compliance tensor $\mathbf{Z}$ has far-reaching implications for the effective elastic properties. To understand them, we substitute equations 9.8 and 9.9 into equation 9.7:

$$\Delta\varepsilon_{ij} = \left[\frac{1}{\mathcal{V}} \sum_{(k)} \left(\mathcal{Z} n_i n_l S \right)^{(k)} \right] \tau_{lj} \, , \quad (i, j = 1, 2, 3) \, . \tag{9.10}$$

Comparison of equations 9.6 and 9.10 helps identify the factor

$$\alpha_{il}^{\text{scal}} = \frac{1}{\mathcal{V}} \sum_{(k)} (\mathcal{Z} n_i n_l S)^{(k)}, \quad (i, l = 1, 2, 3),$$

which contains only the crack-related parameters, as the *crack-density* tensor. It is symmetric, positive-definite, and has second rank. Therefore, the symmetry of an originally isotropic material containing scalar fractures cannot be lower than that of the crack-density tensor itself. Because tensor $\boldsymbol{\alpha}^{\text{scal}}$ has three orthogonal eigenvectors, it possesses orthorhombic symmetry. This observation allowed Kachanov (1980) to draw an important conclusion: a purely isotropic solid with any orientational distribution of scalar cracks is orthorhombic. Also, because any symmetric, positive-definite, second-rank tensor describes an ellipsoid in three-dimensional space, the crack-induced orthotropy is always elliptical and, hence, is characterized by fewer than nine independent constants of general orthorhombic media.

Dry penny-shaped cracks

For circular (or penny-shaped) dry cracks, equation 9.9 is satisfied only approximately, with accuracy dependent on the background Poisson's ratio ν_b. This is clear from the eigenvalues of the tensor $\mathbf{Z}$ (Kachanov, 1992; 1993),

$$Z_N = \frac{16a(1-\nu_b^2)}{3\pi E_b} \quad \text{and} \quad Z_{T1} = Z_{T2} = \frac{Z_N}{1-\nu_b/2} \equiv Z_T, \tag{9.11}$$

where E_b is Young's modulus of the host rock and a is the crack radius. The normal Z_N and tangential Z_T excess fracture compliances are relatively close to each other because the Poisson's ratio ν_b (which is usually positive) satisfies the inequality $\nu_b \leq 1/2$. The difference between Z_N and Z_T leads to the introduction of two fracture-related tensors (Kachanov, 1980),

$$\alpha_{ij} = \frac{1}{\mathcal{V}} \sum_{(k)} \left(a^3 n_i n_j\right)^{(k)}, \quad (i, j = 1, 2, 3) \tag{9.12}$$

and

$$\beta_{ijlm} = -\frac{\nu_b}{2} \frac{1}{\mathcal{V}} \sum_{(k)} \left(a^3 n_i n_j n_l n_m\right)^{(k)}, \quad (i, j, l, m = 1, 2, 3). \tag{9.13}$$

The second-rank crack-density tensor $\boldsymbol{\alpha}$ and the fourth-rank tensor $\boldsymbol{\beta}$ contain all information about the crack distribution over orientations and sizes relevant for obtaining the effective properties in the noninteraction approximation. The tensor $\boldsymbol{\alpha}$ can be viewed as a natural tensorial extension of the scalar crack density,

$$e = \frac{1}{\mathcal{V}} \sum_{(k)} \left(a^3\right)^{(k)} \equiv \alpha_{ii}, \tag{9.14}$$

which was introduced by Bristow (1960). It is important to note that the aspect ratio Θ (i.e., the ratio of the width and diameter) of the cracks does not contribute to equations 9.12 and 9.13, implying that the effective properties of solids with dry fractures are almost independent of Θ (provided that $\Theta \ll 1$). Consequently, for thin dry cracks, the crack-related porosity has virtually no influence on the effective elasticity.

According to equations 9.11, dry circular cracks become scalar (in the terminology of Schoenberg and Sayers, 1995) only when the background Poisson's ratio ν_b goes to zero. Although in reality $\nu_b \neq 0$, the influence of the tensor $\boldsymbol{\beta}$ on the effective properties is still relatively minor because its magnitude is significantly smaller than that of $\boldsymbol{\alpha}$ due to the presence of the multiplier $\nu_b/2 \leq 1/4$ in equation 9.13. Therefore, neglecting $\boldsymbol{\beta}$ and retaining $\boldsymbol{\alpha}$ as the sole crack-density factor constitutes a reasonable approximation. This supports the above-stated conclusion drawn for the (unrealistic) model of scalar fractures: a solid with arbitrarily oriented, circular, dry cracks is nearly orthorhombic. This result is confirmed by the numerical studies of Grechka and Kachanov (2006a; 2006c).

The possibility of describing the effective elasticity in terms of just $\boldsymbol{\alpha}$ results in the following important medium properties:

1. The overall influence of multiple, differently oriented, dry fracture sets is indistinguishable from that of three orthogonal or *principal* sets.

2. The normals to those principal sets coincide with the eigenvectors of the tensor $\boldsymbol{\alpha}$ (equation 9.12). The corresponding principal crack densities e_i ($i = 1, 2, 3$) are the eigenvalues of $\boldsymbol{\alpha}$.

3. The crack-induced orthotropy is elliptical. Furthermore, it is controlled by only four independent quantities [the combinations of E_b, ν_b, and the eigenvalues of $\boldsymbol{\alpha}$; see Kachanov (1980; 1993) for details] rather than nine for general orthorhombic media.

Using the eigenvalues 9.11 of the excess fracture-compliance tensor yields the following expression for Δs_{ijkl} (Kachanov, 1980; Sayers and Kachanov, 1995; Schoenberg and Sayers, 1995):

$$\Delta s_{ijlm} = \frac{8(1-\nu_b^2)}{3E_b(2-\nu_b)} \left(\alpha_{il}\delta_{jm} + \alpha_{im}\delta_{jl} + \alpha_{jl}\delta_{im} + \alpha_{jm}\delta_{il} + 4\beta_{ijlm}\right), \quad (i,\, j,\, l,\, m = 1,\, 2,\, 3)\,. \tag{9.15}$$

Either ignoring the tensor $\boldsymbol{\beta}$ in equation 9.15 or approximating its components as

$$\beta_{ijlm} \approx -\nu_b \left(\alpha_{il}\delta_{jm} + \alpha_{im}\delta_{jl} + \alpha_{jl}\delta_{im} + \alpha_{jm}\delta_{il}\right)/8$$

makes the effective orthorhombic medium elliptical.

Liquid-filled cracks

The results for dry cracks discussed above turn out to be applicable to a wide variety of mixtures of liquids and gases inside the fractures. Indeed, long seismic waves are known to be sensitive to the most compliant portion of rock (e.g., Jaeger et al.,

2007), which would be the gaseous phase if it is present. Hence, as long as the magnitude of elastic dilatation is below the gas saturation, fractures are expected to behave as if they were dry, regardless of the composition of the actual infill. Because strains associated with propagation of seismic waves typically do not exceed 10^{-6} (e.g., Shearer, 2009), a small fraction of gas in otherwise liquid-filled cracks can control the effective compliance. Nevertheless, a significant body of research has been devoted to the effective properties of fractures filled with a *single* liquid phase. We make the same assumption of full saturation with a liquid when referring to "liquid-filled" fractures.

The influence of liquid infill on the overall compliance was first examined by O'Connell and Budiansky (1974) and Budiansky and O'Connell (1976) for cracks with the same aspect ratio. Shafiro and Kachanov (1997) extended their analysis to fractures that have variable aspect ratios and showed that equation 9.15 for liquid-filled fractures takes the form

$$\Delta s_{ijlm} = \frac{8(1-\nu_b^2)}{3E_b(2-\nu_b)}\left(\alpha_{il}\delta_{jm} + \alpha_{im}\delta_{jl} + \alpha_{jl}\delta_{im} + \alpha_{jm}\delta_{il} + 4\beta'_{ijlm}\right), \qquad (9.16)$$

$$(i,\, j,\, l,\, m = 1,\, 2,\, 3)\,,$$

where

$$\beta'_{ijlm} = \beta_{ijlm} - \frac{1}{\mathcal{V}}\left(1 - \frac{\nu_b}{2}\right)\sum_{(k)}\left(\varsigma a^3 n_i n_j n_l n_m\right)^{(k)}, \quad (i,\, j,\, l,\, m = 1,\, 2,\, 3)\,. \qquad (9.17)$$

The dimensionless parameter

$$\varsigma^{(k)} = \frac{1}{1 + \Theta^{(k)}\left[E_b/K_f - 3\left(1 - 2\nu_b\right)\right]} \qquad (9.18)$$

is called the *fluid factor*. The magnitude of $\varsigma^{(k)}$ is governed by the bulk modulus K_f of the fluid and the crack aspect ratio $\Theta^{(k)}$.

The structure of equations 9.17 and 9.18 provides important clues for understanding the influence of fluids on the effective elasticity. The minus sign after β_{ijlm} in equation 9.17 confirms an intuitively obvious fact that the presence of fluids stiffens the fractures. Next, the fluid factors given by equation 9.18 are bounded by $0 \le \varsigma^{(k)} \le 1$. Near-zero values of $\varsigma^{(k)}$ correspond to dry fractures ($K_f/E_b \approx 0$), while $\varsigma^{(k)} \approx 1$ indicate either thin cracks with small $\Theta^{(k)}$ or a relatively stiff fluid infill, or both. Increasing the aspect ratio of cracks with stiff infill reduces the corresponding fluid factor. As is clear from equation 9.17, such "fat" isolated fractures soften the rock and behave similarly to dry ones, pointing to the importance of aspect ratio to the effective elasticity. This analysis shows that the crack-density tensor $\boldsymbol{\alpha}$ alone is no longer sufficient for describing the effective medium properties. The additional (and essential) microstructural parameters are captured by the fourth-rank tensor $\boldsymbol{\beta}'$ (equation 9.17).

Equations 9.12 and 9.17 suggest that the norms of the tensors $\boldsymbol{\alpha}$ and $\boldsymbol{\beta}'$ can be comparable when the fluid factors are not small. Therefore, the tensor $\boldsymbol{\beta}'$ cannot be ignored as could tensor $\boldsymbol{\beta}$ so, in principle, the effective symmetry might be lower than orthorhombic. Nevertheless, the deviation from orthotropy because of the stiffening influence of the fluid infill remains small: the tensor $\boldsymbol{\beta}'$ counteracts the contribution of $\boldsymbol{\alpha}$ and reduces the overall influence of fractures. The net result is that the magnitude of crack-induced anisotropy governing P-wave propagation decreases, and orthotropy still represents a good approximation because of the proximity of the effective elasticity (at least, for P-waves) to isotropy (Grechka and Kachanov, 2006a).

Pore-like cracks

For completeness, we mention an extension of the noninteraction approximation to ellipsoidal pore-like fractures whose aspect ratios are not necessarily small. The contribution $\Delta\mathbf{s}$ of each fracture of this type to the effective compliance (equation 9.4) is given in terms of Eshelby's (1957) tensor $\boldsymbol{\mathcal{S}}$ as (Walpole, 1969; Kachanov et al., 2003)

$$\Delta\mathbf{s} = \phi\left[(\mathbf{s}_i - \mathbf{s}_b)^{-1} + \mathbf{c}_b : (\boldsymbol{\mathcal{J}} - \boldsymbol{\mathcal{S}})\right]^{-1}, \tag{9.19}$$

where ϕ is the crack porosity (the fraction of volume $\mathcal{V}$ occupied by the crack), $\mathbf{s}_i$ is the compliance tensor of the (generally anisotropic) infill material, $\mathbf{c}_b = \mathbf{s}_b^{-1}$ is the background stiffness tensor, and $\boldsymbol{\mathcal{J}} \equiv \mathcal{J}_{ijlm} = (\delta_{il}\,\delta_{jm} + \delta_{im}\,\delta_{jl})/2$ is the fourth-rank identity tensor; the tensor $\boldsymbol{\mathcal{S}}$ is described in Appendix 9A. The colon in equation 9.19 denotes the double dot product (contraction over two indexes) defined as $\mathbf{c}_b : (\boldsymbol{\mathcal{J}} - \boldsymbol{\mathcal{S}}) \equiv c_{b,\,jkrq}\,(\mathcal{J}_{rqlm} - \mathcal{S}_{rqlm})$.

While equation 9.19 is used below to illustrate the influence of a nonzero crack aspect ratio Θ on the effective properties, we caution the reader that the NIA loses its accuracy for relatively "fat" cracks. Specifically, equation 9.19 can give noticeable errors for $\Theta \geq 0.1$.

9.1.4 Linear-slip and Hudson's theories

The noninteraction approximation and the linear-slip theory of Schoenberg (1980) give identical expressions in terms of the **Z**-tensor for dry fractures. The difference is that the former expresses the effective elastic constants through geometric crack-density parameters (at least, for circular and ellipsoidal cracks), whereas the latter lacks a direct link to the microstructure (Schoenberg, 1980; Schoenberg and Sayers, 1995).

For liquid-filled fractures, the linear-slip compliance tensor $\Delta\mathbf{s}$ differs from that given by equation 9.16. Schoenberg and Douma (1988) applied the linear-slip theory to liquid-filled fractures by inverting the effective stiffnesses obtained from Hudson's (1981) formalism. We do not reproduce Schoenberg and Douma's (1988) equations here because the quantitative difference between them and equation 9.16 is small due to the weak overall influence of liquid-filled fractures on the effective properties.

In contrast to the noninteraction approximation that expresses compliances as linear functions of the crack density e, Hudson (1980) operates with the effective stiffnesses $\mathbf{c}_e$. He constructs each stiffness element as a power series in e, which is truncated after either the linear (first-order theory) or quadratic (second-order theory) term.

The first-order theory of Hudson (1980; 1981) has the form

$$\mathbf{c}_e = \mathbf{c}_b + \Delta\mathbf{c}\, e\,. \tag{9.20}$$

It represents the so-called dilute limit obtained by applying the NIA to the compliance tensor and then linearizing the inverted compliances with respect to the crack density. For a single set of penny-shaped fractures with crack density e and normal $\mathbf{n}=\mathbf{x}_1$, the tensor $\Delta\mathbf{c}$ in equation 9.20 is given by

$$\Delta\mathbf{c} = -\frac{1}{\mu_b} \tag{9.21}$$
$$\times \begin{pmatrix} (\lambda_b + 2\mu_b)^2\mathcal{U}_{33} & \lambda_b(\lambda_b + 2\mu_b)\mathcal{U}_{33} & \lambda_b(\lambda_b + 2\mu_b)\mathcal{U}_{33} & 0 & 0 & 0 \\ \lambda_b(\lambda_b + 2\mu_b)\mathcal{U}_{33} & \lambda_b^2\mathcal{U}_{33} & \lambda_b^2\mathcal{U}_{33} & 0 & 0 & 0 \\ \lambda_b(\lambda_b + 2\mu_b)\mathcal{U}_{33} & \lambda_b^2\mathcal{U}_{33} & \lambda_b^2\mathcal{U}_{33} & 0 & 0 & 0 \\ 0 & 0 & 0 & 0 & 0 & 0 \\ 0 & 0 & 0 & 0 & \mu_b^2\mathcal{U}_{11} & 0 \\ 0 & 0 & 0 & 0 & 0 & \mu_b^2\mathcal{U}_{11} \end{pmatrix},$$

where λ_b and μ_b are the Lamé parameters of the host rock, and the factors $\mathcal{U}_{11}$ and $\mathcal{U}_{33}$ are (Hudson, 1980; 1981; Peacock and Hudson, 1990):

$$\mathcal{U}_{11} = \frac{16}{3(3-2g_b)(1+\mathcal{M})}, \qquad \mathcal{U}_{33} = \frac{4}{3(1-g_b)(1+\mathcal{K})}, \tag{9.22}$$

$$\mathcal{M} = \frac{4\mu_i}{\pi\Theta(3-2g_b)\,\mu_b}, \qquad \mathcal{K} = \frac{\lambda_i + 2\mu_i}{\pi\Theta(1-g_b)\,\mu_b}, \tag{9.23}$$

$$g_b = \frac{V_{S,b}^2}{V_{P,b}^2} = \frac{\mu_b}{\lambda_b + 2\mu_b} = \frac{1-2\nu_b}{2(1-\nu_b)}\,. \tag{9.24}$$

Here, $V_{P,b}$ and $V_{S,b}$ are the P- and S-wave velocities of the background, λ_i and μ_i are the Lamé parameters of the isotropic infill, and all cracks are assumed to have the same aspect ratio Θ. If the model includes several differently oriented fracture sets, their stiffness contributions $\Delta\mathbf{c}^{(\ell)}$ are simply added (Hudson, 1981),

$$\Delta\mathbf{c} \to \sum_{(\ell)} \Delta\mathbf{c}^{(\ell)}\,. \tag{9.25}$$

Substitution 9.25 is insensitive to the spatial distribution of fractures, as it should be in the noninteraction approximation.

The second-order Hudson's theory (1980; 1991) extends the linear approximation 9.20 by adding a term quadratic in the crack density:

$$\mathbf{c}_e = \mathbf{c}_b + \Delta\mathbf{c}\, e + \Delta\Delta\mathbf{c}\, e^2 \,, \tag{9.26}$$

where

$$\Delta\Delta c_{ijkl} = \frac{1}{\mu_b}\, \Delta c_{ijmn}\, \chi_{mnqr}\, \Delta c_{qrkl} \,, \tag{9.27}$$

$$\chi_{ijkl} = \frac{1}{15}\Big[\delta_{ik}\delta_{jl}(4+g_b) - (\delta_{il}\delta_{jk} + \delta_{ij}\delta_{kl})\,(1-g_b)\Big], \quad (i,\, j,\, k,\, l = 1,\, 2,\, 3)\,; \tag{9.28}$$

g_b is defined in equation 9.24. Note that the second-order term in equation 9.27 is constructed from the first-order term (equation 9.21) without bringing in any additional information about the fractures.

9.2 Comparison of theoretical predictions

It should be expected that the linear-slip theory and Hudson's first- and second-order theories predict different values for effective properties. Before we compare these theories quantitatively, it is instructive to gain some insight into which of them is likely to be more accurate. For simplicity, we consider a single fracture set with the normal $\mathbf{n}$ parallel to the coordinate axis x_1 and analyze Hudson's (1980) power-series expansion, for instance, of the effective stiffness coefficient $c_{e,11}(e)$:

$$c_{e,11}(e) = c_{b,11} + \left.\frac{dc_{e,11}}{de}\right|_{e=0} e + \frac{1}{2}\left.\frac{d^2c_{e,11}}{de^2}\right|_{e=0} e^2 + \ldots\,, \tag{9.29}$$

where cubic and higher-order terms in e are dropped.

Because the presence of cracks reduces the stiffnesses, $c_{e,11}(e)$ has to decrease. Because the linear term dominates at sufficiently small e, we conclude that

$$\Delta c_{11} = \left.\frac{dc_{e,11}(e)}{de}\right|_{e=0} < 0\,. \tag{9.30}$$

Thus, truncating the series 9.29 after the linear term in accordance with Hudson's (1980) first-order approximation inevitably results in an incorrect, negative $c_{e,11}$ at a certain value of the crack density e. Beyond that crack density, the first-order theory of Hudson violates the elastic stability conditions, such as the requirements $c_{e,11}(e) > 0$ and $c_{e,22}(e) > 0$, and its predictions become physically implausible. For example, for dry fractures, it follows from equations 9.20 – 9.24 [see also equation 21 in Hudson (1981) or equations 20a and 24b in Liu et al. (2000)] that

$$c_{e,11}(e) < 0 \quad \text{when} \quad e > \frac{3}{4}\, g_b(1-g_b)\,. \tag{9.31}$$

Likewise, using the matrix 9.21, we find for $c_{e,22}$:

$$c_{e,22}(e) < 0 \quad \text{when} \quad e > \frac{3g_b(1-g_b)}{4(1-2g_b)^2}\,. \tag{9.32}$$

Inequalities 9.31 and 9.32 indicate that Hudson's first-order scheme encounters problems for small $V_{S,b}/V_{P,b}$ ratios or, equivalently, for large Poisson's ratios ν_b. In fact, the effective stiffnesses $c_{e,11}$ and $c_{e,22}$ vanish for any nonzero crack density in the limit $V_{S,b}/V_{P,b} \to 0$. On the other hand, the first-order predictions for shear moduli $c_{e,44}$, $c_{e,55}$, and $c_{e,66}$, which do not contain potentially small quantity μ_b in the denominator (see equation 9.21), are more accurate.

Hudson's (1980) second-order theory yields a positive coefficient of the quadratic term in expansion 9.29. Therefore, $c_{e,11}(e)$ begins to increase at some value of e, also exhibiting unphysical behavior. In contrast, the NIA (e.g., Bristow, 1960; Schoenberg, 1980), which ignores elastic interactions between the cracks and adds their contributions to the effective compliance, gives a different form for the effective $c_{e,11}(e)$:

$$c_{e,11}^{\text{NIA}}(e) = \frac{c_{b,11}}{1+\kappa_{11}\,e}\,, \tag{9.33}$$

where κ_{11} is a positive coefficient. Therefore, $c_{e,11}^{\text{NIA}}(e)$ is a positive, monotonically decreasing function of e, as expected from the physics of the problem.

Figure 9.1 confirms the above analysis. Hudson's first-order theory (marked with "$\bigtriangledown$") yields unphysical negative values of $c_{e,11}$ and $c_{e,22}$ for the crack density e exceeding 0.07 and 0.12, respectively. The second-order theory of Hudson (marked with "$\bigtriangleup$") results in the obviously incorrect, monotonically growing coefficients $c_{e,11}(e)$ and $c_{e,22}(e)$ for $e \geq 0.05$. Such a behavior implies that adding more fractures stiffens rather than softens the rock. Furthermore, $c_{e,11}(e)$ and $c_{e,22}(e)$ exceed their background values ($c_{b,11} = c_{b,22} = \lambda_b + 2\mu_b = 19.8$ GPa) for $e \geq 0.09$, which indicates that a solid containing fractures is stiffer than the uncracked matrix. The tendency of Hudson's second-order theory to produce unreasonably large effective stiffnesses has been known for quite some time, both for a single fracture set and for randomly oriented cracks (Sayers and Kachanov, 1991; Cheng, 1993).

In contrast, the predictions of the linear-slip theory ("$\star$") and the NIA ("$\circ$") in Figure 9.1 are plausible. They almost coincide with each other, indicating a weak influence of the crack aspect ratio $\Theta = 0.05$ on the effective properties, and are close to the results of numerical modeling (bars; see the next section).

Additional insights into the effective properties can be gained by examining the anisotropy parameters $\epsilon^{(\text{V})}$, $\delta^{(\text{V})}$, and $\gamma^{(\text{V})}$ of the effective HTI medium (the parameter definitions can be found in Appendix 1B). The results in Figure 9.2 allow us to make the following observations:

1. According to the linear-slip theory and NIA, the parameters $\epsilon^{(\text{V})}$ and $\delta^{(\text{V})}$ almost coincide, so anisotropy is nearly elliptical ($\eta^{(\text{V})} \approx \epsilon^{(\text{V})} - \delta^{(\text{V})} \approx 0$).

2. It has been pointed out in the literature (e.g., Bakulin et al., 2000a) that the shear-wave splitting parameter is close to the crack density, $|\gamma^{(\text{V})}| \approx e$. Figure 9.2c

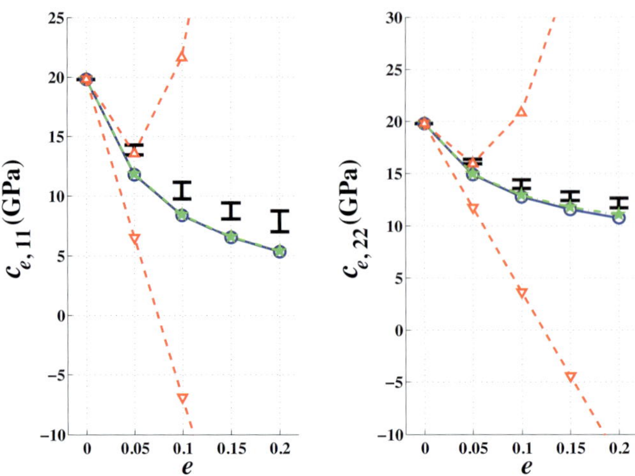

Figure 9.1: Effective stiffness coefficients $c_{e,11}$ and $c_{e,22}$ for a single set of dry cracks (Grechka and Kachanov, 2006b). The background velocities are $V_{P,b}$=3.0 km/s, $V_{S,b}$=1.0 km/s, and the density is ρ_b=2.2 g/cm^3; the corresponding Lamé parameters are λ_b=15.4 GPa and μ_b=2.2 GPa. Symbols mark different theoretical predictions: "$\bigtriangledown$" is Hudson's first-order theory (equations 9.20 and 9.21), "$\bigtriangleup$" is Hudson's second-order theory (equations 9.21 and 9.26 – 9.28), "$\star$" is Schoenberg's linear-slip theory (equations 9.4, 9.12, 9.13, and 9.15), and "$\circ$" is the NIA (equations 9.4 and 9.19) that takes into account nonzero crack aspect ratios (Θ=0.05 for all fractures). The bars correspond to the 95% confidence intervals (the mean values $\pm$ two standard deviations) of the numerically modeled stiffnesses obtained for 100 random realizations of the crack locations.

shows that this conclusion holds mainly for Hudson's theory. Still, although the equality $|\gamma^{(V)}| \approx e$ is not supported by the linear-slip theory and numerical modeling for large crack densities, it represents an acceptable approximation for $e \leq 0.1$.

Overall, we conclude that the linear-slip theory, which is practically equivalent to the noninteraction approximation in compliances, is superior to Hudson's first- and second-order schemes for a single set of dry penny-shaped cracks.

9.3 Numerical modeling of effective elasticity

The numerical results (bars) in Figures 9.1 and 9.2 were obtained by performing static, finite-element modeling on so-called *digital* rocks. The modeling methodology, described in Zohdi and Wriggers (2005), amounts to specifying a desired microstructure (such as that in Figure 9.3) and computing the local stress $\boldsymbol{\tau}(\mathbf{x})$ and strain $\boldsymbol{\varepsilon}(\mathbf{x})$ tensors that correspond to six linearly independent boundary conditions. Then Hooke's law (equation 9.1), written for the components of the volume-averaged tensors $\boldsymbol{\tau}(\mathbf{x})$

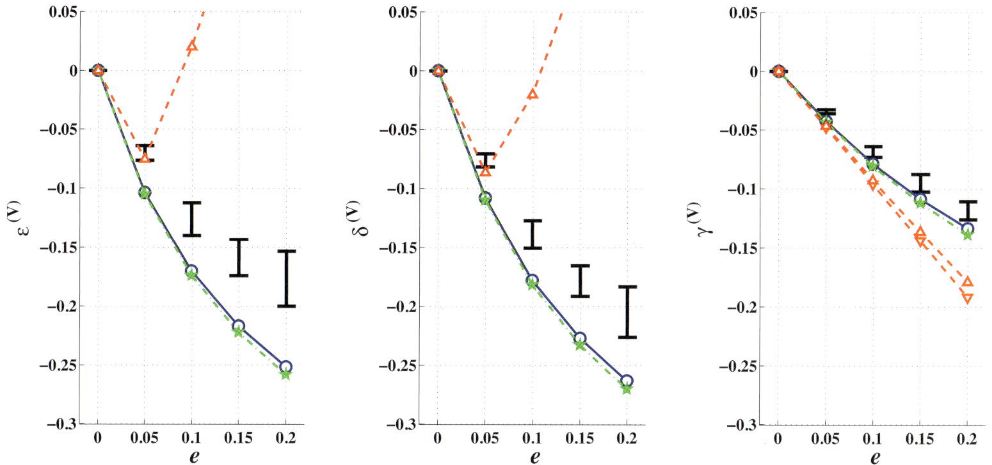

Figure 9.2: Anisotropy parameters $\epsilon^{(\mathrm{V})}$, $\delta^{(\mathrm{V})}$, and $\gamma^{(\mathrm{V})}$ of the effective HTI medium due to a system of dry cracks (Grechka and Kachanov, 2006b). The symbols are the same as those in Figure 9.1.

and $\boldsymbol{\varepsilon}(\mathbf{x})$, is treated as a system of linear equations for the unknown effective stiffness components $c_{e,ijlm}$. There is no need to know or assume the effective symmetry a priori because it can be inferred from the obtained stiffness tensor $\mathbf{c}_e$.

9.3.1 Influence of fracture interaction and intersections

Interaction of fractures

The numerical solution, however, always depends on details of the mutual positions of individual fractures, which causes scatter in the effective parameters for any nonzero crack density. The reason for the scatter is the interaction of the stress fields of adjacent cracks. Figure 9.3 shows the local behavior of the stress component τ_{11}. When the locations of the crack centers (supposed to be random and uncorrelated for the effective parameters to make sense) vary within $\mathcal{V}$ from one model to another, the pattern of interaction changes, introducing variations in the computed effective stiffness tensor $\mathbf{c}_e$ shown as bars in Figures 9.1 and 9.2. Such variations are inevitable for any finite number of cracks (e.g., Zohdi and Wriggers, 2001). The magnitude of these variations and its dependence on the applied boundary conditions allow one to evaluate the proximity of elastic properties of the computational volume to those of the RVE. Huet (1990) and Ostoja-Starzewski (2008) explain how this proximity and the bounds on the effective elasticity for a given crack array can be estimated from the modeling results.

Note that the bars in Figures 9.1 and 9.2 are located above the predictions of the NIA. To understand why this is the case, we need to examine Figure 9.3 more

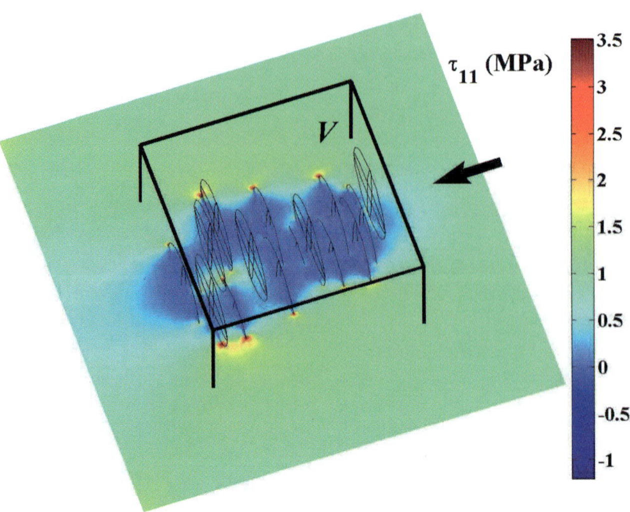

Figure 9.3: Horizontal cross section of the stress component τ_{11} through a model containing dry fractures, which are shown with wire spheroids; some spheroids are cut off by the cross section plane (Grechka and Kachanov, 2006b). The crack density is e=0.15. The arrow indicates the direction of applied remote load with a magnitude of 1 MPa.

closely. It reveals two types of stress disturbances: relatively extended areas of low stress called *shielding* (blue) and smaller areas of elevated stress called *amplification* (yellow and red). The physical reason for shielding is easy to explain. The faces of dry cracks in Figure 9.3 should be traction-free with $\tau_{11} = 0$, regardless of the applied load. Because the stress in the background is continuous, τ_{11} smoothly changes from zero to its far-field value as the distance from a crack face increases. On the other hand, static stress in linear elasticity obeys an analog of the Gauss divergence theorem (e.g., Markov, 1999). According to that theorem, the volume integral of τ_{11} is equal to the surface integral of the applied normal traction T_1. Therefore, areas of higher stress are required to compensate for the low-stress regions in the vicinity of fracture faces. These stress-amplification areas always form at the fracture tips, which primarily support a load applied to the fracture.

Figure 9.3 demonstrates that the shielding areas occupy a major portion of the volume and dominate the amplification areas, even though the local amplification magnitude is greater than that of the shielding. As a result, the shielded volumes that belong to adjacent cracks overlay, making the total shielded volume (or the volume of reduced stress) smaller than it would be in the absence of fracture interaction. This increases the volume integral of the stress tensor $\boldsymbol{\tau}$ compared to that calculated for noninteracting cracks. Hence, the estimated effective model becomes stiffer than the one predicted by the NIA (Figure 9.1). This stiffening naturally translates into a reduction in the magnitude of crack-induced anisotropy, which explains the results in Figure 9.2.

Saenger et al. (2004), in contrast, observed an opposite, softening influence of fracture interaction in their finite-difference wave-propagation experiments. They concluded that the differential effective media scheme (Vavakin and Salganik, 1975) was more appropriate than the NIA for explaining their modeling results. Perhaps further numerical studies and a step change in computing power, which would make it possible to perform computations on finer grids, are necessary to clarify this discrepancy.

Intersecting fractures

Next, we analyze the sensitivity of the effective parameters to the presence of fracture intersections. To this end, we follow Grechka and Kachanov (2006c; 2006d), who examined more than a hundred arrays of both nonintersecting and intersecting cracks, some of which are displayed in Figure 9.4. Intersecting fractures create intricate geometries ranging from relatively simple X- and V-shapes (Figures 9.4b,c) to more complicated ones in Figure 9.4d. Clearly, for such geometries the cracks are no longer penny-shaped or even planar.

Figure 9.4 demonstrates that the presence of crack intersections does not perturb the stress field in the vicinity of fractures. Therefore, fracture intersections are unimportant for modeling the effective medium properties. This conclusion is confirmed by the comparison of the anisotropy parameters for models with nonintersecting and intersecting fractures in Figures 9.5 – 9.7. The confidence interval (bars) in each parameter is computed from finite-element simulations for 40 random realizations of locations of nonintersecting fractures. The triangles mark the same parameters, but for models where the fractures intersect (such as those in Figures 9.4b-d). The numerical results can be summarized as follows:

1. The bars and triangles overlap implying that crack intersections do not change the effective anisotropy in a measurable way.

2. Both the linear-slip theory and NIA tend to slightly overestimate the magnitude of the anisotropy parameters. Similar to models with a single fracture set (Figure 9.2), this is a consequence of stiffening caused by fracture interactions.

3. The linear-slip and NIA predictions in Figures 9.5 – 9.7 do not significantly deviate from one another, again confirming the insensitivity of the effective properties to the aspect ratio of dry cracks.

Because geometric intersections of fractures have virtually no influence on the effective elasticity, the interconnectedness of natural fracture networks (which has a strong impact on permeability) can be hardly established from seismic data.

9.3.2 Noncircular fractures

Natural fractures in rocks are known (e.g., from outcrops) to be notoriously irregular. Because their shapes resemble neither circles nor ellipses (which can be modeled with the semianalytic equation 9.19), it is unclear to what extent the above results are applicable to real cracks. Here, following Grechka et al. (2006), we show that

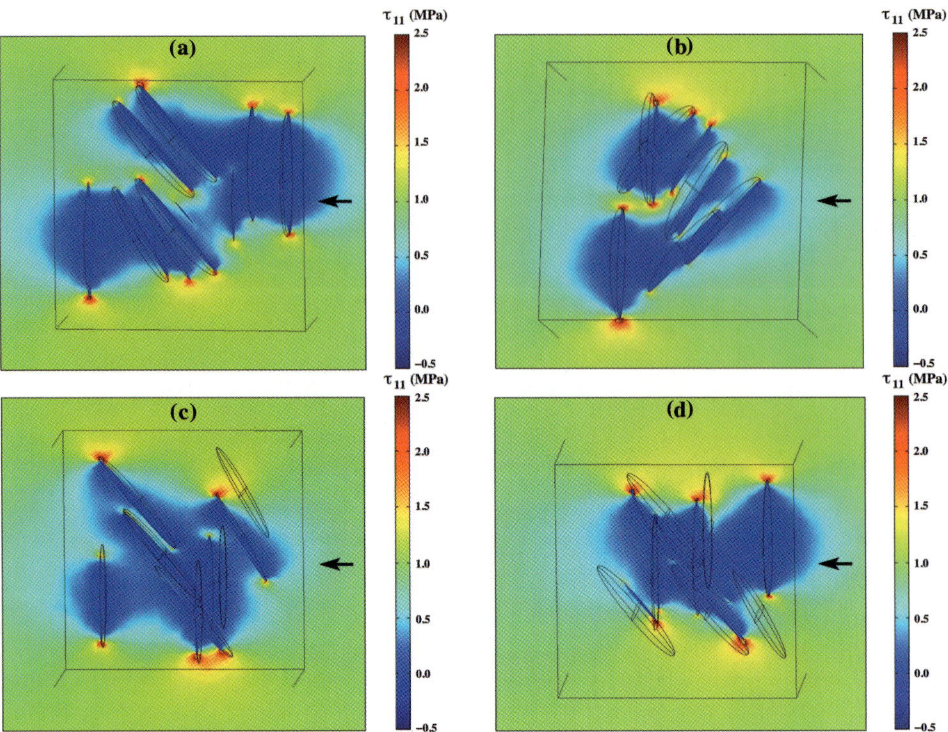

Figure 9.4: Horizontal cross sections of the stress component τ_{11} for arrays of (a) non-intersecting and (b, c, d) intersecting fractures (Grechka and Kachanov, 2006d). The aspect ratios Θ of fractures lie in the range $0.04 \leq \Theta \leq 0.08$. The arrows indicate the direction of applied uniaxial remote load, whose magnitude is 1 MPa.

circular, penny-shaped fractures can legitimately represent flat cracks with random shape irregularities.

Each of the six fracture shapes shown in Figure 9.8 was used to build a single vertical fracture set with random crack locations and orientations (or angles) in the vertical $[x_2, x_3]$-plane. The question is whether or not these irregular fractures can be replaced with penny-shaped cracks for purposes of modeling the effective stiffnesses. After computing the effective tensor $\mathbf{c}_e$, we average the local stiffness values in the $[x_2, x_3]$-plane (this yields $\langle \mathbf{c}_e \rangle$) and calculate the crack contribution by subtracting the background stiffness $\mathbf{c}_b$:

$$\langle \Delta \mathbf{c} \rangle = \langle \mathbf{c}_e \rangle - \mathbf{c}_b \,. \tag{9.34}$$

If this stiffness perturbation can be attributed to penny-shaped cracks, then

$$\langle \Delta \mathbf{c} \rangle = \Delta \mathbf{c}^{\rm ps}(e^{\rm fit},\, \Theta^{\rm fit}) = \mathbf{s}_e^{-1} - \mathbf{c}_b \,, \tag{9.35}$$

where $\Delta \mathbf{c}^{\rm ps}$ is given by equations 9.4 and 9.19. The crack density $e^{\rm fit}$ and aspect ratio $\Theta^{\rm fit}$, which we treat as unknowns, are estimated from nonlinear optimization,

$$\langle \Delta \mathbf{c} \rangle \sim \Delta \mathbf{c}^{\rm ps}(e^{\rm fit},\, \Theta^{\rm fit}) \,, \tag{9.36}$$

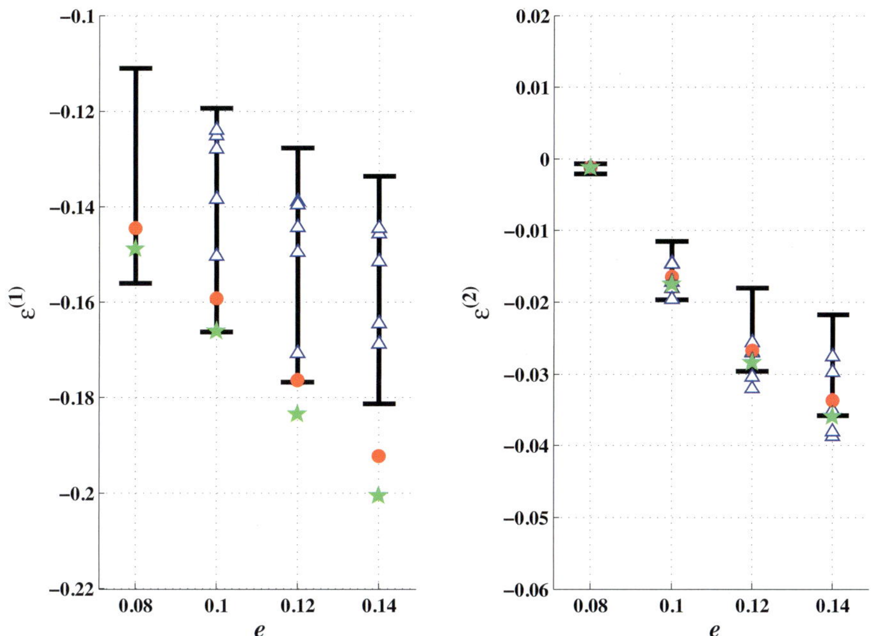

Figure 9.5: Effective anisotropy parameters $\epsilon^{(1)}$ and $\epsilon^{(2)}$ of fractured media (Grechka and Kachanov, 2006d). The bars correspond to the 95% confidence intervals (the mean $\pm$ two standard deviations) of the computed values. The triangles mark the parameters of models with intersecting cracks. The predictions of the linear-slip theory, which ignores the nonzero crack aspect ratios (equations 9.4, 9.12, 9.13, and 9.15), and the noninteraction approximation, which accounts for them (equations 9.4 and 9.19), are shown with "$\star$" and "$\bullet$" (respectively). All effective models are triclinic, but only Tsvankin's parameters, defined for orthorhombic media (Appendix 1B), are displayed.

where we fit $\Delta\mathbf{c}$ with theoretical stiffness contributions $\Delta\mathbf{c}^{\text{ps}}(e^{\text{fit}}, \Theta^{\text{fit}})$ computed with equations 9.4 and 9.19 for a single set of circular, penny-shaped cracks. The quality of the fit is measured by the differences

$$\Delta_c = \langle\Delta\mathbf{c}\rangle - \Delta\mathbf{c}^{\text{ps}} . \tag{9.37}$$

We intentionally operate with the stiffness perturbations $\Delta\mathbf{c}$ rather than the effective stiffness tensors themselves because the former are much more sensitive to fractures. Figure 9.9 shows the magnitudes of the misfits Δ_c calculated as the norm

$$\Delta_c^{\text{nrm}} = \frac{\max|\Delta_c|}{\max|\langle\Delta\mathbf{c}\rangle|} \times 100\% . \tag{9.38}$$

The values of Δ_c^{nrm} do not exceed 0.65% for all fracture geometries. Clearly, the irregular fracture shapes in Figure 9.8 can be accurately represented by circular cracks.

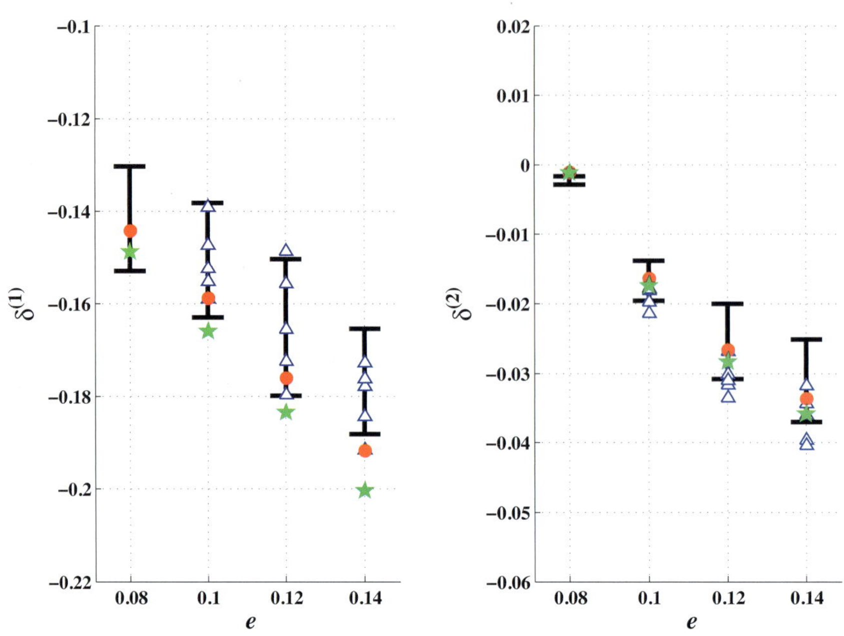

Figure 9.6: Same as Figure 9.5 for the parameters $\delta^{(1)}$ and $\delta^{(2)}$ (Grechka and Kachanov, 2006d).

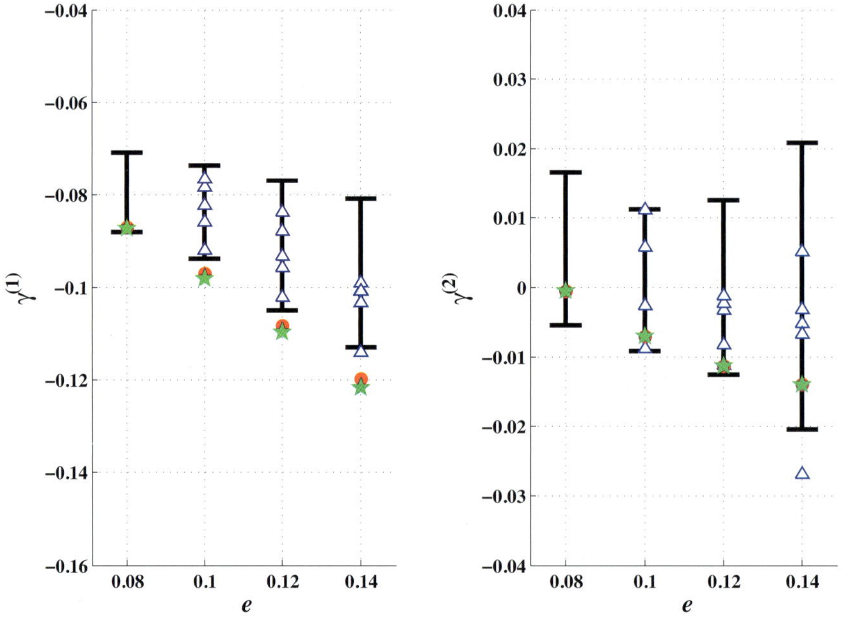

Figure 9.7: Same as Figures 9.5 and 9.6 for the parameters $\gamma^{(1)}$ and $\gamma^{(2)}$ (Grechka and Kachanov, 2006d).

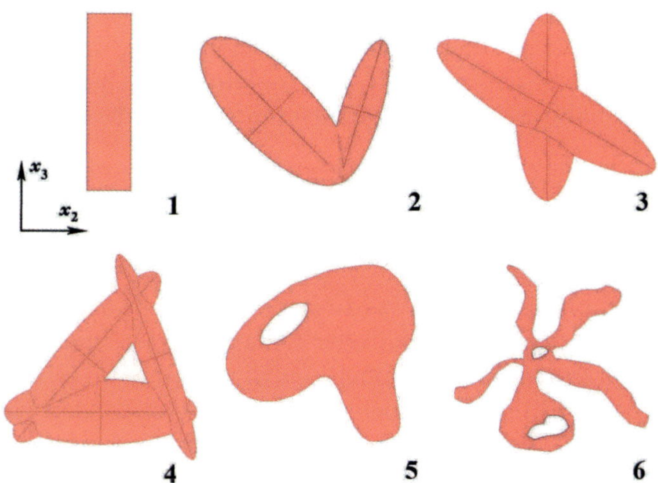

Figure 9.8: Fracture geometries created to study the influence of crack shape on the effective properties (Grechka et al., 2006). All fractures are vertical and planar; their normals are parallel to the x_1-axis. Geometries 4, 5, and 6 contain "rock islands" inside the cracks to model partially closed fractures.

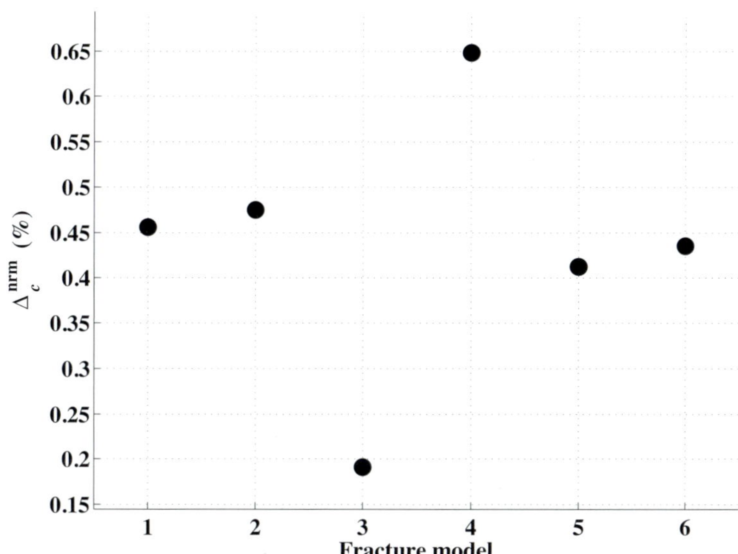

Figure 9.9: Misfits $\Delta_c^{\rm nrm}$ (equation 9.38) for the six fracture shapes in Figure 9.8 (Grechka et al., 2006).

Note that geometries 2, 3, and 4, which are built from ellipsoids, are somewhat rough. They contain sharp edges, where the normals to the crack faces are undefined; Berg et al. (1991) term such fractures *microcorrugated*. The values of Δ_c^{nrm} for those crack models do not stand out in Figure 9.9, implying that the circular-crack approximation performs equally well for all fracture shapes regardless of the smoothness of their faces.

9.3.3 Multiple sets of noncircular fractures

Here, we combine all previously discussed features of cracks and directly examine the effective elasticity of multiple, vertical, noncircular, intersecting fractures with different azimuths. Three representative models out of a total of 20 built for that purpose are displayed in Figure 9.10. The models include three sets of dry, rectangular cracks oriented at azimuths 0°, 40°, and 60° with respect to the coordinate axis x_1. Each set is formed by five cracks rotated around their normals with a 36° increment to remove any preferential in-plane fracture orientation. The locations of the fracture centers are random and uncorrelated with the in-plane crack rotations. The crack densities of the equivalent sets of penny-shaped cracks (computed from the optimization equation 9.36) are $e^{(1)} = 0.04$, $e^{(2)} = 0.03$, and $e^{(3)} = 0.02$; the cumulative crack density $e = \sum_{\ell=1}^{3} e^{(\ell)} = 0.09$.

It is virtually impossible to place all fractures in volume $\mathcal{V}$ randomly and avoid their intersections. For some models, relatively few fractures intersect (Figure 9.10a), while other models include more intersections (Figure 9.10b), or even a single interconnected fracture network (Figure 9.10c). The fracture arrays in Figure 9.10 exhibit a high level of 3D geometric complexity. Specifically, the fracture shapes are noncircular, their faces are not smooth because the cracks often protrude through each other, and there are irregular pieces of host rock between the cracks because of the complicated geometry of their intersections.

Because the models in Figure 9.10 possess no symmetry, the corresponding effective stiffness tensors $\mathbf{c}_e^{(\mathrm{N})}$ computed with the finite-element method are generally triclinic. However, because the fractures are vertical and all directions within fracture planes are equivalent, the effective symmetry is close to monoclinic with a horizontal symmetry plane.

Figure 9.11 shows the effective anisotropy parameters for monoclinic media defined by equations 6.36 – 6.45 and the effective anellipticity parameters $\eta^{(1,2,3)}$ (equations 1.88 – 1.90) computed for 20 random realizations of the fracture locations. We observe that vastly diverse geometries of intersecting fractures do not cause significant variability in the effective anisotropy parameters. This confirms our earlier conclusion, drawn for circular cracks, that crack intersections make a negligible contribution to the effective elasticity. The same result was also obtained by Saenger et al. (2004), who performed wave-propagation rather than static computations for models containing up to 2000 fractures.

Figure 9.11 illustrates two important properties of the effective medium. First, the

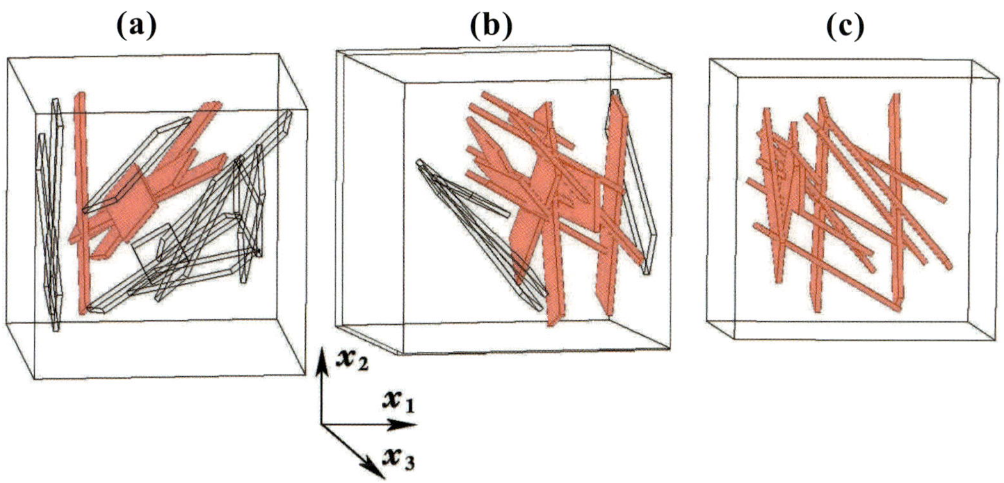

Figure 9.10: Models containing three sets of vertical rectangular cracks (Grechka et al., 2006). Fractures that intersect their neighbors are shaded, while isolated cracks are transparent. The background Poisson's ratio is ν_b=0.44.

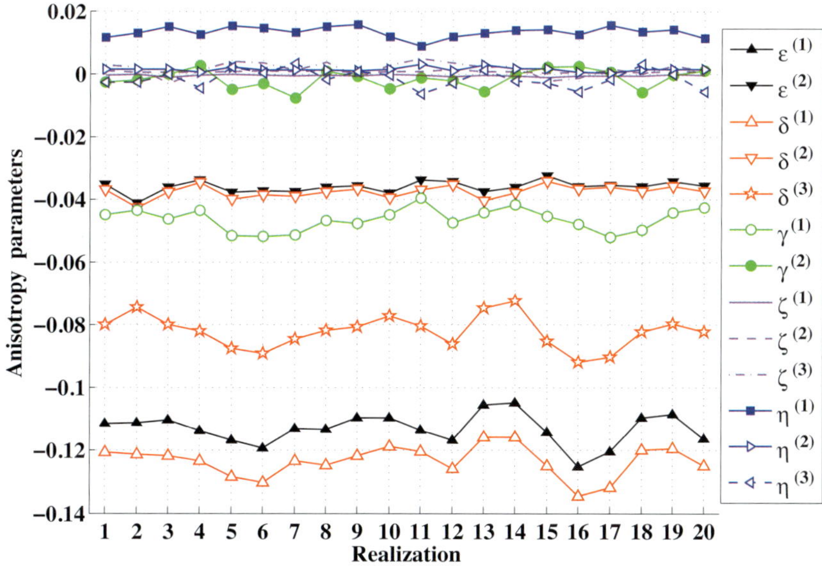

Figure 9.11: Effective anisotropy parameters for three sets of vertical, intersecting, rectangular cracks (Grechka et al., 2006).

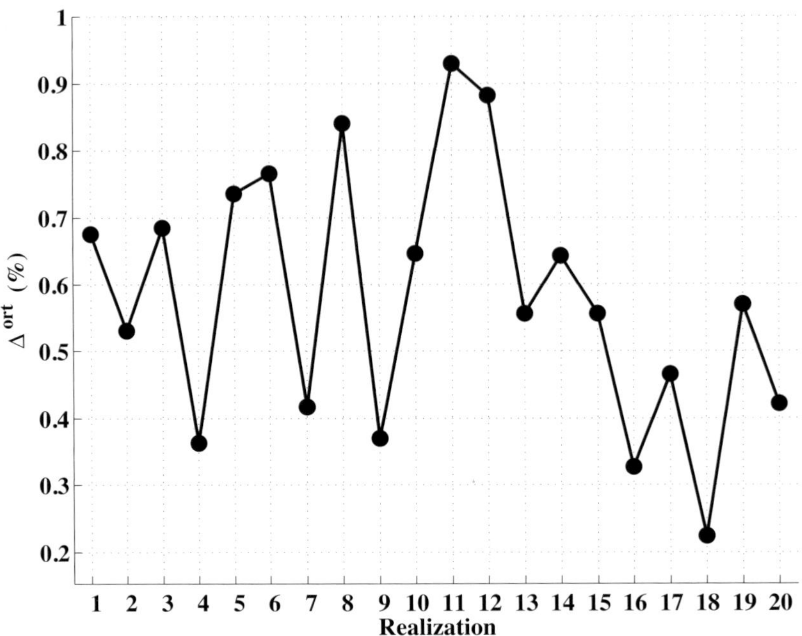

Figure 9.12: Relative deviation Δ^{ort} (equation 9.39) from orthotropy of the effective stiffness tensor for intersecting crack arrays, such as those in Figure 9.10 (Grechka et al., 2006).

anellipticity parameters are small, with $\max|\eta^{(1,2,3)}| = 0.015$. Therefore, for practical purposes the effective anisotropy is elliptical, which is also the case for dry circular cracks (Figures 9.2, 9.5, and 9.6). Second, the magnitude of the ζ-parameters, which quantify the deviation of monoclinic symmetry from orthotropy (see section 6.4), does not exceed 0.005. Clearly, the effective symmetry can be considered orthorhombic, given typical errors in evaluating anisotropy from seismic data.

This last point can be substantiated by approximating the computed effective stiffness tensors $\mathbf{c}_e^{(\text{N})}$ with the best-fit orthorhombic tensors $\mathbf{c}_e^{(\text{ort})}$ (Arts et al., 1991; Dewangan and Grechka, 2003) and calculating the relative misfits:

$$\Delta^{\text{ort}} = \frac{\max|\mathbf{c}_e^{(\text{N})} - \mathbf{c}_e^{(\text{ort})}|}{\max|\mathbf{c}_e^{(\text{N})}|} \times 100\% \,. \tag{9.39}$$

The values of Δ^{ort} for all 20 models are smaller than 1% (Figure 9.12), so the effective crack-induced anisotropy is indeed almost orthorhombic with a horizontal symmetry plane.

As discussed above, in the noninteraction approximation the effective symmetry of solids with multiple sets of scalar cracks is always orthorhombic. The closeness of the effective symmetry of media containing isolated, circular, dry fractures to orthotropy depends on the value of the Poisson's ratio ν_b in the isotropic host rock. According to equations 9.11 and 9.13, the deviation of the effective medium from orthorhombic increases with ν_b. Still, the presented numerical examples demonstrate that the effective symmetry for media with arrays of intersecting, noncircular, dry fractures remains virtually orthorhombic for a fairly large Poisson's ratio (in Figure 9.10 $\nu_b = 0.44$). This conclusion has far-reaching implications for reservoir characterization because it justifies approximating fractured formations with effective orthorhombic models.

9.4 Governing parameters for vertical fractures

Multiple fracture sets embedded in isotropic host rock not only produce orthorhombic (rather than triclinic) symmetry, but also the effective orthotropy for dry cracks has a simplified type. If the fractures are vertical, the effective medium is fully characterized by only five independent parameters instead of the nine needed for general orthorhombic media. Those parameters are the density-normalized Lamé constants λ_b and μ_b of the isotropic background (or, equivalently, the background P- and S-wave velocities $V_{P,b}$ and $V_{S,b}$), the densities e_1 and e_2 (defined as the eigenvalues of the crack-density tensor given by equation 9.12) of the principal orthogonal fracture sets, and the azimuth α_1 of the first principal set.

The effective symmetry remains close to orthotropy for liquid-filled fractures because the anisotropy parameters that control P-wave kinematics are relatively small. When all cracks have the same infill, which is likely to be the case for interconnected fracture networks, the average fluid factor ς (equation 9.18) can be used to incorporate the influence of fluids, bringing the total number of governing parameters to six. Thus, the parameter vector $\mathbf{m}$ describing the effective medium formed by multiple sets of vertical fractures embedded in otherwise isotropic host rock has the form

$$\mathbf{m} = \{V_{P,b}, V_{S,b}, e_1, e_2, \alpha_1, \varsigma\}. \tag{9.40}$$

As long as fractures are vertical, one of the symmetry planes of the effective orthorhombic medium is horizontal.

If the host rock is VTI (e.g., shale), vertical fractures also create nearly orthorhombic effective symmetry. Bakulin et al. (2000b, 2002) analyzed such models containing either a single fracture set or two orthogonal sets. Grechka (2007) extended their results to multiple fracture sets that have arbitrary azimuths. To account for the influence of the background anisotropy on the effective elasticity, he used equation 9.19 with the appropriately modified Eshelby tensor. The excess fracture-compliance tensor $\mathbf{Z}$ for penny-shaped cracks embedded in a VTI background generally lacks rotational invariance and has three different eigenvalues ($Z_N \neq Z_{T1} \neq Z_{T2}$), in contrast to the obvious equality $Z_{T1} = Z_{T2}$ (equation 9.11) for isotropic unfractured rock. Despite this added complexity, the governing fracture-related parameters (e_1, e_2, α_1, and ς) remain the same. Their estimation, however, suffers from the trade-off between the crack-induced and background anisotropy, which can be resolved either by combining wide-azimuth seismic data with borehole information or by assuming the presence of a single fracture set (that is, $e_2 = 0$). If the latter assumption is made, the fracture parameters can be inferred from the differences between seismic signatures in the planes parallel and orthogonal to the cracks, as discussed by Bakulin et al. (2000b).

In the next two sections, we discuss the inverse problem for vertical fractured sets embedded in purely isotropic host rock. If background anisotropy can be ignored, fracture characterization does not require borehole information and can be accomplished using seismic reflection data alone. We begin with the simplest HTI model of a single set of fractures and then proceed to effective media with multiple fracture sets.

9.5 Fracture-characterization methodology

9.5.1 Single fracture set: Effective HTI medium

The simplest fracture model, most often examined in the literature, consists of a single set of vertical, penny-shaped cracks in isotropic host rock. This effective medium is HTI with the symmetry axis perpendicular to the cracks. Wave propagation in HTI models can be described by a subset of Tsvankin's parameters defined for the more general orthorhombic media (see Appendix 1B). As the reference isotropic quantities, we use the vertical velocities of the P-wave (V_{P0}) and slow S-wave (V_{S2} or V_{S0}); at vertical incidence, the S_2-wave is polarized perpendicular to the cracks. The HTI anisotropy parameters $\epsilon^{(V)}$, $\delta^{(V)}$, and $\gamma^{(V)}$ were already analyzed in section 9.2.

Here, following Bakulin et al. (2000a), we discuss which fracture parameters can be estimated from such commonly measured seismic signatures as NMO ellipses and azimuthally varying AVO (amplitude-variation-with-offset) gradients. Because typically seismic data can be inverted for certain combinations of the anisotropy or fracture parameters, our analytic developments involve the following two steps. First, we assume that the crack density e_1 is small and derive linearized expressions for the anisotropy parameters in terms of e_1.[1] Second, the linearized parameters are substituted into equations for reflection seismic signatures to gain insight into the feasibility of fracture characterization.

The result of the first step for a set of dry fractures ($\varsigma = 0$) orthogonal to the x_1-axis is (Bakulin et al., 2000a):

$$\epsilon^{(V)} = \epsilon^{(2)} = -\frac{8}{3}\,e\,, \tag{9.41}$$

$$\delta^{(V)} = \delta^{(2)} = -\frac{8}{3}\,e\left[1 + \frac{g_b(1-2g_b)}{(3-2g_b)(1-g_b)}\right], \tag{9.42}$$

$$\gamma^{(V)} = \gamma^{(2)} = -\frac{8e}{3(3-2g_b)}\,, \tag{9.43}$$

[1] In this section, we denote e_1 by e to maintain consistency with the HTI notation used in sections 9.1 and 9.2 (because there is no second fracture set, $e_2 = 0$).

$$\eta^{(\mathrm{V})} = \epsilon^{(\mathrm{V})} - \delta^{(\mathrm{V})} = \frac{8}{3} e \left[\frac{g_b(1 - 2g_b)}{(3 - 2g_b)(1 - g_b)} \right], \tag{9.44}$$

where

$$g_b = \left(\frac{V_{S,b}}{V_{P,b}} \right)^2 \tag{9.45}$$

is the squared ratio of the S- and P-wave background velocities. Although equations 9.41 – 9.44 were originally derived in Bakulin et al. (2000a) using relatively inaccurate Hudson's (1980) theory, exactly the same equations follow from the linear-slip theory of Schoenberg (1980). Their analytic equivalence implies that both schemes can be used to describe effective elastic properties of fractured media in the limit of a vanishingly small crack density e.

The linearized anisotropy parameters for cracks fully filled with a stiff liquid ($\varsigma = 1$) are

$$\epsilon^{(\mathrm{V})} = 0, \tag{9.46}$$

$$\delta^{(\mathrm{V})} = -\frac{32 g_b e}{3(3 - 2g_b)}, \tag{9.47}$$

$$\gamma^{(\mathrm{V})} = -\frac{8e}{3(3 - 2g_b)}, \tag{9.48}$$

$$\eta^{(\mathrm{V})} = -\delta^{(\mathrm{V})}. \tag{9.49}$$

The influence of fluid saturation on the anisotropy parameters is illustrated in Figure 9.13. In accordance with equations 9.43 and 9.48, the shear-wave splitting parameter $|\gamma^{(\mathrm{V})}|$ is close to e for both dry and fluid-filled cracks and can be used as a proxy for the crack density. In contrast, the values of $\epsilon^{(\mathrm{V})}$ and $\delta^{(\mathrm{V})}$ strongly depend on fluid infill. For dry cracks, $\epsilon^{(\mathrm{V})}$ is independent of the ratio g_b (equation 9.41), while the influence of g_b on $\delta^{(\mathrm{V})}$ is weak; hence, the crack density can be found directly from either $\epsilon^{(\mathrm{V})}$ or $\delta^{(\mathrm{V})}$. In contrast, knowledge of the background velocity ratio $V_{S,b}/V_{P,b}$ is needed to estimate e from $\delta^{(\mathrm{V})}$ for liquid-filled fractures (the corresponding $\epsilon^{(\mathrm{V})} = 0$).

Even though the linearized equations 9.41 – 9.44 and 9.46 – 9.49 lose accuracy for realistic values of e (compare, for instance, equation 9.41 with the numerical results in Figure 9.2a), they still provide useful insights for assessing the feasibility of fracture characterization from seismic data. Next, we describe several options available for estimating the fracture azimuth, the crack density e, and the type of fracture infill.

Reflection moveout and AVO gradient of P-waves

P-wave NMO ellipse. If the layer of interest is sufficiently thick, information about fracturing can be obtained from the azimuthal variation of the interval NMO velocity (i.e., from the NMO ellipse). After estimating the effective NMO ellipses for the top and bottom of the layer, the interval NMO ellipse can be computed from

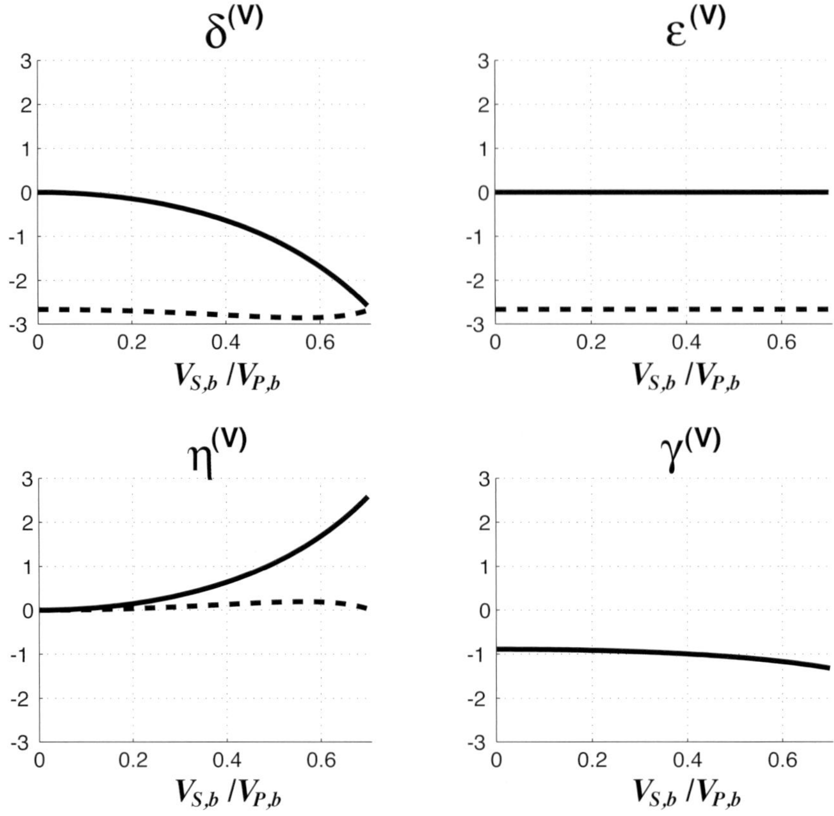

Figure 9.13: HTI anisotropy parameters computed from equations 9.41 – 9.44 and 9.46 – 9.49 and normalized by the crack density (Bakulin et al., 2000a). The solid lines correspond to fluid-filled cracks, the dashed lines to dry cracks. The vertical scale has the units of the crack density e.

the generalized Dix equations discussed in detail in Chapters 1 and 2. The P-wave NMO ellipse in a horizontal, homogeneous HTI layer can be found as a special case of equations 2.61 and 2.62 derived for orthorhombic media. Setting $\delta^{(1)}$ in equation 2.62 to zero and replacing $\delta^{(2)}$ with $\delta^{(\mathrm{V})}$ yields:

$$V^2_{P,\,\mathrm{nmo}}(\beta) = V^2_{P0}\,\frac{1+2\delta^{(\mathrm{V})}}{1+2\delta^{(\mathrm{V})}\sin^2\beta}\,, \tag{9.50}$$

where β is the azimuth of the common-midpoint line with respect to the symmetry axis (normal to the fractures). Because $\delta^{(\mathrm{V})} \le 0$, the major axis of the ellipse lies in the fracture plane. Therefore, P-wave NMO-velocity measurements along three or more well-separated azimuths can be inverted for the fracture azimuth α_1, the velocity V_{P0}, and the anisotropy parameter $\delta^{(\mathrm{V})}$. If the fractures are dry, their crack density can be obtained as $e \approx -\delta^{(\mathrm{V})}/2.8$ (see Figure 9.13). For liquid-filled cracks, estimation of e requires knowledge of the ratio g_b.

P-wave AVO gradient. As discussed in Chapter 8, reflection coefficients measured at oblique incidence angles over fractured formations vary with the azimuth of the source-receiver line. This azimuthal dependence can be used in AVO analysis to obtain information about the orientation, density, and content of the fractures. Suppose that a horizontal plane interface separates isotropic (incidence) and weakly anisotropic HTI (reflecting) halfspaces, and the elastic contrasts across the reflector is small. Then the difference B^{ani} between the P-wave AVO gradients in the directions perpendicular and parallel to the cracks is approximately given by equation 8.9 (Rüger, 1997; Rüger and Tsvankin, 1997):

$$B^{\text{ani}} = \frac{\delta^{(V)}}{2} - 4g\gamma^{(V)}, \tag{9.51}$$

where g has the meaning of the average ratio V_S^2/V_P^2 for the two halfspaces.[2]

Substituting the approximate anisotropy parameters for dry cracks (equations 9.42 and 9.43) into equation 9.51, we find

$$B^{\text{ani}}_{\text{dry}} = \frac{4(-8g^2 + 12g - 3)}{3(3 - 2g)(1 - g)}\, e\,. \tag{9.52}$$

Likewise, using equations 9.47 and 9.48 for liquid-filled cracks leads to

$$B^{\text{ani}}_{\text{wet}} = \frac{16g}{3(3 - 2g)}\, e\,. \tag{9.53}$$

As expected, the magnitude of the azimuthal AVO response in weakly anisotropic HTI media is proportional to the crack density. In addition, B^{ani} is sensitive to the V_S/V_P ratio of the host rock (Figure 9.14). If V_S/V_P takes a typical value of 0.55, $B^{\text{ani}}_{\text{wet}}$ is positive, while $B^{\text{ani}}_{\text{dry}}$ is close to zero. This result is in agreement with the computations of Strahilevitz and Gardner (1995), Rüger and Tsvankin (1997), and Sayers and Rickett (1997), who noticed that the AVO gradient for dry cracks is almost independent of azimuth for models with $V_S/V_P \approx 0.55$. The relationship between $B^{\text{ani}}_{\text{wet}}$ and $B^{\text{ani}}_{\text{dry}}$, however, changes substantially for higher or lower values of V_S/V_P.

Potentially, the factor B^{ani}, combined with an estimate of the V_S/V_P ratio, may serve as an indicator of fluid content. Reliable interpretation of the azimuthal AVO response, however, requires additional information about the crack density or anisotropy parameters. In particular, if the cracks are dry and V_S/V_P is between 0.5 and 0.6, the P-wave AVO gradient does not vary much with azimuth, irrespective of the degree of fracturing (e.g., the medium might be isotropic). If the crack density e can be obtained, for example, from the shear-wave splitting at vertical incidence, B^{ani} can be used to evaluate fluid saturation.

Reflection moveout and AVO gradient of PS-waves

When multicomponent data generated with P-wave sources are available, it is advantageous to supplement PP-wave NMO ellipses with reflection traveltimes of mode-converted PS-waves, or carry out joint inversion of PP- and PS-wave AVO gradients.

[2] The splitting parameter $\gamma^{(S)}$ in equation 8.9 is equal to $-\gamma^{(V)}$.

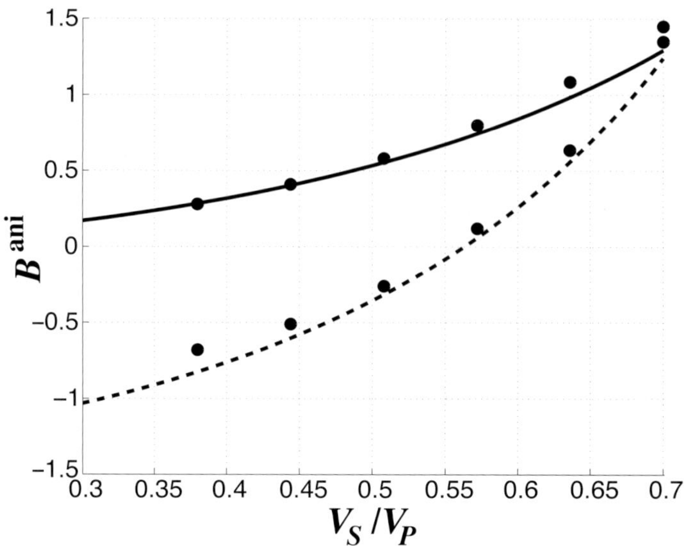

Figure 9.14: Azimuthal variation of the AVO gradient for P-waves reflected from a horizontal boundary between isotropic (incidence) and HTI (reflecting) halfspaces (Bakulin et al., 2000a). The difference B^{ani} between the AVO gradients in the directions perpendicular and parallel to the cracks was calculated for dry (equation 9.52, dashed line) and fluid-filled cracks (equation 9.53, solid line). The vertical scale has the units of the crack density e. The dots mark B^{ani} estimated from the exact reflection coefficient for the crack density e=0.07. The upper halfspace and the host rock in the fractured lower halfspace have the same P- and S-wave velocities: V_P=6 km/s, and V_S changes in accordance with the V_S/V_P ratio on the horizontal axis. The densities of the upper and lower media are 2.8 g/cm^3 and 2.3 g/cm^3, respectively.

PS-wave traveltimes. In Chapter 4 we introduced the PP+PS=SS method designed to produce the traveltimes of pure SS-wave reflections from PP and PS data. If the range of source-receiver offsets and the average V_S/V_P ratio are sufficient for computing the conventional-spread moveout function of the constructed SS reflections (see Chapters 4 and 5 for details), it should be possible to estimate SS-wave NMO velocities (in 2D) or ellipses (in 3D). Alternatively, NMO ellipses of SS-waves for laterally homogeneous models with a horizontal symmetry plane can be obtained directly from the corresponding PP- and PS-wave NMO ellipses using equation 1.66.

As explained in sections 6.1 and 8.1, P-waves propagating in vertical symmetry planes of azimuthally anisotropic media are coupled only to the in-plane polarized shear waves. Therefore, by combining PP and PS data in the isotropy (fracture) plane of HTI media, we can find the NMO velocity of the fast shear wave S_1 polarized parallel to the fractures. Because velocities in the isotropy plane are independent of the propagation angle, V_{nmo} of the S_1-wave is equal to the vertical velocity V_{S1}. Similarly, PP and PS data acquired in the symmetry-axis plane (orthogonal to the fractures) can be used to determine the NMO velocity of the slow shear wave S_2 given by equation 6.12 (V_{S2} is the vertical velocity of the S_2-wave):

$$V_{\text{nmo},S2} = V_{S2}\sqrt{1+2\sigma^{(\text{V})}}\,; \tag{9.54}$$

$$\sigma^{(\text{V})} \equiv \left(\frac{V_{P0}}{V_{S2}}\right)^2 (\epsilon^{(\text{V})} - \delta^{(\text{V})})\,. \tag{9.55}$$

If the vertical velocities V_{P0}, V_{S1}, and V_{S2} and the parameter $\delta^{(\text{V})}$ were determined from the P-wave NMO ellipse (equation 9.50) and the vertical traveltimes, $V_{\text{nmo},S2}$ yields the parameter $\epsilon^{(\text{V})}$. We conclude that multicomponent reflection data are sufficient for estimating all HTI parameters and, therefore, the crack density e and the fluid content of the fractures (assuming that there is no fluid flow between the fractures and pore space). Note that the crack density can be found directly from the fractional difference between the vertical velocities of the split shear waves, which is close to the parameter $\gamma^{(\text{V})}$.

PS-wave AVO gradient. For relatively thin target layers and surveys acquired without shear-wave sources, supplementing the P-wave AVO response with prestack reflection amplitudes of PS-waves might be the only option for fracture characterization. As before, we analyze azimuthally varying reflection coefficients for an interface between isotropic (incidence) and HTI media. For PS-waves, the reflection coefficient at small offsets is proportional to $\sin\theta$, where θ is the P-wave phase incidence angle (Rüger, 2002). Therefore, the PS-wave AVO gradient is usually defined as the coefficient of the $\sin\theta$-term. Rüger (2002) provides the weak-contrast, weak-anisotropy approximation for the AVO gradients of the converted waves propagating in the vertical symmetry planes of the model. The difference between the PS-wave AVO gradients measured across the fractures and parallel to them can be expressed through the anisotropy parameters as

$$B_{\text{PS}}^{\text{ani}} = \frac{\delta^{(\text{V})}}{2(1+\sqrt{g})} - 2\sqrt{g}\,\gamma^{(\text{V})}\,. \tag{9.56}$$

Despite the apparent similarity between equation 9.56 and the corresponding P-wave expression 9.51, the factor B^{ani} for PP- and PS-waves is controlled by different combinations of $\delta^{(V)}$ and $\gamma^{(V)}$. Hence, joint azimuthal AVO analysis of the two modes might help constrain both the crack density and fluid content of fractures, provided that the squared vertical-velocity ratio g was determined from PP and PS traveltimes.

Figure 9.15 reproduces for converted waves the numerical example from Figure 9.14. As is the case for compressional waves, the factor $B_{\text{PS}}^{\text{ani}}$ stays positive when the cracks are liquid-filled and changes sign with the increasing V_S/V_P ratio for dry cracks. In the latter case, however, the PS-wave AVO gradient becomes azimuthally invariant for a smaller V_S/V_P ratio than does the gradient of PP-waves. Hence, if V_S/V_P is close to a typical value of 0.55 and the cracks are dry (so that B^{ani} for P-waves is negligible), the addition of PS-waves may help identify fracturing and, moreover, discriminate between dry and fluid-filled cracks.

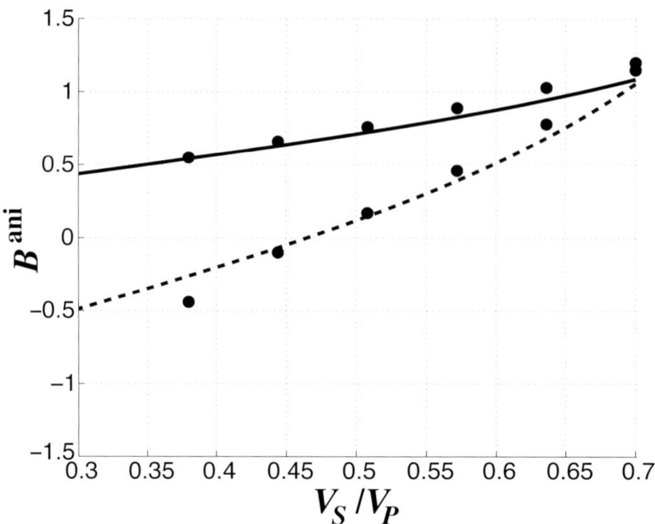

Figure 9.15: Difference between the converted-wave AVO gradients in the directions perpendicular and parallel to the cracks for the isotropic/HTI boundary (Bakulin et al., 2000a). The dashed line corresponds to dry cracks, the solid line to liquid-filled cracks; both curves are computed from the weak-anisotropy approximation 9.56. The dots mark $B_{\text{PS}}^{\text{ani}}$ estimated from the exact reflection coefficient for the crack density e=0.07. The model parameters are the same as those in Figure 9.14.

9.5.2 Multiple fracture sets: Effective orthotropy

As discussed above, the parameters of a single set of vertical fractures can be obtained from several different combinations of seismic signatures. In the presence of multiple fracture sets embedded in isotropic host rock, the effective medium is close to orthorhombic and can be described by the parameter vector $\mathbf{m}$ in equation 9.40. Despite the increased model complexity, Grechka and Kachanov (2006a) showed that for a horizontal, homogeneous fractured layer all components of $\mathbf{m}$, including the principal crack densities e_1 and e_2, can be estimated from the following input data:

$$\mathbf{d}(\mathbf{m}) = \left\{ \frac{V_{S1}}{V_{P0}}, \frac{V_{S2}}{V_{P0}}, \mathbf{W}^P, \mathbf{W}^{S1}, \mathbf{W}^{S2} \right\}, \tag{9.57}$$

where V_{P0}, V_{S1}, and V_{S2} are the vertical velocities of P-, S$_1$-, and S$_2$-waves, and the matrices $\mathbf{W}$ (see section 1.1) define the NMO ellipses of the three modes. The velocity ratios in $\mathbf{d}$ can be computed from the zero-offset traveltimes after establishing P-to-S event correspondence. The weak-anisotropy (or small-crack-density) approximations in Appendices 9B and 9C indicate that all elements of the vector $\mathbf{m}$ are well constrained by the vector $\mathbf{d}$. Clearly, this algorithm requires acquisition of wide-azimuth, multicomponent reflection data. Note that azimuthally varying reflection traveltimes of S$_1$- and S$_2$-waves can be obtained by applying the 3D version of the PP+PS=SS method to PP and PS (PS$_1$ and PS$_2$) data.

9.6 Case study of fracture characterization

Vasconcelos and Grechka (2007) tested the fracture-characterization methodology suggested by Grechka and Kachanov (2006a) on a 3D, 9C seismic data set acquired by the Reservoir Characterization Project (Colorado School of Mines) at Rulison field in Colorado, USA. The wide-azimuth P-wave data from that survey were used in the case studies presented in sections 3.4 and 8.3. The reservoir at Rulison field is sufficiently thick (Figure 3.28) and reflector dips throughout the section are small (Figure 3.29), which allows application of the algorithm based on the input data in equation 9.57. Also, the reservoir produces gas from fractured sand lenses, so the model of dry cracks embedded in isotropic matrix should be adequate.

Effective NMO ellipses of P-, S_1, and S_2-waves were computed over the entire survey area. The azimuthal velocity variations for reflections from the bottom of the producing interval are strong and clearly visible on both P- and S-wave data (Figure 9.16). In contrast, the eccentricity of the P-wave NMO ellipses in the overburden is consistently smaller than 3%, suggesting that the observed azimuthal anisotropy is caused primarily by fractures in the producing interval. The interval NMO ellipses in the reservoir needed for the inversion (equation 9.57) were estimated by generalized Dix differentiation (equation 1.31).

The laterally varying background velocity fields $V_{P,b}$ and $V_{S,b}$ and the principal crack densities e_1 and e_2 (by definition, $e_1 \geq e_2$), inverted using exact equations rather than those in Appendices 9B and 9C, are displayed in Figure 9.17. For most of the area, the values of e_1 (Figure 9.17c) are substantially larger than those of e_2 (Figure 9.17d; the color scale is the same), implying that the dominant fracture direction is west-northwest – east-southeast. Figures 9.17c,d also indicate that the western part of the area is controlled mostly by a single fracture set with the density e_1, while the eastern part has a nonnegligible contribution of differently oriented fractures that manifest themselves as the second principal set with the density e_2. The estimated fluid factor ς (not shown) does not exceed 0.01, which is expected for a gas reservoir.

The survey area contains a borehole (the star in Figure 9.17), where a Formation MicroImager (FMI) log was used to count fractures for the entire reservoir interval. The FMI fracture count is compared in Figure 9.18 with the orientation of the dominant fracture set estimated from the seismic inversion. (According to Figure 9.17d, the density e_2 of the second principal fracture set at the borehole location is almost zero.) The two fracture sets identified by the FMI log are oriented at approximately N70W and N73E. Although these sets are not perpendicular, their influence on propagation of long (compared to the fracture size) seismic waves is equivalent to that of two principal orthogonal sets. Evidently, the azimuth of the dominant equivalent fracture set was accurately estimated from seismic reflection data. Note that the average azimuth of the largest (by magnitude) P-wave AVO gradient for the reservoir bottom is close to N70W (Figure 8.10).

Figure 9.18 illustrates the resolution of seismic fracture characterization: while

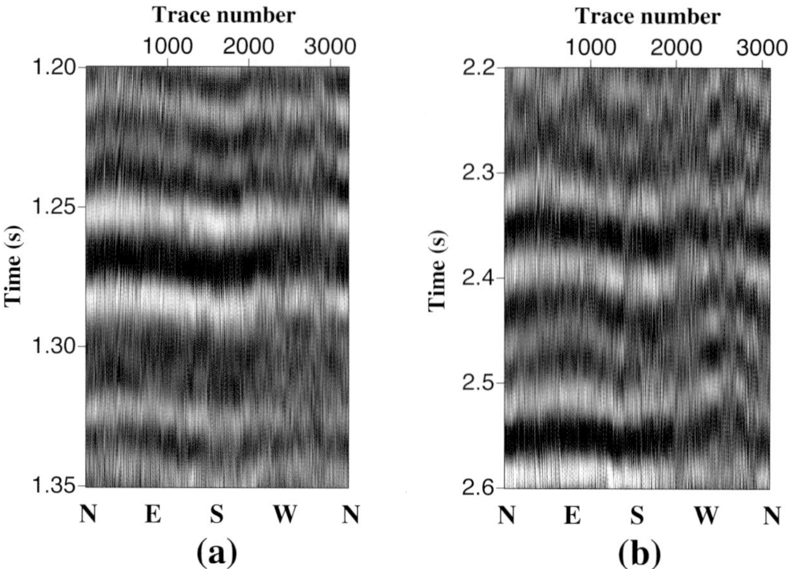

Figure 9.16: CMP gathers of the (a) P-wave and (b) S_1-wave at Rulison field after azimuthally-invariant (isotropic) NMO correction (Vasconcelos and Grechka, 2007). The traces are sorted by azimuth; the apparent sinusoidal variation of the residual moveout is indicative of azimuthal anisotropy. The offsets for different azimuths are not exactly the same, which causes the jitter. The reflections from the bottom of the reservoir are at approximately 1.27 s (a) and 2.35 s (b). The offset-to-depth ratio is close to 1.

long seismic waves cannot resolve each fracture set, the properties of waves are determined by the effective model of the fractured formation as a whole. The cumulative influence of fractures on wave propagation is described by the crack-density tensor $\boldsymbol{\alpha}$, which can be represented in terms of the contributions of mutually orthogonal (or principal) fracture sets. It is those principal sets that control seismic signatures and can be unambiguously estimated from seismic data. Clearly, the knowledge of principal crack densities e_1 and e_2 is insufficient to uniquely reconstruct the individual fracture sets displayed in Figure 9.18.

The inversion results in Figure 9.17 (along with the fluid factor $\varsigma \approx 0$) are sufficient for computing an orthorhombic depth model of the reservoir (Figures 9.19 and 9.20). It should be emphasized that this model was reconstructed from reflection data without using any borehole information. As discussed in Chapter 6, building orthorhombic subsurface models in depth generally requires well-log or check-shot data. Here, however, we considered a specific crack-induced orthorhombic medium with no background anisotropy, so just five out of nine model parameters were independent. This reduction in the number of unknowns makes it possible to rely solely on surface seismic data in depth-domain inversion.

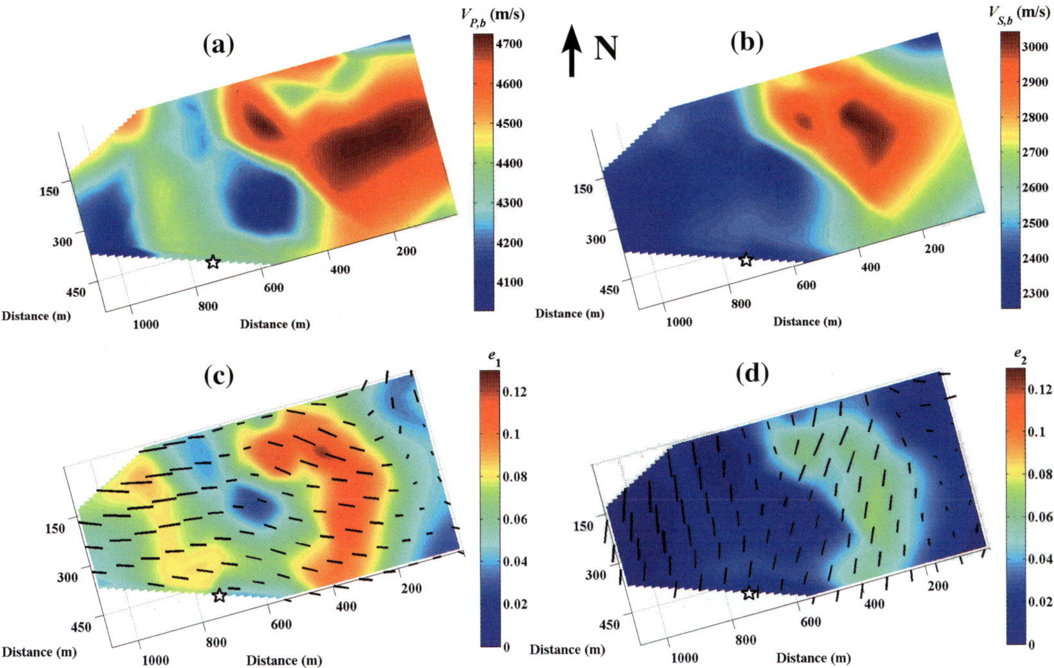

Figure 9.17: Results of parameter estimation at Rulison field (Vasconcelos and Grechka, 2007). The background velocities of (a) P-waves ($V_{P,b}$) and (b) S-waves ($V_{S,b}$), and the principal crack densities (c) e_1 and (d) e_2. The ticks mark the directions of the principal fracture sets. The star marks a borehole with an available FMI log (see Figure 9.18).

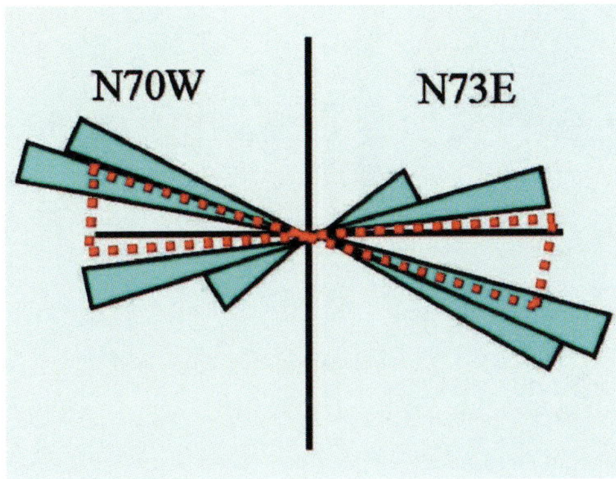

Figure 9.18: Fracture count (shaded in blue) in the borehole marked by the star in Figure 9.17 and the 90% confidence interval (dashed red line) for the azimuth of the fracture set with the density e_1 (Vasconcelos and Grechka, 2007).

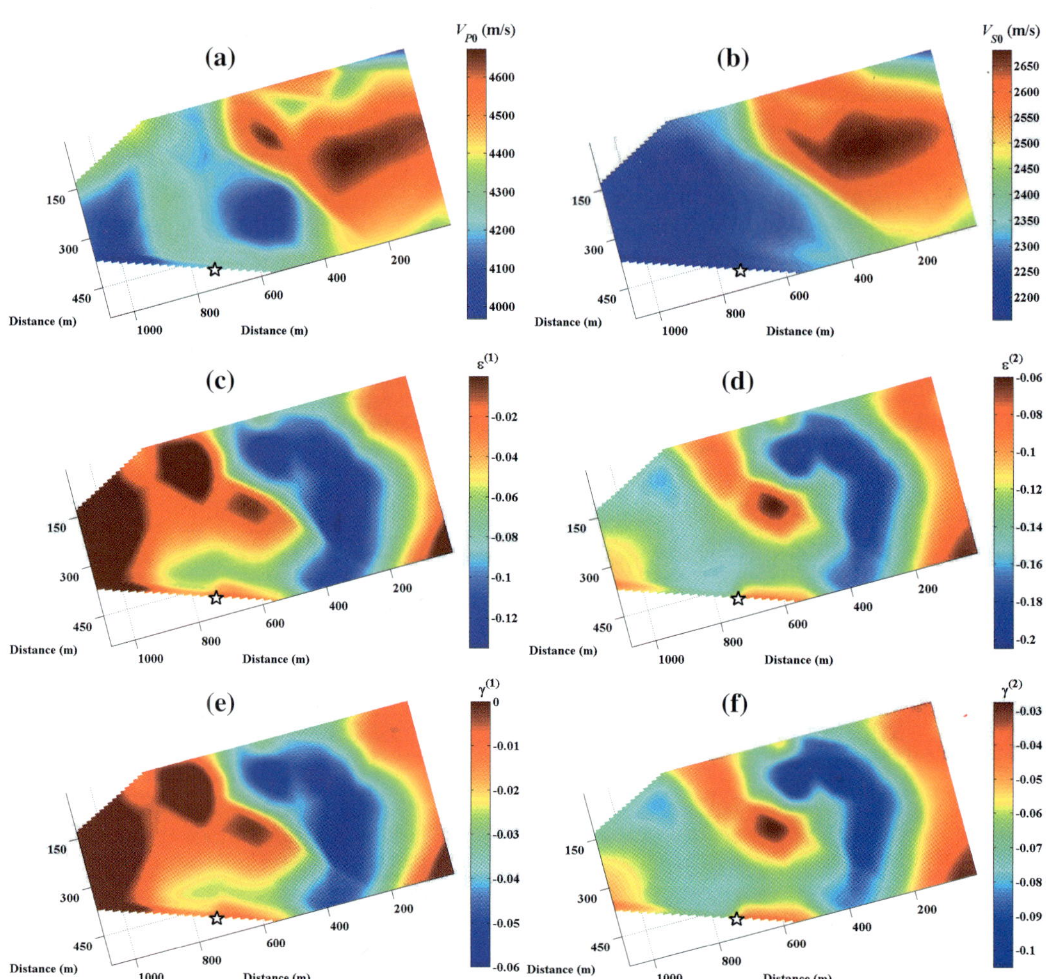

Figure 9.19: Estimated orthorhombic model of the Rulison reservoir (Vasconcelos and Grechka, 2007). The vertical velocities (a) V_{P0} and (b) V_{S0}, and the anisotropy parameters (c)$\epsilon^{(1)}$, (d) $\epsilon^{(2)}$, (e) $\gamma^{(1)}$, and (f) $\gamma^{(2)}$.

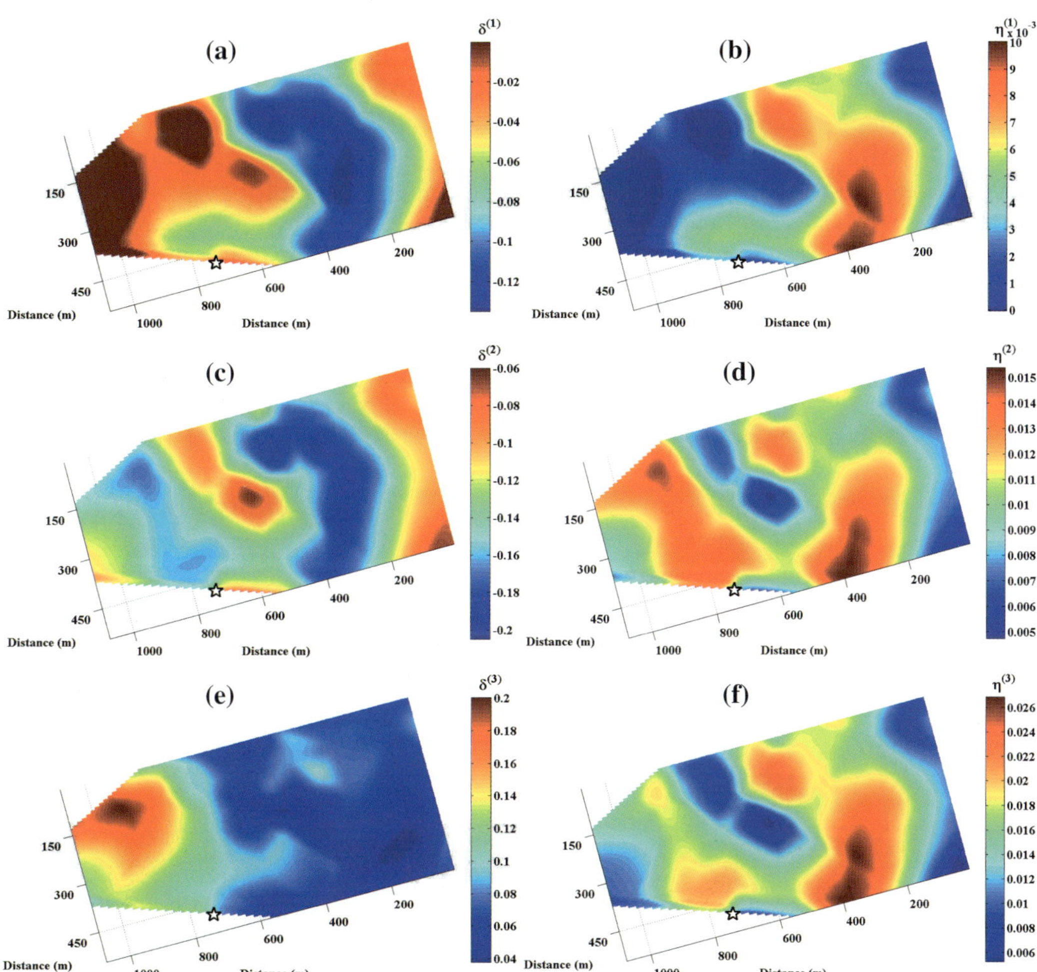

Figure 9.20: Estimated anisotropy parameters (a, c, e) $\delta^{(1,2,3)}$ and (b, d, f) $\eta^{(1,2,3)}$ of the Rulison reservoir (Vasconcelos and Grechka, 2007).

The strength of velocity anisotropy in the reservoir is substantial, with the magnitude of the ϵ-parameters reaching 0.2 (Figures 9.19c,d). As expected, the effective medium is close to elliptical because the parameters $\eta^{(1)}$, $\eta^{(2)}$, and $\eta^{(3)}$ are small for the whole area (Figures 9.20b,d,f).

This case study confirms that 3D, wide-azimuth, multicomponent seismic data provide sufficient information to establish the presence of multiple sets of vertical fractures in an otherwise isotropic host rock. The obtained principal crack densities are indicative of an interconnected fracture network in the eastern part of the study area. This conclusion follows from a straightforward geometric consideration: it is extremely difficult to place into a rock volume multiple sets of fractures that have random locations and the principal crack densities $e_1 = 0.11$ and $e_2 = 0.06$ without letting some fractures intersect each other (see Figures 9.4 and 9.10).

Despite the successful reconstruction of the effective fracture model, the spatially varying crack densities e_1 and e_2 are not well correlated with the estimated ultimate recovery (EUR) of wells. One reason for this discrepancy might be that EUR is known to be influenced by several operational factors unrelated to formation permeability and fracturing, such as injection rates and pressures used in hydraulic stimulation, flow-back schedule, and the choice of proppants and well-treatment fluids. Still, it should be mentioned that crack-related permeability of low-porosity rocks depends primarily on the fracture width (e.g., Kachanov and Sevostianov, 2005; Jaeger et al., 2007), whereas seismic signatures are governed by the crack density (i.e., by the fracture length). Without a spatially consistent relationship between the fracture width and length, the effective elasticity cannot serve as a reliable predictor of permeability. Also, the areas of the largest estimated crack density do not coincide with the pronounced P-wave AVO anomalies for the top (Figure 8.7) and bottom (Figure 8.8) of the Rulison reservoir. As discussed in section 8.3, weak correlation between the AVO response and the kinematic signatures used in this section is likely due to the reservoir heterogeneity.

9.7 Summary

In this chapter, we examined essential features of seismic anisotropy induced by the presence of multiple fracture sets. The theoretical development was based on the assumption that the fracture size is much smaller than the length of seismic waves. Under this assumption, the macroscopic properties of cracked solids can be derived using effective media theories. Successful modeling and inversion of fracture-related anisotropy requires evaluating the accuracy of a particular theory and understanding the information content of low-frequency data collected over a microheterogeneous medium.

To address the first issue, we analyzed and compared two popular effective media schemes proposed by Schoenberg (1980) and Hudson (1980). The main difference between them is that the linear-slip theory of Schoenberg (1980) adds the excess fracture compliances to the compliance of the background, while Hudson's (1980) theory does the same in stiffnesses. Although the difference might seem minor because compliance and stiffness are reciprocal, the linear-slip theory turns out to be significantly more accurate.

The mathematical reason for the inaccuracy of Hudson's theory lies in the presence of the background shear modulus μ_b in the denominator of the fracture-related stiffness contribution $\Delta \mathbf{c}$ (equation 9.21). Because $\lim_{\mu_b \to 0} ||\Delta \mathbf{c}|| = \infty$, Hudson's predictions might be physically implausible even for a vanishingly small crack density e. For realistic nonzero values of μ_b, the maximum eigenvalue of the tensor $(e\,\Delta \mathbf{c})$ might still be greater than unity, implying that the entire Hudson's formalism is based on a potentially diverging Taylor series. The tendency of Hudson's theory to predict inaccurate effective properties can also be understood on a qualitative level. It is clear that the softer the background, the smaller change in the stiffnesses is produced by a certain incremental increase in the crack density. Because the first- and second-order Hudson's equations contain expansions of the stiffness tensor at $e = 0$, they operate with the largest possible derivatives $||d\mathbf{c}/de||$ and $||d^2\mathbf{c}/de^2||$. Evidently, distortions of these derivatives for finite crack densities lead to either under- or overestimation of the effective parameters.

Operating in compliances, the theory of Schoenberg (1980) avoids the above problems and always results in plausible effective properties. The main disadvantage of the linear-slip theory, however, is the absence of a direct link to the fracture geometry. To provide such a connection, we combined the linear-slip predictions with a suite of theories developed by Kachanov (1980; 1992; 1993) for applications in materials science.

In addition, we discussed finite-element modeling of the effective properties of fractured media using so-called digital rocks. We introduced such realistic features as crack intersections, random shape irregularities, and partial contacts of the fracture faces and found that their influence on the effective elasticity is well approximated by isolated, penny-shaped cracks.

While comparing the predictions of semianalytic theories with the results of numerical experiments, this chapter also verified Kachanov's (1980) finding that the effective symmetry of initially isotropic solids containing multiple fracture sets is close to orthorhombic. This conclusion led to identification of the *principal* crack densities and orientations as the proper microstructural parameters that should be targeted in seismic fracture characterization. The case study from Rulison field confirmed the feasibility of obtaining the densities and orientations of the principal fracture sets from wide-azimuth, multicomponent seismic reflection data. As a by-product of the inversion, we built a depth-domain orthorhombic model of the Rulison reservoir without using any borehole information. This became possible because orthotropy induced only by fractures is governed by fewer independent parameters than general orthorhombic media, and reflection traveltimes can be inverted for the vertical velocities and reflector depth.

Appendices for Chapter 9

9A Eshelby's tensor

Here we present exact and approximate equations for calculating Eshelby's (1957) tensor $\boldsymbol{\mathcal{S}}$ used in equation 9.19. A more general representation of that tensor can be found in Mura (1987).

Eshelby (1957) examined static deformation of elastic solids caused by a constant eigenstrain $\boldsymbol{\varepsilon}^*$ imposed inside a homogeneous ellipsoidal inclusion embedded in a homogeneous isotropic medium. The ellipsoid has the semiaxes $\{a_1, a_2, a_3\}$, and its interior Ω is described by the inequality

$$\frac{x_1^2}{a_1^2} + \frac{x_2^2}{a_2^2} + \frac{x_3^2}{a_3^2} \le 1 , \tag{9.58}$$

where x_i are the Cartesian coordinates. The eigenstrain is defined as any nonelastic strain; it can be caused, for example, by thermal expansion, phase transformation, or pre-existing plastic strain. The details of the physics behind $\boldsymbol{\varepsilon}^*$ are unimportant for the purposes of this appendix.

The eigenstrain $\boldsymbol{\varepsilon}^*$ in the ellipsoid Ω deforms its exterior $\bar{\Omega}$, which, in turn, creates a nonzero elastic strain $\boldsymbol{\varepsilon}^{\Omega}(\mathbf{x})$ in Ω. Perhaps the most unexpected finding of Eshelby (1957) is that the strain $\boldsymbol{\varepsilon}^{\Omega}$ is independent of $\mathbf{x}$. Hence, $\boldsymbol{\varepsilon}^{\Omega}$ can be represented as

$$\varepsilon_{ij}^{\Omega} = \mathcal{S}_{ijkl}\, \varepsilon_{kl}^{*} , \quad (i, j = 1, 2, 3) , \tag{9.59}$$

where $\boldsymbol{\mathcal{S}}$ is the Eshelby tensor; summation from 1 to 3 with respect to repeated indices is implied.

A general expression for the tensor $\boldsymbol{\mathcal{S}}$ is given by Mura (1987):

$$\begin{aligned} \mathcal{S}_{ijkl} = \; & \frac{1}{8\pi(1-\nu_b)} \Big\{ \delta_{ij}\,\delta_{kl} \left(2\nu_b\, \mathcal{I}_i - \mathcal{I}_k + a_i^2\, \mathcal{I}_{ik} \right) \\ & + \left(\delta_{ik}\delta_{jl} + \delta_{jk}\delta_{il} \right) \left[a_i^2\, \mathcal{I}_{ij} - \mathcal{I}_j + (1-\nu_b) \left(\mathcal{I}_k + \mathcal{I}_l \right) \right] \Big\} , \\ & (i, j, k, l = 1, 2, 3) . \end{aligned} \tag{9.60}$$

Here ν_b is Poisson's ratio in the exterior $\bar{\Omega}$, δ_{ij} is Kronecker's delta, and there is no summation with respect to repeated indices.

The integrals $\mathcal{I}_i$ and $\mathcal{I}_{ij}$ in equation 9.60 are defined as (Eshelby, 1957)

$$\mathcal{I}_i = 2\pi a_1 a_2 a_3 \int_0^{\infty} \frac{dz}{P(z)\,(a_i^2 + z)} , \quad (i = 1, 2, 3) \tag{9.61}$$

and

$$\mathcal{I}_{ij} = \mathcal{I}_{ji} = 2\pi a_1 a_2 a_3 \int_0^{\infty} \frac{dz}{P(z)\,(a_i^2 + z)\,(a_j^2 + z)} , \quad (i, j = 1, 2, 3) , \tag{9.62}$$

where

$$P(z) = \sqrt{\prod_{i=1}^{3} (a_i^2 + z)}\ . \tag{9.63}$$

Because we examine penny-shaped fractures in the main text, here we consider only oblate spheroids, that is, ellipsoids with two equal semiaxes and a smaller third semiaxis:

$$a_2 = a_3 \geq a_1 \geq 0\,. \tag{9.64}$$

The equality $a_2 = a_3$ allows one to evaluate integrals 9.61 and 9.62 in elementary functions:

$$\mathcal{I}_1 = \frac{2\pi a_2^2}{a_\Delta^3}\Big(2a_\Delta - b a_1\Big), \tag{9.65}$$

$$\mathcal{I}_2 = \mathcal{I}_3 = -\frac{\pi a_1}{a_\Delta^3}\Big(2a_\Delta a_1 - b a_2^2\Big), \tag{9.66}$$

$$\mathcal{I}_{11} = \frac{2\pi a_2^2}{3a_\Delta^5 a_1^2}\Big[2a_\Delta\left(a_2^2 - 4a_1^2\right) + 3ba_1^3\Big], \tag{9.67}$$

$$\mathcal{I}_{12} = \mathcal{I}_{13} = \frac{\pi}{a_\Delta^5}\Big[2a_\Delta\left(a_1^2 + 2a_2^2\right) - 3ba_1 a_2^2\Big], \tag{9.68}$$

$$\mathcal{I}_{22} = \mathcal{I}_{23} = \mathcal{I}_{33} = \frac{\pi a_1}{4a_\Delta^5 a_2^2}\Big[2a_\Delta a_1\left(2a_1^2 - 5a_2^2\right) + 3ba_2^4\Big], \tag{9.69}$$

where

$$a_\Delta = \sqrt{a_2^2 - a_1^2} \quad \text{and} \quad b = \pi - 2\tan^{-1}\left(\frac{a_1}{a_\Delta}\right). \tag{9.70}$$

Useful approximations of integrals $\mathcal{I}_i$ and $\mathcal{I}_{ij}$ can be derived for thin spheroids, whose aspect ratios $\Theta \equiv a_1/a_2$ are much smaller than unity. Equations 9.65 – 9.70 then reduce to

$$\mathcal{I}_1 = 4\pi - 2\pi^2\Theta + \mathcal{O}\left(\Theta^2\right), \tag{9.71}$$

$$\mathcal{I}_2 = \mathcal{I}_3 = \pi^2\Theta + \mathcal{O}\left(\Theta^2\right), \tag{9.72}$$

$$\mathcal{I}_{11} = \frac{4\pi}{3a_1^2} - \frac{8\pi}{3a_2^2} + \frac{2\pi^2\Theta}{a_2^2} + \mathcal{O}\left(\Theta^2\right), \tag{9.73}$$

$$\mathcal{I}_{12} = \mathcal{I}_{13} = \frac{4\pi}{a_2^2} - \frac{3\pi^2\Theta}{a_2^2} + \mathcal{O}\left(\Theta^2\right), \tag{9.74}$$

$$\mathcal{I}_{22} = \mathcal{I}_{23} = \mathcal{I}_{33} = \frac{3\pi^2\Theta}{4a_2^2} + \mathcal{O}\left(\Theta^2\right). \tag{9.75}$$

9B Effective anisotropy parameters for small crack density

Here, using the work of Vasconcelos and Grechka (2007), we present approximations for the effective properties of rocks containing multiple fracture sets. To obtain general but tractable expressions, we assume the principal crack densities to be small ($\{e_1,\, e_2\} \ll 1$) and linearize the medium parameters and then the vertical and NMO velocities in e_1 and e_2. The results of this appendix generalize equations 9.41 – 9.49, which are valid only for a single set of fractures (i.e., $e_2 = 0$).

One of the main conclusions of the theoretical part of Chapter 9 is that the effective symmetry of initially isotropic rocks containing multiple fracture sets is nearly orthorhombic. As discussed throughout this volume, seismic signatures of orthorhombic media can be conveniently described in terms of Tsvankin's parameters defined in Appendix 1B. Linearizing those parameters in the principal crack densities e_1 and e_2 yields (Vasconcelos and Grechka, 2007)

$$V_{P0} = V_{P,b} \left[1 + \frac{2\lambda_b^2\,(\varsigma - 1)\,(e_1 + e_2)}{3\mu_b\,(\lambda_b + \mu_b)} \right], \tag{9.76}$$

$$V_{S0} = V_{S,b} \left(1 - \frac{8}{3}\, e_1\, \frac{\lambda_b + 2\mu_b}{3\lambda_b + 4\,\mu_b} \right), \tag{9.77}$$

$$\epsilon^{(2)} = \frac{8}{3}\, e_1\,(\varsigma - 1), \tag{9.78}$$

$$\epsilon^{(1)} = \frac{8}{3}\, e_2\,(\varsigma - 1), \tag{9.79}$$

$$\delta^{(2)} = \frac{8}{3}\, e_1 \left[\frac{(\varsigma - 1)\,\lambda_b}{\lambda_b + \mu_b} - \frac{4\mu_b}{3\lambda_b + 4\mu_b} \right], \tag{9.80}$$

$$\delta^{(1)} = \frac{8}{3}\, e_2 \left[\frac{(\varsigma - 1)\,\lambda_b}{\lambda_b + \mu_b} - \frac{4\mu_b}{3\lambda_b + 4\mu_b} \right], \tag{9.81}$$

$$\gamma^{(2)} = -\frac{8}{3}\, e_1\, \frac{\lambda_b + 2\mu_b}{3\lambda_b + 4\mu_b}, \tag{9.82}$$

$$\gamma^{(1)} = -\frac{8}{3}\, e_2\, \frac{\lambda_b + 2\mu_b}{3\lambda_b + 4\mu_b}. \tag{9.83}$$

Here

$$V_{P,b} = \sqrt{\frac{\lambda_b + 2\mu_b}{\rho}} \tag{9.84}$$

and

$$V_{S,b} = \sqrt{\frac{\mu_b}{\rho}} \tag{9.85}$$

are the P- and S-wave velocities in the isotropic background rock, ρ is the density, λ_b and μ_b are the background Lamé coefficients, and ς is the average fluid factor. The anisotropy parameters $\epsilon^{(2)}$, $\delta^{(2)}$, and $\gamma^{(2)}$ are defined in the $[x_1, x_3]$-plane, which contains the normal (x_1) to the fracture set with the density e_1. Therefore, the parameters with the superscript "(2)" are proportional to e_1 and independent of e_2. Likewise, the parameters $\epsilon^{(1)}$, $\delta^{(1)}$, and $\gamma^{(1)}$ are controlled solely by e_2. The same observations were made by Bakulin et al. (2000b), who studied the properties of orthorhombic models formed by two orthogonal fracture sets embedded in isotropic host rock.

9C Approximations for the vertical and NMO velocities

Next, we present the small-crack-density approximations for the components of the data vector in equation 9.57. Assuming that the slow shear wave S_2 at vertical incidence is polarized in the x_1-direction, the ratios of the vertical velocities can be written as

$$\frac{V_{S1}}{V_{P0}} = \frac{V_{S0}}{V_{P0}} \sqrt{\frac{1+2\gamma^{(1)}}{1+2\gamma^{(2)}}} = \frac{V_{S,b}}{V_{P,b}} \left\{1 - \frac{2}{3\mu_b(\lambda_b+\mu_b)} \left[\lambda_b^2 (\varsigma - 1)\, e_1 + \frac{(3\lambda_b + 2\mu_b)(\lambda_b^2 \varsigma + 4\mu_b^2) - \lambda_b^2 (3\lambda_b - 2\mu_b \varsigma)}{3\lambda_b + 4\mu_b} e_2 \right]\right\}, \tag{9.86}$$

$$\frac{V_{S2}}{V_{P0}} = \frac{V_{S0}}{V_{P0}} = \frac{V_{S,b}}{V_{P,b}} \left\{1 - \frac{2}{3\mu_b(\lambda_b+\mu_b)} \left[\lambda_b^2 (\varsigma - 1)\, e_2 + \frac{(3\lambda_b + 2\mu_b)(\lambda_b^2 \varsigma + 4\mu_b^2) - \lambda_b^2 (3\lambda_b - 2\mu_b \varsigma)}{3\lambda_b + 4\mu_b} e_1 \right]\right\}. \tag{9.87}$$

The pure-mode NMO ellipses for a horizontal reflector beneath a homogeneous orthorhombic layer are given by equation 1.20:

$$\frac{1}{V_{\mathrm{nmo},Q}^2(\beta)} = W_{11,Q} \cos^2\beta + W_{22,Q}\sin^2\beta = \frac{\cos^2\beta}{\left[V_{\mathrm{nmo},Q}^{(1)}\right]^2} + \frac{\sin^2\beta}{\left[V_{\mathrm{nmo},Q}^{(2)}\right]^2}, \tag{9.88}$$

where Q (P, S_1, or S_2) denotes the wave type, β is the azimuth measured from the x_1-axis, $W_{11,Q}$ and $W_{22,Q}$ are the diagonal elements of the matrix $\mathbf{W}$ that describes the NMO ellipse, and $V_{\mathrm{nmo},Q}^{(i)}$ are the semiaxes of the ellipse (i.e., the NMO velocities in the vertical symmetry planes).

Exact expressions for the symmetry-plane NMO velocities of P-, S_1, and S_2-waves in an orthorhombic layer can be found in section 6.1. Substituting the medium parameters obtained in Appendix 9B (equations 9.76 – 9.83) into equations 6.1, 6.2, and 6.9 – 6.12 and applying further linearization in e_1 and e_2, we find

$$V_{\mathrm{nmo},P}^{(1)} = V_{P,b}\left\{1 + \frac{2}{3\mu_b(\lambda_b+\mu_b)}\left\{\lambda_b^2(\varsigma-1)\, e_1 + \frac{1}{3\lambda_b+4\mu_b}\left\{3\lambda_b^3(\varsigma-1) + 16\mu_b\left[\lambda_b^2(\varsigma-1) + \lambda_b\mu_b(\varsigma-2) - \mu_b^2\right]\right\} e_2\right\}\right\}, \tag{9.89}$$

$$
\begin{aligned}
V_{\mathrm{nmo},P}^{(2)} &= V_{P,b}\left\{1+\frac{2}{3\mu_b(\lambda_b+\mu_b)}\left\{\lambda_b^2(\varsigma-1)\,e_2+\frac{1}{3\lambda_b+4\mu_b}\left\{3\lambda_b^3(\varsigma-1)\right.\right.\right. \\
&\left.\left.\left.+\,16\mu_b\left[\lambda_b^2(\varsigma-1)+\lambda_b\,\mu_b(\varsigma-2)-\mu_b^2\right]\right\}e_1\right\}\right\}, \qquad (9.90)
\end{aligned}
$$

$$
V_{\mathrm{nmo},S1}^{(1)} = V_{S,b}\left\{1+\frac{8(\lambda_b+2\mu_b)}{3(\lambda_b+\mu_b)}\,e_2\left[\varsigma-\frac{\mu_b}{3\lambda_b+4\mu_b}\right]\right\}, \qquad (9.91)
$$

$$
V_{\mathrm{nmo},S1}^{(2)} = V_{\mathrm{nmo},S2}^{(1)} = V_{S,b}\left[1-\frac{8(\lambda_b+2\mu_b)}{3(3\lambda_b+4\mu_b)}(e_1+e_2)\right], \qquad (9.92)
$$

$$
V_{\mathrm{nmo},S2}^{(2)} = V_{S,b}\left\{1+\frac{8(\lambda_b+2\,\mu_b)}{3(\lambda_b+\mu_b)}\,e_1\left[\varsigma-\frac{\mu_b}{3\lambda_b+4\mu_b}\right]\right\}. \qquad (9.93)
$$

To construct the corresponding Fréchet-derivative matrix, the quantities given by equations 9.86, 9.87, and 9.89 – 9.93 should be differentiated with respect to the elements of the parameter vector **m** (equation 9.40). None of the singular values of the Fréchet matrix vanishes, which means that all model parameters can be resolved from the input data.

References

Adam L., K. van Wijk, and T. Davis, 2003, Multi-level 3D VSP travel time inversion in VTI media, Weyburn field, Canada: 73rd Annual International Meeting, SEG, Expanded Abstracts, 753–756.

Akbar, N., J. Dvorkin, and A. Nur, 1993, Relating P-wave attenuation to permeability: Geophysics, **58**, 20–29.

Aki, K., and P. G. Richards, 2002, Quantitative seismology: University Science Books.

Al-Dajani, A., and I. Tsvankin, 1998, Nonhyperbolic reflection moveout for horizontal transverse isotropy: Geophysics, **63**, 1738–1753.

Al-Dajani, A., I. Tsvankin, and M. N. Toksöz, 1998, Nonhyperbolic reflection moveout for azimuthally anisotropic media: 68th Annual International Meeting, SEG, Expanded Abstracts, 1479–1482.

Alford, R. M., 1986, Shear data in the presence of azimuthal anisotropy: 56th Annual International Meeting, SEG, Expanded Abstracts, 476–479.

Alkhalifah, T., 1997, Velocity analysis using nonhyperbolic moveout in transversely isotropic media: Geophysics, **62**, 1839–1854.

Alkhalifah, T., and I. Tsvankin, 1995, Velocity analysis for transversely isotropic media: Geophysics, **60**, 1550–1566.

Alkhalifah, T., I. Tsvankin, K. Larner, and J. Toldi, 1996, Velocity analysis and imaging in transversely isotropic media: Methodology and a case study: The Leading Edge, **15**, 371–378.

Angerer, E., S. A. Horne, J. E. Gaiser, R. Walters, S. Bagala, and L. Vetri, 2002, Characterization of dipping fractures using PS mode-converted data: 72nd Annual International Meeting, SEG, Expanded Abstracts, 1010–1013.

Antich P., and S. Mehta, 1997, Ultrasound critical-angle reflectometry (UCR): A new modality for functional elastometric imaging: Physics in Medicine and Biology, **42**, 1763–1777.

Arts, R. J., K. Helbig, and N. J. P. Rasolofosaon, 1991, General anisotropic elastic tensor in rocks – Approximation, invariants, and particular direction: 61st Annual International Meeting, SEG, Expanded Abstracts, ST 2.4.

Ayres, A., and F. Theilen, 1999, Relationship between P- and S-wave velocities and geological properties of near-surface sediments of the continental slope of the Barents Sea: Geophysical Prospecting, **47**, 431–441.

Bachrach, R., M. Sengupta, A. Salama, and P. Miller, 2009, Reconstruction of the layer anisotropic elastic paraments and high-resolution fracture characterization from P-wave data: A case study using seismic inversion and Bayesian rock physics parameter estimation: Geophysical Prospecting, **57**, 253–262.

Backus, G. E., 1965, Possible form of seismic anisotropy of the uppermost mantle under oceans: Journal of Geophysical Research, **70**, 3429–3439.

Bakulin, A., V. Grechka, and I. Tsvankin, 2000a, Estimation of fracture parameters from reflection seismic data – Part I: HTI model due to a single fracture set: Geophysics, **65**, 1788–1802.

——, 2000b, Estimation of fracture parameters from reflection seismic data – Part II: Fractured models with orthorhombic symmetry: Geophysics, **65**, 1803–1817.

——, 2000c, Estimation of fracture parameters from reflection seismic data – Part III: Fractured models with monoclinic symmetry: Geophysics, **65**, 1818–1830.

Bakulin, A., V. Grechka, and I. Tsvankin, 2002, Seismic inversion for the parameters of two orthogonal fracture sets in a VTI background medium: Geophysics, **67**, 292–299.

Banik, N. C., 1987, An effective parameter in transversely isotropic media: Geophysics, **52**, 1654–1664.

Behura, J., and I. Tsvankin, 2009a, Role of the inhomogeneity angle in anisotropic attenuation analysis: Geophysics, **74**, no. 5, WB177–WB191.

——, 2009b, Reflection coefficients in attenuative anisotropic media: Geophysics, **74**, no. 5, WB193–WB202.

——, 2009c, Estimation of interval anisotropic attenuation from reflection data: Geophysics, **74**, no. 6, A69–A74.

Ben-Menahem, A., and S. J. Singh, 1981, Seismic waves and sources: Springer-Verlag.

Berg, E., J. Hood, and G. Fryer, 1991, Reduction of the general fracture compliance matrix Z to only five independent elements: Geophysical Journal International, **107**, 703–707.

Berthet, P., J.-P. Dunand, P. Julien, and J. Arnaud, 2004, 3D azimuthal velocity analysis on OBC data: SEG/EAGE Summer Research Workshop.

Bóna, A., and M. A. Slawinski, 2008, Comparison of two inversions for elasticity tensor: Journal of Applied Geophysics, **65**, 6–9.

Bóna, A., I. Bucataru, and M. A. Slawinski, 2008, Inversion of ray velocity and polarization for elasticity tensor: Journal of Applied Geophysics, **65**, 1–5.

Brajanovski, M., B. Gurevich, and M. Schoenberg, 2005, A model for P-wave attenuation and dispersion in a porous medium permeated by aligned fractures: Geophysical Journal International, **163**, 372–384.

Bristow, J. R., 1960, Microcracks and the static and dynamic elastic constants of annealed and heavily cold-worked metals: British Journal of Applied Physics, **11**, 81–85.

Brown, R. J., D. C. Lawton, and S. P. Cheadle, 1991, Scaled physical modelling of anisotropic wave propagation: Multioffset profiles over an orthorhombic medium: Geophysical Journal International, **107**, 693–702.

Budiansky, B., and R. J. O'Connell, 1976, Elastic moduli of a cracked solid: International Journal of Solids and Structures, **12**, 81–97.

Byun, B. S., D. Corrigan, and J. E. Gaiser, 1989, Anisotropic velocity analysis for lithology discrimination: Geophysics, **54**, 1564–1574.

Carcione, J. M., 2000, A model for seismic velocity and attenuation in petroleum source rocks: Geophysics, **65**, 1080–1092.

——, 2001, Wave fields in real media: Wave propagation in anisotropic, anelastic, and porous media: Pergamon Press.

Cardona, R., 2002, Fluid substitution theories and multicomponent seismic characterization of fractured reservoirs: Ph.D. thesis, Colorado School of Mines.

Castle, R. J., 1994, A theory of normal moveout: Geophysics, **59**, 983–999.

Červený, V., 1961, The amplitude curves of reflected harmonic waves around the critical point: Studia Geophysica et Geodaetica, **5**, no. 4, 319–351.

——, 1962, On the position of the maximum of the amplitude curves of reflected waves: Studia Geophysica et Geodaetica, **6**, no. 3, 215–233.

——, 1972, Seismic rays and ray intensities in inhomogeneous anisotropic media: Geophysical Journal of the Royal Astronomical Society, **29**, 1–13.

——, 2001, Seismic ray theory: Cambridge University Press.

Červený, V., and I. Pšenčík, 2005a, Plane waves in viscoelastic anisotropic media – I. Theory: Geophysical Journal International, **161**, 197–212.

——, 2005b, Plane waves in viscoelastic anisotropic media – II. Numerical examples: Geophysical Journal International, **161**, 213–228.

Červený, V., I. A. Molotkov, and I. Pšenčík, 1977, Ray method in seismology: University of Karlova.

Chang, H., and G. McMechan, 2009, 3D 3-C full-wavefield elastic inversion for estimating anisotropic parameters: A feasibility study with synthetic data: Geophysics, **74**, no. 6, WCC159–WCC175.

Chapman, M., 2003, Frequency-dependent anisotropy due to mesoscale fractures in the presence of equant porosity: Geophysical Prospecting, **51**, 369–379.

——, 2009, Modeling the effect of multiple sets of mesoscale fractures in porous rock on frequency-dependent anisotropy: Geophysics, **74**, no. 6, D97–D103.

Cheadle, S. P., R. J. Brown, and D. C. Lawton, 1991, Orthorhombic anisotropy: A physical seismic modeling study: Geophysics, **56**, 1603–1613.

Cheng, C. H., 1993, Crack models for a transversely isotropic medium: Journal of Geophysical Research, **98**, 675–684.

Chernjak, V. S., and S. A. Gritsenko, 1979, Interpretation of effective common-depth-point parameters for a spatial system of homogeneous beds with curved boundaries: Soviet Geology and Geophysics, **20**, no. 12, 91–98.

Chichinina, T., V. Sabinin, and G. Ronquillo-Jarillo, 2006, QVOA analysis: P-wave attenuation anisotropy for fracture characterization: Geophysics, **71**, no. 3, C37–C48.

Chichinina, T., I. Obolentseva, L. Gik, B. Bobrov, and G. Ronquillo-Jarillo, 2009, Attenuation anisotropy in the linear-slip model: Interpretation of physical-modeling data: Geophysics, **74**, no. 5, WB165–WB176.

Chin, R., G. W. Hedstrom, and L. Thigpen, 1984, Matrix methods in synthetic seismograms: Geophysical Journal of the Royal Astronomical Society, **77**, 483–502.

Clark, R., P. Benson, A. Carter, and C. Guerrero Moreno, 2009, Anisotropic P-wave attenuation measured from a multi-azimuth surface seismic reflection survey: Geophysical Prospecting, **57**, 835–845.

Cohen, J. K., 1998, A convenient expression for the NMO velocity function in terms of ray parameter: Geophysics, **63**, 275–278.

Contreras, P., V. Grechka, and I. Tsvankin, 1999, Moveout inversion of P-wave data for horizontal transverse isotropy: Geophysics, **64**, 1219–1229.

Corrigan, D., R. Withers, J. Darnall, and T. Skopinski, 1996, Fracture mapping from azimuthal velocity analysis using 3D surface seismic data: 66th Annual International Meeting, SEG, Expanded Abstracts, 1834–1837.

Crampin, S., 1991, Effects of point singularities on shear-wave propagation in sedimentary basins: Geophysical Journal International, **107**, 531–543.

Crampin, S., and M. Yedlin, 1981, Shear-wave singularities of wave propagation in anisotropic media: Journal of Geophysics, **49**, 43–46.

Cumella, S. P., and D. B. Ostby, 2003, Geology of the basin-centered gas accumulation, Piceance Basin, Colorado: Piceance Basin 2003 Guidebook, Rocky Mountain Association of Geologists, 171–193.

Daley, P., and F. Hron, 1977, Reflection and transmission coefficients for transversely isotropic media: Bulletin of the Seismological Society of America, **67**, 661–675.

de Bazelaire, E., and J. R. Viallix, 1994, Normal moveout in focus: Geophysical Prospecting, **42**, 477–499.

de Parscau, J., 1991, P- and SV-wave transversely isotropic phase velocity analysis from VSP data: Geophysical Journal International, **107**, 629–638.

del Monte, A. A., E. Angerer, C. Reiser, and C. Glass, 2004, Integrated approach for seismic fracture characterization: SEG/EAGE Summer Research Workshop.

Dellinger, J., and F. Muir, 1988, Imaging reflections in elliptically anisotropic media: Geophysics, **53**, 1616–1618.

Dewangan, P., and V. Grechka, 2003, Inversion of multicomponent, multiazimuth, walkaway VSP data for the stiffness tensor: Geophysics, **68**, 1022–1031.

Dewangan, P., and I. Tsvankin, 2006a, Modeling and inversion of PS-wave moveout asymmetry for tilted TI media: Part 1 – Horizontal TTI layer: Geophysics, **71**, no. 4, D107–D122.

——, 2006b, Modeling and inversion of PS-wave moveout asymmetry for tilted TI media: Part 2 – Dipping TTI layer: Geophysics, **71**, no. 4, D123–D134.

——, 2006c, Velocity-independent layer stripping of PP and PS reflection traveltimes: Geophysics, **71**, no. 4, U59–U65.

Dewangan, P., I. Tsvankin, M. Batzle, K. van Wijk, and M. Haney, 2006, PS-wave moveout inversion for tilted TI media: A physical-modeling study: Geophysics, **71**, no. 4, D135–D143.

DiSeina, J. P., J. E. Gaiser, and D. Corrigan, 1984, Horizontal components and shear wave analysis of three-component VSP data: in M. N. Toksöz and R. R. Stewart, eds., Vertical Seismic Profiling, Part B: Advanced Concepts, Geophysical Press, 177–188.

Dix, C. H., 1955, Seismic velocities from surface measurements: Geophysics, **20**, 68–86.

Douma, H., and A. Calvert, 2006, Nonhyperbolic moveout analysis in VTI media using rational interpolation: Geophysics, **71**, no. 3, D59–D71.

Dunne, J., and G. Beresford, 1998, Improving seismic data quality in the Gippsland Basin (Australia): Geophysics, **63**, 1496–1506.

Epstein, M., and M. A. Slawinski, 1999, On rays and ray parameters in inhomogeneous isotropic media: Canadian Journal of Exploration Geophysics, **35**, 7–19.

Eshelby, J. D., 1957, The determination of the elastic field of an ellipsoidal inclusion and related problems: Proceedings of the Royal Society, **A241**, 376–396.

Farra, V., 2001, High-order perturbations of the phase velocity and polarization of qP and qS waves in anisotropic media: Geophysical Journal International, **147**, 93–104.

Fomel, S., 2002, Applications of plane-wave destruction filters: Geophysics, **67**, 1946–1960.

——, 2004, On anelliptic approximations for qP velocities in VTI media: Geophysical Prospecting, **52**, 247–259.

Fowler, P. J., A. Jackson, J. Gaffney, and D. Boreham, 2008, Direct nonlinear traveltime inversion in layered VTI media: 78th Annual International Meeting, SEG, Expanded Abstracts, 3028–3032.

Franco, G., T. L. Davis, and V. Grechka, 2007, Seismic anisotropy of tight-gas sandstones, Rulison Field, Piceance Basin, Colorado: 77th Annual International Meeting, SEG, Expanded Abstracts, 1461–1464.

Fryer, G. J., and L. N. Frazer, 1984, Seismic waves in stratified anisotropic media: Geophysical Journal of the Royal Astronomical Society, **78**, 691–710.

Gaiser, J. E., 1990, Transversely isotropic phase velocity analysis from slowness estimates: Journal of Geophysical Research, **95**, 11241–11254.

——, 2000, Advantages of 3-D PS-wave data to unravel S-wave birefringence for fracture detection: 70th Annual International Meeting, SEG, Expanded Abstracts, 1201–1204.

Gaiser, J. E., and R. Van Dok, 2005, Borehole calibration of PS-waves for fracture characterization: Pinedale field, Wyoming: 75th Annual International meeting, SEG, Expanded Abstracts, 873–876.

Gajewski, D., and I. Pšenčík, 1987, Computation of high-frequency seismic wavefields in 3-D laterally inhomogeneous anisotropic media: Geophysical Journal of the Royal Astronomical Society, **91**, 383–411.

——, 1990, Vertical seismic profile synthetics by dynamic ray tracing in laterally varying layered anisotropic structures: Journal of Geophysical Research, **95**, 11301–11315.

Gal'perin, E. I., 1971, Vertical seismic profiling: Nedra (in Russian). English translation by A. J. Hermont and J. E. White, eds., 1974, SEG.

Gautam, K., M. Batzle, and R. Hofmann, 2003, Effect of fluids on attenuation of elastic waves: 73rd Annual International Meeting, SEG, Expanded Abstracts, 1592–1595.

Gibson, R. L., Jr., and S. Theophanis, 1996, Ultrasonic and numerical modeling of reflections from azimuthally anisotropic media: 66th Annual International Meeting, SEG, Expanded Abstracts, 1025–1028.

Goldin, S. V., 1986, Seismic traveltime inversion: SEG.

Graebner, M., 1992, Plane-wave reflection and transmission coefficients for a transversely isotropic solid: Geophysics, **57**, 1512–1519.

Granli, J. R., B. Arntsen, A. Sollid, and E. Hilde, 1999, Imaging through gas-filled sediments using marine shear-wave data: Geophysics, **64**, 668–677.

Gray, F. D., and D. Todorovic-Marinic, 2004, Fracture detection using 3D azimuthal AVO: CSEG Recorder, **29**, 5–8.

Gray, F. D., G. Roberts, and K. J. Head, 2002, Recent advances in determination of fracture strike and crack density from P-wave seismic data: The Leading Edge, **21**, 280–285.

Grechka, V., 1998, Transverse isotropy versus lateral heterogeneity in the inversion of P-wave reflection traveltimes: Geophysics, **63**, 204–212.

——, 2007, Multiple cracks in VTI rocks: Effective properties and fracture characterization: Geophysics, **72**, no. 5, D81–D91.

——, 2009, Applications of seismic anisotropy in the oil and gas industry: EAGE.

Grechka, V., and P. Dewangan, 2003, Generation and processing of pseudo-shear-wave data: Theory and case study: Geophysics, **68**, 1807–1816.

Grechka, V., and M. Kachanov, 2006a, Seismic characterization of multiple fracture sets: Does orthotropy suffice?: Geophysics, **71**, no. 3, D93–D105.

——, 2006b, Effective elasticity of fractured rocks: A snapshot of the work in progress: Geophysics, **71**, no. 6, W45–W58.

——, 2006c, Effective elasticity of rocks with closely spaced and intersecting cracks: Geophysics, **71**, no. 3, D85–D91.

——, 2006d, Effective elasticity of fractured rocks: The Leading Edge, **25**, 152–155.

Grechka, V., and A. Mateeva, 2007, Inversion of P-wave VSP data for local anisotropy: Theory and a case study: Geophysics, **72**, no. 4, D69–D79.

Grechka, V., and G. A. McMechan, 1997, Analysis of reflection traveltimes in 3-D transversely-isotropic heterogeneous media: Geophysics, **62**, 1884–1895.

Grechka, V., and I. Obolentseva, 1993, Geometrical structure of shear wave surfaces near singularity directions in anisotropic media: Geophysical Journal International, **115**, 609–616.

Grechka, V., and I. Tsvankin, 1998a, 3-D description of normal moveout in anisotropic inhomogeneous media: Geophysics, **63**, 1079–1092.

——, 1998b, Feasibility of nonhyperbolic moveout inversion in transversely isotropic media: Geophysics, **63**, 957–969.

——, 1999a, 3-D moveout inversion in azimuthally anisotropic media with lateral velocity variation: Theory and a case study: Geophysics, **64**, 1202–1218.

——, 1999b, 3-D moveout velocity analysis and parameter estimation for orthorhombic media: Geophysics, **64**, 820–837.

——, 2000, Inversion of azimuthally dependent NMO velocity in transversely isotropic media with a tilted axis of symmetry: Geophysics, **65**, 232–246.

——, 2002a, NMO-velocity surfaces and Dix-type formulas in anisotropic heterogeneous media: Geophysics, **67**, 939–951.

——, 2002b, Processing-induced anisotropy: Geophysics, **67**, 1920–1928.

——, 2002c, PP + PS = SS: Geophysics, **67**, 1961–1971.

——, 2002d, The joint nonhyperbolic moveout inversion of PP and PS data in VTI media: Geophysics, **67**, 1929–1932.

——, 2003, Feasibility of seismic characterization of multiple fracture sets: Geophysics, **68**, 1399–1407.

——, 2004, Characterization of dipping fractures in a transversely isotropic background: Geophysical Prospecting, **52**, 1–10.

Grechka, V., A. Bakulin, and I. Tsvankin, 2003, Seismic characterization of vertical fractures described as general linear-slip interfaces: Geophysical Prospecting, **51**, 117–130.

Grechka, V., P. Contreras, and I. Tsvankin, 2000, Inversion of normal moveout for monoclinic media: Geophysical Prospecting, **48**, 577–602.

Grechka, V., A. Pech, and I. Tsvankin, 2002a, P-wave stacking-velocity tomography for VTI media: Geophysical Prospecting, **50**, 151–168.

——, 2002b, Multicomponent stacking-velocity tomography for transversely isotropic media: Geophysics, **67**, 1564–1574.

——, 2005, Parameter estimation in orthorhombic media using multicomponent wide-azimuth reflection data: Geophysics, **70**, no. 2, D1–D8.

Grechka, V., S. Theophanis, and I. Tsvankin, 1999a, Joint inversion of P- and PS-waves in orthorhombic media: Theory and a physical-modeling study: Geophysics, **64**, 146–161.

Grechka, V., I. Tsvankin, and J. K. Cohen, 1999b, Generalized Dix equation and analytic treatment of normal-moveout velocity for anisotropic media: Geophysical Prospecting, **47**, 117–148.

Grechka, V., I. Vasconcelos, and M. Kachanov, 2006, The influence of crack shape on the effective elasticity of fractured rocks: Geophysics, **71**, no. 5, D153–D160.

Grechka, V., A. Pech, I. Tsvankin, and B. Han, 2001, Velocity analysis for tilted transversely isotropic media: A physical-modeling example: Geophysics, **66**, 904–910.

Grechka, V., I. Tsvankin, A. Bakulin, J. O. Hansen, and C. Signer, 2002c, Joint inversion of PP and PS reflection data for VTI media: A North Sea case study: Geophysics, **67**, 1382–1395.

Grechka, V., I. Tsvankin, A. Bakulin, C. Signer, and J. O. Hansen, 2002d, Anisotropic inversion and imaging of PP and PS reflection data in the North Sea: The Leading Edge, **21**, 90–97.

Grechka, V., A. Mateeva, G. Franco, C. Gentry, P. Jorgensen, and J. Lopez, 2007, Estimation of seismic anisotropy from P-wave VSP data: The Leading Edge, **26**, 756–759.

Grimm, R., H. Lynn, C. Bates, D. Phillips, K. Simon, and W. Beckham, 1999, Detection and analysis of naturally fractured reservoirs: Multiazimuth seismic surveys in the Wind River Basin, Wyoming: Geophysics, **64**, 1277–1292.

Hale, D., N. R. Hill, and J. Stefani, 1992, Imaging salt with turning wave seismic waves: Geophysics, **57**, 1453–1462.

Hall, S., and J. M. Kendall, 2003, Fracture characterization at Valhall: Application of P-wave amplitude variation with offset and azimuth (AVOA) analysis to a 3-D ocean-bottom data set: Geophysics, **68**, 1150–1160.

Helbig, K., 1994, Foundations of anisotropy for exploration seismics: Pergamon Press.

Helbig, K., and L. Thomsen, 2005, 75-plus years of anisotropy in exploration and reservoir seismics: A historical review of concepts and methods: Geophysics, **70**, no. 6, 9ND–23ND.

Horne, S., and S. Leaney, 2000, Short note: Polarization and slowness component inversion for TI anisotropy: Geophysical Prospecting, **48**, 779–788.

Hosten, B., M. Deschamps, and B. R. Tittmann, 1987, Inhomogeneous wave generation and propagation in lossy anisotropic solids: Application to the characterization of viscoelastic composite materials: Journal of the Acoustical Society of America, **82**, 1763–1770.

Hsu, K., M. Schoenberg, and J. Walsh, 1991, Anisotropy from polarization and moveout: 61st Annual International Meeting, SEG, Expanded Abstracts, 1526–1529.

Hubral, P., and T. Krey, 1980, Interval velocities from seismic reflection measurements: SEG.

Hudson, J. A., 1980, Overall properties of a cracked solid: Mathematical Proceedings of Cambridge Philosophical Society, **88**, 371–384.

——, 1981, Wave speeds and attenuation of elastic waves in material containing cracks: Geophysical Journal of the Royal Astronomical Society, **64**, 133–150.

Huet, C., 1990, Application of variational concepts to size effects in elastic heterogeneous bodies: Journal of the Mechanics and Physics of Solids, **38**, 813–841.

Ikelle, L. T., 1996, Anisotropic migration velocity based on inversion of common azimuthal sections: Journal of Geophysical Research, **101**, 22461–22484.

Isaac, J. H., and D. C. Lawton, 1999, Image mispositioning due to dipping TI media: A physical seismic modeling study: Geophysics, **64**, 1230–1238.

Iversen, E., H. Gjøystdal, and J. O. Hansen, 2000, Prestack map migration as an engine for parameter estimation in TI media: 70th Annual International Meeting, SEG, Expanded Abstracts, 1004–1007.

Jaeger, J. C., N. G. W. Cook, and R. W. Zimmerman, 2007, Fundamentals of rock mechanics: Blackwell Publishing.

Jansen, K. J., 2005, Seismic investigation of wrench faulting and fracturing at Rulison field, Colorado: M.S. thesis, Colorado School of Mines.

Jenner, E., 2001, Azimuthal anisotropy of 3D compressional wave seismic data, Weyburn Field, Saskatchewan, Canada: Ph.D. thesis, Colorado School of Mines.

Jenner, E., M. Williams, and T. Davis, 2001, A new method for azimuthal velocity analysis and application to a 3D survey, Weyburn field, Saskatchewan, Canada: 71st Annual International Meeting, SEG, Expanded Abstracts, 102–105.

Jílek, P., 2002a, Converted PS-wave reflection coefficients in weakly anisotropic media: Pure and Applied Geophysics, **159**, 1527–1562.

——, 2002b, Modeling and inversion of converted-wave reflection coefficients in anisotropic media: A tool for quantitative AVO analysis: Ph.D. thesis, Colorado School of Mines.

Jílek, P., B. Hornby, and A. Ray, 2003, Inversion of 3D VSP P-wave data for local anisotropy: A case study: 73rd Annual International Meeting, SEG, Expanded Abstracts, 1322–1325.

Johnston, D. H., and M. N. Toksöz, 1981, Seismic wave attenuation: SEG.

Kachanov, M., 1980, Continuum model of medium with cracks: Journal of the Engineering Mechanics Division, ASCE, **106** (EM5), 1039–1051.

——, 1992, Effective elastic properties of cracked solids: Critical review of some basic concepts: Applied Mechanics Review, **45**, 304–335.

——, 1993, Elastic solids with many cracks and related problems: in J. W. Hutchinson and T. Wu, eds., Advances in Applied Mechanics, **30**, 259–445.

Kachanov, M., and I. Sevostianov, 2005, On quantitative characterization of microstructures and effective properties: International Journal of Solids and Structures, **42**, 309–336.

Kachanov, M., B. Shafiro, and I. Tsukrov, 2003, Handbook of elasticity solutions: Kluwer Academic Publishers.

Kashtan, B. M., 1982, Calculation of geometrical spreading in piecewise homogeneous anisotropic media: Problems of Dynamic Theory of Seismic Wave Propagation, **22**, 14–24 (in Russian).

Kendall, J-M., and C. J. Thomson, 1989, A comment on the form of the geometrical spreading equations, with some numerical examples of seismic ray tracing in inhomogeneous, anisotropic media: Geophysical Journal International, **99**, 401–413.

Kochetov, M., and M. A. Slawinski, 2009a, Estimating effective elasticity tensors from Christoffel equations: Geophysics, **74**, no. 5, WB67–WB73.

——, 2009b, On obtaining effective orthotropic elasticity tensors: Quarterly Journal of Mechanics and Applied Mathematics, **62**, 149–166.

——, 2009c, On obtaining effective transversely isotropic elasticity tensors: Journal of Elasticity, **94**, 1–13.

Krebes, E. S., and L. H. T. Le, 1994, Inhomogeneous plane waves and cylindrical waves in anisotropic anelastic media: Journal of Geophysical Research, **99**, 23899–23919.

Krebes, E. S., and M. A. Slawinski, 1991, On raytracing in an elastic–anelastic medium: Bulletin of the Seismological Society of America, **81**, 667–686.

Landrø, M., and I. Tsvankin, 2007, Seismic critical-angle reflectometry: A method to characterize azimuthal anisotropy?: Geophysics, **72**, no. 3, D41–D50.

Laubach, S. E., J. E. Olson, and J. F. W. Gale, 2004, Are open fractures necessarily aligned with maximum horizontal stress?: Earth and Planetary Science Letters, **222**, 191–195.

Le Rousseau, J., 1997, Depth migration in heterogeneous, transversely isotropic media with the phase-shift-plus-interpolation method: 67th Annual International Meeting, SEG, Expanded Abstracts, 1703–1706.

Le Stunff, Y., and D. Grenié, 1998, Taking into account a priori information in 3D tomography: 68th Annual International Meeting, SEG, Expanded Abstracts, 1875–1878.

Le Stunff, Y., V. Grechka, and I. Tsvankin, 2001, Depth-domain velocity analysis in VTI media using surface P–wave data: Is it feasible?: Geophysics, **66**, 897–903.

Leslie, J. M., and D. C. Lawton, 1996, Structural imaging below dipping anisotropic layers: Predictions from seismic modeling: 66th Annual International Meeting, SEG, Expanded Abstracts, 719–722.

——, 1998, Anisotropic pre-stack depth migration: CSEG Recorder, **23**, no. 10, 23–36.

Levin, F. K., 1971, Apparent velocity from dipping interface reflections: Geophysics, **36**, 510–516.

Li, X.-Y., and J. Yuan, 1999, Converted-wave moveout and parameter estimation for transverse isotropy: 61st EAGE Conference, Extended Abstracts, 4–35.

Liu, E., J. A. Hudson, and T. Pointer, 2000, Equivalent medium representation of fractured rock. Journal of Geophysical Research, **105**, no. B2, 2981–3000.

Liu, E., S. Crampin, J. H. Queen, and W. D. Rizer, 1993, Velocity and attenuation anisotropy caused by microcracks and microfractures in a multiazimuth reverse VSP: Canadian Journal of Exploration Geophysics, **29**, 177–188.

Lynn, H., D. Campagna, K. Simon, and W. Beckham, 1999a, Relationship of P-wave seismic attributes, azimuthal anisotropy, and commercial gas pay in 3-D P-wave multiazimuth data, Rulison Field, Piceance Basin, Colorado: Geophysics, **64**, 1293–1311.

Lynn, H., K. Simon, C. Bates, and R. Van Dok, 1996, Azimuthal anisotropy in P-wave 3-D (multiazimuth) data: The Leading Edge, **15**, 923–928.

Lynn, H., W. Beckham, K. Simon, C. Bates, M. Layman, and M. Jones, 1999b, P-wave and S-wave azimuthal anisotropy at a naturally fractured gas reservoir, Bluebell-Altamont Field, Utah: Geophysics, **64**, 1312–1328.

Lynn, W., 2007, Uncertainty implications in azimuthal velocity analysis: 77th Annual International Meeting, SEG, Expanded Abstracts, 84–87.

MacBeth, C., 1999, Azimuthal variation in P-wave signatures due to fluid flow: Geophysics, **64**, 1181–1192.

——, 2002, Multi-component VSP analysis for applied seismic anisotropy: Pergamon Press.

Mallick, S., K. Craft, L. Meister, and R. Chambers, 1998, Determination of the principal directions of azimuthal anisotropy from P-wave seismic data: Geophysics, **63**, 692–706.

Markov, K. Z., 1999, Elementary micromechanics of heterogeneous media: in K. Z. Markov and L. Preziosi, eds., Heterogeneous media: Modeling and simulations, Birkhauser.

Martinez, R. D., 1993, Wave propagation effects on amplitude variation with offset measurements: A modeling study: Geophysics, **58**, 534–543.

Mattocks, B., 1998, Borehole geophysical investigation of seismic anisotropy at Vacuum Field, Lea County, New Mexico: Ph.D. thesis, Colorado School of Mines.

Maultzsch, S., M. Chapman, E. Liu, and X. Y. Li, 2003a, Modeling frequency-dependent seismic anisotropy in fluid-saturated rock with aligned fractures: Implication of fracture size estimation from anisotropic measurements: Geophysical Prospecting, **51**, 381–392.

Maultzsch, S., S. Horne, S. Archer, and H. Burkhardt, 2003b, Effects of an anisotropic overburden on azimuthal amplitude analysis in horizontal transverse isotropic media: Geophysical Prospecting, **51**, 61–74.

Mavko, G., and A. Nur, 1979, Wave attenuation in partially saturated rocks: Geophysics, **44**, 161–178.

Mehdizadeh, H., M. Landrø, B. A. Mythen, N. Vedanti, and R. Srivastava, 2005, Time lapse seismic analysis using long-offset PS data: 75th Annual International Meeting, SEG, Expanded Abstracts, TL 3.7.

Mensch, T., and P. Rasolofosaon, 1997, Elastic-wave velocities in anisotropic media of arbitrary symmetry – generalization of Thomsen's parameters ϵ, δ, and γ: Geophysical Journal International, **128**, 43–64.

Michaud, G., 2001, Multicomponent borehole seismic monitoring of a pilot CO_2 flood: Ph.D. thesis, Colorado School of Mines.

Miller, D. E., and C. Spencer, 1994, An exact inversion for anisotropic moduli from phase slowness data: Journal of Geophysical Research, **99**, B11, 21651–21657.

Milton, G. M., 2002, The theory of composites: Cambridge University Press.

Minsley, B. J., M. E. Willis, M. Krasovec, D. R. Burns, and M. N. Toksöz, 2004, Investigation of a fractured reservoir using P-wave AVOA analysis: A case study of the Emilio Field with support from synthetic examples: SEG/EAGE Summer Research Workshop.

Mura, T., 1987, Micromechanics of defects in solids: Martinus Nijhoff Publishers.

Musgrave, M. J. P., 1970, Crystal acoustics: Holden Day.

Muyzert, E., 2000, Scholte wave velocity inversion for a near surface S-velocity model and PS-statics: 70th Annual International Meeting, SEG, Expanded Abstracts, 1197–1200.

Narr, W., D. W. Schechter, and L. B. Thompson, 2006, Naturally fractured reservoir characterization: SPE.

Nemat-Nasser, S., and M. Hori, 1999, Micromechanics: Overall properties of heterogeneous materials: Elsevier Science Publ. Co., Inc.

Neves, F., A. Al-Marzoug, J. J. Kim, and E. L. Nebrija, 2003, Fracture characterization of deep tight gas sands using azimuthal velocity and AVO seismic data in Saudi Arabia: The Leading Edge, **22**, 469–475.

Nolte, B., D. V. Sukup, P. M. Krail, B. O. Temple, and B. Cafarelli, 2000, Anisotropic 3D prestack depth imaging of the Donald Field with converted waves: 70th Annual International Meeting, SEG, Expanded Abstracts, 1158–1161.

O'Connell, R. J., and B. Budiansky, 1974, Seismic velocities in dry and saturated cracked solids: Journal of Geophysical Research, **79**, 5412–5426.

Ostoja-Starzewski, M., 2008, Microstructural randomness and scaling in mechanics of materials: Chapman & Hall.

Pankhurst, D., K. Marfurt, C. Sullivan, H. Zhou, F. Hilterman, and J. Gallaghar, 2002, Long offset AVO in a mid-continent tight gas sand reservoir: 72nd Annual International Meeting, SEG, Expanded Abstracts, 297–299.

Peacock, S., and J. A. Hudson, 1990, Seismic properties of rocks with distributions of small cracks: Geophysical Journal International, **102**, 471–484.

Pech, A., and I. Tsvankin, 2004, Quartic moveout coefficient for a dipping azimuthally anisotropic layer: Geophysics, **69**, 699–707.

Pech, A., I. Tsvankin, and V. Grechka, 2003, Quartic moveout coefficient: 3D description and application to tilted TI media: Geophysics, **68**, 1600–1610.

Pérez, M. A., V. Grechka, and R. J. Michelena, 1999, Fracture detection in a carbonate reservoir using a variety of seismic methods: Geophysics, **64**, 1266–1276.

Pointer, T., E. Liu, and J. A. Hudson, 2000, Seismic wave propagation in cracked porous media: Geophysical Journal International, **142**, 199–231.

Prasad, M., and A. Nur, 2003, Velocity and attenuation anisotropy in reservoir rocks: 73rd Annual International Meeting, SEG, Expanded Abstracts, 1652–1655.

Press, W. H., B. P. Flannery, S. A. Teukolsky, and W. T. Vetterling, 1987, Numerical recipes: The art of scientific computing: Cambridge University Press.

Pšenčík, I., and D. Gajewski, 1998, Polarization, phase velocity, and NMO velocity of qP-waves in arbitrary weakly anisotropic media: Geophysics, **63**, 1754–1766.

Rathore, J. S., E. Fjaer, R. M. Holt, and L. Renlie, 1995, P- and S-wave anisotropy of a synthetic sandstone with controlled crack geometry: Geophysical Prospecting, **43**, 711–728.

Rommel, B. E., and I. Tsvankin, 2000, Analytic description of P-wave ray direction and polarization in orthorhombic media: *in* L. Ikelle and A. Gangi, eds., Anisotropy 2000: Fractures, converted waves and case studies, Proceedings of the 9th International Workshop on Seismic Anisotropy, SEG.

Rüger, A., 1997, P-wave reflection coefficients for transversely isotropic models with vertical and horizontal axis of symmetry: Geophysics, **62**, 713–722.

——, 1998, Variation of P-wave reflectivity with offset and azimuth in anisotropic media: Geophysics, **63**, 935–947.

——, 2002, Reflection coefficients and azimuthal AVO analysis in anisotropic media: SEG.

Rüger, A., and I. Tsvankin, 1997, Using AVO for fracture detection: Analytic basis and practical solutions: The Leading Edge, **16**, 1429–1434.

Saenger, E. H., O. S. Krüger, and S. A. Shapiro, 2004, Effective elastic properties of randomly fractured soils: 3D numerical experiments: Geophysical Prospecting, **52**, 183–195.

Sarkar, D., R. T. Baumel, and K. Larner, 2002, Velocity analysis in the presence of amplitude variation: Geophysics, **67**, 1664–1672.

Sayers, C. M., 1998, Misalignment of the orientation of fractures and the principal axes for P and S-waves in rocks containing multiple non-orthogonal fracture sets: Geophysical Journal International, **133**, 459–466.

Sayers, C. M., and D. A. Ebrom, 1997, Seismic traveltime analysis for azimuthally anisotropic media: Theory and experiment: Geophysics, **62**, 1570–1582.

Sayers, C. M., and M. Kachanov, 1991, A simple technique for finding effective elastic constants of cracked solids for arbitrary crack orientation statistics: International Journal of Solids and Structures, **27**, 671–680.

——, 1995, Microcrack-induced elastic wave anisotropy of brittle rocks, Journal of Geophysical Research, **100**, B3, 4149–4156.

Sayers, C. M., and J. E. Rickett, 1997, Azimuthal variation in AVO response for fractured gas sands: Geophysical Prospecting, **45**, 165–182.

Schoenberg, M., 1980, Elastic wave behavior across linear slip interfaces: Journal of the Acoustical Society of America, **68**, 1516–1521.

Schoenberg, M., and J. Douma, 1988, Elastic wave propagation in media with parallel fractures and aligned cracks: Geophysical Prospecting, **36**, 571–590.

Schoenberg, M., and K. Helbig, 1997, Orthorhombic media: Modeling elastic wave behavior in a vertically fractured earth: Geophysics, **62**, 1954–1974.

Schoenberg, M., and C. Sayers, 1995, Seismic anisotropy of fractured rock: Geophysics, **60**, 204–211.

Scuta, M., 1997, 3D geological characterization of Central Vacuum unit, Lea County, New Mexico: Ph.D. thesis, Colorado School of Mines.

Sena, A. G., 1991, Seismic traveltime equations for azimuthally anisotropic and isotropic media: Estimation of interval elastic properties: Geophysics, **56**, 2090–2101.

Seriff, A. J., and K. P. Sriram, 1991, P-SV reflection moveouts for transversely isotropic media with a vertical symmetry axis: Geophysics, **56**, 1271–1274.

Sevostianov, I., and M. Kachanov, 2002, On elastic compliances of irregularly shaped cracks: International Journal of Fracture, **114**, 245–257.

Shafiro, B., and M. Kachanov, 1997, Materials with fluid-filled pores of various shapes: Effective elastic properties and fluid pressure polarization: International Journal of Solids and Structures, **34**, 3517–3540.

Shah, P. M., 1973, Use of wavefront curvature to relate seismic data with subsurface parameters: Geophysics, **38**, 812–825.

Shearer, P. M., 2009, Introduction to seismology: Cambridge University Press.

Shuey, R. T., 1985, A simplification of the Zoeppritz equations: Geophysics, **50**, 609–614.

Signer, C., J. O. Hansen, G. Hutton, M. Nickel, B. Reymond, J. Schlaf, L. Sønneland, B. Tjøstheim, and H. H. Veire, 2000, Reservoir characterization using 4C seismic and calibrated 3D AVO: *in* K. Ofstad, J. E. Kittilsen, and P. Alexander-Marrack, eds., Improving the exploration process by learning from the past, Elsevier Science Publ. Co., Inc.

Siliqi, R., and N. Bousqué, 2000, Anelliptic time processing based on a shifted hyperbola approach: 70th Annual International Meeting, SEG, Expanded Abstracts, 2245–2248.

Slawinski, M. A., 2010, Waves and rays in elastic continua: World Scientific.

Slawinski, M. A., and P. S. Webster, 1999, On generalized ray parameters for vertically inhomogeneous and anisotropic media: Canadian Journal of Exploration Geophysics, **35**, 28–31.

Slotnick, M. M., 1959, Lessons in seismic computing: SEG.

Stovas, A., and B. Ursin, 2003, Reflection and transmission responses of layered transversely isotropic viscoelastic media: Geophysical Prospecting, **51**, 447–477.

Strahilevitz, R., and G. H. F. Gardner, 1995, Fracture detection using P-wave AVO: 65th Annual International Meeting, SEG, Expanded Abstracts, 589–591.

Sun, Z., 1994, Seismic anisotropy of salt from theoretical study, modeling, and field experiments: M.S. thesis, University of Calgary.

Taner, M. T., and F. Koehler, 1969, Velocity spectra – digital computer derivation and applications of velocity functions: Geophysics, **34**, 859–881.

Tarantola, A., 1987, Inverse problem theory: Methods for data fitting and model parameter estimation: Elsevier Science Publ. Co., Inc.

Tessmer, G., and A. Behle, 1988, Common reflection point data-stacking technique for converted waves: Geophysical Prospecting, **36**, 671–688.

Thomsen, L., 1986, Weak elastic anisotropy: Geophysics, **51**, 1954–1966.

——, 1988, Reflection seismology over azimuthally anisotropic media: Geophysics, **53**, 304–313.

——, 1993, Weak anisotropic reflections: in J. Castagna and M. Backus, eds., Offset dependent reflectivity, SEG, 103–114.

——, 1999, Converted-wave reflection seismology over inhomogeneous, anisotropic media: Geophysics, **64**, 678–690.

——, 2002, Understanding seismic anisotropy in exploration and exploitation: SEG/EAGE Distinguished Instructor Series.

Thomsen, L., I. Tsvankin, and M. C. Mueller, 1999, Coarse-layer stripping of vertically variable azimuthal anisotropy from shear-wave data: Geophysics, **64**, 1126–1138.

Todorovic-Marinic, D., G. Larson, D. Gray, G. Soule, Y. Zheng, and J. Pelletier, 2004, Identifying vertical productive fractures in the Narraway gas field using the envelope of seismic anisotropy: SEG/EAGE Summer Research Workshop.

Toldi, J., T. Alkhalifah, P. Berthet, J. Arnaud, P. Williamson, and B. Conche, 1999, Case study of estimation of anisotropy: The Leading Edge, **18**, 588–594.

Tsvankin, I., 1995a, Normal moveout from dipping reflectors in anisotropic media: Geophysics, **60**, 268–284.

——, 1995b, Body-wave radiation patterns and AVO in transversely isotropic media: Geophysics, **60**, 1409–1425.

——, 1995c, Seismic wavefields in layered isotropic media: Samizdat Press, Colorado School of Mines.

——, 1996, P-wave signatures and notation for transversely isotropic media: An overview: Geophysics, **61**, 467–483.

——, 1997a, Anisotropic parameters and P-wave velocity for orthorhombic media: Geophysics, **62**, 1292–1309.

——, 1997b, Reflection moveout and parameter estimation for horizontal transverse isotropy: Geophysics, **62**, 614–629.

——, 1997c, Moveout analysis in transversely isotropic media with a tilted symmetry axis: Geophysical Prospecting, **45**, 479–512.

——, 2005, Seismic signatures and analysis of reflection data in anisotropic media, 2nd ed.: Elsevier Science Publ. Co., Inc.

Tsvankin, I., and E. M. Chesnokov, 1990, Synthesis of body wave seismograms from point sources in anisotropic media: Journal of Geophysical Research, **95**, 11317–11331.

Tsvankin, I., and L. Thomsen, 1994, Nonhyperbolic reflection moveout in anisotropic media: Geophysics, **59**, 1290–1304.

——, 1995, Inversion of reflection traveltimes for transverse isotropy: Geophysics, **60**, 1095–1107.

Tsvankin, I., J. Gaiser, V. Grechka, M. van der Baan, and L. Thomsen, 2010, Seismic anisotropy in exploration and reservoir characterization: An overview: Geophysics, **75**, no. 5, 75A15–75A29.

Uren, N. F., G. N. F. Gardner, and J. A. McDonald, 1990, Normal moveout in anisotropic media: Geophysics, **55**, 1634–1636.

Ursin, B., and K. Hokstad, 2003, Geometrical spreading in a layered transversely isotropic medium with vertical symmetry axis: Geophysics, **68**, 2082–2091.

Ursin, B., Tygel, M., and E. Iversen, 2009, SS traveltime parameters from PP and PS reflections: Geophysics, **74**, no. 4, R35–R48.

van der Baan, M., 2004, Processing of anisotropic data in the τ–p domain: I – Geometric spreading and moveout corrections: Geophysics, **69**, 719–730.

——, 2005, Processing of anisotropic data in the τ–p domain: II – Common-conversion-point sorting: Geophysics, **70**, no. 4, D29–D36.

van der Baan, M., and J.-M. Kendall, 2002, Estimating anisotropy parameters and traveltimes in the τ–p domain: Geophysics, **67**, 1076–1086.

——, 2003, Traveltime and conversion point computations and parameter estimation in layered anisotropic media by τ–p transform: Geophysics, **68**, 210–224.

Vanelle, C., and D. Gajewski, 2003, Determination of geometrical spreading from traveltimes: Journal of Applied Geophysics, **54**, 391–400.

Vasconcelos, I., and V. Grechka, 2007, Seismic characterization of multiple fracture sets at Rulison Field, Colorado: Geophysics, **72**, no. 2, B19–B30.

Vasconcelos, I., and E. Jenner, 2005, Estimation of azimuthally varying attenuation from wide-azimuth P-wave data: 75th Annual International Meeting, SEG, Expanded Abstracts, 123–126.

Vasconcelos, I., and I. Tsvankin, 2006, Nonhyperbolic moveout inversion of wide-azimuth P-wave data for orthorhombic media: Geophysical Prospecting, **54**, 535–552.

Vavakin, A. S., and R. L. Salganik, 1975, Effective characteristics of nonhomogeneous media with isolated inhomogeneities: Mechanics of Solids, Allerton Press, 58–66 (English translation of Izvestia AN SSSR, Mekhanika Tverdogo Tela, **10**, 65–75).

Vavryčuk, V., 2007, Ray velocity and ray attenuation in homogeneous anisotropic viscoelastic media: Geophysics, **72**, No. 6, D119–D127.

Vavryčuk, V., and I. Pšenčík, 1998, PP-wave reflection coefficients in weakly anisotropic elastic media: Geophysics, **63**, 2129–2141.

Vernik, L., and X. Liu, 1997, Velocity anisotropy in shales: A petrophysical study: Geophysics, **62**, 521–532.

Vestrum, R. W., D. C. Lawton, and R. Schmid, 1999, Imaging structures below dipping TI media: Geophysics, **64**, 1239–1246.

Virieux, J., and S. Operto, 2009, An overview of full-waveform inversion in exploration geophysics: Geophysics, **74**, no. 6, WCC1–WCC26.

Walpole, L. J., 1969, On the overall elastic moduli of composite materials: Journal of the Mechanics and Physics of Solids, **17**, 235–251.

Wang, D. Y., and D. W. McCowan, 1989, Spherical divergence correction for seismic reflection data using slant stacks: Geophysics, **54**, 563–569.

Wang, X., and I. Tsvankin, 2009, Estimation of interval anisotropy parameters using velocity-independent layer stripping: Geophysics, **74**, no. 5, WB117–WB127.

Wang, Z., 2002, Seismic anisotropy in sedimentary rocks, part 2: Laboratory data: Geophysics, **67**, 1423–1440.

Wapenaar, K., D. Draganov, and J. O. A. Robertsson (Eds.), 2008, Seismic interferometry: History and present status: Geophysics reprint series, SEG.

White, J. E., L. Martineau-Nicoletis, and C. Monash, 1983, Measured anisotropy in Pierre shale: Geophysical Prospecting, **31**, 709–725.

Wild, P., and S. Crampin, 1991, The range of effects of azimuthal isotropy and EDA anisotropy in sedimentary basins: Geophysical Journal International, **107**, 513–529.

Williamson, P., and E. Maocec, 2001, Estimation of local anisotropy using polarizations and travel times from the Oseberg 3D VSP: *in* L. Ikelle and A. Gangi, eds., Anisotropy 2000: Fractures, converted waves and case studies, Proceedings of the 9th International Workshop on Seismic Anisotropy, SEG, 339–348.

Willis, M., R. Rao, D. Burns, J. Byun, and L. Vetri, 2004, Spatial orientation and distribution of reservoir fractures from scattered seismic energy: 74th Annual International Meeting, SEG, Expanded Abstracts, 1535–1538.

Winterstein, D. F., and M. A. Meadows, 1991, Shear-wave polarizations and subsurface stress directions at Lost Hills field: Geophysics, **56**, 1331–1348.

Withers, R., and D. Corrigan, 1997, Fracture detection using wide azimuth 3D seismic surveys: 59th EAGE Conference, Extended Abstracts, Paper E003.

Xu, X., and I. Tsvankin, 2006a, Anisotropic geometrical-spreading correction for wide-azimuth P-wave reflections: Geophysics, **71**, no. 5, D161–D170.

——, 2006b, Azimuthal AVO analysis with anisotropic spreading correction: A synthetic study: The Leading Edge, **25**, 1336–1342.

——, 2007, A case study of azimuthal AVO analysis with anisotropic spreading correction: The Leading Edge, **26**, 1552–1561.

——, 2008, Moveout-based geometrical-spreading correction for PS-waves in layered anisotropic media: Journal of Geophysics and Engineering, **5**, 195–202.

Xu, X., I. Tsvankin, and A. Pech, 2005, Geometrical spreading of P-waves in horizontally layered, azimuthally anisotropic media: Geophysics, **70**, no. 5, D43–D54.

Yan, J., and I. Tsvankin, 2008, AVO-sensitive semblance analysis for wide-azimuth data: Geophysics, **73**, no. 2, U1–U11.

Zheng, X., and I. Pšenčík, 2002, Local determination of weak anisotropy parameters from qP-wave slowness and particle motion measurements: Pure and Applied Geophysics, **159**, 1881–1905.

Zhu, Y., and I. Tsvankin, 2006, Plane-wave propagation in attenuative transversely isotropic media: Geophysics, **71**, no. 2, T17–T30.

——, 2007, Plane-wave attenuation anisotropy in orthorhombic media: Geophysics, **72**, no. 1, D9–D19.

Zhu, Y., I. Tsvankin, and I. Vasconcelos, 2007a, Effective attenuation anisotropy of thin-layered media: Geophysics, **72**, no. 5, D93–D106.

Zhu, Y., I. Tsvankin, P. Dewangan, and K. van Wijk, 2007b, Physical modeling and analysis of P-wave attenuation anisotropy in transversely isotropic media: Geophysics, **72**, no. 1, D1–D7.

Zohdi, T. I. and P. Wriggers, 2001, Aspects of the computational testing of the mechanical properties of microheterogeneous material samples: International Journal for Numerical Methods in Engineering, **50**, 2573–2599.

——, 2005, Introduction to computational micromechanics: Springer.

Author Index

Subject Index